Scroggie's Foundations of Wireless and Electronics

Eleventh Edition

S. W. Amos
R. S. Amos

Newnes
An imprint of Butterworth-Heinemann
Linacre House, Jordan Hill, Oxford OX2 8DP
A division of Reed Educational and Professional Publishing Ltd

 A member of the Reed Elsevier plc group

OXFORD BOSTON JOHNANNESBURG
MELBOURNE NEW DELHI SINGAPORE

First published as *Foundations of Wireless*, 1936, by Iliffe Books
Eighth edition, rewritten and published as *Foundations of Wireless and Electronics*, 1971
Ninth edition, 1975 by Newnes Technical Books
Tenth edition, 1984
Eleventh edition, 1997

© Reed Educational and Professional Publishing Ltd 1997

All rights reserved. No part of this publication may be
reproduced in any material form (including photocopying
or storing in any medium by electronic means and whether
or not transiently or incidentally to some other use of
this publication) without the written permission of the
copyright holder except in accordance with the provisions
of the Copyright, Designs and Patents Act 1988 or under
the terms of a license issued by the Copyright Licensing
Agency Ltd, 90 Tottenham Court Road, London, England W1P 9HE.
Applications for copyright holder's written permission
to reproduce any part of this publication should be
addressed to the publishers.

British Library Cataloguing in Publication Data
A catalogue record for this book is available from the British Library

ISBN 0 7506 3430 8

Library of Congress Cataloguing in Publication Data
A catalogue record for this book is available from the Library of Congress

Typeset by Tekoa Graphics
Printed in Great Britain by Scotprint Ltd, Musselburgh

Contents

Preface		vii

1 General View of a System — 1

1.1	What Wireless Does	1
1.2	Nature of Sound Waves	1
1.3	Characteristics of Sound Waves	2
1.4	Frequency	3
1.5	Wavelength	3
1.6	The Transmitter	4
1.7	The Receiver	4
1.8	Electrical Communication by Wire	4
1.9	Electric Waves	6
1.10	Why High Frequencies are Necessary	6
1.11	Radio Telegraphy	7
1.12	Tuning	7
1.13	Radio Telephony	7
1.14	Recapitulation	8

2 Electricity and Circuits — 10

2.1	Electrons	10
2.2	Electric Charges and Currents	10
2.3	Conductors and Insulators	11
2.4	Electromotive Force	11
2.5	Electrical Units	12
2.6	Ohm's Law	12
2.7	Larger and Smaller Units	13
2.8	Circuit Diagrams	13
2.9	Resistances in Series and in Parallel	15
2.10	Series-Parallel Combinations	16
2.11	Resistance Analysed	17
2.12	Conductance	17
2.13	Kirchhoff's Laws	17
2.14	P.D. and E.M.F.	18
2.15	Electrical Effects	19
2.16	Instruments for Measuring Electricity	19
2.17	Electrical Power	20
2.18	A Broader View of Resistance	21

3 Capacitance — 22

3.1	Charging Currents	22
3.2	Capacitance: What It Is	22
3.3	Capacitance Analysed	23
3.4	Capacitors	24
3.5	Charge and Discharge of a Capacitor	25
3.6	Where the Power Goes	26
3.7	Recapitulation	26
3.8	Displacement Currents	27

4 Inductance — 28

4.1	Magnets and Electromagnets	28
4.2	Interacting Magnetic Fields	29
4.3	Induction	29
4.4	Self-inductance	30
4.5	Lenz's Law	30
4.6	Inductance Analysed	30
4.7	Practical Considerations	31
4.8	Growth of Current in an Inductive Circuit	32
4.9	Power During Growth	33
4.10	More Comparison and Contrast	33
4.11	Mutual Inductance	33

5 Alternating Currents — 35

5.1	Frequencies of Alternating Current	35
5.2	The Sine Wave	35
5.3	Circuit with Resistance Only	36
5.4	R.M.S. Values	36
5.5	A.C. Meters	37
5.6	Phase	38
5.7	Phasor Diagrams	38
5.8	Adding Alternating Voltages	39
5.9	Direction Signs	39
5.10	Subscript Notation	40
5.11	Current Phasors	41

6 Capacitance in A.C. Circuits — 42

6.1	Current Flow in a Capacitive Circuit	42
6.2	Capacitive Current Waveform	43
6.3	The 'Ohm's Law' for Capacitance	43
6.4	Capacitances in Parallel and in Series	44
6.5	Power in a Capacitive Circuit	45
6.6	Phasor Diagram for Capacitive Circuit	45
6.7	Capacitance and Resistance in Series	46
6.8	Impedance	46
6.9	Power in Mixed Circuits	47
6.10	Capacitance and Resistance in Parallel	47

7 Inductance in A.C. Circuits — 49

7.1	Current Flow in an Inductive Circuit	49
7.2	The 'Ohm's Law' for Inductance	50
7.3	Inductances in Series and in Parallel	50
7.4	Power in an Inductive Circuit	50
7.5	Phasor Diagram for Inductive Circuit	51
7.6	Inductance and Resistance in Series	51
7.7	Inductance and Resistance in Parallel	51
7.8	Transformers	52
7.9	Load Currents	52
7.10	Transformer Losses	53

7.11	Impedance Transformation	54

8 The Tuned Circuit — 56

8.1	Inductance and Capacitance in Series	56
8.2	L, C and R all in Series	56
8.3	The Series Tuned Circuit	56
8.4	Magnification	57
8.5	Resonance Curves	57
8.6	Selectivity	58
8.7	Frequency of Resonance	59
8.8	L and C in Parallel	59
8.9	The Effect of Resistance	60
8.10	Dynamic Resistance	60
8.11	Parallel Resonance	61
8.12	Frequency of Parallel Resonance	61
8.13	Series and Parallel Resonance Compared	62
8.14	The Resistance of the Coil	62
8.15	Dielectric Losses	63
8.16	H.F. Resistance	63
8.17	Cavity Resonators	64

9 Diodes — 65

9.1	Electronic Devices	65
9.2	Diodes	65
9.3	Thermionic Emission of Electrons	65
9.4	The Vacuum Diode Valve	66
9.5	Semiconductors	67
9.6	Holes	67
9.7	Intrinsic Conduction	69
9.8	Effects of Impurities	69
9.9	P-N Junctions	71
9.10	The Semiconductor Diode	72
9.11	Diode Characteristics	73
9.12	Recapitulation	74

10 Triodes — 75

10.1	The Vacuum Triode Valve	75
10.2	Amplification Factor	75
10.3	Mutual Conductance	76
10.4	Anode Resistance	76
10.5	Alternating Voltage at the Grid	77
10.6	Grid Bias	77
10.7	The Transistor	77
10.8	Transistor Characteristic Curves	79
10.9	Transistor Parameters	80
10.10	Field-effect Transistors	81
10.11	Insulated-gate F.E.T.s	82
10.12	Light-sensitive Diodes and Triodes	83

11 The Triode at Work — 84

11.1	Input and Output	84
11.2	Source and Load	84
11.3	Feeds and Signals	84
11.4	Load Lines	85
11.5	Voltage Amplification	86
11.6	An Equivalent Generator	86
11.7	Calculating Amplification	87
11.8	The Maximum-power Law	88
11.9	Transistor Load Lines	89
11.10	Class A, B and C Operation	89

12 Transistor Equivalent Circuits — 91

12.1	The Equivalent Current Generator	91
12.2	Duality	92
12.3	Voltage or Current Generator?	93
12.4	Transistor Equivalent Output Circuit	93
12.5	Some Box Tricks	94
12.6	Input Resistance	95
12.7	Complete Transistor Equivalent Circuit	96
12.8	A Simpler Equivalent	97
12.9	Other Circuit Configurations: Common Collector	98
12.10	Common Base	100
12.11	A Summary of Results	101
12.12	F.E.T.s and Valves	104

13 The Working Point — 106

13.1	Feed Requirements	106
13.2	Effect of Amplification Factor Variations	107
13.3	Influence of Leakage Current	107
13.4	Methods of Base Biasing	108
13.5	Biasing the F.E.T.	110
13.6	Valve Biasing	110
13.7	Biasing by Diodes	111

14 Oscillation — 112

14.1	Generators	112
14.2	The Oscillatory Circuit	112
14.3	Frequency of Oscillation	113
14.4	Damping	113
14.5	Maintaining Oscillation	114
14.6	Practical Oscillator Circuits	115
14.7	Resistance-Capacitance Oscillators	117
14.8	Negative Resistance	117
14.9	Amplitude of Oscillation	118
14.10	Automatic Biasing of Oscillators	120
14.11	Distortion of Oscillation	120
14.12	Constancy of Frequency	120
14.13	Crystal Oscillators	121
14.14	Microwave Oscillators	121

15 Modulation — 123

15.1	The Need for Modulation	123
15.2	Amplitude Modulation	123
15.3	Methods of Amplitude Modulation	124
15.4	Frequency Modulation	124
15.5	Telegraphy and Keying	125
15.6	Methods of Frequency Modulation	125
15.7	Sources of Modulating Signal	125
15.8	Theory of Sidebands	126

15.9	Channel Separation	128		18.17	Quadrature Detector	160
15.10	Multiplex	129		18.18	Phase-locked-loop Detector	160
15.11	Pulse Modulation Systems	130		18.19	Counter Discriminator	161

16 Transmission Lines — 131

16.1	Feeders	131
16.2	Circuit Equivalent of a Line	131
16.3	Characteristic Resistance	132
16.4	Waves along a Line	133
16.5	Wave Reflection	134
16.6	Standing Waves	135
16.7	Line Impedance Variations	135
16.8	The Quarter-wave Transformer	136
16.9	Fully Resonant Lines	136
16.10	Lines as Tuned Circuits	137
16.11	Waveguides or Radio-wave Plumbing	137

17 Radiation and Antennas — 139

17.1	Bridging Space	139
17.2	The Quarter-wave Resonator Again	139
17.3	A Rope Trick	140
17.4	Electromagnetic Waves	141
17.5	Radiation	141
17.6	Polarisation	142
17.7	Antennas	142
17.8	Radiation Resistance	143
17.9	Directional Characteristics	143
17.10	Reflectors and Directors	144
17.11	Antenna Gain	145
17.12	Choice of Frequency	145
17.13	Influence of the Atmosphere	146
17.14	Earthed Antennas	146
17.15	Feeding the Antenna	147
17.16	Tuning	148
17.17	Effective Height	148
17.18	Microwave Antennas	149
17.19	Inductor Antennas	149

18 Detection — 150

18.1	The Need for Detection	150
18.2	The Detector	150
18.3	Rectifiers	151
18.4	Linearity of Rectification	152
18.5	Rectifier Resistance	152
18.6	Action of Reservoir Capacitor	152
18.7	Choice of Component Values	153
18.8	The Diode Detector in Action	154
18.9	The Detector as a Load	154
18.10	Filters	155
18.11	Detector Distortion	156
18.12	Shunt-diode Detector	157
18.13	Television Diode Detector	158
18.14	F.M. Detectors	158
18.15	Foster-Seeley Discriminator	158
18.16	Ratio Detector	159

19 Audio Frequency Amplification — 162

19.1	Recapitulation	162
19.2	Decibels	162
19.3	Gain/Frequency Distortion	163
19.4	Non-linearity Distortion	164
19.5	Generation of Harmonics	164
19.6	Intermodulation	165
19.7	Allowable Limits of Non-linearity	166
19.8	Phase Distortion	166
19.9	Loudspeakers	167
19.10	The Output Stage	168
19.11	Class B Amplification	169
19.12	Distortion with Class B	170
19.13	Class B Circuits	171
19.14	Compound Transistors	173
19.15	Negative Feedback	173
19.16	Phase Shift with Feedback	174
19.17	Input and Output Resistance	176
19.18	Linearising the Input	177
19.19	Some Circuit Details	178
19.20	Noise	179
19.21	Another Box Trick	180

20 Selectivity and Tuning — 181

20.1	Need for R.F. Amplification	181
20.2	Selectivity and Q	181
20.3	Over-sharp Tuning	182
20.4	A General Resonance Curve	182
20.5	More Than One Tuned Circuit	184
20.6	Coupled Tuned Circuits	185
20.7	Effects of Varying Coupling	187
20.8	Practical Tuning Difficulties	188

21 The Superheterodyne Principle — 189

21.1	A Difficult Problem Solved	189
21.2	The Frequency-changer	189
21.3	Frequency-changers as Modulators	191
21.4	Types of Frequency-changer	191
21.5	Conversion Conductance	192
21.6	Ganging the Oscillator	193
21.7	Whistles	194

22 Radio-frequency and Intermediate-frequency Amplification — 196

22.1	Amplification So Far	196
22.2	Active-device Interelectrode Capacitances	196
22.3	Miller Effect	197
22.4	High-frequency Effects in Transistors	197
22.5	Transistor Phasor Diagrams	198
22.6	Limiting Frequencies	199

22.7	An I.F. Amplifier	200
22.8	R.F. Amplification	201
22.9	Screening	203
22.10	Cross-modulation	204
22.11	Automatic Gain Control	205
22.12	The Antenna Coupling	206
22.13	Klystrons and Travelling-wave Tubes	208

23 Electronic Imaging: Video 210

23.1	Description of Cathode-Ray Tube	210
23.2	Electric Focusing and Deflection	211
23.3	Magnetic Deflection and Focusing	212
23.4	Oscilloscopes	213
23.5	Time Bases	213
23.6	Application to Television	214
23.7	Characteristics of Television Signals	216
23.8	Television Receivers	217
23.9	Synchronising Circuits	218
23.10	Scanning Circuits	220
23.11	Colour Television	222
23.12	Television Camera Tubes	224
23.13	Application to Radar	225
23.14	Application to Computing	227
23.15	Alternatives to the C.R. Tube	227

24 Non-Sinusoidal Signal Amplification 229

24.1	Waveforms	229
24.2	Frequency and Time	230
24.3	Frequency Range	231
24.4	Importance of Phase Response	231
24.5	Obtaining the High-frequency Response	232
24.6	Operational Amplifiers	234
24.7	Integrated Circuits	236
24.8	Analogue Computers	236

25 Electronic Waveform Generators and Switches 237

25.1	Differentiating and Integrating Circuits	237
25.2	Waveform Shapers	239
25.3	Waveform Generators: Relaxation Oscillators	240
25.4	The Blocking Oscillator	241
25.5	The Multivibrator	242
25.6	Bistables	243
25.7	Monostables	244
25.8	Clampers and Clippers	244
25.9	Gates and Choppers	246
25.10	Design	246

26 Digital Techniques 247

26.1	Analogue and Digital	247
26.2	The Binary Scale	247
26.3	Logic Operations and Circuits	248
26.4	Adders	250
26.5	Shift Registers	251
26.6	Counters	252
26.7	Digital Computers	253
26.8	Stores	253
26.9	Computer Languages	254
26.10	Input and Output Devices	254
26.11	Microprocessors and Microcomputers	255
26.12	Digital-to-Analogue and Analogue-to-Digital Conversion	255

27 Electronic Data Storage 258

27.1	The Need for Storage	258
27.2	Lessons from History	258
27.3	Magnetic Storage Systems—Analogue	259
27.4	Magnetic Storage Systems—Digital	261
27.5	Optical Systems—Compact Disc (CD)	262
27.6	Other Optical Storage Systems	263
27.7	Solid-State Systems	263
27.8	The Future	265

28 Power Supplies 266

28.1	The Power Required	266
28.2	Batteries	266
28.3	D.C. from A.C.	267
28.4	Types of Rectifier	268
28.5	Rectifier Circuits	268
28.6	Filters	270
28.7	Decoupling	270
28.8	Bias Supplies	271
28.9	Stabilisation	271
28.10	E.H.T.	273
28.11	Switch-mode Power Suppliers	274
28.12	Cathode Heating	274

Appendix A: Algebraic Symbols 275

A.1	Letter Symbols	275
A.2	What Letter Symbols Really Mean	275
A.3	Some Other Uses of Symbols	276
A.4	Abbreviations	276
A.5	How Numbers are Used	276

Appendix B: Graphs 278

B.1	What is a Graph?	278
B.2	Scales	278
B.3	What a 'Curve' Signifies	279
B.4	Three-dimensional Graphs	279
B.5	Significance of Slope	280
B.6	Non-uniform Scales	281

Appendix C: Alternative Technical Terms 282

Appendix D: Symbols and Abbreviations **283**

D.1 General Abbreviations 283
D.2 Greek Letters 283
D.3 Quantities and Units (S.I.) 283
D.4 Unit Multiple and Submultiple Prefixes 284
D.5 Transistor Abbreviations and Symbols 284
D.6 Valve Abbreviations and Symbols 284
D.7 Special Abbreviations used in this Book 284

Appendix E: Decibel Table **285**

Index **286**

Preface

Preface to Ninth Edition

Wireless—or radio—has branched out and developed so tremendously that very many books would be needed to describe it all in detail. In ordinary conversation, 'radio' has come to mean sound broadcasting, as distinct from 'TV' which gives pictures as well as sound, but of course the pictures come by radio just as much as the sound. Besides broadcasting sound and vision, radio is used for communication with and between ships, cars, aircraft, satellites and spacecraft; for direction-finding and radar (radiolocation), photograph and 'facsimile' transmission, telegraph and telephone links, meteorological probing of the upper atmosphere, astronomy, and other things. Very similar techniques are applied on an increasing scale to industrial control ('automation') and scientific and financial computation. All of these are based on the same foundational principles now generally comprised in the word *electronics*. The purpose of this book is to start at the beginning and lay these foundations, on which more detailed knowledge can then be built.

At the beginning ... If you had to tell somebody about a cricket match you had seen, your description would depend very much on whether or not your hearer was familiar with the jargon of the game. If he wasn't and you assumed he was, he would be puzzled. If it was the other way about, he would be irritated. There is the same dilemma with electronics. It takes much less time to explain it if the reader is familiar with methods of expression, such as symbols, that are taken for granted in technical discussion but not in ordinary conversation. This book assumes hardly any special knowledge. But if the use of graphs and symbols had to be completely excluded, or else accompanied everywhere by digressions explaining them, it would be very boring for the initiated. So the methods of technical expression are explained separately, as indicated below.

Then there are the technical terms. They are explained one by one as they occur and their first occurrence is distinguished by printing in italics; but in case any are forgotten they can be looked up in the Index; and the symbols and abbreviations in Appendix D. These references are there to be used whenever the meaning of anything is not understood.

Most readers find purely abstract principles very dull; it is more stimulating to have in mind some application of those principles. As it would be confusing to have all the applications of electronics or even of radio in mind at once, broadcasting of sound is mentioned most often because it directly affects nearly everyone. But the same principles apply more or less to all the other things. The reader who is interested in the communicating of something other than sound has only to substitute the appropriate word.

Glancing through this book, you can see numerous strange signs and symbols. Most of the diagrams consist of little else, while the occasional appearance of what looks like algebra may create a suspicion that this is a Mathematical Work and therefore quite beyond a beginner. Yet these devices are not, as might be supposed, for the purpose of making the book look more learned or difficult. Quite the contrary. Experience has shown them to be the simplest, clearest and most compact ways of conveying the sort of information needed.

Two of these devices are Algebraic Symbols and Graphs. Explanations of these for readers who are not quite used to them are given in Appendices A and B. Circuit Diagrams constitute a third device and these are described in Chapter 2.

1975   M. G. SCROGGIE

Preface to Tenth Edition

Since the ninth edition was published in 1975 there have been a number of significant developments in electronics. For example small thermionic valves are no longer used in new electronic equipment and manufacture of replacement types has ceased. The miniaturisation of integrated circuits has continued and a single i.c. can now contain more than 100 000 components, making even complex equipment very compact, as instanced by digital watches and domestic computers no larger than a book. Nearly all homes in this country now have colour television sets, but sales of plain (i.e. black-and-white) receivers continue mainly for use as second or portable sets.

To bring the book up to date most references to valves have been removed but sufficient has been left to provide an introduction to cathode-ray tubes and microwave tubes. Some information on senders has been omitted but more space has been devoted to modulation, including modern pulse methods. Accounts of up-to-date f.m. detectors have been added. Information on colour television has been expanded and now covers all the basic processes of transmission and reception. A new chapter on non-sinusoidal signal amplification has been introduced, and the chapter on computers has been updated.

All the circuit diagrams have been redrawn using the graphical symbols recommended by BSI.

This new edition was prepared by S. W. Amos who collaborated with M. G. Scroggie to ensure continuity of approach and style.

1983   S. W. AMOS

Preface to Eleventh Edition

Few technologies have advanced so fast during the past few decades as electronics and it was not surprising, therefore, to find that there was a need to update the tenth edition which was published in 1984.

Since then, the original author, M. G. Scroggie, has died and the preparation of this new edition has been undertaken by S. W. Amos and his son, R. S. Amos.

The early chapters dealing with basic electronics theory have needed very little revision, but the chapters on digital techniques and their applications have been extensively rewritten by R. S. Amos. The former chapter on digital computers has been thoroughly revised and a new chapter on data storage has been added. Modifications and additions have been made to include information on video recorders, video cameras, digital-to-analogue and analogue-to-digital conversion, liquid crystal displays and other recent developments in electronics.

1997   S. W. AMOS, R. S. AMOS

Chapter 1
General View of a System

Most of the chapters of this book are about subjects the usefulness of which, taken by themselves, might not be at all obvious. Proverbially, one might not be able to see the wood for the trees. So before examining these things in detail we may find it helpful to take a general view of a complete electronic system. The choice is a broadcasting and receiving system, because it is so familiar in everyday life and it happens to embody most of the principles of electronics.

1.1 What Wireless Does

People who remark on the wonders of wireless seldom seem to consider the fact that most people can broadcast speech and song merely by using their voices. We can instantly communicate our thoughts to others, without wires or any other visible lines of communication and without even any sending or receiving apparatus outside of ourselves. If anything is wonderful, that is. Wireless, or radio, is merely a device for increasing the range.

There is a fable about a dispute among the birds as to which could fly the highest. The claims of the wren were derided until, when the great competition took place and the eagle was proudly outflying the rest, the wren took off from his back and established a new altitude record.

It has been known for centuries that communication over vast distances of space is *possible*—it happens every time we look at the stars and detect light coming from them. We know, then, that a long-range medium of communication exists. Radio simply uses this medium for the carrying of sound and vision.

Simply?

Well, there actually is quite a lot to learn about it. But the same basic principles can be applied to many other electronic systems.

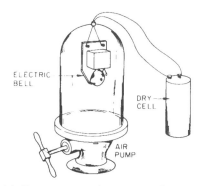

Figure 1.1 Experiment to demonstrate that sound waves cannot cross empty space. As the air is removed from the glass vessel, the sound of the bell fades away

1.2 Nature of Sound Waves

Radio (if, for convenience, we adopt the illogical convention that it excludes television) is, as we have just observed, essentially a means of extending natural communication beyond its limits of range. It plays the part of the swift eagle in carrying the wren of human speech to remote distances. So first of all let us see what is required for 'natural' communication.

There are three essential parts: the *speaker*, who generates sound; the *hearer* who receives it; and the *medium*, which connects one with the other.

The last of these is the key to the other two. By suspending a source of sound in a container and extracting the air (Figure 1.1) one can demonstrate that sound cannot travel without air (or some other physical substance) all the way between sender and receiver.

Anyone who has watched a cricket match will recall that the smack of bat against ball is heard a moment after they are seen to meet; the sound of the impact has taken an appreciable time to travel from the pitch to the viewpoint. If the pitch were 335 metres away, the time delay would be one second. The same speed of travel is found to hold good over longer or shorter distances. Knowing the damage that air does when it travels at even one-tenth of that speed (hurricane force) we conclude that sound does not consist of the air itself leaping out in all directions.

When a stone is thrown into a pond it produces ripples, but the ripples do not consist of the water which was directly struck by the stone travelling outwards until it reaches the banks. If the ripples pass a cork floating on the surface they make it bob up and down; they do not carry it along with them.

What *does* travel, visibly across the surface of the pond, and invisibly through the air, is a *wave*, or more often a succession of waves. While ripples or waves on the surface of water do have much in common with waves in general, they are not of quite the same type as sound waves *through* air or any other medium (including water).

So let us forget about the ripples now they have served their purpose, and consider a number of boys standing in a queue. If the boy at the rear is given a sharp push he will bump into the boy in front of him, and he into the next boy; and so the push may be passed on right to the front. The push has travelled along the line although the boys—representing the medium of communication—have all been standing still except for a small temporary displacement.

In a similar manner the cricket ball gives the bat a sudden push; the bat pushes the air next to it; that air gives the air around it a push; and so a wave of compression travels outwards in all directions like an expanding bubble. Yet all that any particular bit of air does is to move a small fraction of an inch away from the cause of the disturbance and back again. As the original impulse is spread over an ever-expanding wavefront the extent of this tiny air movement becomes less and less and the sound fainter and fainter.

1.3 Characteristics of Sound Waves

The generation of sound is only incidental to the functioning of cricket bats, but the human vocal organs make up a generator capable of emitting a great variety of sounds at will. What surprises most people, when the nature of sound waves is explained to them, is how such a variety of sound can be communicated by anything so simple as to-and-fro tremors in the air. Is it really possible that the subtle inflexions of the voice by which an individual can be identified, and all the endless variety of music, can be represented by nothing more than that?

Examining the matter more closely we find that this infinite variety of sounds is due to four basic characteristics, which can all be illustrated on a piano. They are:

(a) *Amplitude*. A single key can be struck either gently or hard, giving a soft or loud sound. The harder it is struck, the more violently the piano string vibrates and the greater the tremor in the air, or, to use technical terms, the greater the amplitude of the sound wave. Difference in amplitude is shown graphically in Figure 1.2a. (Note that to-and-fro displacement of the air is represented here by up-and-down displacement of the curve, because horizontal displacement is usually used for time.)

(b) *Frequency*. Now strike a key nearer the right-hand end of the piano. The resulting sound is distinguishable from the previous one by its higher pitch. The piano string is tighter and lighter, so vibrates more rapidly and generates a greater number of sound waves per second. In other words, the waves have a higher frequency, which is represented in Figure 1.2b.

(c) *Waveform*. If now several keys are struck simultaneously to give a chord, they blend into one sound which is richer than any of the single notes. The first three waves in Figure 1.2c represent three such notes singly; when the displacements due to these separate notes are added together they give the more complicated waveform shown on the last line. Actually any one piano note itself has a complex waveform, which enables it to be distinguished from notes of the same pitch played by other instruments.

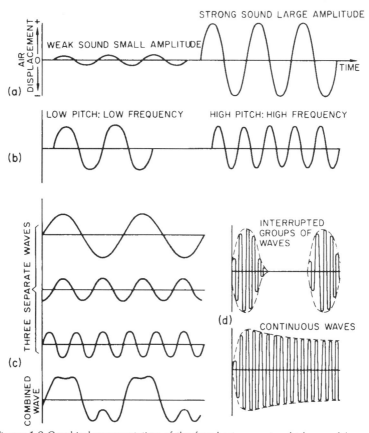

Figure 1.2 Graphical representation of the four basic ways in which sound (or any other) waves can differ: (a) amplitude; (b) frequency; (c) waveform; and (d) envelope

(d) *Envelope*. A fourth difference in sound character can be illustrated by first 'pecking' at a key, getting brief (staccato) notes, and then pressing it firmly down and keeping it there, producing a sustained (legato) note. This can be shown as in Figure 1.2d. The dashed lines, called envelopes, drawn around the wave peaks, have different shapes. Another example of this type of difference is the contrast between the flicker of conversational sound and the steadiness of a long organ note.

Every voice or musical instrument, and in fact everything that can be heard, is something that moves; its movements cause the air to vibrate and radiate waves; and all the possible differences in the sounds are due to combinations of the four basic differences. Mathematically, envelope is covered by waveform, so may not always be listed separately.

The details of human and other sound generators are outside the scope of this book, but the sound waves themselves are the things that have to be carried by long-range communication systems such as radio. Even if something other than sound is to be carried (pictures, for instance) the study of sound waves is not wasted, because the basic characteristics illustrated in Figure 1.2 apply to all waves—including radio waves.

The first (amplitude) is simple and easy to understand. The third and fourth (waveform and envelope) lead us into complications that are too involved for this preliminary survey, and will have to be gone into later. The second (frequency) is very important indeed and not too involved to understand at this stage.

1.4 Frequency

Let us consider the matter as exemplified by the vibrating string or wire shown in Figure 1.3. The upper and lower limits of its vibration are indicated (rather exaggeratedly,

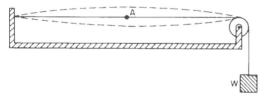

Figure 1.3 A simple sound generator, consisting of a string held taut by a weight. When plucked it vibrates as indicated by the dashed lines

perhaps) by the dashed lines. If a point on the string, such as A, were arranged to record its movements on a strip of paper moving rapidly from left to right, it would trace out a wavy line which would be its displacement/time graph, like those in Figure 1.2. Each complete up-and-down-and-back-again movement is called a *cycle*. In Figure 1.4, where several cycles are shown, one has been picked out in heavy line. The time it takes, marked T, is known as its *period*.

We have already noted that the rate at which the vibrations take place is called the *frequency*, and determines the pitch of the sound produced. Its symbol is f, and it used to be reckoned in cycles per second

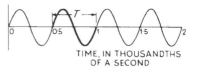

Figure 1.4 The extent of one cycle in a series of waves is shown in heavy line. The time occupied by its generation, or by its passing a given point, is marked T

(abbreviated c/s) but the cycle per second is now called the hertz (abbr. Hz) in honour of Heinrich Hertz for a reason to be given very soon. Middle C in music has a frequency of 261 Hz. The full range* of a piano is 27 to 3516 Hz. The frequencies of audible sounds—about 20–20 000 Hz for young people with normal hearing but much lower for old people—are distinguished as *audio frequencies*.

You may have noticed in passing that T and f are closely related. If the period of each cycle is one-hundredth of a second, then the frequency is obviously one hundred cycles per second. Putting it in general terms, $f = 1/T$. So if the time scale of a waveform graph is given, it is quite easy to work out the frequency. In Figure 1.4 the time scale shows that T is 0.0005 s, so the frequency must be 2000 Hz.

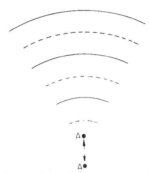

Figure 1.5 End view of the vibrating string at A in Figure 1.3. As it moves up and down over the distance AA it sends out air waves which carry some of its energy of vibration to the listener's ear

1.5 Wavelength

Next, consider the air waves set up by the vibrating string. Figure 1.5 shows an end view, with point A vibrating up and down. It pushes the air alternately up and down, and these displacements travel outwards from A. Places where the air has been temporarily moved a little farther from A than usual are indicated by full lines, and places where it is nearer by dashed lines; these lines should be imagined as expanding outwards at a speed of about 335 metres per second. So if, for example, the string is vibrating at a frequency of 20 Hz, at one second from the start the first air wave will be 335 metres away, and there will be 20

---

* In radio the word 'range' means distance that can be covered, so to avoid confusion a range of frequencies is more often referred to as a *band*.

complete waves spread over that distance (Figure 1.6). The length of each wave is therefore one-twentieth of 335 metres, which is 16.75 metres.

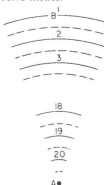

Figure 1.6 Twenty successive waves from the string shown in Figures 1.3 and 1.5

Now it is a fact that sound waves of practically all frequencies and normal amplitudes travel through the air at the same speed, so the higher the frequency the shorter the wavelength. This relationship can be expressed by the equation

$$\lambda = \frac{v}{f}$$

where the Greek letter λ (pronounced 'lambda') stands for wavelength, v for the velocity of the waves, and f for the frequency. If v is given in metres per second and f in hertz, then λ will be in metres. The letter v has been put here instead of 335, because the exact velocity of sound waves in air depends on its temperature; moreover the equation applies to waves in water, wood, rock, or any other substance, if the v appropriate to the substance is filled in.

It should be noted that it is the frequency of the wave which determines the pitch, and that the wavelength is a secondary matter depending on the speed of the wave. That this is so can be shown by sending a sound of the same frequency through water, in which the velocity is 1432 metres per second; the wavelength is therefore more than four times as great as in air, but the pitch, as judged by the ear, remains the same as that of the shorter air wave.

1.6   The Transmitter

The device shown in Figure 1.3 is a very simple sound transmitter or sender. Its function is to generate sound waves by vibrating and stirring up the surrounding air. In some instruments, such as violins, the stirring-up part of the business is made more effective by attaching the vibrating parts to surfaces which increase the amount of air disturbed.

The pitch of the note can be controlled in two ways. One is to vary the weight of the string, which is conveniently done by varying the length that is free to vibrate, as a violinist does with his fingers. The other is to vary its tightness, as a violinist does when tuning, or can be done in Figure 1.3 by altering the weight W. To lower the pitch, the string is made heavier or slacker. As this is not a book on sound there is no need to go into this further, but there will be a lot to say (especially in Chapter 8) on the electrical equivalents of these two things, which control the frequency of radio waves.

1.7   The Receiver

What happens at the receiving end? Reviving our memory of the stone in the pond, we recall that the ripples made corks and other small flotsam bob up and down. Similarly, air waves when they strike an object try to make it vibrate with the same characteristics as their own. That is how we can hear sounds through a door or partition; the door is made to vibrate, and its vibrations set up a new lot of air waves on our side of it. When they reach our ears they strike the ear-drums; these vibrate and stimulate a very remarkable piece of receiving mechanism, which sorts out the sounds and conveys its findings via a multiple nerve to the brain.

1.8   Electrical Communication by Wire

To extend the very limited range of sound-wave communication it was necessary to find a carrier. Nothing served this purpose very well until the discovery of electricity. At first electricity could only be controlled rather crudely, by switching the current on and off; so spoken messages had to be translated into a code of signals before they could be sent, and then translated back into words at the receiving end. A simple electric telegraph consisted of a battery and a switch or 'key' at the sending end, some device for detecting the current at the receiving end, and a wire between to carry the current.

Most of the current detectors made use of the discovery that when some of the electric wire was coiled round a piece of iron the iron became a magnet so long as the current flowed, and would attract other pieces of iron placed near it. If the neighbouring iron was in the form of a flexible diaphragm held close to the ear, its movements towards and away from the iron-cored coil when current was switched on and off caused audible clicks even when the current was very weak.

To transmit speech and music, however, it was necessary to make the diaphragm vibrate in the same complicated way as the sound waves at the sending end. Since the amount of attraction varied with the strength of current, this could be done if the current could be made to vary in the same way as the distant sound waves. In the end this problem was solved by quite a simple device—the carbon microphone, which, with only details improved, was until recently still the type used in telephones.

In its simplest form, then, a telephone consists of the equipment shown in Figure 1.7. Although this invention extended the range of speech from metres to kilometres, every kilometre of line weakened the electric currents and set a limit to clear communication. So it was not until the invention of the electronic valve, making it possible to

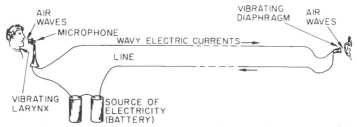

Figure 1.7 An early form of one-way electric telephone, showing how the characteristics of the original air waves are duplicated in electric currents, which travel farther and quicker, and which are then transformed back into air waves to make them perceptible by the listener.

amplify the complicated current variations, that telephoning could be done over hundreds and even thousands of kilometres.

So far we have concentrated on the transmission of sound. Now let us consider how moving images can be sent instantaneously over distances and we shall see that the problem has something in common with sound transmission.

For the purpose of transmitting an image, no matter what its content, it is regarded as made up of a very large number of tiny areas called picture elements (or just elements) arranged in regularly spaced horizontal rows or lines. Each element has its particular brightness, some being white, others black and still others with intermediate shades of grey. The brightnesses are so distributed among the elements that they determine the detail, the highlights and the lowlights of the image we wish to transmit. To give good definition several hundred lines of elements are needed and the individual elements are so small that, at normal viewing distances, they cannot be seen. This idea of dividing an image into elements is also used in the reproduction of photographs in newspapers; these, too, if examined closely, can be seen to be composed of a very large number of elements of different brightnesses (*tones* or *tonal values* are the terms normally used).

To transmit the image the lines of elements are 'read' in the same order as a printed page is read, i.e., the elements in a line are examined in order from left to right after which the reader moves rapidly back to the left to read the next line of elements below as shown in Figure 1.8. In this way all the lines of elements are read (*scanned* is the technical term) and the scanner then returns to the top of the image to begin the scanning process again. At each element the scanner generates an electrical signal proportional to the tonal value of that element. Thus, as scanning proceeds, the scanner output consists of an electrical signal whose strength rises and falls in accordance with the tonal values of the elements being scanned; this signal therefore represents the detail in the image and, not surprisingly, is known as the *picture signal*. Figure 1.9 shows a very simple image and the form of the corresponding picture signal.

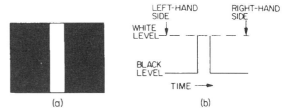

Figure 1.9 (a) A simple picture consisting of a vertical white bar on a black background and (b) the picture signal corresponding to (a) for all scanning lines

At the receiving end a scanner, moving over a screen in the same pattern as that of the transmitter scanner, generates light where it strikes the screen, the intensity of the light being determined by the instantaneous magnitude of the picture signal. In this way a reproduction of the original image is built up on the screen. Synchronising signals are added to the picture signal to make the receiver scanner always precisely in step with that at the transmitter. The combination of picture signal and synchronising signal is known as a *video* signal.

In the cinema the impression of smooth movement in the picture is obtained by projecting 24 pictures per second, persistence of vision ensuring that movement seems natural. The same technique is used in television; in Europe 25 pictures are transmitted per second and the speed of scanning must be such that all the elements composing the image are examined in 1/25th second. It would be impossible to achieve such a scanning speed by mechanical means and in practice electron beams are used for scanning at transmitting and receiving ends of the system. In television camera tubes, for example, the electron beam scans a target on which an image of the scene to be televised is focused. The equipment used for transmitting recorded scenes, i.e., slides and cinema film, similarly employs a cathode-ray tube in which a target is scanned by an electron beam. For reproduction of television pictures a

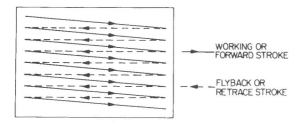

Figure 1.8 Simplified representation of scanning pattern used in television. In practice there are hundreds of lines

cathode-ray tube, known as a picture tube, is again used, the density of the beam which scans the screen being controlled by the video signal.

The foregoing information refers to signals for plain television (i.e., without colour). The main modifications needed for colour come in Section 23.11. By the way 'black and white' is a clumsy and inaccurate description for pictures without colour, and 'monochrome' is wholly wrong because it means 'one colour', whereas white is composed of every colour! So, recalling the old advertising slogan, 'penny plain, twopence coloured', why not 'plain'?

What has been said about plain television applies equally to colour television because in all colour television systems the reproduced image is basically a plain one with areas filled in by colour. In fact all the fine detail in a colour television picture is reproduced in black and white. The colour information is transmitted by additional signals incorporated in the video signal. More information about colour television is given in Section 23.11.

To summarise the last few paragraphs, moving and coloured pictures can be reproduced at a distant point by sending to that point the video signal representing the original scene. The video signal thus corresponds to the audio signal of a sound system, but is more complex and, as we shall see later, contains much higher frequencies. Video signals can be sent to a distant fixed receiver by telephone lines just as audio signals can, but some other means of transmission is needed for sending video and audio signals to ships, aircraft, man-made satellites or the moon.

1.9 Electric Waves

At this stage we can profitably think again of the process of unaided voice communication. When you talk to another person you do not have to transfer the vibrations of your vocal organs directly to the ear-drum of your hearer by means of a rod or other 'line'. What you do is to stir up waves in the air, and these spread out in all directions, shaking anything they strike, including ear-drums.

'Is it possible', experimenters might have asked, 'to stir up *electric* waves by any means, and if so would they travel over greater distances than sound waves?' Gradually scientists supplied the answer. Clerk Maxwell showed mathematically that electric waves were theoretically possible, and indeed that in all probability light, which could easily be detected after travelling vast distances, was an example of such waves. But they were waves of unimaginably high frequency, about 5×10^{14} Hz. A few years later the German experimenter Hertz actually produced and detected electric waves of much lower (but still very high) frequency, and found that they shared with sound the useful ability to pass through things that are opaque to light.

Unlike sound, however, they are not carried by the air. They travel equally well (if anything, better) where there is no air or any other material substance present. So the broadcasting expression 'on the air' (or, worse still, 'air waves' for radio waves) is misleading. If the experiment of Figure 1.1 is repeated with a source of electric waves instead of sound waves, extracting the air makes no appreciable difference. We know, of course, that light and heat waves travel to us from the sun across 149 668 620 kilometres of empty space. What does carry them is a debatable question. It was named ether (or aether, to avoid confusion with the anaesthetic liquid ether), but experiments which ought to have given results if there had been any such thing just didn't. So scientists deny its existence. But it is very hard for lesser minds to visualise how electric waves, whose behaviour corresponds in so many respects with waves through a material medium, can be propagated by nothing.

There is no doubt about their speed, however. Light waves travel through space at about 300 000 (more precisely 299 792) kilometres per second. Other sorts of electric waves, such as X-rays, ultra-violet and radio waves, travel at the same speed, and differ only in frequency. Their length is generally measured in metres; so filling in the appropriate value of v in the formula in Section 1.5 we have, very nearly,

$$\lambda = \frac{300\ 000\ 000}{f}$$

If, for example, the frequency is 1 000 000 Hz, the wavelength is 300 metres.

1.10 Why High Frequencies are Necessary

Gradually it was discovered that electric waves of these frequencies, called radio waves, are capable of travelling almost any distance, even round the curvature of the earth to the antipodes.

Their enormous speed was another qualification for the duty of carrying messages; the longest journey in the world takes less than a tenth of a second. But attempts to stir up radio waves of the same frequencies as sound waves were not very successful. To see why this is so we may find it helpful to consider again how sound waves are stirred up.

If you try to radiate air waves by waving your hand you will fail, because the highest frequency you can manage is only a few hertz, and at that slow rate the air has time to rush round from the front to the back of your hand and vice versa instead of piling up and giving the surrounding air a push. If you could wave your hand at the speed of an insect's (or even a humming bird's) wing, then the air would have insufficient time to equalise the pressure and would be alternately compressed and rarefied at the front or back of your hand and so would generate sound waves. Alternatively, if your hand were as big as the side of a house, it could stir up air waves even if waved only a few times per second, because the air would have too far to go from front to back every time. It is true that the frequency of these air waves would be too low to be heard, but they could be detected by the rattling of windows and doors all over the neighbourhood.

The same principle applies to radio waves. But because of the vastly greater speed with which they can rush round, it is necessary for the aerial or antenna—which corresponds to the waving hand—to be correspondingly vaster. To radiate radio waves at the lower audible frequencies the antenna would have to be miles high, so one might just as well (and more conveniently) use it on the ground level as a telephone line.

Even if there had not been this difficulty, another would have arisen when large numbers of transmitters had started to radiate waves in the audible frequency band. There would have been no way of picking out the one wanted; it would have been like a babel of giants.

That might have looked like the end of any prospects of radio telephony, but human ingenuity was not to be beaten. The solution arose by way of the simpler problem of radio telegraphy, so let us follow that way.

1.11 Radio Telegraphy

Hertz discovered how to stir up radio waves; we will not bother about exactly how, because his method is obsolete. Their frequency was what we would now call ultra-high, in the region of hundreds of millions of hertz, so their length was only a few centimetres. Such frequencies are used nowadays for radar. He also found a method of detecting them over distances of a few yards. With the more powerful transmitters and more sensitive receivers developed later, ranges rapidly increased, until in 1901 Marconi is believed to have signalled across the Atlantic. The various sorts of detectors that were invented from time to time worked by causing an electric current to flow in a local receiving circuit when radio waves impinged on the receiving antenna. Human senses are unable to respond to these 'wireless' waves directly, and even if they could, the frequency is usually too high for the ear and far too low for the eye—but the electric currents resulting from their detection can be used to produce audible or visible effects.

At this stage we have the radio counterpart of the simple telegraph. Its diagram, Figure 1.10, more or less explains itself. The sending key turns on the transmitter, which generates a rapid succession of radio waves, radiated by an antenna. When these reach the receiver they operate the detector and cause an electric current to flow, which can make a noise in a telephone earpiece. By working the sending key according to the Morse code, messages can be transmitted. That is roughly the basis of radio telegraphy to this day, though some installations have been elaborated almost out of recognition, and print the messages on paper faster than the most expert typist.

1.12 Tuning

One important point to note is that the current coming from the detector does not depend on the frequency of the waves, within wide limits. (The frequency of the current is actually the frequency with which the waves are turned on and off at the transmitter.) So there is a wide choice of wave frequency. In Section 17.12 we shall consider how the choice is made; but in the meantime it will be sufficient to remember that if the frequency is low, approaching audibility, the antennas have to be immense; while if the frequency is much higher than those used by Hertz it is difficult to generate them powerfully, and they are easily obstructed, like light waves. The useful limits of these *radio* frequencies are about 15 000 to 50 000 000 000 Hz.

The importance of having many frequencies to choose from was soon apparent, when it was found to be possible—and highly advantageous—to make the receiver respond to the frequency of the transmitter and reject all others. This invention of *tuning* was essential to effective radio communication, and is the subject of a large part of this book. So it may be enough just now to point out that it has its analogy in sound. We can adjust the frequency of a sound wave by means of the length and tightness of the sender. A piano contains nearly a hundred strings of various fixed lengths and tightnesses, which we can select by means of the keys. If there is a second piano in the room, with its sustaining pedal held down so that all the strings are free to vibrate, and we strike a loud note on the first piano, the string tuned to the same note in the second can be heard to vibrate. If we had a single adjustable string, we could tune it to respond in this way to any one note of the piano and ignore those of substantially different frequency. In a corresponding manner the electrical equivalents of length (or weight) and tightness are adjusted by the tuning controls of a radio receiver to select the desired sending station.

1.13 Radio Telephony

The second important point is that (as one would expect) the strength of the current from the detector increases with

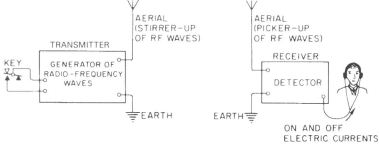

Figure 1.10 Elementary radio telegraph. Currents flow in the receiving headphones and cause sound whenever the transmitter is radiating waves

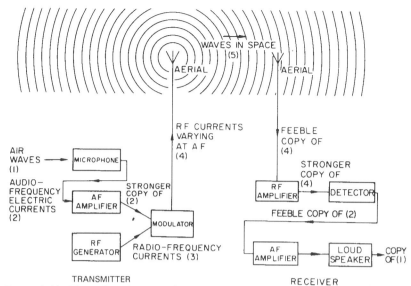

Figure 1.11 The radio telegraph of Figure 1.10 converted into a radio telephone by controlling the output of the transmitter by sound waves instead of just turning it on and off. In practice it is also generally necessary to use amplifiers where shown

the strength of the waves transmitted. So, with Figure 1.7 in mind, we need no further hint to help us to convert the radio telegraph of Figure 1.10 into a radio telephone or a broadcasting system, by substituting for the Morse key a device for varying the strength of the waves, just as the microphone varied the strength of the line current in Figure 1.7. Such devices are called *modulators*, and will be referred to in more detail in Chapter 15. The microphone is still needed, because the output of the radio transmitter is not controlled directly by the impinging sound waves but by the wavily varying currents from the microphone.

A practical system needs amplifiers in various places, firstly to amplify the feeble sound-controlled currents from the microphone until they are strong enough to modulate a powerful wave transmitter; then to amplify the feeble radio-frequency currents stirred up in a distant receiving antenna by the far flung waves; then another to amplify the audio-frequency currents from the detector so that they are strong enough to work a loudspeaker instead of the less convenient headphones. This somewhat elaborated system is indicated—but still only in broadest outline—by the block diagram, Figure 1.11.

We have already shown the analogy between audio and video signals and in order to obtain a simple block diagram of a television broadcasting system we can, in Figure 1.11, replace the microphone by a television camera, the a.f. amplifier by a video amplifier and the loudspeaker by a picture tube. We must remember, however, that the camera tube and picture tube require sources of power for vertical and horizontal deflection of their electron beams and that a television system includes a sound system. So another system such as that illustrated in Figure 1.11 is required to convey the sound associated with the picture.

1.14 Recapitulation

At this stage it may be as well to review the results of our survey. We realise that in all systems of communication the signals must, in the end, be detectable by the human senses. Of these, hearing and seeing are the only ones that matter. Both are wave-operated. Our ears respond to air waves between frequency limits of about 20 and at most 20 000 Hz; our eyes respond to waves within a narrow band of frequency centred on 5×10^{14} Hz—waves not carried by air but by space, and which have been proved to be electrical in nature.

Although our ears are remarkably sensitive, air-wave (i.e. sound) communication is very restricted in range. Eyes, it is true, can see over vast distances, but communication is cut off by the slightest wisp of cloud. If, however, we can devise means for 'translating' audible or visible waves into and out of other kinds of waves, we have a far wider choice, and can select those that travel best. It has been found that these are electric waves of lower frequencies than the visible ones, but higher than audible frequencies. The lowest of the radio frequencies require excessively large antennas to radiate them, and the highest are impeded by the atmosphere. All travel through space at the same speed of nearly 300 000 000 metres per second, so the wavelength in metres can be calculated by dividing that figure by the frequency in Hz.

The frequency of the transmitter is arranged by a suitable choice and adjustment of the electrical equivalents of weight and tightness (or, more strictly, slackness) in the tuning of radio-frequency generators. In the same way the receiver can be made to respond to one frequency and reject others. There are so many applicants for frequencies—broadcasting authorities, telegraph and telephone authorities, naval, army, and air forces, space organisations, weather stations, police, the merchant navies, airlines, research scientists, amateurs and others—that even with such a wide band to choose from it is difficult to find enough for all, especially as only a small part of the whole band is generally suitable for any particular kind of service.

The change-over from the original sound waves to radio waves of higher frequency is done in two stages: first by a microphone into electric currents having the same frequencies as the sound waves; these currents are then used in a modulator to control the radio-frequency waves in such a way that the original sound characteristics can be extracted at the receiver. Here the change-back is done in the same two stages reversed: first by a detector, which yields electric currents having the original sound frequencies; then by earphones or loudspeakers which use these currents to generate sound waves.

Since the whole thing is an application of electricity, our detailed study must obviously begin with the general principles of electricity. And it will soon be necessary to pay special attention to electricity varying in a wavelike manner at both audio and radio frequencies. This will involve the electrical characteristics that form the basis of tuning. To generate the radio-frequency currents in the transmitter, as well as to amplify and perform many other services, use is made of various types of electronic devices, such as valves and transistors, so it is necessary to study these in some detail. The process of causing the r.f. currents to stir up waves, and the reverse process at the receiver, demands some knowledge of antennas and radiation. Other essential matters that need elucidation are modulation and detection. Armed with this knowledge we can then see how they are applied in typical receivers. By then, most of the basic principles of electronics in general will have been covered, but a few more chapters will be needed to fill in the special requirements in things such as radar, electronic instruments, computers and microprocessors.

Chapter 2 Electricity and Circuits

2.1 Electrons

The exact nature of electricity is a mystery that may never be fully cleared up, but from what is known it is possible to form a sort of working model or picture which helps us to understand how it produces the results it does, and even to think out how to produce new results. The reason why the very existence of electricity went unnoticed until comparatively recent history was that there are two opposite kinds which exist in equal quantities and, unless separated in some way, cancel one another out. This behaviour reminded the investigating scientists of the use of positive (+) and negative (−) signs in arithmetic: the introduction of either +1 or −1 has a definite effect, but +1−1 equals just nothing. So, although at that time hardly anything was known about the two kinds of electricity, they were called positive and negative. Both positive and negative electricity produce very remarkable effects when they are separate, but a combination of equal quantities shows no signs of electrification.

Further research led to the startling conclusion that all matter consists largely of electricity. All the thousands of different kinds of matter, whether solid, liquid, or gaseous, consist of atoms which contain only a very few different basic components. Of these we need take note of only one kind—particles of negative electricity, called *electrons*. For a simple study it is convenient to imagine each atom as a sort of ultra-microscopic solar system in which a number of electrons are distributed around a central nucleus, somewhat as our earth and the other planets around the sun (Figure 2.1). (The subject is treated in much more detail in M. G. Scroggie's *The Electron in Electronics*). The nucleus is generally a more or less composite structure; it makes up nearly all the weight of the atom, and the number and type of particles it contains determine which element it is— oxygen, carbon, iron, etc. It is possible to change some elements into others by breaking up the nucleus under very intense laboratory treatment. For our present purpose, however, the only things we need remember about the nucleus are that it is far heavier than the electrons, and that it has a surplus of positive electricity.

In the normal or unelectrified state of the atom this positive surplus is exactly neutralised by the planetary electrons. But it is a comparatively easy matter to dislodge one or more of these electrons from each atom. One method is by simply rubbing; for example, if a glass rod is rubbed vigorously with silk some of the electrons belonging to the glass are transferred to the silk. Doing this produces no change in the material itself—glass is still glass and the silk remains silk. As separate articles they show no obvious evidence of the transfer. But if they are brought close together it is found that they attract one another. And the glass rod can pick up small scraps of paper. It is even possible, in a dry atmosphere, to produce sparks. These curious phenomena gradually fade away, and the materials become normal again.

2.2 Electric Charges and Currents

The surplus of electrons on the silk is called a negative electric *charge*. The glass has a corresponding deficiency of electrons, so its positive nuclear electricity is incompletely neutralised, and the result is a positive charge. It is a pity that the people who decided which to call positive did not know anything about electrons, because if they had they would certainly have called electrons positive and so spared us a great deal of confusion. As it is, however, one just has to remember that a *surplus* of electrons is a *negative* charge.

Any unequal distribution of electrons is a condition of stress. Forcible treatment of some kind is needed in order to create a surplus or deficiency anywhere, and if the electrons get a chance they move back to restore the balance; i.e., from negatively to positively charged bodies. This tendency shows itself as an attraction between the bodies themselves. That is what is meant by saying that opposite charges attract.

The space between opposite charges, across which this attraction is exerted, is said to be subject to an *electric field*. The greater the charges, the more intense is the field and the greater the attractive force.

If a number of people are forcibly transferred from town A to town B, the pressure of overcrowding and the intensity of desire to leave B depend, not only on the number of displaced persons, but also on the size of B and possibly on whether there is a town C close by with plenty of accommodation. The electrical pressure also depends on other things than the charge, reckoned in displaced electrons; so it is more convenient to refer to it by another term, namely *difference of potential*, often abbreviated to

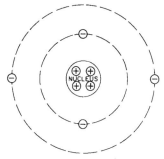

Figure 2.1 Diagrammatic picture of an atom, showing the nucleus surrounded by a number of electrons

p.d. We shall consider the exact relationship between charge and p.d. in the next chapter. The thing to remember now is that wherever there is a difference of potential between two points there is a tendency for electrons to move from the point of lower (or negative) potential to that of higher (or positive) potential in order to equalise the distribution.* If they are free to move they will do so; and their movement is what we call an *electric current*.

This is where we find it so unfortunate that the names 'positive' and 'negative' were allocated before anybody knew about electrons. For it amounted to a guess that the direction of current flow was the opposite to that in which we now know electrons flow. By the time the truth was known this bad guess had become so firmly established that reversing it would have caused worse confusion. Moreover, when the positively charged atoms (or positive ions, as they are called) resulting from the removal of electrons are free to do so they also move, in the direction + to −, though much more slowly owing to their greater mass. So this book follows the usual custom of talking about current flowing from + to − or high to low potential, but it must be remembered that most often this means electrons moving from − to +.

2.3 Conductors and Insulators

Electrons are not always free to move. Except in special circumstances (such as high temperature, which we shall consider in Chapter 9) atoms do not allow their electrons to fly off completely on their own, even when they are surplus. The atoms of some substances go so far as to allow frequent exchanges, however, like dancers in a Paul Jones, and in fact such exchanges go on all the time, even in the normal unelectrified state. The directions in which the electrons flit from one atom to another are then completely random, because there is nothing to influence them one way or another.

But suppose the whole substance is pervaded by an electric field. The electrons feel an attraction towards the positive end; so between partnerships they tend to drift that way. The drift is what is known as an electric current, and substances that allow this sort of thing to go on are called *conductors* of electricity. All the metals are more or less good conductors; hence the extensive use of metal wire. Carbon and some liquids are fairly good conductors.

Note that although electrons start to go in at the negative end and electrons start to arrive at the positive end the moment the p.d. is set up, this does not mean that these are the same electrons, which have instantaneously travelled the whole length of the conductor. An electron drift starts almost instantaneously throughout miles of wire, but the speed at which they drift is seldom much more than an inch a minute.

---

* Note that +1 is positive with respect to −1 or even to 0, but it is negative with respect to +2; −1 or 0 are both positive with respect to −2. There is nothing inconsistent in talking about two oppositely charged bodies one minute, and in the next referring to them both as positive (relative to something else).

Other substances keep their electrons, as it were, on an elastic leash which allows them a little freedom of movement, but never 'out of sight' of the atoms. If such a substance occupies the space between two places of different potential, the electrons strain at the leash in response to the positive attraction, but a continuous steady drift is impossible. Materials of this kind, called *insulators*, can be used to prevent charges from leaking away, or to form boundaries restricting currents to the desired routes. Dry air, glass, polythene, rubber, and paraffin wax are among the best insulating materials. None is absolutely perfect, however; electrified glass rods, for example, gradually lose their charges.

We shall see in Chapter 9 that there is a third class of substances called *semiconductors*, of immense importance in electronics.

2.4 Electromotive Force

If electrons invariably moved from − to +, as described, they would in time neutralise all the positive charges in the world, and that, for all practical purposes, would be the end of electricity. But fortunately there are certain appliances, such as batteries and dynamos (or generators) which can force electrons to go from + to −, contrary to their natural inclination. In this way they can continuously replenish a surplus of electrons. Suppose A and B in Figure 2.2 are two insulated metal terminals, and a number of electrons have been transferred from A to B, so that A is at a higher potential than B. If now they are joined by a wire, electrons will drift along it, and, if that were all, the surplus would soon be used up and the potential difference would disappear. But A and B happen to be the terminals of a battery, and as soon as electrons leave B the battery provides more, while at the same time it withdraws electrons arriving at A. So while electrons are moving from B to A through the wire, the battery keeps them moving from A to B through itself. The battery is, of course, a conductor; but if it were only that it would be an additional path for electrons to go from B to A and dissipate the charge all the quicker. It is remarkable, then, for being able to make electrons move against a p.d. This ability is called *electromotive force* (usually abbreviated to e.m.f.).

Figure 2.2 shows what is called a *closed circuit*, there being a continuous endless path for the current. This is invariably necessary for a continuous current, because if the circuit were opened, anywhere, say by disconnecting the

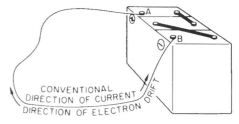

Figure 2.2 A very simple electrical circuit, round which the electromotive force of a battery makes a current of electrons flow continuously

wire from one of the terminals, continuation of the current would cause electrons to pile up at the break, and the potential would increase without limit, which is impossible. So the electron flow ceases throughout the conducting path the moment the path is broken at any point.

2.5 Electrical Units

The amount or strength of an electric current might reasonably be reckoned as the number of electrons passing any point in the circuit each second, but the electron is so extremely small that such a unit would lead to inconveniently large numbers. For practical purposes it has been agreed to base the unit on the metric system. It is called the *ampere* (or more colloquially the amp), and happens to be about 6 240 000 000 000 000 000 (or 6.24×10^{18}) electrons per second.

As one would expect, the number of amps caused to flow in given circuit depends on the strength of the electromotive force operating in it. The practical unit of e.m.f. is the *volt*.

Obviously the strength of current depends also on the circuit—whether it is made up of good or bad conductors. This fact is expressed by saying that circuits differ in their electrical *resistance*. The resistance of a circuit or of any part of it can be reckoned in terms of the e.m.f. required to drive a given current through it. For convenience the unit of resistance is made numerically equal to the number of volts required to cause one amp to flow, and to avoid the cumbersome expression 'volts per amp' this unit of resistance has been named the *ohm*.

By international agreement the following letter symbols have been allocated to denote these electrical quantities and their units, and will be used from now on:

Quantity	Symbol for quantity	Unit of quantity	Symbol for unit
E.M.F.	E	Volt	V
Current	I	Ampere	A
Resistance	R	Ohm	Ω

(Ω is a Greek capital letter omega)

These three important quantities are not all independent, for we have just defined resistance in terms of e.m.f. and current. Expressing it in symbols:

$$R = \frac{E}{I}$$

Note that sloping letters (italics) are used to denote quantities, and upright ones (Roman) for the units. This is always done; see Section A.4.

2.6 Ohm's Law

The question immediately arises: does R depend only on the circuit, or does it depend also on the current or voltage? This was one of the first and most important investigations into electric currents. We can investigate it for ourselves if we have an instrument for measuring current, called an *ammeter*, a battery of identical cells, and a length of thin wire or other resistive conductor. A closed circuit is formed of the wire, the ammeter, and a varying number of the cells; and the ammeter reading corresponding to each number of cells is noted (Figure 2.7). The results can be plotted in the form of a graph (Figure 2.3). If the voltage of a cell is known, the current can be plotted against voltage instead of merely against number of cells.

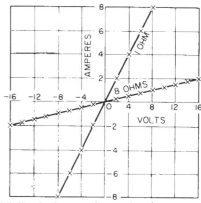

Figure 2.3 Graph showing the results of an experiment on the relationship between current in amps and e.m.f. in volts, for two different circuits

When all the points are joined up by a line it will probably be found that the line is straight and inclined at an angle and passes through the origin (O). (If the line is not quite straight, this may be for the reason given at the end of Section 2.11.) If a different resistance is tried, its line will slope at a different angle. Figure 2.3 shows two possible samples resulting from such an experiment. In this case each cell gave 2 volts, so points were plotted at 2, 4, 6, 8, etc., volts. The points at –2, –4, etc., volts were obtained by reversing the battery. A current of –2 A means a current of the same strength as +2 A, but flowing in the opposite direction.

Looking at the steep line, we see that an e.m.f. of 2 V caused a current of 2 A, 4 V caused 4 A, and so on. The result of dividing the number of volts by the number of amps is always 1. And this, according to our definition, is the resistance in ohms. So there is no need to specify the current at which the resistance must be measured, or to make a graph; it is only necessary to measure the voltage required to cause any one known current (or the current caused by any one voltage). The amount (or, as one says in technical language, the value) of resistance so obtained can be used to find the current at any other voltage, or voltage at any other current. For these purposes it is more convenient to write the relationship $R = E/I$ in the form

$$I = \frac{E}{R} \quad \text{or} \quad E = IR$$

Thus if any two of these quantities are known, the third can be found.

Conductors whose resistance does not vary with the amount of current flowing through them are described as *linear*, because the line representing them on a graph of the Figure 2.3 type (called their *current/voltage characteristic*) is straight. Although the relationship $I = E/R$ is true for any resistance, its greatest usefulness lies in the fact, discovered by Dr Ohm, and known as Ohm's law, that ordinary conductors are linear. From Chapter 9 onwards we shall come across some exceptions. And often the relationship $I = E/R$ is itself rather loosely called Ohm's law.

As an example of the use of Ohm's law, we might find, in investigating the value of an unknown resistance, that when it was connected to the terminals of a 9 V battery a current of 0.01 A was driven through it. Using Ohm's law in the form $R = E/I$, we get for the value of the resistance $9/0.01 = 900\,\Omega$.

Alternatively, we might know the value of the resistance and find that an old battery, nominally of 9 V, could only drive a current of 0.007 A through it. We could deduce, since $E = I \times R$, that the voltage of the battery had fallen to $900 \times 0.007 = 6.3$ V. Note that in driving current through a resistance an e.m.f. causes a p.d. of the same voltage between the ends of the resistance. This point is elaborated in Section 2.14.

2.7 Larger and Smaller Units

It is unusual to describe a current as 0.007 ampere, as was done just now; one speaks of '7 milliamperes', or, more familiarly, '7 milliamps'. A milliampere is one-thousandth part of an ampere. Several other prefixes are used; they are tabulated in Appendix D, the commonest being

Prefix	Meaning	Symbol
milli-	one thousandth of	m
micro-	one millionth of	μ
nano-	one thousandth-millionth of	n
pico-	one billionth of (million millionth)	p
kilo-	one thousand	k
mega-	one million	M
giga-	one thousand million	G

The prefixes can be put in front of any unit; one speaks commonly of milliamps, microamps, megohms, and so on. 'Half a megohm' is easier to say than 'Five hundred thousand ohms', and '½ MΩ' is quicker to write than '500 000 Ω'.

It must be remembered, however, that Ohm's law in the forms given on p.12 assumes volts, ohms, and amps. If a current of 5 mA is flowing through 15 000 Ω, the voltage across that resistance will not be 75 000 V. But since most of the currents we shall be concerned with are of the milliamp order it is worth noting that there is no need to convert them to amps if R is expressed in thousands of ohms (kΩ). So in the example just given one can get the correct answer, 75 V, by multiplying 5 by 15.

2.8 Circuit Diagrams

At this stage it will be as well to start getting used to circuit diagrams. In a book like this, concerned with general principles rather than with constructional details, what matters about (say) a battery is not its shape or the design of its label, but its voltage and how it is connected in the circuit. So it is a waste of time drawing a picture as in Figure 2.2. All one needs is the symbol as shown in Figure 2.4a, which represents one cell. An accumulator cell gives 2 V, and a dry cell about 1.4 V. The longer stroke represents the + terminal. To get higher voltages, several cells are connected one after the other, as in Figure 2.4b, and if so many are needed that it is tedious to draw them all one can represent all but the end ones by a dashed line, as in Figure 2.4c, which also shows how to indicate the voltage.

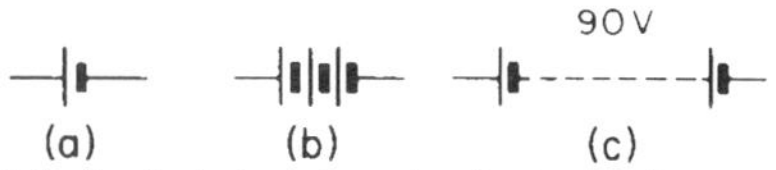

Figure 2.4 A first instalment of circuit-diagram symbols, representing (a) a cell, (b) a battery of three cells and (c) a battery of many cells

It is easier to identify the politicians depicted in our daily papers in the cartoons than in the news photos. There is a sense in which the distorted versions of those individuals presented by the caricaturist are more like them than they are themselves; the distinguishing features are picked out and reduced to conventional forms that can be recognised at a glance.

In a circuit diagram each component or item that has a significant electrical effect is represented by a conventional symbol, and the wires connecting them up are shown as lines. Fortunately, except for a few minor differences or variations, these symbols are recognisably the same all over the world. Most of them indicate their functions so clearly that one could guess what they mean. However, they are introduced as required in this book.

One convention about which there are differences of opinion is the way in which crossing wires should be distinguished from connected wires. Figure 2.5a shows the commonest method. Crossing wires are assumed to be unconnected unless marked by a blob, as at b. There is obviously a risk that either the blob may be omitted where needed or may form unintentionally (especially when the lines are drawn in ink) where not wanted. Even when correctly drawn the difference is not very conspicuous. Method b ought never to be used, especially along with a. Connecting crossings should always be staggered, as at c. To distinguish unconnected crossings more clearly they are sometimes drawn as at d or e. To make quite sure, the author uses methods a and c respectively to denote unconnected and connected crossings. This is in accordance with the recommendations of BS 3939: *Graphical Symbols for Electrical and Electronics Diagrams*.

Another variation is that some people omit to draw a ring round the symbols that represent the 'innards' of a valve or transistor. In this book they are drawn, because they make

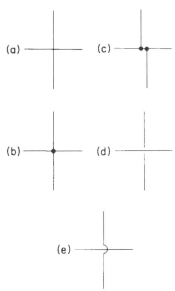

Figure 2.5 Alternative methods of distinguishing crossing from connected wires in circuit diagrams

the transistors or valves, which are usually key components, stand out distinctly.

There is a good deal more in the drawing of circuit diagrams than just showing all the right symbols and connections. If the letters composing a message were written at random all over the paper, nobody would bother to read it even if the correct sequence were shown by a maze of connecting lines—unless it was a clue to valuable buried treasure. And a circuit diagram, though perfectly accurate, is difficult to 'read' if it is laid out in an unfamiliar manner. The layout has by now become largely standardised, though again there are some variations. It is too soon in the book to deal with layout in detail, but here are some general principles for later reference.

The first is that the diagram should be arranged for reading from left to right. In a radio receiver, for example, the antenna should be on the left and the final item, the loudspeaker, on the right.

Next, there is generally an 'earthed' or 'low-potential' connection. This should be drawn as a thicker line, usually straight across the diagram. The most positive potential connections should be at the top and the most negative at the bottom. This helps one to visualise the potential distribution and the direction of current flow (normally downwards).

Long runs of closely parallel wires are difficult to follow and should be avoided.

As with written symbols, facility in drawing and reading circuit diagrams soon comes with practice.

One warning is needed in connection with circuit diagrams, especially when going on to more advanced work. They are an indispensably convenient aid to thought, but it is possible to allow one's thoughts to be moulded too rigidly by them. The diagram takes for granted that all electrical properties are available in separate lumps, like chemicals in bottles, and that one can make up a circuit like a prescription. Instead of which they are more like the smells from the said chemicals when the stoppers have been left open—each one strongest near its own bottle, but pervading the surrounding space and mixing inextricably with the others. This is particularly true at very high frequencies, so that the diagram must not always be taken as showing the whole of the picture. Experience tells one how far a circuit diagram can be trusted.

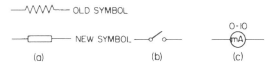

Figure 2.6 An instalment of circuit-diagram symbols representing (a) a resistor, (b) a switch and (c) a milliammeter

Figure 2.6 shows a group of much-used symbols: *a* is a circuit element called a *resistor*, because resistance is its significant feature; *b* a switch shown 'open' or 'off'; and *c* a measuring instrument, the type of which can be shown by 'mA' for milliammeter, 'V' for voltmeter, etc., and the range of measurement marked as here. The original resistor symbol, the zigzag, is still in use but the alternative rectangular symbol, widely used on the Continent, is now recommended by the British Standards Institution and is used throughout this book.

A simple line represents an electrical connection of negligible resistance, often called a lead (rhyming with 'feed').

The circuit diagram for the Ohm's law experiment can therefore be shown as in Figure 2.7. The arrow denotes a movable connection, used here for including anything from one to eight cells in circuit.

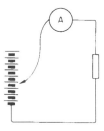

Figure 2.7 Circuit diagram of the apparatus used for obtaining Figure 2.3

Figure 2.7 is a simple example of a *series* circuit. Two elements are said to be connected *in series* when, in tracing out the path of the current, we encounter them one after the other. With steadily flowing currents the electrons do not start piling up locally, so it is clear that the same current flows through both elements. In Figure 2.7 for instance, the current flowing through the ammeter is the same as that through the resistor and the part of the battery 'in circuit'.

Two elements are said to be *in parallel* if they are connected so as to form alternative paths for the current; for example, the resistors in Figure 2.10a. Since a point cannot be at two different potentials at once, it is obvious that the same potential difference exists across both of them (i.e., between their ends or terminals). One element in parallel with another is often described as *shunting* the other.

The method of connection can sometimes be regarded as either series or parallel, according to circumstances. In Figure 2.8, R_1 and R_2 are in series with battery B_1, and in parallel with B_2.

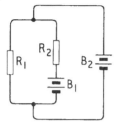

Figure 2.8 R_1 is in series with R_2 and B_1, but in parallel with it and B_2

Looking at the circuit diagrams of radio sets—especially television receivers—one often sees quite complicated networks of resistors. Fortunately Ohm's law can be applied to every part of a complicated system of e.m.f.s and resistances as well as to the whole. Beginners often seem reluctant to make use of this fact, being scared by the apparent difficulty of the problem. So let us see how it works out with more elaborate circuits.

2.9 Resistances in Series and in Parallel

Complicated circuit networks can be tackled by successive stages of finding a single element that is equivalent (in the quantity to be found) to two or more. Consider first two resistances R_1 and R_2, in series with one another and with a source of e.m.f. E (Figure 2.9a).

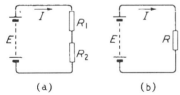

Figure 2.9 Resistances in series. The circuit (b) is equivalent to the circuit (a), in the sense that both take the same current from the battery, if $R = R_1 + R_2$

To bring this circuit as a whole within the scope of calculation by Ohm's law we need to find the single resistance R (Figure 2.9b) equivalent to R_1 and R_2 together.

We know that the current in the circuit is everywhere the same; call it I. Then by Ohm's law, the voltage across R_1 is IR_1, and that across R_2 is IR_2. The total voltage across both must therefore be $IR_1 + IR_2$, or $I(R_1 + R_2)$, which must be equal to the voltage E.

In the equivalent circuit, E is equal to IR, and since to make the circuits truly equivalent, the currents must be the same in both for the same battery voltage, we see that

$$R = R_1 + R_2$$

Generalising from this result, we conclude that: *The equivalent resistance of several resistances in series is equal to the sum of their individual resistances.* It follows that if a number of resistances are connected in series, the effective resistance is always greater than any individual resistance.

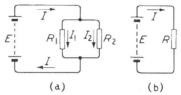

Figure 2.10 Resistances in parallel. The circuit (b) is equivalent to the circuit (a), in the sense that both take the same current from the battery, if $1/R = 1/R_1 + 1/R_2$

Turning to the parallel-connected resistances of Figure 2.10a, we see that they have the same voltage across them; in this case the e.m.f. of the battery. Each of these resistances will take a current depending on its own resistance and on the e.m.f.—the simplest case of Ohm's law. Calling the currents respectively I_1 and I_2 we therefore know that $I_1 = E/R_1$ and $I_2 = E/R_2$. The total current drawn is the sum of the two: it is

$$I = \frac{E}{R_1} + \frac{E}{R_2} = E\left(\frac{1}{R_1} + \frac{1}{R_2}\right)$$

In the equivalent circuit of Figure 2.10b the current is E/R, which may also be written $E(1/R)$. Since, for equivalence between the circuits, the current must be the same for the same battery voltage, we see that

$$\frac{1}{R} = \frac{1}{R_1} + \frac{1}{R_2}$$

Generalising from this result, we conclude that: *The reciprocal of the equivalent resistance of several resistances in parallel is equal to the sum of the reciprocals of their individual resistances.* It follows from this that the effective resistance of a number of resistances in parallel is always less than the smallest of them.

If the resistances of Figure 2.10a were 100 Ω and 200 Ω, the single resistance R which, connected in their place, would draw the same current is given by $1/R = 1/100 + 1/200 = 0.01 + 0.005 = 0.015$. Hence, $R = 1/0.015 = 66.67$ Ω. This could be checked by adding together the individual currents through 100 Ω and 200 Ω, and comparing the total with the current taken from the same voltage source by 66.67 Ω. In both cases the result is 15 mA per volt of battery.

We can summarise the two rules in symbols:

1. Series Connection: $R = R_1 + R_2 + R_3 + R_4 + \ldots$
2. Parallel Connection: $1/R = 1/R_1 + 1/R_2 + 1/R_3 \ldots$

For only two resistances in parallel, a more convenient form of the same rule can be obtained by multiplying above and below by R_1R_2:

$$R = \frac{1}{1/R_1 + 1/R_2} = \frac{R_1 R_2}{R_1 + R_2}$$

$$= \frac{\text{product of resistances}}{\text{sum of resistances}}$$

2.10 Series-Parallel Combinations

How the foregoing rules, derived directly from Ohm's law, can be applied to the calculation of more complex circuits can perhaps best be illustrated by a thorough working-out of one fairly elaborate network, Figure 2.11. We will find the total current flowing, the equivalent resistance of the whole circuit, and the voltage and current of every resistor individually.

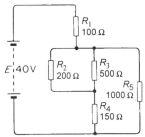

Figure 2.11 The current through and voltage across each resistor in this complicated network can be calculated by applying the two simple rules derived from Figures 2.9 and 2.10

The policy is to look for any resistances that are in simple series or parallel. The only two in this example are R_2 and R_3. Writing R_{23} to symbolise the combined resistance of R_2 and R_3 taken together, we know that $R_{23} = R_2 R_3/(R_2 + R_3) = (200 \times 500)/700 = 142.8\,\Omega$. This gives us the simplified circuit of Figure 2.12a. If R_{23} and R_4 were one resistance, they and R_5 in parallel would make another simple case, so we proceed to combine R_{23} and $R_4 = R_{23} + R_4 = 142.8 + 150 = 292.8\,\Omega$. Now we have the circuit of Figure 2.12b. Combining R_{234} and R_5, $R_{2345} = (292.8 \times 1000)/1292.8 = 226.5\,\Omega$. This brings us within sight of the end; Figure 2.12c shows us that the equivalent resistance of the network now is simply the sum of the two remaining resistances; that is, R in Figure 2.12d is $R_{2345} + R_1 = 226.5 + 100 = 326.5\,\Omega$.

From the point of view of current drawn from the 40 V source the whole system of Figure 2.11 is equivalent to a single resistor of this value. The current taken from the battery will therefore be $40/326.5 = 0.1225\,\text{A} = 122.5\,\text{mA}$.

To find the current through each resistor individually now merely means applying Ohm's law to some of our previous results. Since R_1 carries the whole current of 122.5 mA, the potential difference across it will be $100 \times 0.1225 = 12.25\,\text{V}$. R_{2345} also carries the whole current (Figure 2.12c); the p.d. across it will again be the product of resistance and current, in this case $226.5 \times 0.1225 = 27.75\,\text{V}$. This same voltage also exists, as comparison of the various diagrams will show, across the whole complex system $R_2 R_3 R_4 R_5$ in Figure 2.11. Across R_5 there exists the whole of this voltage; the current through this resistor will therefore be $27.75/1000\,\text{A} = 27.75\,\text{mA}$.

The same p.d. across R_{234} of Figure 2.12b, or across the system $R_2 R_3 R_4$ of Figure 2.11, will drive a current of $27.75/292.8 = 94.75\,\text{mA}$ through this branch.

The whole of this flows through R_4 (Figure 2.12a), the voltage across which will accordingly be $150 \times 0.09475 = 14.21\,\text{V}$. Similarly, the p.d. across R_{23} in Figure 2.12a, or across both R_2 and R_3 in Figure 2.11, will be $0.09475 \times 142.8 = 13.54\,\text{V}$, from which we find that the currents through R_2 and R_3 will be respectively $13.54/200$ and $13.54/500\,\text{A}$, or 67.68 and $27.07\,\text{mA}$, making up the required total of 94.75 mA for this branch.

This completes an analysis of the entire circuit; we can now collect our scattered results in the form of the following table.

	Resistance (Ω)	Current through it (mA)	Potential difference across it (V)
R_1	100	122.5	12.25
R_2	200	67.68	13.54
R_3	500	27.07	13.54
R_4	150	94.75	14.21
R_5	1000	27.75	27.75
R	326.5	122.5	40

Note that by using suitable resistors any potential intermediate between those given by the terminals of the battery can be obtained. For instance, if the lower and

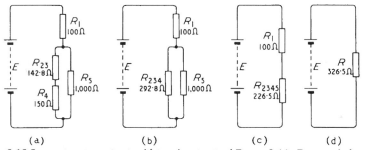

Figure 2.12 Successive stages in simplifying the circuit of Figure 2.11. R_{23} stands for the single resistance equivalent to R_2 and R_3; and so on. R represents the whole system

upper ends of the battery in Figure 2.11 are regarded as 0 and +40 (they could equally be reckoned as −40 and 0, or −10 and +30, with respect to any selected level of voltage), the potential of the junction between R_3 and R_4 is 14.21 V. The arrangement is therefore called a *potential divider*, and is often employed for obtaining a desired potential not given directly by the terminals of the source. If a sliding connection is provided on a resistor, to give a continuously variable potential, it is generally known—though not always quite justifiably—as a *potentiometer*.

2.11 Resistance Analysed

So far we have assumed the possibility of almost any value of resistance without considering exactly what determines the resistance of any particular resistor or part of a circuit. We understand that different materials vary widely in the resistance they offer to the flow of electricity, and can guess that with any given material a long piece will offer more resistance than a short piece, and a thin piece more than a thick piece. We have indeed actually proved as much and more; by the rule for resistances in series the resistance of a uniform wire is exactly proportional to its length—doubling the length is equivalent to adding another equal resistance in series, and so on (Figure 2.13a). Similarly, putting two equal pieces in parallel, which halves the resistance, is equivalent to doubling the thickness; so resistance is proportional to the reciprocal of thickness, or, in more precise language, to the cross-section area (Figure 2.13b). Altering the shape of the cross-section has no effect on the resistance—with steady currents, at least. Figure 2.13c shows how doubling the *diameter* (i.e. thickness) of a piece of wire divides the resistance by *four*.

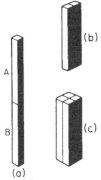

Figure 2.13 The rule for resistances in series proves that doubling the length of a uniform conductor doubles its resistance (a). By the rules for resistances in parallel, doubling its cross-section area halves its resistance (b); and doubling its diameter quarters its resistance (c)

To compare the resistances of different materials it is necessary to bring them all to the same standard length and cross-section area, which, in the system of units known as SI (Système Internationale d'Unités), is one metre long by one square metre, a metre cube, in fact (*not* a cubic metre!). This comparative resistance for any material is called its *resistivity* (symbol: ρ pronounced 'roe'). Knowing the resistivity (and it can be found in almost any electrical reference book*) one can easily calculate the resistance of any wire or piece having uniform section by multiplying by the length in metres (l) and dividing by the cross-section area in square metres (A). In symbols:

$$R = \frac{\rho l}{A}$$

Such calculations are seldom necessary, because there are tables showing the resistance per metre of all the different gauges of wire, both for copper (usually used for parts of the circuit where resistance should be as low as possible) and for the special alloys intended to have a high resistance.

Resistance varies to some extent with temperature, so the tables show the temperature at which they apply, and also the proportion by which the resistance increases with rise of temperature.

2.12 Conductance

It is often more convenient to work in terms of the *ease* with which a current can be made to flow, rather than the difficulty; in other words, in *conductance* rather than resistance. The symbol for conductance is G, and its unit (called the siemens, S; formerly, the mho) is equal to the number of amps caused to flow by one volt. So $I = EG$ is an alternative form of $I = E/R$; and $G = I/R$.

It is not difficult to see that corresponding to the equation for calculating resistance there will be one for conductance:

$$G = \frac{\sigma A}{l}$$

where σ (Greek 'sigma') stands for conductivity (siemens per metre).

The advantage of conductance appears chiefly in parallel circuits, for which it is easy to see that the formula is $G = G_1 + G_2 + \ldots$.

2.13 Kirchhoff's Laws

We have already taken it as obvious that no one point can be at more than one potential at the same time. (If it is not obvious, try to imagine a point with a difference of potential within itself; then electrons will flow from itself to itself, i.e., a journey of no distance and therefore non-existent.) So if you start at any point on a closed circuit and go right round it, adding up all the voltages on the way, with due regard for + and −, the total is bound to be zero. If it isn't, a mistake has been made somewhere; just as a surveyor must have made a mistake if he started from a certain spot, measuring his ascents as positive and his descents as negative, and his figures told him that when he returned to his starting-point he had made a net ascent or descent.

---

* In the older books, resistivity is given in ohms per *centimetre* cube, or ohm-cm. Resistivity in metre units is a number 100 times smaller.

This simple principle is the modern form of one of what are known as Kirchhoff's laws, and is a great help in tackling networks of resistances where the methods already described fail because there are no two that can be simply combined. One writes down an equation for each closed loop or mesh, by adding together all the voltages and equating the total to zero, and one then solves the resulting simultaneous equations. This can be left for more advanced study, but the same law is a valuable check on any circuit calculations. Try applying it to the several closed loops in Figure 2.11, such as that formed by E, R_1, and R_5. If we take the route clockwise we go 'uphill' through the battery, becoming more positive by 40 V. Coming down through R_1 we move from positive to negative so adding -12.25 V, and, through R_5, -27.75 V, reaching the starting point again. Check: $40 - 12.25 - 27.75 = 0$.

Taking another clockwise route via R_3, R_5, and R_4, we get $13.54 - 27.75 + 14.21 = 0$. And so on for any closed loop.

The other Kirchhoff law is equally obvious; it states that if you count all currents arriving at any point in a circuit as positive, and those leaving as negative, the total is zero. This can be used as an additional check, or as the basis of an alternative method of solving circuit problems.

2.14 P.D. and E.M.F.

The fact that both e.m.f. and p.d. are measured in volts can be confusing. So it will be time well spent if, before going any further, we look into the matter rather carefully.

We have just been likening difference of potential to difference in height. This analogy is implied whenever we talk about low and high potential. But although we measure height in feet or metres what we have in mind is not just a *distance*; we are really concerned with the force of gravity close to a massive body, usually the earth. ('Height' has no meaning in outer space.) This force tends to make things move downwards, or fall. Corresponding to this gravitational field is the electric field between opposite electric charges, which tends to make mobile charges move in the direction of that field, towards the low-potential or negative end if the mobile charges are positive ones, but 'upwards' if they are electrons. Just as an effort, which we might call a weight-motive force, is required to lift a weight against the force of a gravitational field, so an electromotive force is needed to lift positive electric charges to a point of higher potential; that is to say, against the force of an electric field. And just as the amount of weight-motive force could be reckoned as the height it could lift a weight—any weight—so e.m.f. is specified as the p.d. it can overcome. If water has to be supplied to the top floor of a 60 m block of flats, then a water-motive force which could be specified as not less than 60 m must be provided by a pump or other active device. The difference in height that the pressure of the water source can overcome is the most directly informative way of stating it. Similarly the p.d. in volts that a source of electricity can overcome is the most helpful way of stating its e.m.f., which indeed is sometimes referred to as its pressure.

Another reason for reckoning e.m.f. in volts is that it cannot be measured as such, but only in terms of the p.d. it can overcome or create.

Note that the p.d. between two points is something quite definite, but the potential of a point (like the height of a hill) is relative, depending on what one takes as the reference level. Heights are usually assumed to be relative to sea level, unless it is obvious that some other 'zero' is used (such as street level for the height of a building). The potential of the earth is the corresponding zero for electrical potentials, where no other is implied. To ensure definiteness of potential, apparatus is often connected to earth.

The p.d. across a resistor (or, as it is sometimes called, the voltage drop, or just 'drop') is, as we know, equal to its resistance multiplied by the current through it; and the positive end is that at which the current enters (or electrons leave). The positive end of a source of e.m.f., on the other hand, is that at which current would leave if its terminals were joined by a passive conductor. That is because an e.m.f. has the unique ability of being able to drive a current against a p.d. Note, however, that the positive terminal of a source of e.m.f. is not always the one at which current actually does leave; there may be a greater e.m.f. opposing it in the circuit.

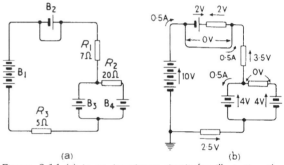

(a) (b)

Figure 2.14 (a) is an imaginary circuit for illustrating the meanings of e.m.f. and p.d. The voltages and currents in different parts of it are marked in (b). Arrows alongside voltages show the direction of rise of potential

Figure 2.14a shows a circuit invented to illustrate questions that can arise in the early stages of studying these matters. Let us assume that each cell in it has an e.m.f. of 2 V.

All five cells in the battery B_1 are connected in series and all tend to drive current clockwise, so the total clockwise e.m.f. of B_1 is $5 \times 2 = 10$ V, and we can show this as in Figure 2.14b.

B_2 seems to be a flat contradiction of Kirchhoff's voltage law. Here we have an e.m.f. of 2 V between two points which must be at the same potential because they are joined by a wire of negligible resistance! The cell is, as we say, *short-circuited*, or 'dead-shorted'. Part of the answer to this apparent impossibility is that no real source of e.m.f. can be entirely devoid of resistance. Often it is not enough to matter, so is not shown in diagrams. But where the source is short-circuited it is vital, because it is the only thing to limit the current. To represent a source of e.m.f. completely, then, we must include between its terminals not only a symbol for e.m.f. but also one for resistance. When a source of e.m.f. is open-circuited, so that no current flows, there is no voltage drop in its internal resistance, so the terminal voltage is equal to the e.m.f. When current is drawn the terminal voltage falls, and if the source is dead-shorted the internal drop must be equal to the e.m.f., as shown in Figure 2.14b. The other part of the answer is that when the source has a very low resistance (e.g., a large accumulator or generator) such an enormous current flows that even a wire of normally negligible resistance does not reduce the terminal voltage to zero, and the wire is burnt out, and perhaps the source too.

In our example we shall assume that all the resistance is internal, so the external effect of B_2 is nil.

If the loop containing B_3, B_4, and R_2 is considered on its own as a series circuit, we see two e.m.f.s each of 4 V in opposition, so the net e.m.f. is zero, and no current due to them flows through R_2. As regards the main circuit, B_3 and B_4 are in parallel, and contribute an e.m.f. of –4 V (because in opposition to B_1, which we considered positive). This can be indicated either by '–4 V' and a clockwise arrow, or by '4 V' and an anticlockwise arrow as shown.

It would be wise to stop for a moment and re-read that last sentence. So often there are disputes about which is the 'right' direction for a voltage or current direction arrow, when all the time either direction could be right. The important thing is to say what the arrow is intended to mean—direction of positive voltage rise or current flow, or direction of actual voltage rise or current flow (Section 5.9).

We see that putting a second battery in parallel does not increase the main circuit e.m.f., but it does reduce the internal resistance by putting two in parallel, so is occasionally done for that reason. In our example the 20 Ω in series with B_4 is probably far greater than the internal resistance of B_3, so practically all the current will flow through B_3, and the effect of B_4 will be negligible.

We have, then (ignoring B_2), 10 – 4 = 6 V net clockwise e.m.f. in the circuit, and (neglecting the internal resistances of B_1 and B_3) a total resistance of 7 + 5 = 12 Ω. The current round the circuit is therefore 6/12 = 0.5 A, as marked, and the voltage drops across R_1 and R_3 are 0.5 × 7 = 3.5 V and 0.5 × 5 = 2.5 V respectively.

Finally, the symbol at the bottom left-hand corner means that the negative end of B_1 is connected to earth (or 'earthed'). So its potential is reckoned as zero, and the potential of every other point in the circuit is thereby defined.

To distinguish between e.m.f. and p.d. the symbol E is sometimes reserved for e.m.f., p.d.s being denoted by V. But often there is no need to make any distinction, and we shall see in Section 5.11 that attempting to do so can lead to controversy and confusion, so there is much to be said for reckoning in voltage rises (falls being negative rises), as for example, those indicated in Figure 2.14b.

2.15 Electrical Effects

It is time now to see what electricity can *do*. One of the most familiar effects of an electric current is heat. It is as if the jostling of the electrons in the conductor caused a certain amount of friction. But whereas it makes the wire in an electric fire red hot, and in an electric light bulb white hot, it seems to have no appreciable effect on the flex and other parts of the circuit—in a reputable installation, anyway. Reasoning of the type we used in connection with Figure 2.13 shows that the rate at which heat is generated in a conductor containing no e.m.f.s is proportional to the product of the current flowing through it and the voltage across it (i.e., the p.d. between its ends); in symbols, EI. Since both E and I are related to the resistance of the conductor, we can substitute IR for E or E/R for I, getting I^2R or E^2/R as alternative measures of the heating effect. Thus for a given current the heating is proportional to the resistance. The wires connecting electric lamps and heaters to the generator are chosen to have a resistance low compared with that of the electric devices themselves. I^2R shows that the heating increases very rapidly as the current rises. Advantage is taken of this in fuses which are short pieces of wire that melt (or 'blow') and cut off the current if, owing to a short-circuit somewhere, it rises dangerously above normal.

Another effect of an electric current is to magnetise the space around it. This effect is particularly strong when the wire is coiled round a piece of iron. As we shall see, it is of far more significance then merely being a handy way of making magnets.

An effect that need not concern us much is the production of chemical changes, especially when the current is passed through watery liquids. A practical example is the charging of batteries. The use of the word 'charging' for this process is unfortunate, for no surplus or deficiency of electrons is accumulated by it.

All these three effects are reversible; that is to say, they can be turned into causes, the effect in each case being an e.m.f. The production of an e.m.f. by heating the junction of two different metals (a *thermocouple*) is of only minor importance, but magnetism is the basis of all electrical generating machinery, and chemical changes are very useful on a smaller scale in batteries.

2.16 Instruments for Measuring Electricity

All three effects can be used for indicating the strength of currents; but although thermocouple ammeters (Section 5.5) are used in radio transmitters, the vast majority of current meters are based on the magnetic effect. There are two main kinds: the moving-iron instrument, in which the current coil is fixed, and a small piece of iron with a pointer attached moves to an extent depending on the amperage; and the moving-coil (generally preferred) in which the coil is deflected by a fixed permanent magnet. Besides these current effects there is the potential effect—the attraction between two bodies at different potential. The force of attraction is seldom enough to be noticeable, but if the p.d. is not too small it can be measured by a delicate instrument. This is the principle of what is called the electrostatic voltmeter, in which a rotatable set of small metal vanes is held apart from an interleaving fixed set by a hairspring. When a voltage is applied between the two sets the resulting attraction turns the moving set through an angle depending on the voltage. This type of instrument, which is a true voltmeter because it depends directly on potential and draws no current, is practical and convenient for full-scale readings from about 1000 V to 10 000 V, but not for low voltages. ('Full-scale reading' means the reading when the pointer is deflected to the far end of the scale.)

Since V or $E = IR$, voltage can be measured indirectly by measuring the current passing through a known resistance. If this additional conducting path were to draw a substantial current from a source of voltage to be measured, it might give misleading results by lowering that voltage appreciably; so R is made relatively large. The current meter (usually of the moving-coil type) is therefore a milliammeter, or better still a microammeter. For example, a voltmeter for measuring up to 1 V could be made from a milliammeter reading up to 1 mA by adding sufficient resistance to that of the moving coil to bring the total up to 1000 Ω. The same instrument could be adapted to read up to 10 V (all the scale readings being multiplied by 10) by adding another resistance, called a *multiplier*, of 9000 Ω, bringing the total to 10 000 Ω. Such an arrangement, shown in Figure 2.15a, can obviously be extended.

The same milliammeter can of course be used for measuring currents up to 1 mA by connecting it in series in the current path; but to avoid increasing the resistance of the circuit more than need be, the voltage resistances would have to be short-circuited or otherwise cut out. Care must be taken never to connect a current meter directly across a voltage source, for its low resistance might pass sufficient current to destroy it. Higher currents can be measured by diverting all but a known fraction of the current through a bypass resistance called a *shunt*. If, for example, the resistance of the 0–1 milliammeter is 75 Ω, shunting it with 75/9 = 8.33 Ω would cause nine times as much current to flow through the shunt as passes through the meter, thereby multiplying the range by 10 (Figure 2.15b).

Thus a single moving-coil instrument can be made to cover many ranges of current and voltage measurement, simply by connecting known resistances as shunts and multipliers. The operation can be reversed by incorporating a battery sufficient to give full-scale deflection; then, when an *unknown* resistance is connected, either as a shunt or multiplier, according to whether it is small or large, the extent to which the reading is reduced depends on the value of that resistance. It is therefore possible to graduate the scale in ohms and we have an *ohmmeter* (Figure 2.15c).

2.17 Electrical Power

When everyday words have been adapted for scientific purposes by giving them precise and restricted meanings, they are more likely to be misunderstood than words that were specially invented for use as scientific terms—'electron', for example. *Power* is a word of the first kind. It is commonly used to mean force, or ability, or authority. In technical language it has only one meaning: rate of doing work. *Work* here is also a technical term, confined to purely physical activity such as lifting weights or forcing things to move against pressure or friction—or potential difference. If by exerting muscular force you lift a 10 kg weight 0.5 m against the opposing force of gravity, you do $10 \times 0.5 = 5$ kg m of work. And if you take 1 second to do it, your rate of doing work—your power output—is 5 kg m per second, which is about 1/20th of a horse-power.

Correspondingly, if the electromotive force of a battery raises the potential of 10 electrons by 3 V (i.e., causes them to flow against an opposing p.d. of 3 V) it does 30 electron-volts (eV) of work; and if it takes one second to do it the output of power is 30 volt-electrons per second. The electron per second is, as we saw, an extremely small unit of electric current, and it is customary to use one more than 6×10^{18} times larger—the ampere. So the natural choice for the electrical unit of power might be called the volt-ampere. Actually, for brevity (and for another reason which appears in Chapter 6), it has been given the name *watt*.

We now have the following to add to the table which was given in Section 2.5:

Quantity	Symbol for quantity	Unit of quantity	Symbol for unit
Electrical power	P	Watt	W

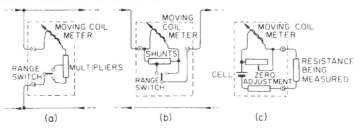

Figure 2.15 Showing how a single moving-coil indicating instrument can be used for measuring (a) voltage, (b) current, (c) resistance. The moving coil and pointer attached are shown diagrammatically

We also have the equation $P = EI$. So when your 3 V torch battery is lighting a 0.2 A bulb it is working at the rate of $0.2 \times 3 = 0.6$ W. In this case the battery e.m.f. is occupied in forcing the current through the resistance of the filament in the bulb, and the resulting heat is visible. We have already noted that the rate at which heat is produced in a resistance is proportional to EI—the product of the e.m.f. applied and the current flowing—and now we know this to be a measure of the power expended.

Alternative forms of the power relationship can be found in Section 2.15:

$$P = I^2R = \frac{E^2}{R}$$

Thus if any two of the four quantities E, I, R, and P are given, all are known. For example, if a 1000 Ω resistor is to be used in a circuit where 50 V will be maintained across it, the power dissipated as heat will be $50^2/1000 = 2.5$ W. In choosing such a resistor one must take care not only that the resistance is correct but also that it is large enough to get rid of heat at the rate equivalent to 2.5 W without reaching such a high temperature as to damage itself or anything near it.

To familiarise oneself with power calculations it would be a good exercise to add another column to the table in Section 2.10, headed 'Power (watts)', working out the figures for each resistor and checking that they are the same whether derived from EI, I^2R, or E^2/R. A further check is to add up all the wattages for R_1 to R_5 and see that the total is the same as that for R.

It is often necessary to be able to tell whether a certain part of a circuit is acting as a giver (or *generator*, or *source*) of power, or as a receiver (or *load*). The test is whether the current is going through it in the direction of rising potential, in which case power is being generated there, or of falling potential, in which case power is being absorbed. In Figure 2.14 the current is going through B_1 from – to +, so B_1 is a generator, as one would expect of a battery. But it is going through B_3 from + to –, so although in a different circuit B_3 could be a generator, in this one it is a load, into which power is being delivered, perhaps for charging it. Through *any* resistance alone, current can only flow from + to –, because a resistance is essentially a user-up of power.

2.18 A Broader View of Resistance

It will be found later on in this book that there are other ways in which electrical power can be 'lost' than by heating up the circuit through which the current flows. This leads to a rather broader view of resistance; instead of defining it as E/I, resulting from an experiment such as that connected with Figure 2.3, one defines it as P/I^2 (derived from $P = I^2R$), P being the expenditure of power in the part of the circuit concerned, and I the current through it. The snag is that P is often difficult to measure. But as a definition to cover the various sorts of 'resistance' encountered in radio it is very useful.

A feature of electrical power expended in resistance—however defined—is that it leaves the circuit for good. Except by some indirect method it is not possible to recover any of it in the form of electrical power. But there are ways in which electrical power can be employed to create a store of energy that can be released again as electrical energy. *Energy* here is yet another technical term, meaning the amount of work that such a store can do. It can be reckoned in the same units as work, and so power can be defined alternatively as rate of release of energy. To return to our mechanical analogy, 5 kg m of energy expended by exerting a force of 10 kg to push a box across the floor is all dissipated as heat caused by friction.* But 5 kg m used in raising a 10 kg weight 0.5 m is stored as what is called potential energy. If the weight is allowed to descend, it delivers up the 5 kg m of energy, which could be used to drive a grandfather clock for a week. Another way of storing the energy would be to push a heavy truck mounted on ball bearings along rails. In this case very little of the pressure would be needed to overcome the small amount of friction; most of it would have the effect of giving it momentum which would keep it going long after the push had finished. A heavy truck in motion is capable of doing work because the force setting it in motion has been used to store energy in it; this is called *kinetic* energy.

In the next chapter we shall begin the story of how electrical energy can be stored and released in two ways, corresponding to potential and kinetic energy, and how this makes it possible to tune in to different stations on the radio or television.

---

* Electrical units are defined in terms of mechanical forces, etc., in metric units. Because of the relationships between metric and British units, a power of 746 watts is equal to 550 ft-lb per second (1 horse-power). It is confirmed by experiment that the rate of heating is the same, whether produced by dissipation of 1 mechanical HP or 746 electrical watts.

Chapter 3 Capacitance

3.1 Charging Currents

Although most of the last chapter was about steady e.m.f.s driving currents through resistances, you may remember that the whole thing began with the formation of an electric charge. We likened it to the deportation of unwilling citizens from town A to town B. Transferring a quantity of electrons from one place to another and leaving them there sets up a stress between the two places, which is only relieved when the same quantity of electrons has been allowed to flow back. The total amount of the stress is the p.d., measured in volts. And the place with the surplus of electrons is said to be negatively charged. The original method of charging—by friction—is inconvenient for most purposes and has been generally superseded by other methods. If an e.m.f. of, say, 100 V is used, electrons will flow until the p.d. builds up to 100 V; then the flow will cease because the charging and discharging forces are exactly equal.

It was at this stage that we provided a conductive path between the negatively and positively charged bodies, allowing the electrons to flow back at a rate proportional to the p.d.; and, as this p.d. is being maintained by a constant e.m.f., a continuous steady current is set up. This state of affairs could be represented as in Figure 3.1. Our attention then became attached to this continuous current and the circuit through which it flowed. The electron surplus at the negative end—and the deficit at the positive end—seemed to have no bearing on this, and dropped out of the picture. So long as e.m.f. and current are quite steady, the charges can be ignored.

But not during the process of charging. While that is going on there must be more electrons entering B than are leaving it; and vice versa during discharging. Assuming that there are parts of a circuit requiring a substantial

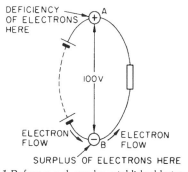

Figure 3.1 Before a p.d. can be established between any two parts of a circuit (such as A and B) they must be charged, and when the p.d. is rapidly varied the charging current may be important

number of electrons to charge them to a potential equal to the applied e.m.f., then it is clear that the ordinary circuit principles we have been studying become more complicated whenever the e.m.f. varies. For one thing, it seems as if the current is no longer the same in all parts of a series circuit. The moment the e.m.f. is applied, a charging current starts to flow, over and above any current through conducting paths. If the e.m.f. is increased at any time, a further charging current flows to raise the p.d. correspondingly. Reducing the e.m.f. causes a discharging current. Since telecommunications and most other applications of electronics involve e.m.f.s that are continually varying, it is evidently important to know how far the charging and discharging currents affect matters.

3.2 Capacitance: What It Is

The first thing is to find what decides the number of electrons needed to charge any part of a circuit. The number of people that could be transferred to another town would probably depend on (*a*) the pressure brought to bear on them, (*b*) the size of the town and (*c*) its remoteness. The number of electrons that are transferred as a charge certainly depends on the electrical pressure applied—the e.m.f. As one would expect, it is proportional to it. If the battery in Figure 3.1 gave 200 V instead of 100, the surplus of electrons forced into B would be just twice as great. The quantity of electrons required to charge any part of a circuit to 1 V is called its *capacitance*.

We have already noted in Section 2.5 that as a practical unit of electric current one electron-per-second would be ridiculously small and the number of electrons per second that fits the metric system (SI) is rather more than 6×10^{18} (= 1 A). So it is natural to take the same number of electrons as the unit of electric charge or quantity of electricity (symbol: Q). It is therefore equal to the number that passes a given point every second when a current of 1 A is flowing, and the name for it is the *coulomb*. So the unit of capacitance is that which requires 1 coulomb to charge it to 1 volt. It is named the *farad* (abbreviation: F) in honour of Michael Faraday, who contributed so much to electrical science. In symbols, the relationship between electric charge, voltage and capacitance is

$$Q = VC$$

which should be compared with the relationship between electric current, voltage, and conductance (Section 2.12). V is used instead of E because fundamentally it is the p.d. that is involved rather than the charging e.m.f. If the conducting path and e.m.f. in Figure 3.1 were both removed, leaving A and B well insulated, their charge and the p.d. would remain. In practice one need not bother to distinguish between e.m.f. and p.d. every time (Section 2.14). All that matters is the voltage, and E and V do equally well.

As capacitance is equal to the charge required to set up a given potential *difference* between two parts of a circuit or other objects, one has to speak of the capacitance *between* those objects, or of one to another.

3.3 Capacitance Analysed

We saw that when any object is being charged the current going into it is greater than that coming out. So if Kirchhoff's law about currents is true in all circumstances, the object must be more than just a point; it must have some size. We would expect the capacitance of objects to one another to depend in some way on their size; the question is, what way?

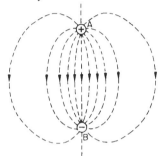

Figure 3.2 The dashed lines indicate the direction and relative intensity of electric field between two oppositely charged terminals, A and B

When analysing resistance (Section 2.11) we found that it depended on the dimensions of the conductor and what it was made of. So, of course, does conductance, though in inverse ratio (Section 2.12). Capacitance, however, has to do with the space between the conductors; it results from the force of attraction that can be set up in that space, or from what is called the electric field. Suppose A and B in Figure 3.2 are our oppositely charged terminals. The directions along which the force acts are indicated by dashed lines, called *lines of electric force*. These show the paths electrons would take if they were free to move slowly under the force of attraction to +. (But to agree with the convention for direction of electric currents the arrows point in the opposite direction.) To represent a more intense field, lines are drawn more closely together. Each can be imagined to start from a unit of positive charge and end on a unit of negative charge. It must be understood that it is a matter of convenience what size of unit is taken for this purpose; and that the lines are not meant to convey the idea that the field itself is in the form of lines, with no field in the spaces between.

The fact that the space between two small, widely separate objects such as these is not confined to a uniform path makes the capacitance between them a more difficult thing to calculate than the resistance of a wire. But consider two large flat parallel plates, shown edge-on in Figure 3.3a. Except for a slight leakage around the edges, the field is confined to the space directly between the plates, and this space, where the field is uniformly distributed, is a rectangular slab.

Suppose, for the sake of example, the plates are charged to 100 V. Since the field is uniform, the potential changes at a uniform rate between one plate and the other. Reckoning the potential of the lower plate as zero, the potential half way between the plates is therefore +50 V. A third plate could be inserted here, and we would then have two capacitances in series, each with a p.d. of 50 V. Provided that the thickness of the third plate was negligible, its presence would make no difference to the amount of the charge required to maintain 100 V across the outer plates. Each of the two capacitances in series would therefore have the same charge but only half the voltage. To raise their p.d. to 100 V each, the amount of charge would have to be doubled (Figure 3.3b). In other words (remembering that $C = Q/V$) halving the spacing doubles the capacitance. In general, C is inversely proportional to the distance between the plates (which we can denote by t, for thickness of the space).

Next, imagine a second pair of plates, exactly the same as those in Figure 3.3a. Obviously they would require the same quantity of electricity to charge them to 100 V. If the two pairs were joined together as in Figure 3.3c, they would form one unit having twice the cross-section area and requiring twice the charge to set up a given p.d., in other words, twice the capacitance. In general, C is directly proportional to the cross-section area of the space between the plates (which we can denote by A).

So the capacitance between two plates separated by a slab of space in which the electric field is uniform depends on the dimensions of the slab in the same way as does the conductance of a piece of material (Section 2.12):

$$C = \frac{\varepsilon A}{t}$$

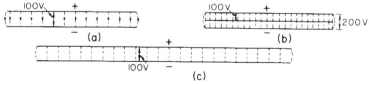

Figure 3.3 Consideration of the electric field between closely spaced parallel plates leads to the relationship between dimensions and capacitance

Here ε ('epsilon') is the 'constant', corresponding to conductivity, needed to link the chosen system of units with the experimentally observed facts. Formerly the symbol κ was used. One would expect it to be called 'capacitivity', but actually the name is *absolute permittivity*. When, as in SI units (Section 2.11), C is reckoned in farads, t in metres, and A in square metres, and the space is completely empty, ε turns out to be 8.854×10^{-12}. This particular value is sometimes called the 'electric space constant' and is given the distinctive symbol ε_0.

Filling the space with air makes very little difference; but if you use solid or liquid insulating materials such as glass, paper, plastics or oil, the capacitance is considerably increased. The amount by which the capacitance is multiplied in this way by filling the whole of the space with such a material is called the *relative permittivity* of the material, or often just *permittivity* (symbol: ε_r).

Insulating materials in this role are called *dielectrics*, and another name for relative permittivity is *dielectric constant*. For most solids and liquids it lies between 2 and 10, but for special materials it may be as much as several thousands.

Just as the embodiment of resistance is a resistor, a circuit element designed for its capacitance is called a *capacitor*; the former name was *condenser*. Usually it consists of two or more parallel plates spaced so closely that most of the field is directly between them, as in Figure 3.3.

The capacitance of a capacitor thus depends on three factors: area and thickness of the space between the plates, and permittivity of any material there. For convenience, the permittivity specified for materials is the relative value, ε_r; and since ε is ε_r times ε_0 the formula is usually adapted for ε_r by filling in the value of ε_0:

$$C = \frac{8.854 \times 10^{-12} \varepsilon_r A}{t}$$

This practice is so common that the little r is often omitted, 'ε' being understood to be the relative value.

Supposing, for example, the dielectric had a permittivity of 5 and was 0.1 cm thick, the area needed to provide a capacitance of 1 F would be about 23 square kilometres. In practice the farad is far too large a unit, so is divided by a million into microfarads (μF), or even by 10^{12} into picofarads (pF). Adapting the above formula to the most convenient units for practical purposes, we have:

$$C = \frac{\varepsilon A}{11.3 t} \text{ pF}$$

the dimensions being in *centimetres*.

3.4 Capacitors

The general symbol for a capacitor in circuit diagrams is Figure 3.4a. The capacitors themselves appear in great variety in circuits, according to the required capacitance and other circumstances. Some actually do consist of a single sheet of insulating material sandwiched between a pair of metal plates, but sometimes there are a number of plates on each side as indicated in Figure 3.4b. The effective area of one dielectric is of course multiplied by their number—in this case four.

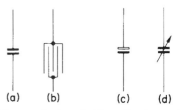

Figure 3.4 The standard symbol for a fixed capacitor is at *a*: it often has many plates interleaved as at *b*. Electrolytic capacitors are distinguished as at *c*, in which the white plate is positive. *d* denotes a variable capacitor

A very common form of capacitor consists of two long strips of aluminium foil separated by polyester film and rolled up into a compact block. Above a few microfarads it is often more economical to separate the aluminium electrodes by chemically impregnated paper or other material which causes an extremely thin insulating film to form on one of them. Apart from this filmforming, the chemical solution also acts as a conductor constituting one plate of the capacitor. The aluminium foil is the other plate and the two plates are separated by the film which acts as a dielectric. These *electrolytic* capacitors must always be connected and used so that the terminal marked + never becomes negative relative to the other. Even when correctly used there is always a small leakage current. To help the wireman, the symbol Figure 3.4c, in which the + plate is distinguished by outlining, is used for electrolytic types.

Air is seldom used as the dielectric for fixed capacitors, but is quite common in variable capacitors. These, indicated in diagrams as 3.4d, consist of two sets of metal vanes which can be progressively interleaved with one another, so increasing the effective area, to obtain any desired capacitance up to the maximum available, seldom more than 500 pF. Plastic film dielectric is used in variable capacitors where compactness is essential.

Besides the capacitance, an important thing to know about a capacitor is its maximum working voltage. In Section 2.3 we visualised insulators as substances in which the electrons are unable to drift through the material in large numbers under the influence of an electric field but instead they 'strain at the leash', being slightly displaced from their normal positions in the atoms. If the intensity of the electric field is increased beyond a certain limit the elastic leashes tethering the electrons snap under the strain, and the freed electrons rush uncontrollably to the positive plate, just as if the dielectric were a good conductor. It has, in fact, been broken down or punctured by the excessive voltage. For a given thickness, mica stands an exceptionally high voltage, or, as one says, has a relatively high dielectric strength. Air is less good, but has the advantage that the sparking across resulting from breakdown does it no permanent harm.

In our concentration on intentional capacitance between plates, we should not forget that, whether we like it or not, every part of a circuit has capacitance to surrounding parts.

When the circuit e.m.f.s are varying rapidly, these stray capacitances may be very important.

Semiconductor diodes can also behave as capacitors and these are described in Chapter 9.

3.5 Charge and Discharge of a Capacitor

It is instructive to consider the charging process in greater detail. In Figure 3.5 the capacitor can be charged by moving the switch to A, connecting it across a battery; and discharged by switching to B. R might represent the resistance of the wires and capacitor plates, but as that is generally very small it will be easier to discuss the matter if we assume we have added some resistance, enough to bring the total up to, say, $200\,\Omega$. Suppose the capacitance C is $5\,\mu F$ and the battery e.m.f. 100 V.

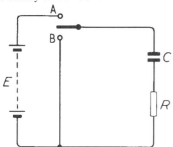

Figure 3.5 Circuit used for obtaining the charge and discharge curves shown in Figure 3.6

At the exact moment of switching to A the capacitor is as yet uncharged, so there can be no voltage across it (Section 3.2); the whole of the 100 V of the battery must therefore be occupied in driving current through R, and by Ohm's law that current is found to be 0.5 A (extreme left of Figure 3.6a). Reckoning the potential of the negative end of the battery as zero, we have at this stage the positive end of the battery, the switch, the upper plate of the capacitor, and (because the capacitor has no p.d. across it) the lower plate also, all at +100 V. The lower end of the resistor is at zero volts, so we have a 100 V drop across the resistor, as already stated. Note that *immediately* the switch is closed the potential of the lower plate of the capacitor as well as that of the upper plate jumps from zero to +100. The use of a capacitor to transfer a sudden change of potential to another point, without a conducting path, is very common.

We have already seen that the number of coulombs required to charge a capacitance of C farads to V volts is VC. In this case, it is $100 \times 5 \times 10^{-6}$, which is 0.000 5. As a coulomb is an ampere-second, the present charging rate of 0.5 A, *if maintained*, would fully charge the capacitor in 0.001 s. The capacitor voltage would rise steadily as shown by the sloping dashed line in Figure 3.6b. Directly it starts to do so, however, there are fewer volts available for driving current through R, and so the current becomes less and the capacitor charges more slowly. When it is charged to 50 V, 50 V are left for the resistor, and the charging rate is 0.25 A. When C is charged to 80 V, 20 are left for R, and the current is 0.1 A and so on, as shown by the curves in Figure 3.6.

Curves of this type are known as *exponential*, and are characteristic of many growth and decay phenomena. Note that the current curve's downward slope, showing the rate at which the current is decreasing, is at every point proportional to the current itself. It can be shown mathematically that in the time a capacitor would take to charge at the starting rate it actually reaches 63.2% of its full charge. By using the principles we already know we

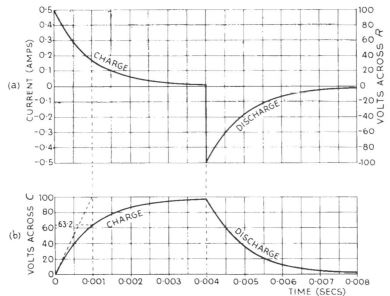

Figure 3.6 These curves show the current and voltage in the circuit of Figure 3.5 during charge and discharge.

have calculated that this time in the present example is 0.001 s. Try working it out with letter symbols instead of numbers, so that it covers all charging circuits, and you should find that it is CR s, regardless of the voltage E. CR—equal to the time in seconds taken to reach 63.2% of final charge—is of course a characteristic of the circuit, and is called its *time constant*. C in this example being 5×10^{-6} F and R 200 Ω, CR is 0.001 s, confirming Figure 3.6.

Theoretically, the capacitor is never completely charged, because the charging depends on there being a difference between the applied e.m.f. and the capacitor voltage, to drive current through the resistance. If the current scale is multiplied by 200 (the value of R), the upper curve is a graph of this voltage across R. Note that the voltages across C and R at all times up to 0.004 s add up to 100, just as Kirchhoff's voltage law says they ought (Section 2.13).

For practical purposes the capacitor may be as good as fully charged in a very short time. Having allowed 0.004 s for this to happen in the present case, we move the switch to B. Here the applied voltage is zero, so the voltages across C and R must now add up to zero. We know that the capacitor voltage is practically 100, so the voltage across R must be -100, and the current -0.5 A; that is to say, it is flowing in the opposite direction, discharging the capacitor. And so we get the curves in Figure 3.6 from 0.004 s onwards.

A point to remember is that the voltages set up across R at the instants of switching to A and B are short-lived; some idea of their duration in any particular case can be found by multiplying C by R in farads and ohms respectively, giving the answer in seconds. These temporary changes are called *transients*.

3.6 Where the Power Goes

The height of the charging curve above the zero line in Figure 3.6a represents the current supplied by the battery. Multiplied by its e.m.f. (100 V) it represents the power. When first switched on, the battery is working at the rate of $0.5 \times 100 = 50$ W, but at the end of 0.004 s this has fallen off almost to nothing. At first the whole 50 W goes into the resistance, so all its energy is lost as heat. But the effort of the e.m.f. immediately becomes divided, part being required to drive the current through R and the rest to charge C. It is clear that energy is going into the capacitance (Section 2.18); what happens to it? The answer is that it is stored as an electric field, in a similar way to the storage of mechanical energy in a spring when it is compressed by an applied force. During the charging process as a whole, then, part of the release of energy from the battery is lost as heat in the resistor, and part is stored in the capacitor.

When we come to the discharge we find that the voltage across R and the current through R follow exactly the same curves as during the charge, except that they are both reversed. But multiplying two negatives together gives a positive, so the energy expended in R is the same amount as during charge. The capacitor voltage and current during the discharge are the same as for the resistor, so the amount of energy is equal; in fact, it is the same energy, for what goes into the resistor has to come out of the capacitor. Applying the test described at the end of Section 2.17, we can check that the capacitor is a source of energy; the current that is going into the resistor at its positive terminal is coming out from the capacitor's positive terminal.

At the end of the discharge, all the energy supplied by the battery during the charging process has been dissipated in the resistance as heat. And as the amounts dissipated therein during charge and discharge are equal, it follows that during charge the energy from the battery was shared equally by R and C.

Summing up: during the charging period half the energy supplied by the battery is stored as an electric field in the capacitor and half is dissipated in the resistor; during discharge the half that was stored is given back at the expense of the collapsing field and is dissipated in the resistor.

It may be as well to repeat the warning that the 'charging' and 'discharging' of accumulator batteries is an entirely different process, accompanied by only minor variations in voltage. The energy put into the battery is not stored as an electric charge; it causes a reversible chemical change in the plates.

3.7 Recapitulation

By this time you may be getting a little confused about the meanings and relationships of the various quantities we have met. So let us recapitulate. First, those that do not necessarily depend on time. There is electric charge (Q), reckoned in coulombs (C), which can be thought of as surpluses or deficiencies of electrons. Transferring electrons from A to B charges A positively and B negatively (as we say), and set up a p.d. (V) between them, which is a kind of stress and is reckoned in volts (V). Q/V is called the capacitance (C) between A and B, and its unit is the farad (F). To make the electrons move from A to B against the p.d. created by the separation of the charges we need an e.m.f. (E) which is reckoned in terms of the p.d. it is able to counteract, so its unit also is the volt. The p.d. is always tending to restore the charge to the normal uncharged state, and if it is allowed to do so the electrical energy stored in C is released. That energy is equal to the effort (called work) that the source of e.m.f. had to do in order to charge C. Because energy and work, like income and expenditure, are the same thing looked at from different points of view, they are both denoted by W and have the same unit—the *joule* (J).

We know (Section 3.2) that if a charge Q is imparted to a capacitance C the resulting p.d., V, is equal to Q/C. But as the starting p.d. was zero and increased in proportion to the charge, the average p.d.—and therefore the average e.m.f. needed to transfer the charge—must be half as much, $Q/2C$. The work done by the e.m.f., and therefore the energy stored in C, is Q multiplied by the average e.m.f., that is to say

$$W = Q \times Q/2C = Q^2/2C = V^2C^2/2C = \tfrac{1}{2}V^2C$$

The important thing to remember about energy is that it never ceases to exist. Even when the switch in Figure 3.5 is moved to B the electrical energy stored in C is converted into an equal amount of heat energy in R.

Time comes into the equation, in the *rate* of the process. The rate at which charge is moved is called current (I), reckoned in amperes (A). So if a current of I amps continues steadily for t s the charge moved past a fixed point en route is It coulombs and $I = Q/t$. If the charge is moved against a constant p.d. V, energy equal to VQ is transferred every second, and an equal amount of work has to be done by a source of e.m.f. The rate at which energy is transferred and work is done is called power (P) and is reckoned in watts (W). So $P = VQ/t = VI$ (Section 2.17).

3.8 Displacement Currents

If any readers are worrying about reconciling capacitance with Kirchhoff's current law, which implies that current in a series circuit is the same throughout, they deserve a note on the subject.

There should be no difficulty so far as the extra capacitance due to dielectrics is concerned. A charging or discharging current in the circuit connected to the capacitor plates can be regarded as continued through the dielectric by the displacements of electrons as they 'strain at the leash'. The fact that this movement is very restricted is no problem, for it is consistent with the very limited quantity of electricity that can be allowed to flow in one direction as a charging current. To take some practical figures, 1 nF ($= 10^{-9}$ F) is typical of capacitors in radio circuits, and the voltages across them are usually quite small. But let us go up to 1000 V as a typical breakdown voltage. Then using the basic equation $Q = VC$ we find the charge needed to bring it to that point is only 10^{-6} C, which would be delivered in 1 second at the rate of only $1\,\mu$A. The enormous number of electrons in the dielectric would not have to move very far to match such a small drift of their colleagues in the charging circuit.

The real problem concerns the capacitance left when the dielectric is removed and even the air between the plates is pumped out, so that no electrons or any other charges are there. It looks as if Kirchhoff's current law then breaks down completely.

Clerk Maxwell, who has already come into this story (Section 1.9), was well aware of this difficulty, and got himself out of it by imagining what he called a *displacement current* in the empty space, equal to the charging or discharging current in the rest of the circuit. (The name connected it with a quantity already well known in the theory of electricity as displacement.) This seems rather a desperate expedient for a respectable scientist to fall back on merely in order to save Kirchhoff's law from failing in a particular case, and so perhaps it would be if that were all. But as we saw earlier (Section 1.9) Maxwell was concerned with a much more important matter than that: the possibility of electric waves in space. These could not exist, and neither could we, if something equivalent to electric currents did not function in empty space. After all, we do not have to understand everything fully in order to use it.

Chapter 4 Inductance

4.1 Magnets and Electromagnets

If a piece of paper is laid on a straight bar magnet, and iron filings are sprinkled on the paper, they are seen to arrange themselves in a pattern something like Figure 4.1*a*. Gently tapping the paper helps the filings to take up the pattern. The lines show the paths along which the attraction of the magnet exerts itself. (Compare the lines of electric force in Figure 3.2.) As a whole, they map out the *magnetic field*, which is the sphere of influence, as it were, of the magnet. The field is most concentrated around two regions, called *poles*,* at the ends of the bar. The lines may be supposed to continue right through the magnet, emerging at the end marked N and returning at S. This direction, indicated by the arrows, is (like the direction of an electric current) purely conventional (representing the direction in which a north pole would be moved), and the lines themselves are an imaginary representation of a condition occupying the whole space around the magnet.

still found to be surrounded by a magnetic field. Though very much less concentrated than in the coil, if the current is strong enough it can be demonstrated with filings as in Figure 4.2.

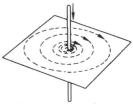

Figure 4.2 Conventional direction of magnetic field around a straight wire carrying current

The results of these and other experiments in electromagnetism are expressed by saying that wherever an electric current flows it surrounds itself with a field which

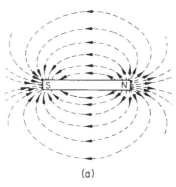 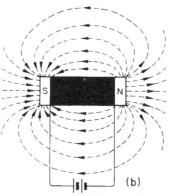

(a) (b)

Figure 4.1 The dashed lines indicate the direction and distribution of magnetic field around (*a*) a permanent magnet, and (*b*) a coil of wire carrying current

The same result can be obtained with a previously quite ordinary and unmagnetised piece of iron by passing an electric current through a wire coil round the iron—an arrangement known as an *electromagnet*. It is not even necessary to have the iron core (as it is called); the coil alone, so long as it is carrying current, is interlinked with a magnetic field having the same general pattern as that due to a bar magnet, as can be seen by comparing Figure 4.1*a* and *b*. Without the iron core, however, it is considerably weaker. Finally, if the wire is unwrapped, every inch of it is

sets up magnetic *flux* (symbol: Φ, the Greek capital phi). This flux and the circuit link one another like two adjoining links in a chain. By winding the circuit into a compact coil, the field due to each turn of wire is made to operate in the same space, producing a concentrated magnetic flux (Figure 4.1*b*). Finally, certain materials, notably iron and its alloys, described as *ferromagnetic*, are found to have a very large multiplying effect on the flux, this property being called the *relative permeability* of the material, denoted by the symbol μ_r, or often just μ. The permeabilities of most things, such as air, water, wood, brass, etc., have practically no effect on the flux and are therefore generally reckoned as 1. Certain alloys, not all containing iron, have permeabilities running into many thousands. Unlike resistivity and permittivity, however, the

---

* The terminals of a battery or other source of e.m.f., between which electric field lines are imagined, are sometimes also called poles. The distinction between + and − in such sources, and N and S in magnets, is called its *polarity*.

permeability of ferromagnetic materials is not even approximately constant but varies greatly according to the flux density (i.e., the flux passing through unit area at right angles to its direction).

It is now known that all magnets are really electromagnets. Those that seem not to be (called *permanent magnets*) are magnetised by the movements of the electrons in their own atomic structure. The term 'electromagnet' is, however, understood to refer to one in which the magnetising currents flow through an external circuit.

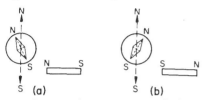

Figure 4.3 Deflection of a compass needle by a magnet, proving that like poles repel and unlike poles attract

4.2 Interacting Magnetic Fields

The 'N' and 'S' in Figure 4.1 stand for North-seeking and South-seeking respectively. Everybody knows that a compass needle points to the north. The needle is a magnet, and turns because its own field interacts with the field of the earth, which is another magnet. Put differently, the north magnetic pole of the earth attracts the north-seeking pole of the needle, while the earth's south pole attracts its south-seeking pole. The poles of a magnet are often called just north and south, but strictly, except in referring to the earth itself, this is incorrect. By bringing the two poles of a bar magnet—previously identified by suspending it as a compass—in turn towards a compass needle it is very easy to demonstrate, as in Figure 4.3, that unlike poles attract one another. This reminds one of the way electric charges behave.

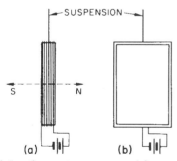

Figure 4.4 A coil—carrying current and free to rotate—sets itself with its axis pointing N and S, (a) is a side view and (b) an end view

Now suppose we hang a coil in such a way that it is free to turn, as suggested in Figure 4.4. So long as no current is passing through the coil it will show no tendency to set itself in any particular direction, but if a battery is connected to it the flow of current will transform the coil into a magnet. Like the compass needle, it will then indicate the north, turning itself so that the plane in which the turns of the coil lie is east and west, the axis of the coil pointing north. If now the current is reversed the coil will turn through 180 degrees, showing that what was the N pole of the coil is now, with the current flowing the opposite way, the S pole.

The earth's field is weak, so the force operating to turn the coil is small. When it is desired to make mechanical use of the magnetic effect of the current in a coil it is better to provide an artificial field of the greatest possible intensity by placing a powerful magnet as close to the coil as possible. This is the basis of all electric motors.

In electronics, however, we are more interested in other applications of the same principle, such as the moving-coil measuring instruments described very briefly in Section 2.16. The coil is shaped somewhat as in Figure 4.4, and is suspended in the intense field of a permanent magnet. Hairsprings or taut metal ribbons at top and bottom of the coil serve the double purpose of conducting the current to the coil and keeping the coil normally in the position where the pointer attached to it indicates zero. When current flows through it the coil tends to rotate against the restraint of the springs, and the angle through which it moves is proportional to the magnetic interaction, which in turn is proportional to the value of the current.

4.3 Induction

In Section 4.1 we noted that the magnetic circuit formed by the imaginary flux lines is linked with the electric circuit that causes them, like two adjoining links in a chain (Figure 4.5). Now suppose the electric circuit includes no source of e.m.f. and consequently no current, but that some magnetic

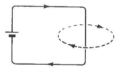

Figure 4.5 An electric current and the magnetic flux set up thereby are linked with one another as indicated in this diagram, which shows the relative directions established by convention

flux is made to link with it. There are various ways in which this can be done: a permanent magnet or an electromagnet—or any current-carrying circuit—could be moved towards it, or current could be switched on in a stationary circuit close to it. Whatever the method, the result will be the same: so long as the flux linked with the circuit is changing, an e.m.f. is generated in the circuit. The magnitude and direction of this induced e.m.f. can be found by means of a suitable voltmeter, as in Figure 4.6. The amount of e.m.f. is proportional to the rate at which the amount of flux linked with the circuit is changing. So a constant flux linkage yields no e.m.f. When the flux is removed or decreased, a reverse e.m.f. is induced. In Figure 4.6 the increasingly linked flux has the same direction as in Figure 4.5; the e.m.f. induced thereby would tend to drive current in the direction shown; i.e., in the opposite direction to the current in Figure 4.5.

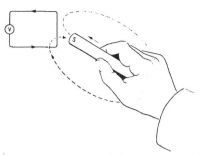

Figure 4.6 When the magnetic linkage with any closed path is varied, say by moving a magnet near it, an e.m.f. is generated around the path

These results, first discovered by Michael Faraday in 1831, are the basis of all electrical engineering, including of course radio. The electrical generators in power stations are devices for continuously varying the magnetic flux linked with a circuit, usually by making electromagnets rotate past fixed coils. The alternative method of varying the flux linkages—by varying the current in adjacent fixed coils—is adopted in transformers (Section 4.11) and, as we shall see, in many kinds of radio equipment.

The SI unit of flux, the *weber* (Wb), is such that the induced e.m.f. in volts is equal to the rate at which the linked flux is changing, in webers per second.

4.4 Self-inductance

We have just been considering an e.m.f. being induced in a circuit by the varying of magnetic flux due to current in some other circuit. But in Figure 4.5 we see a dashed loop representing the flux due to current flowing in that circuit itself. Before the current flowed there was no flux. So, directly the current was switched on, the flux linked with the circuit must have been increasing from zero to its present value, and in the process it must have been inducing an e.m.f. in its own circuit. This effect is—very naturally—known as *self-induction*.

Since Figure 4.5 can be taken to represent any circuit carrying current, and because a flow of current inevitably results in some linked magnetic flux, it follows that an e.m.f. is induced in any circuit in which the current is varying.

The amount of e.m.f. induced in a circuit (or part of a circuit) when the current is varied at some standard rate is called its *self-inductance*, or often just inductance (symbol: L). Its unit, the henry (abbreviation: H), was so defined that the inductance of a circuit in henries is equal to the number of volts self-induced in it when the current is changing at the rate of 1 ampere per second.

Actually it is the change of *flux* that induces the e.m.f., but since information about the flux is less likely to be available than that about the current, inductance is defined in terms of current. What it depends on, then, is the amount of flux produced by a given current, say 1 A. The more flux, the greater the e.m.f. induced when that current takes one second to grow or die, and so the greater the inductance. The flux produced by a given current depends, as we have seen, on the dimensions and arrangement of the circuit or part of the circuit, and on the permeability.

4.5 Lenz's Law

In Section 4.3 we noted the direction of the induced e.m.f. in relation to the direction of the flux inducing it. By comparing Figure 4.6 with Figure 4.5 we see that a self-induced e.m.f. would tend to send current round the circuit in the opposite direction to the current producing the flux that is inducing it. In more general terms, *any* induced e.m.f. tends to oppose its cause. This statement is known as Lenz's law, and is really inevitable, because if induced e.m.f.s aided their cause there would be no limit to their growth. So it is easy to remember.

When you try to stop the clockwise current in Figure 4.5 by switching off, the flux in the direction shown begins to *decrease*, and in doing so it induces an e.m.f. in the *same* direction as that of the battery, tending to oppose the switch-off and keep the current going.

This is a good point at which to learn another easily remembered rule—the corkscrew rule. Imagine that the current in Figure 4.2 is a corkscrew being driven downwards. The direction in which the corkscrew has to be turned is the direction of the resulting flux. This can be seen in Figure 4.5 too. And it holds good if the roles are interchanged; turning the corkscrew in the direction of the current around the circuit, the direction in which it moves, in or out, is the direction of the resulting flux through the circuit. With this rule and Lenz's law we are fully equipped to deal with induction polarity questions. But we must realise that the directions are conventions, not physical facts.

4.6 Inductance Analysed

We had no great difficulty in finding a formula expressing resistance in terms of resistivity and the dimensions of the circuit element concerned, because currents are usually confined to wires or other conductors of uniform cross-section. Capacitance was not so easy, because it depends on the dimensions of space between conductors, and the distribution of the electric field therein is likely to be anything but uniform, but for the particular and practically important case of a thin space between parallel plates we found a formula for capacitance in terms of permittivity and the dimensions of the space. Inductance is even more difficult, because the magnetic fields of many of the circuit forms used for providing it, known in general as *inductors*, are not even approximately uniform or easily calculated. And some of the flux may link with only part of the circuit. In coils, for example, the flux generated by one turn usually links with some of the others but not all.

However, Figure 4.7 shows a type of inductor in which the coil is linked with an iron core of nearly uniform cross-section area. The permeability of the iron is usually so high that the core carries very nearly all the magnetic flux; the remainder—called leakage flux—that strays from it through the surrounding air or the material of the coil itself, can be

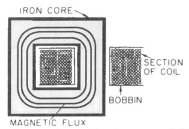

Figure 4.7 Showing how a closed iron core links with the turns of the coil and carries nearly all the flux set up by the coil

neglected. So it is nearly true to assume that all the flux links all the turns, and that it is uniformly distributed through the whole core. On that assumption, the inductance of a coil is related to the dimensions in the same way as for conductance and capacitance:

$$L = \frac{\mu A}{l}$$

where A is the cross-section area of the core, l is its length, and μ is its absolute permeability. As with permittivity, the 'absolute' value is the one that gives the answer in the desired units without the assistance of other numbers ('constants'). In SI units (Section 2.11), in which L is in henries, A in square metres and l in metres, its value for empty space (sometimes called the 'magnetic space constant'), denoted by μ_0, is 1.257×10^{-6}. The same value is practically correct for air and most other substances except the ferromagnetic materials. Although strictly the permeability (μ) is this μ_0 multiplied by the relative permeability μ_r (Section 4.1), the values quoted for materials are invariably relative, so the formula is usually adapted for them by incorporating μ_0:

$$L = \frac{1.257 \times 10^{-6} \mu_r A}{l}$$

And, as with permittivity, μ_r is then usually written 'μ' and called just 'permeability'. For air, etc., it can be omitted altogether, being practically 1. If the dimensions are in centimetres, 10^{-8} must be substituted for 10^{-6}.

The above equation is for only one turn. If there are N turns the same current flows N times around the area A, so the flux (assuming μ to be constant) is N times as much. Moreover, this N-fold flux links the circuit N times, inducing N times as much e.m.f. when it varies as in one turn. So the inductance is N^2 times as great:

$$L = \frac{1.257 \times 10^{-6} \mu_r A N^2}{l}$$

4.7 Practical Considerations

In practice this formula is not of much use for calculating the inductance of even iron-cored coils. For one thing, in order to get the core into position around the coil it is necessary to have it in more than one piece, and the joint is never so magnetically perfect as continuous metal. Even a microscopic air gap makes an appreciable difference, because it is equivalent to an iron path perhaps tens of thousands of times as long. Then the permeability of iron depends very much on the flux density and therefore on the current in the coil. So one cannot look it up in a list of materials as one can resistivity or permittivity. It is usually plotted as a graph against the strength of the magnetising field, which has the symbol H and is reckoned as the total current encircling the iron (i.e., the number of ampere-turns) per unit length of iron carrying the flux. For many purposes it is even more helpful to have graphs of the flux density per square metre, which has the symbol B and is reckoned in teslas (i.e., webers of flux per square metre) against H, as in Figure 4.8. The permeability, equal to B/H, can be found from this for any particular value of H or B. It is represented by the steepness of the slope of the graph. A characteristic of ferromagnetic materials is that increasing the ampere-turns yields progressively smaller increases in flux density. In other words, the permeability gets less and less. This effect is called *magnetic saturation*.

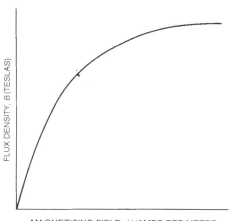

Figure 4.8 The flux density in many magnetic materials such as iron varies with the magnetising field in this sort of way.

If you are finding this section rather heavy going, with so many new quantities cropping up, comfort yourself with the assurance that they will not be found essential within the scope of this book. But if you are keen enough to pursue the magnetic relationships further you may care to note, starting from the fact that $\mu = B/H$, that one can derive the first formula for inductance in Section 4.6:

$$\mu = \frac{B}{H} = \frac{\Phi l}{A I}$$

So

$$\frac{\mu A}{l} = \frac{\Phi}{I} = L$$

With 'air-core' coils, even when all the turns are close together it is not correct to assume that all the flux due to each turn links with the whole lot; and the error is of course much greater if the turns are widely distributed as in Figure 4.1*b*. An even greater difficulty is that the flux does not all follow a path of constant area A and equal length l.

A general formula for inductance has nevertheless been given, to bring out the basic similarity in form to those for conductance and capacitance, but for practical purposes it is necessary either to measure inductance or calculate it from the various formulae and graphs that have been worked out for coils of various shapes and straight wires and published in radio reference books.

Between a short length of wire and an iron-cored coil with thousands of turns there is an enormously large range of inductance. The iron-cored coil might have an inductance of many henries; the short wire a small fraction of a microhenry. For most purposes the inductance of connecting wires is small enough to ignore. But not always. The maximum current may be much less than 1 A, but if it comes and goes a thousand million times a second its rate of change may well be millions of amps per second and the induced voltage quite large.

It is, of course, this induced voltage that makes inductance significant. Just how significant will appear in later chapters, but we can make a start now by comparing and contrasting it with the voltage that builds up across a capacitance while it is being charged.

4.8 Growth of Current in an Inductive Circuit

What happens when current is switched on in an inductive circuit can be studied with the arrangement shown in Figure 4.9, which should be compared with Figure 3.5. To make comparison of the results easy, let us assume that the inductance, for which the scallops marked L is the standard symbol, is 0.2 H, and that R (which includes the resistance of the coil) is 200 Ω and E is 100 V as before.

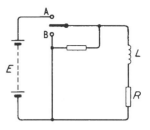

Figure 4.9 Circuit used for studying the rise and fall of current in an inductive circuit

With the switch as shown, no current is flowing through L and there is no magnetic field. At the instant of switching to A, the full 100 V is applied across L and R. The current cannot instantly leap to 0.5 A, the amount predicted by Ohm's law, for that would mean increasing the current at an infinite rate, which would induce an infinite voltage opposing it. So the current must rise gradually, and at the exact moment of closing the circuit it is zero. There is therefore no voltage drop across R, and the battery e.m.f. is opposed solely by the inductive e.m.f., which must be 100 V (Figure 4.10*a*). That enables us to work out the rate at which the current will grow. If L were 1 H, it would have to grow at 100 A per second to induce 100 V. As it is only 0.2 H it must grow at 500 A/s.

In the graph (Figure 4.10*b*) corresponding to Figure 3.6*b* for the capacitive circuit, the dashed line represents a steady current growth of 500 A/s. If it kept this up, it would reach the full 0.5 A in 0.001 s. But directly the current starts to flow it causes a voltage drop across R. And as the applied voltage remains 100, the induced voltage must diminish. The only way this can happen is for the current to grow less rapidly. By the time it has reached 0.25 A there are 50 V across R, therefore only 50 across L, so the current must be growing at half the rate. The full line shows how it

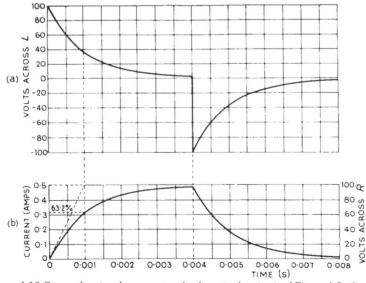

Figure 4.10 Curves showing the current and voltage in the circuit of Figure 4.9 when the switch is moved first to A and then to B

grows; here is another exponential curve and, as in the capacitive circuit, it theoretically never quite finishes. In the time that the current would take to reach its full Ohm's law value if it kept up its starting pace, it actually reaches 63.2% of it. This time is, as before, called the time constant of the circuit. A little elementary algebra based on the foregoing shows it to be L/R seconds.

The voltage across L is also shown. Compared with Figure 3.6, current and voltage have changed places. The voltage across R is of course proportional to the current and therefore its curve is the same shape as Figure 4.10b. Added to the upper, it must equal 100 V as long as that is the voltage applied.

When the current has reached practically its full value, we flick the switch instantaneously across to B. The low resistance across the contacts is merely to prevent the circuit from being completely interrupted in this process. At the moment the switch is operated, the magnetic field is still in existence. It can only cease when the current stops, and the moment it begins to do so an e.m.f. is induced which tends to keep it going. At first the full 0.5 A is going, which requires 100 V to drive it through R; and as the battery is no longer in circuit this voltage must come from L, by the current falling at the required rate: 500 A/s. As the current wanes, so does the voltage across R, and so must the induced voltage, and therefore the current dies away more slowly, as shown in the continuation of Figure 4.10. In L/R (= 0.001 s) it has been reduced by the inevitable 63.2%.

4.9 Power During Growth

In reckoning how the power in the circuit varies during these operations, we note firstly that the output from the battery (calculated by multiplying the current at each instant by its 100 V) instead of starting off at 50 W and then tailing off almost to nothing, as in the capacitive example, starts at nothing and works up towards 50 W. If there had been no C in Figure 3.5 there would have been no flow of power at any time; the effect of C was to make possible a temporary flow, during which half the energy was dissipated in R and half stored in C. In Figure 4.9, on the other hand, the absence of L would have meant power going into R at the full 50 W rate all the time: the effect of L was to withhold part of that expenditure from R for a short time. Not all this energy withheld was a sheer saving to the battery, represented by the temporarily sub-normal current; part of it—actually a half—went into L, for the current had to be forced into motion through it against the voltage induced by the growing magnetic field. The energy was in fact stored in the magnetic field, in a similar way to the storage of mechanical energy when force is applied to set a heavy vehicle in motion.

Just as the vehicle gives up its stored energy when it is brought to a standstill, the result being the heating of the brakes, so the magnetic field in L gave up its energy when the current was brought to a standstill, the result being the heating of R. Self-inductance is therefore analogous to mechanical inertia.

The amount of energy stored when a current I is flowing is again equal to the charge Q transferred during the magnetising process, multiplied by the p.d. it has been transferred against. Supposing the current grew at the starting rate in Figure 4.10 it would reach its full value I in time L/R, so $Q = IL/R$. The average p.d. across the coil is clearly half the maximum value IR. So

$$W = \frac{IL}{R} \times \tfrac{1}{2}IR = \tfrac{1}{2}I^2L$$

If we write this in the form $\tfrac{1}{2}LI^2$ it compares directly with $\tfrac{1}{2}mv^2$, the expression for kinetic energy. Just as $\tfrac{1}{2}LI^2$ expresses the energy stored in the magnetic field of an inductor, so $\tfrac{1}{2}mv^2$ gives the energy stored in a body of mass m moving at a velocity v. Thus inductance is analogous to mass and current to velocity.

4.10 More Comparison and Contrast

In the previous chapter we saw that besides the conducted currents, reckoned by dividing voltage by resistance, there is an extra current whenever the voltage is varying. This current is made up of movements of charges caused by the electric field adjusting itself to the new voltage. It is proportional not only to the rate at which the voltage is changing but also the capacitance (intentional or otherwise) of the part of the circuit considered. In this chapter we have seen that in addition to the resistive voltage, reckoned by multiplying current by resistance, there is an extra voltage whenever the current is varying. This voltage is caused by the magnetic field adjusting itself to the new current. It is proportional not only to the rate at which the current is changing but also to the inductance (intentional or otherwise) of the part of the circuit considered.

4.11 Mutual Inductance

If a coil with the same number of turns as that in Figure 4.9 were wound so closely around it that practically all the magnetic flux due to the original *primary winding* embraced also this *secondary winding*, then an equal voltage would at all times be induced in the secondary. If an appreciable part of the flux is not linked with the secondary winding, the voltage induced in it is correspondingly less. But by giving the secondary more turns than the primary, it is possible to obtain a greater voltage in the secondary than in the primary. Remember that this voltage depends on the *varying* of the primary current. A device of this kind, for stepping voltages up or

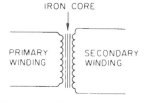

Figure 4.11 How a transformer is represented in circuit diagrams

down, or for inducing voltages into circuits without any direct connection, is named (rather inappropriately) a *transformer*. It is represented in circuit diagrams by two or more coils drawn closely together. An iron core is shown by one or more straight lines drawn between them (Figure 4.11).

The effect that one coil can exert, through its magnetic field, on another, is called *mutual inductance* (symbol: M), and like self-inductance is measured in henries. The definition is similar too: the mutual inductance in henries between two coils is equal to the number of volts induced in one of them when the current in the other is changing at the rate of 1 A per second.

Transformers are an immensely useful application of the inductive effect, and in fact are the main reason why alternating current (a.c.) has almost superseded direct current (d.c.) for electricity supply. So we now proceed to study a.c.

Chapter 5 Alternating Currents

5.1 Frequencies of Alternating Current

In our first chapter we saw that speech and music are conveyed from one person to another by sound waves, which consist of rapid vibrations or alternations of air pressure. And that to transmit them over longer distances by telephone it is necessary for electric currents to copy these alternations. And further, that for transmitting them across space, by radio, it is necessary to use electric currents alternating still more rapidly. Of course we must never forget that sound waves and radio waves are quite different in nature, sound waves being mechanical vibrations of material substances (solid, liquid or gas) whereas radio waves are oscillations of electric and magnetic fields which can cross interstellar space.

We have now just noted the fact that the great advantage of being able to step voltages up and down as required is obtainable only when the electricity supply is continually varying, which is most conveniently done by arranging for it to be alternating. Public electricity supplies which light and heat our houses and provide the power that works our television receivers and many other facilities therefore use mostly alternating current.

The only essential difference between all these alternating currents is the number of double-reversals they make per second; in a word, the *frequency*. We have already gone fairly fully into the matter of frequency (Section 1.4), so there is no need to repeat it; but it may be worth recalling the main divisions of frequency.

There is no hard-and-fast dividing line between one lot of frequencies and another; but those below 100 Hz are used for power (the standard in Britain is 50 Hz); those between about 20 and 20 000 Hz are audible, and therefore are classed as audio frequencies (a.f.); while those above about 15 000 are more or less suitable for carrying signals across space, and are known as radio frequencies (r.f.). Certain of these, such as 525 000 to 1 605 000, are allocated for broadcasting. All those above about 1000 MHz (wavelengths shorter than about 30 cm) are often lumped together as *microwave frequencies*.

For convenience the enormous range of radio frequencies is divided into subranges (decades), each with a 10 : 1 ratio of maximum to minimum frequency (and wavelength), as shown in the table below which also gives the terminology and the abbreviations in common use for the subranges.

What has been said about currents applies to voltages, too; it requires an alternating voltage to drive an alternating current.

The last two chapters have shown that when currents and voltages are varying there are two circuit effects—capacitance and inductance—that have to be taken into account as well as resistance. We shall therefore be quite right in expecting a.c. circuits to be a good deal more complicated and interesting than d.c. Before going on to consider these circuit effects there are one or two things to clear up about a.c. itself.

5.2 The Sine Wave

An alternating current might reverse abruptly or it might do so gradually and smoothly. In Figure 3.6a we have an example of abrupt reversal. There is, in fact, an endless variety of ways in which current or voltage can vary with time, and when graphs are drawn showing how they do so we get a corresponding variety of wave shapes. The steepness of the wave at any point along the time scale shows the rapidity with which the current is changing at that time.

Fortunately for the study of alternating currents, all repetitive wave shapes, however complicated, can be regarded as built up of waves of one fundamental shape, called the *sine wave* (adjective: sinusoidal). Figure 1.2c shows examples of this. Note the smooth, regular alternations of the component sine waves, like the swinging of a pendulum. Repetitive waves of even such a spiky

Frequency range	Wavelength range	Name	Abbreviation
less than 30 kHz	greater than 10 000 m	very low frequencies	v.l.f.
30 kHz–300 kHz	1000 m–10 000 m	low frequencies	l.f.
300 kHz–3 MHz	100 m–1000 m	medium frequencies	m.f.
3 MHz–30 MHz	10 m–100 m	high frequencies	h.f.
30 MHz–300 MHz	1 m–10 m	very high frequencies	v.h.f.
300 MHz–3 GHz	10 cm–1 m	ultra high frequencies	u.h.f.
3 GHz–30 GHz	1 cm–10 cm	super high frequencies	s.h.f.
30 GHz–300 GHz	1 mm–1 cm	extremely high frequencies	e.h.f.

appearance as those in Figure 3.6 can be analysed into a number of sine waves of different frequencies.

Most circuits used in radio and other systems of telecommunications consist basically of sources of alternating e.m.f. (often called generators) connected to combinations of resistance, inductance and/or capacitance (usually called loads). With practice one can reduce even very complicated-looking circuits to standard generator-and-load combinations.

We have just seen that although there is no end to the variety of waveforms that generators can produce, they are all combinations of simple sine waves. So it is enough to consider the sine-wave generator. Sources of more complicated waveforms can be imitated by combinations of sine-wave generators. Theoretically this dodge is always available, but in practice (as, for example, with square waves) there are sometimes easier alternatives, as we shall see in Chapter 25. Unless the contrary is clear, it is assumed from now on that a generator means a sine-wave generator.

5.3 Circuit with Resistance Only

The simplest kind of a.c. circuit is one that can be represented as a generator supplying current to a purely resistive load (Figure 5.1). We have already found (Section 2.6) how to calculate the current any given e.m.f. will drive through a resistance. Applying this method to an alternating

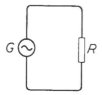

Figure 5.1 Circuit consisting of an a.c. generator (G) feeding a purely resistive load (R)

e.m.f. involves nothing new. Suppose, for example, that the voltage at the peak of each wave is 200, and that its frequency is 50 Hz. Figure 5.2 shows a graph of this e.m.f. during rather more than one cycle. And suppose that R is 200 Ω. Then we can calculate the current at any point in the cycle by dividing the voltage at that point by 200, in accordance with Ohm's law. At the peak it is, of course, 200/200 = 1 A; half-way up the wave it is 0.5 A. Plotting a number of such points we get the current wave, shown dashed. It is identical in shape, though different in vertical scale.

5.4 R.M.S. Values

How is an electricity supply that behaves in this fashion to be rated? Can one fairly describe it as a 200 V supply, seeing that the actual voltage is changing all the time and is sometimes zero? Or should one take the average voltage?

The answer that has been agreed upon is based on comparison with d.c. supplies. It is obviously very convenient for a lamp or heater intended for a 200 V d.c. system to be equally suited to a.c. mains of the same nominal voltage. This condition will be fulfilled if the average power taken by the lamp or heater is the same with both types of supply, for then the element will reach the same temperature and the light or heat will be the same in both cases.

We know (Section 2.17) that the power in watts is equal to the voltage multiplied by the amperage. Performing this calculation for a sufficient number of points in the alternating cycle in Figure 5.2, we get a curve showing how the power varies. To avoid confusion it has been drawn below the E and I curves. A significant feature of this power curve is that although during half of every cycle the voltage and current are negative, their negative half-cycles always exactly coincide, so that even during these half-cycles the power is (by the ordinary rules of arithmetic) positive. This mathematical convention agrees with and

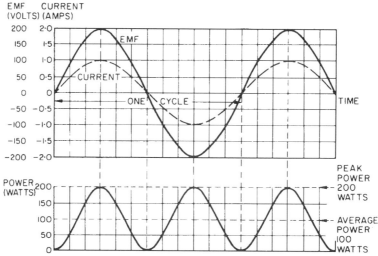

Figure 5.2 Time graphs of e.m.f., current and power in the circuit of Figure 5.1. Note that the frequency of the power waves is twice that of the e.m.f. and current

represents the physical fact, checked by the test given at the end of Section 2.17, that when both current and voltage reverse their direction the flow of power is always in the same direction, namely, out of the generator.

Another thing one can see by looking at the power curve is that its average height is half the height of the peaks. The peak height is $200 \times 1 = 200$ W; so the average power must be 100 W. This can be checked by cutting out the areas of wave above 100 W and finding that they exactly fill the empty troughs below 100 W, giving a steady 100 W of d.c. The next step is to find what direct voltage would be needed to dissipate 100 W in 200 Ω; the answer is again in Section 2.17: $P = E^2/R$, from which $E^2 = PR$, which in this case is $100 \times 200 = 20\,000$. So $E = \sqrt{20\,000} = 141$ approximately.

Judged on the basis of equal power, then, an a.c. supply with maximum or peak voltage of 200 is equivalent to a d.c. supply of 141 V.

Generalising this calculation, we find that when an alternating voltage is adjusted to deliver the same power to a given resistance as a direct voltage, its peak value is √2 times the direct voltage, or about 1.414 times as great. Put the other way round, its nominal or equivalent or effective voltage—called its *r.m.s.* (root-mean-square) value—is 1/√2 or about 0.707 times its peak value. Since the resistance is the same in both cases, the same ratio exists between r.m.s. and peak values of the current. What is called a 240 V a.c. supply therefore alternates between + and $-\sqrt{2} \times 240 = 339$ V peak.

Since the r.m.s. values of voltage and current are 0.707 times the peak values, the power dissipated in a resistance is given by $0.707 \times E_{max} \times 0.707 \times I_{max} = \frac{1}{2} E_{max} I_{max}$, i.e., one-half the product of the peak values. This is important because, when we come to the output power of certain amplifier types (Section 19.10), we shall be dealing in peak-to-peak values of voltage and current. These are respectively double the single-peak values E_{max} and I_{max}, and so power is equal to one-eighth the product of the peak-to-peak values.

The r.m.s. value is *not* the same as the average voltage or current (which is actually 0.637 times the peak value, if averaged over a half cycle, or zero over a whole cycle). If you have followed the argument carefully you will see that this is because we have been comparing the a.c. and d.c. on a power basis, and power is proportional to voltage or current *squared*.

Remember that the figures given above apply only to the sine waveform. With a square wave (Figure 6.2b) for example, it is obvious that peak, r.m.s., and average (or mean) values are all the same.

There is one other recognised 'value'—the instantaneous value, changing all the time in an a.c. system. It is the quantity graphed in Figure 5.2. It can be found for any stage of a cycle by dividing the cycle into 360 degrees of angle as in Section 5.7 and then looking up the sine of that angle in a trigonometrical table and multiplying the peak value by it.

As regards symbols, plain capital letters such as E, V and I are generally understood to mean r.m.s. values unless the contrary is obvious. Instantaneous values, when it is necessary to distinguish them, are denoted by small letters, such as i; peak values by I_{max}; and average (or mean) values by I_{ave}.

Using r.m.s. values for voltage and current we can forget the rapid variations in instantaneous voltage, and, *so long as our circuits are purely resistive*, carry out all a.c. calculations according to the rules discussed in Chapter 2.

5.5 A.C. Meters

If alternating current is passed through a moving-coil meter (Section 4.2) the coil is pushed first one way and then the other, because the current is reversing in a steady magnetic field. The most one is likely to see is a slight vibration about the zero mark. Certainly it will not indicate anything like the r.m.s. value of the current.

If, however, the direction of magnetic field is reversed at the same times as the current, the double-reversal makes the force act in the same direction as before, and the series of pushes will cause the pointer to take up a position that will indicate the value of current. The obvious way to obtain this reversing magnetic field is to replace the permanent magnet by a coil and pass through it the current being measured. When the current is small the field also is very weak and the deflection too small to be read, so this principle is seldom used, except in *wattmeters*, in which the main current is passed through one coil, and the other—the volt coil—is connected across the supply.

A more usual type is that in which there is only one coil, which is fixed. Inside are two pieces of iron, one fixed and the other free to move against a hairspring. When either d.c. or a.c. is passed through the coil, both irons are magnetised with the same polarity, and so repel one another, to a distance depending on the strength of current. These *moving-iron* meters are useful when there is plenty of power to spare in the circuit for working them, but they tend to use up too much in low-power circuits.

Another method is to make use of the heating effect of the current. When a junction of two different metals is heated, a small unidirectional e.m.f. is generated, which can be measured by a moving-coil meter. Instruments of this kind, called *thermojunction* or *thermocouple* types, if calibrated on d.c. will obviously read r.m.s. values of a.c. regardless of waveform. They are particularly useful for much higher frequencies than can be measured with instruments in which the current to be measured has to pass through a coil.

The electrostatic instrument (Section 2.16) can be used for alternating voltage and responds to r.m.s. values.

But perhaps the most popular method of all is to convert the alternating current into direct by means of a *rectifier*—a device that allows current to pass through it in one direction only—so that it can be measured with an ordinary moving-coil meter. The great advantage of this is that by adding a rectifier a multi-range d.c. instrument can be used on a.c. too. Most of the 'multimeters' used in radio and electronics are of this kind. Because the moving-coil instrument measures the average current, which in general is not the same as the r.m.s. current, the instrument is arranged to take account of the factor necessary to convert

38 Scroggie's Foundations of Wireless and Electronics

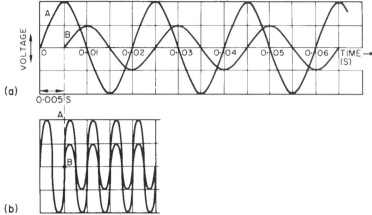

Figure 5.3 The time lag between the start of A and the start of B is the same in (a) and (b), but in (a) the two voltages are quarter of a cycle out of phase and in (b) are in phase. Time is therefore not the basic measure of phase

one to the other. This, as we have seen, is approximately 0.707/0.638 (= 1.11) for sine waves. It is different for most other waveforms, so the rectifier type of meter reads them incorrectly.

5.6 Phase

Looking again at the current graph in Figure 5.2 we see that it not only has the same shape as the e.m.f. that causes it but it is also exactly in step with it. Consideration of Ohm's law proves that this must be so in any purely resistive circuit. The technical word for being in step is being *in phase*. The idea of phase is very important, so we had better make sure we understand it.

Suppose we have two alternating generators, A and B. The exact voltages they give will not matter, but to make it easier to distinguish their graphs we shall suppose A gives double the voltage of B. If we start drawing the voltage graph of A just as it begins a cycle the result will be something like waveform A in Figure 5.3a. Next, consider generator B, which has the same frequency, but is out of step with A. It starts each of its cycles, say, a quarter of a cycle later than A, as shown by waveform B. This fact is stated by saying that voltage B lags voltage A by a phase difference of quarter of a cycle or 90°. It may alternatively be expressed by saying that the two waveforms are in *quadrature*. Seeing that it is a time graph, it might seem more natural to say that B lagged A by a phase difference of 0.005 s. The incorrectness of doing so is shown in Figure 5.3b, where the frequency of A and B is four times as great. The time lag is the same as before, but the two voltages are now *in phase*. So although phase is usually very closely related to time, it is not advisable to think of it as time. The proper basis of reckoning phase is in fractions of a cycle.

5.7 Phasor Diagrams

The main advantage of a time graph of alternating voltage, current, etc., is that it shows the waveform. But on those very many occasions when the form of the wave is not in question (because we have agreed to stick to sine waves) it is a great waste of effort to draw a number of beautifully exact waves merely in order to show the relative phases. Having examined Figures 1.2, 5.2 and 5.3, we ought by now to be able to take the sine waveform for granted, and be prepared to accept a simpler method of indicating phases.

Figure 5.4 Starting position of an alternative method of representing sinusoidal variation

In Figure 5.4, imagine the line OP to be pivoted at O and to be capable of rotating steadily in the direction shown (anticlockwise), which by mathematical convention is the positive direction of rotation. Then the height of the end P above a horizontal line through O varies in exactly the same way as a sine wave. Assuming OP starts, as shown in Figure 5.4, at '3 o'clock', P is actually on the horizontal line, so its height above it is, of course, nil. That represents the starting point of a sine-wave cycle. As OP rotates, its height above the line increases at first rapidly and then more slowly, until, after a quarter of a revolution, it is at right angles to its original position. That represents the first quarter of a sine-wave cycle. During the third and fourth quarter of the revolution, the height of P is negative, corresponding to the same quarters of a cycle.

Readers with any knowledge of trigonometry will know that the ratio of the height of P to its distance from O (i.e., PQ/PO in Figure 5.5) is called the sine of the angle that OP

Figure 5.5 The Figure 5.4 phasor after having turned through an angle, θ

has turned through; hence the name given to the waveform we have been considering and which is none other than a time graph of the sine of the angle θ (abbreviated 'sin θ') when θ (pronounced 'theta') is increasing at a steady rate.

We now have a much more easily drawn diagram for representing sine waves. The length of a line such as OP represents the peak value of the voltage, etc., and the angle it makes with the '3 o'clock' position represents its phase. The line itself is called a *phasor*. The angle it turns through in one whole revolution (which, of course, brings it to the same position as at the start) is 360°, and that corresponds to one whole cycle of the voltage. So a common way of specifying a phase is in angular degrees. Quarter of a cycle is 90°, and so on. Usually even a waveform diagram such as Figure 5.3 is marked in degrees.

Besides the common angular measure, in which one whole revolution is divided into 360°, there is the mathematical angular measure in which it is divided into 2π *radians*. Seeing that π is a very odd number (Section A.2 of Appendix A) this may seem an extremely odd way of doing things. It arises because during one revolution P travels 2π times its distance from O. Even this may not seem a good enough reason, but the significance of it will appear very soon in the next chapter.

As an example of a simple phasor diagram, Figure 5.6 is the equivalent of Figure 5.3a. In the position shown it is

Figure 5.6 Phasor diagram corresponding to Figure 5.3a

equivalent to it at the beginning or end of each cycle of A, but since both phasors rotate at the same rate the 90° phase difference shown by it between A and B is the same at any stage of any cycle. If the voltages represented had different frequencies, their phasors would rotate at different speeds, so the phase difference between them would be continually changing.

It should be stated here that the subject of phasor diagrams is currently treated in a number of different ways, which are examined in M. G. Scroggie's book *Phasor Diagrams*. Phasors are still often called vectors, but vectors are the subject of a branch of mathematics dealing with vector quantities, which current and voltage are not. The likelihood of misunderstanding through using the same name for things that are essentially different is increased by the common practice of attaching an arrow head to the end of each phasor, making it look exactly like a vector. The arrow head is quite unnecessary, and can be confusing until it is realised that it does not indicate a direction of motion.

Worthy of more respect is the doctrine that phasors should not be supposed to rotate. Some of the reasons are rather too subtle for the present level of study, and in any case the difference between the two schools of thought is theoretical rather than practical, because the diagrams that even rotating-phasor exponents draw on paper are inevitably fixed there. The effort of imagining them to be

rotating at 3000 rev/min to represent a frequency of 50 Hz (let alone the speeds corresponding to radio frequencies!) might well induce a feeling of dizziness. So long as all the phasors in a diagram represent quantities having the same frequency, the shape of the diagram does not change. Nevertheless, in the early stages of study we may find it helpful to turn the diagram around in order to see 'in slow motion' how each voltage and current in an a.c. circuit changes relative to the others as the cycle progresses.

It is an essential part of the rotating phasor concept that the length of the phasor represents the peak value. But without forgetting this there is really no reason why we should not make use of the fact that the ratio between peak and r.m.s. values is constant ($\sqrt{2}$) to regard the phasors alternatively as representing r.m.s. values. It is merely a matter of choice of scale. This is convenient and is often done, especially when one has become used to interpreting stationary diagrams.

5.8 Adding Alternating Voltages

A great advantage of the phasor diagram is the way it simplifies adding and subtracting alternating voltages or currents that are not in phase. To add voltages A and B in Figure 5.3a, for example, one would have to add the heights of the A and B waves at close intervals of time and plot them as a third wave. This would be a very laborious way of finding how the peak value and phase of the combined voltage (or *resultant*, as it is called) compared with those of its component voltages. In a phasor diagram it is quickly done by 'adding' phasor B to the rotating end of phasor A, as in Figure 5.7, and drawing a straight line from

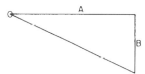

Figure 5.7 How to find the phasor representing the resultant of A and B (Figure 5.3a)

O to the 'loose' end of B. This line is the phasor of the combined voltage, A + B. Its length represents its peak value (to the same scale as A and B) and its angle represents its relative phase. Alternatively of course A can be added to B. Note that the result of adding two voltages that are not in phase is less than the result of adding them by ordinary arithmetic, and its phase is somewhere between those of the components. Since it is representable by a phasor, it is sinusoidal, like its components.

5.9 Direction Signs

Have you ever been directed to a destination by someone who supposed that it was being approached from the opposite direction? It is a very perplexing and frustrating experience. Yet quite a common one for students of a.c., because much of the literature on the subject is ambiguous in its direction signs.

All our talk about voltages and currents will do us no practical good unless we can relate them to circuits. Already we have been drawing graphs and phasor diagrams of voltages A and B without any reference to where they could be found. So let us suppose that voltage A (call it V_A) occurs between terminals c and d in a circuit, and V_B between terminals e and f in the same or another circuit. As these circuits might be very complicated, in Figure 5.8 only

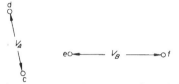

Figure 5.8 If V_B is known to lag quarter of a cycle behind V_A, which of its terminals first becomes positive after d?

the terminals are shown. A very common way of indicating the voltages between them is as shown there, with two-way arrows. Now let us try to relate this to the waveform graph, Figure 5.3a, which tells us that the positive peak of V_B occurs quarter of a cycle later than the positive peak of V_A. The phasor diagram, Figure 5.6, tells us the same thing in a simpler way. Suppose that at a certain instant d is at positive peak relative to c. Then we know that a quarter of a cycle later either e or f will be at peak positive relative to the other. Which one? The answer we have is like that of the man who was asked 'Will Oxford or Cambridge win the boat race?' and replied 'Yes'. Quite true, but quite useless. So are two-way arrows as voltage signposts.

One solution to the difficulty is to insist on arrows pointing one way only, like any sensible signpost. The same applies of course to currents. When we were dealing with d.c. circuits, such as Figure 2.14, one point that had to be made clear was that there needed to be agreement on whether the arrow heads marked what we knew were (say) the positive ends, or merely the ends that were to be taken as positive, so that if any voltage, at first unknown, turned out to be negative we would know that the arrow head would mark what was actually the negative end. With a.c. the voltages and currents are continually reversing, so no terminal or direction is any more positive than negative; but one-way arrows or even + and − signs can be used for a.c. to show the polarity that is arbitrarily taken as positive at a given instant. Half a cycle later everything will be reversed, but things that were opposite in sign will still be so and like signs will still be like. Provided that all phasors are correctly marked to correspond (hence the arrow-head custom) this method can be made to work. M. G. Scroggie's *Phasor Diagrams* mentions eight disadvantages, however, and advises a system that is free from all of them. A brief outline follows.

5.10 Subscript Notation

If the terminals are labelled, as in Figure 5.8, there is no need to mark the voltages or their directions on the circuit diagram at all: V_A is automatically V_{cd} or V_{dc}, and so on. In practice there is no need even to write the V; it is much easier to write (and especially to type) just cd or dc. Similarly there is no need to mark the phasor with an A, or V_A, as in Figure 5.7; nor with an arrow. Marking its ends c and d makes it correspond completely with the circuit. We do not then have to make an arbitrary choice of polarity; both sets of half-cycles are equally represented. If necessary, we can distinguish between a voltage and its phasor by following the usual convention of italic type for a voltage (e.g., cd) and roman letters for its phasor (cd).

Only one thing remains to be settled. When cd is positive, which terminal is the positive one? Some people who use a subscript voltage notation write the letters in the order that makes a voltage rise negative. It seems more natural, and fits in with other conventions better, to regard a rise as positive, and that is what will be done here. So whenever terminal d is positive relative to c, voltage cd will be taken as positive, signifying in fact the change in potential on moving from c to d. It follows that dc would be negative.

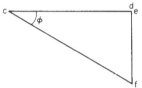

Figure 5.9 Figure 5.7 redrawn, using double-subscript convention related to Figure 5.8

Figure 5.9 shows how Figure 5.7 would be redrawn according to the foregoing conventions. It tells us that if terminals d and e were connected together, the voltage between c and f would be correctly represented by the phasor between c and f. The angle by which it lags cd is marked φ (Greek small phi), the usual symbol for a phase difference. It answers all questions unambiguously. For example, when d is at peak positive with respect to c, what is the potential of f relative to c, and when will it reach its positive peak? If we turn Figure 5.9 quarter of a turn anticlockwise, so that cd is bolt upright and therefore appropriately representing cd at positive peak, d being at its farthest above c, corresponding to Figure 5.3a 0.005 s from the origin, cf shows that cf is nearing its positive peak and will reach it in a fraction φ/360° of a cycle.

If instead of connecting e to d we had connected it to c, Figure 5.10 would be the appropriate diagram. It shows the potential at f a full quarter-cycle behind that at d (both relative to c). This agrees with Figure 5.3a, but until we had a proper notation we were unable to tell which was the right connection, and in fact went astray in Figure 5.7.

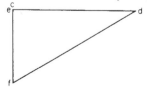

Figure 5.10 If one of the circuit connections was reversed, this would apply instead of Figure 5.9

One of the advantages of not using arrow heads in phasor diagrams is that we are not tied to any one point as the centre about which the thing is supposed to rotate. Whichever point in the circuit we choose to regard as the reference point, or zero potential, with respect to which all other voltages are reckoned, is the appropriate centre, and we can mark it with a circle or stick a pin through it to serve as an axis.

5.11 Current Phasors

Attempts to extend subscript notation to currents usually run into trouble, because people write I_{cd} to mean the current flowing from c to d. But whereas there can be only one voltage at a time between two points, there can be as many currents as there are parallel paths. This difficulty can be avoided by labelling the *meshes* or loops of the circuit. There are excellent theoretical reasons for doing this, but it also works in practice. To distinguish currents from voltages, capital letters can be used for the meshes. All we need is to agree on which order of letters to use in order to denote the positive direction of current. Either would do, but to agree with M. G. Scroggie's *Phasor Diagrams* let us say that I_{AB} (which we can conveniently write as AB) in Figure 5.11a means positive current moving up through the generator and down through the resistor. Note that although this simplest of circuits appears to have only one mesh we must mark the external one (A) too. Note also that I_B, or B, means the current circulating clockwise around mesh B, and AB means the net current crossed in moving from A to B.

For such a simple circuit the phasor diagram is very simple. There are only two terminals, so only one voltage, ab or ba, between them and therefore one voltage phasor (Figure 5.11b). When ab is peak positive, as shown, we know (Section 5.3) that the current AB, which according to our convention is the one flowing upward through the generator, is also peak positive, so is represented by the vertically upward phasor AB.

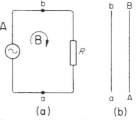

Figure 5.11 Voltage and current phasor diagrams relating to the very simple circuit shown

It can be said (and many of those who use other conventions probably will say) that there are two voltages in this circuit. And in a sense there are, but they are really the same voltage looked at from two opposite points of view. The only way of finding two voltages with a voltmeter is to measure it once and then measure it again with reversed leads. If we think of Kirchhoff's voltage law and start at a, going clockwise around the circuit with the current, at the instant represented by the phasors, we find that when we have reached b the potential has risen. In other words, ab is positive. Going now through the resistor R to b we find the potential falls. In other words, ba is negative. This may seem too obvious to have to explain, but it is surprising how confused people who use arrows can get about it. And if the confusion is possible with this simplest of circuits the need to be quite clear before going on to more complicated ones needs no emphasis.

One objection that is sometimes raised against the foregoing notation is that it introduces a minus sign into Ohm's law, since the voltage across R in the direction of AB is ba. But this is not so in the form of the law which says that the current is proportional to the voltage applied. (This is usually implied by using E for the voltage, in $I = E/R$.) In Figure 5.11a the voltage applied is the generator voltage, ab.

Chapter 6

Capacitance in A.C. Circuits

6.1 Current Flow in a Capacitive Circuit

The next type of basic circuit to consider is the one shown in Figure 6.1. If the generator gave an unvarying e.m.f. no steady current could flow, because there is a complete break in the circuit. The most that could happen would be a momentary current when switching on, as shown in Figure 3.6.

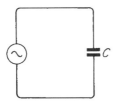

Figure 6.1 Circuit consisting of an a.c. generator with a purely capacitive load, for comparison with Figure 5.1

Some idea of what is likely to happen when an alternating e.m.f. is applied can be obtained by an experiment similar to Figure 3.5 but with an inverted battery in the B path, as shown in Figure 6.2a. Moving the switch alternately to A and B at equal time intervals will then provide an alternating voltage of square waveform (Figure 6.2b).

Whenever the switch is moved to A there is a momentary charging current in one direction, and moving it to B causes a similar current in the reverse direction.

We know that for a given voltage the quantity of electricity transferred at each movement of the switch is proportional to the capacitance (Section 3.2). And it is obvious that the more rapidly the switch is moved to and fro (that is to say, the higher the frequency of the alternating e.m.f.) the more often this quantity of electricity will surge to and fro in the circuit (that is to say, the greater the quantity of electricity that will move in the circuit per second, or, in other words, the greater the current).

So although the circuit has no conductance an alternating e.m.f. causes an alternating current. It does not flow conductively *through* the capacitor, whose dielectric is normally a very good insulator, but can be visualised as flowing in and out of it. Alternatively, Section 3.8 suggested how the current can be regarded as existing even between the capacitor plates.

The fact that an alternating e.m.f. can make a current flow in a circuit blocked to d.c. by a capacitance in series is easily demonstrated by bridging the open contacts of an electric light switch by a capacitor (of suitable a.c. rating!); say 2 µF for a 40 W lamp (Figure 6.3). The lamp will light and stay alight as long as the capacitor is connected. But it will not be as bright as usual. The less the capacitance, the

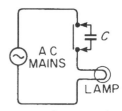

Figure 6.3 Demonstration of current without conduction

dimmer the light. The amount of current due to a given voltage thus depends on two things: the capacitance and the frequency.

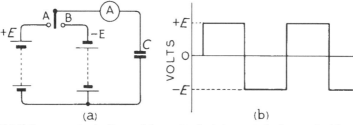

Figure 6.2 If the generator in Figure 6.1 consisted of the periodically switched batteries as shown at (a), the waveform would be as at (b), and the current (assuming a certain amount of resistance) could be calculated as discussed in connection with Figure 3.6

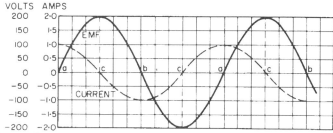

Figure 6.4 E.m.f. and current diagram for Figure 6.1, drawn to correspond with Figure 5.2

6.2 Capacitive Current Waveform

Let us now consider the action of Figure 6.1 in greater detail, by drawing the graph of instantaneous e.m.f. (Figure 6.4) exactly as for the resistive circuit, in which we assumed a sine waveform. But unlike the resistive case there is no Ohm's law to guide us in plotting the current waveform. We have, however, a rather similar relationship (Section 3.2):

$$V = Q/C$$

where V is the p.d. across C, and Q is the charge in C. This is true at every instant, so we can rewrite it $v = q/C$, to show that we mean instantaneous values. In Figure 6.1, v is always equal in magnitude to e, the instantaneous e.m.f. So we can say with confidence that at the moments when e is zero the capacitor is completely uncharged, and at all other moments the charge is exactly proportional to e. If we knew the right scale, the voltage wave in Figure 6.4 would do also as a charge curve. But we are not so much interested in the charge as in the current; that is to say, the rate at which charging takes place. At points marked a, although the voltage and charge are zero they are growing faster than at any other stage in the cycle. So we may expect the current to be greater than at any other times. At points marked b, the charge is decreasing as fast as it was growing at a, so the current is the same in magnitude but opposite in direction and sign. At points marked c, the charge reaches its maximum, but just for an instant it is neither growing nor waning; its rate of increase or decrease is zero, so at that instant the current must be zero. At intermediate points the relative strength of current can be estimated from the steepness of the voltage (and charge) curve. Joining up all the points gives a current curve shaped like the dashed line.

If this job is done carefully the shape of the current curve is also sinusoidal, as in the resistive circuit, but out of phase, being a quarter of a cycle (or 90°) ahead of the voltage.

On seeing this, students are sometimes puzzled and ask how the current *can* be ahead of the voltage. 'How does the current know what the voltage is going to be, quarter of a cycle later?' This difficulty arises only when it is supposed, quite wrongly, that the current maximum at a is caused by the voltage maximum a quarter of a cycle later, at c; whereas it is actually caused by the voltage at a increasing at its maximum rate.

6.3 The 'Ohm's Law' for Capacitance

What we want to know is how to predict the actual magnitude of the current, given the necessary circuit data. Since, as we know (Section 3.2), when there is 1 V between the terminals of a 1 F capacitor its charge must be 1 C, we can see that if the voltage were increasing at the rate of 1 V/s the charge would be increasing at 1 C/s. But we also know that 1 C/s is a current of 1 A; so from our basic relationship, $Q = VC$, we can derive

$$\text{Amps} = \text{Volts-per-second} \times \text{Farads}$$

The problem is how to calculate the volts-per-second in an alternating e.m.f., especially as it is varying all the time. All we need do, however, is find it at the point in the cycle where it is greatest—at a in Figure 6.4. That will give the peak value of current, from which all its other values follow (Section 5.4).

Going back to Figure 5.5, you may remember that one revolution of P about O represents one cycle of alternating voltage, the fixed length OP represents the peak value of the voltage, and the length of PQ (which, of course, is varying all the time) represents the instantaneous voltage. When P is on the starting line, as in Figure 5.4, the length of PQ is zero, *but at that moment it is increasing at the rate at which P is revolving around O*. Now the distance travelled by P during one cycle is 2π times the length of OP. And if the frequency is 50 Hz, it does this distance 50 times per second. Its rate is therefore $2\pi \times 50$ times OP. More generally, if f stands for the frequency in hertz, the rate at which P moves round O is $2\pi f$ times OP. And as OP represents E_{max}, we can say that P's motion represents $2\pi f E_{max}$ volts-per-second (Figure 6.5). But we have just seen that it also represents the maximum rate at which PQ, the instantaneous voltage, is growing. So, fitting this information into our equation, we have

$$I_{max} = 2\pi f E_{max} C$$

where I is in amps, E in volts, and C in farads. Since I_{max} and E_{max} are both $\sqrt{2}$ times I and E (the r.m.s. values) respectively, it is equally true to say

$$I = 2\pi f E C$$

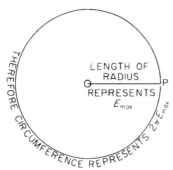

Figure 6.5 From this diagram it can be inferred that the maximum rate at which E increases is $2\pi f E_{max}$ volts-per-second

And if we rearrange it like this:

$$\frac{E}{I} = \frac{1}{2\pi f C}$$

and then take another look at one of the forms of expressing Ohm's law:

$$\frac{E}{I} = R$$

we see they are the same except that $1/2\pi f C$ takes the place of R in the role of limiting the current in the circuit. So, although $1/2\pi f C$ is quite a different thing physically from resistance, for purposes of calculation it is of the same kind, and it is convenient to reckon it in ohms. To distinguish it from resistance it is named *reactance*. To avoid having to repeat the rather cumbersome expression $1/2\pi f C$, the special symbol X has been allotted. The 'Ohm's law' for the Figure 6.1 circuit can therefore be written simply as

$$\frac{E}{I} = X \text{ or } I = \frac{E}{X} \text{ or } E = IX$$

For example, the reactance of a 2 µF capacitor to a 50 Hz supply is $1/(2\pi \times 50 \times 2 \times 10^{-6}) = 1592\,\Omega$, and if the voltage across it were 240 V the current would be $240/1592 = 0.151$ A.

The factor group $2\pi f$ occurs so often in connection with a.c. circuits that it is commonly denoted by the Greek small omega, ω. So an alternative way of writing the reactance is $1/\omega C$. But in this elementary book the formulae will be used mainly for numerical examples, so it may be clearer if the separate factors are shown.

6.4 Capacitances in Parallel and in Series

The argument that led us to conclude that the capacitance between parallel plates is proportional to the area across which they face one another (Section 3.3) leads also to the conclusion that capacitances in parallel add up just like resistances in series.

We can arrive at the same conclusion by simple algebraic reasoning based on the behaviour of capacitors to alternating voltage. Figure 6.6 represents an a.c. generator connected to two capacitors in parallel. The separate

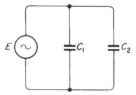

Figure 6.6 In this circuit the capacitance of C_1 and C_2 together is equal to $C_1 + C_2$

currents in them are respectively $E \times 2\pi f C_1$ and $E \times 2\pi f C_2$. As these currents are in phase with one another (because both lead E by the same angle), the total current is $E \times 2\pi f (C_1 + C_2)$, which is equal to the current that would be taken by a single capacitance equal to the sum of the separate capacitances of C_1 and C_2. Thus the effective capacitance of a number of capacitors in parallel is always greater than that of the largest (c.f. resistances in series, Section 2.9).

We also saw that if a single capacitor is divided by a plate placed midway between its two plates, the capacitance between the middle plate and either of the others is twice that of the original capacitor. In other words, the capacitance of two equal capacitors in series is half that of each of them. Let us now consider the more general case of any two capacitances in series.

If X_1 and X_2 are respectively the reactances of C_1 and C_2 in Figure 6.7, their combined reactance X is $(X_1 + X_2)$, as in the case of resistances in series. By first writing down the

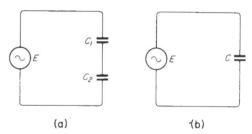

Figure 6.7 If the capacitor C in (b) is to take the same current as the two in series at (a), its capacitance must be equal to $1/(1/C_1 + 1/C_2)$

equation $X = X_1 + X_2$, and then replacing each X by its known value, of form $1/2\pi f C$, we deduce that

$$\frac{1}{C} = \frac{1}{C_1} + \frac{1}{C_2}$$

That is, the sum of the reciprocals of the separate capacitances is equal to the reciprocal of the total capacitance.

So the rule for capacitances in series is identical with that for resistances or reactances in parallel. From the way it was derived it is evidently not limited to two capacitances only, but can be applied to any number. It implies that if capacitors are connected in series, the capacitance of the

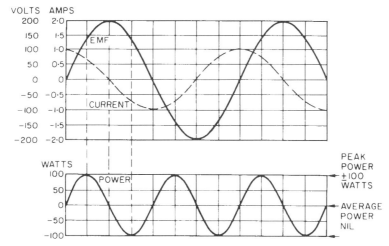

Figure 6.8 The current and voltage graphs for a capacitance load (Figure 6.4) are here repeated, with a power graph derived from it plotted below

combination is always less than that of the smallest (c.f. resistances in parallel, Section 2.9).

Applied to two capacitances only, the rule can, as with resistances, be put in the more easily workable form

$$C = \frac{C_1 C_2}{C_1 + C_2} = \frac{\text{product}}{\text{sum}}$$

6.5 Power in a Capacitive Circuit

Figure 6.4 was drawn to match Figure 5.2 as regards peak voltage and current. The only difference is the current's 90° phase lead. Let us now complete the picture by calculating the wattage at a sufficient number of points to draw the power curve (Figure 6.8). Owing to the phase difference between E and I, there are periods in the cycle when they are of opposite sign, giving a negative power. In fact, the power is alternately positive and negative, just like E and I except for having twice the frequency. The net power, taken over a whole cycle, is therefore zero. Since we are applying the sign '+' to positive current going through the generator towards positive potential, positive power means power going out of the generator, and negative power is power returned to the generator by the load (end of Section 2.17).

The power curve in Figure 6.8 represents the fact that although a pure capacitance draws current from an alternating generator it does not permanently draw any power—it only borrows some twice per cycle, repaying it in full with a promptitude that might be commended to human borrowers.

A demonstration of this fact is given by the capacitor in series with the lamp (Figure 6.3). The lamp soon warms up, showing that electrical power is being dissipated in it, but although the capacitor is carrying the same current it remains quite cold.

6.6 Phasor Diagram for Capacitive Circuit

In dealing with such a simple circuit as Figure 6.1 we are not likely to become uncertain about the directions of e.m.f. and current represented in graphs such as Figure 6.8, especially as we have just stated them fully in words in order to check the directions of power flow. But in order to get into practice for more complicated circuits let us draw the phasors for Figure 6.1, as we did for the simple resistive circuit in Figure 5.11.

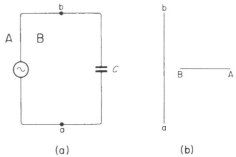

Figure 6.9 Simplest capacitive circuit and corresponding phasor diagrams

Figure 6.9a shows the circuit suitably labelled. We can always choose the angle at which to draw the first phasor, so to make it easy to compare with Figure 5.11 let us draw the voltage phasor in the same way, indicating that terminal b is at peak positive potential with respect to a. At that moment, therefore, the voltage across C has just stopped increasing and is about to decrease; so C is charged to the maximum extent, positive on the upper plate. The current, which has been flowing clockwise to deliver that charge, is now at zero and about to discharge anticlockwise. The current phasor must therefore be drawn horizontal, to signify zero current. And as anticlockwise current, which our convention denotes by *BA*, is the one about to become positive, it must be drawn left to right as shown (Figure 6.9b). As *ab* is the generator voltage applied to *C*, one would usually refer to the clockwise current *AB*, which the diagram shows to be a quarter of a cycle ahead of *ab*. This confirms the rule for sinusoidal a.c., well known to electrical engineers, that the current into a pure capacitance leads the applied voltage by 90°.

6.7 Capacitance and Resistance in Series

Figure 6.10 shows the next type of circuit to be considered: *C* and *R* in series. We know how to calculate the relationship (magnitude and phase) between voltage and current for each of these separately. When they are in series the same current will flow through both, and this current cannot at one and the same time be in phase with the generator voltage and 90° ahead of it. But it is quite possible—and in fact essential—for it to be in phase with the voltage across the resistance and 90° ahead of the voltage across the capacitance. These two voltages, which must therefore be 90° out of phase with one another, must add up to equal the generator voltage. We have already

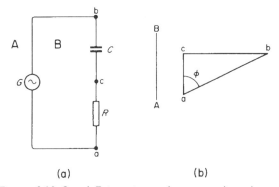

Figure 6.10 C and R in series, and corresponding phasor diagrams

added out-of-phase voltages by two different methods (Figure 5.3 and Figure 5.7), so that part of it should not be difficult. Because the current, and not the voltage, is common to both circuit elements, it will be easier to reverse the procedure we have adopted until now, and, starting with a current, find the e.m.f. required to drive it. When we have in that way discovered the key to this kind of circuit we can easily use it to calculate the current resulting from a given e.m.f.

We could, of course, represent the two voltages by drawing waveform graphs, and, by adding them, plot the graph of the total generator voltage and compare it for phase and magnitude with the current graph. But seeing we have taken the trouble to learn the much quicker phasor method of arriving at the same result, we might as well use it. (If you prefer not to, there is nothing to stop you drawing the waveforms.)

The first thing, then, is to draw a current phasor AB at any angle we please. In Figure 6.10 it is vertical. *AB* being a clockwise current, we know that it is in phase with the voltage applied clockwise, via *G* and *C*, to the resistor R; that is to say *ac*. So we draw ac parallel to AB to show that they are in phase. We know that *AB* leads by 90° the voltage applied clockwise, via *R* and *G*, to *C*; that is to say *cb*. So cb should be drawn to lag AB by 90°. The generator e.m.f., *ab*, is then indicated in magnitude and phase. The current is leading it by the angle φ, less than 90°.

You may ask how we know how long to draw these phasors. The current one has to be drawn first, and at that stage we do not know how much current there is, even if we know *ab*, *f*, *C* and *R*. Whatever *AB* may turn out to be, however, any length of phasor will be right if a suitable scale is chosen. As for *ac* and *cb*, those voltages are proportional to *R* and *X* (= $1/2\pi fC$), so we can draw their phasor lengths in the ratio *R* : *X*. Then if we know the generator e.m.f., ab represents it, so the voltage scale is found and hence the values of *ac* and *cb*. Lastly, *AB* can be calculated by Ohm's law as *ac/R*.

Most of the time, however, phasor diagrams are drawn just to show the principle without bothering about the exact values they represent. At low frequencies at least, it is likely that C would impede current more than *R* and so take the major share of the available e.m.f., so cb has been drawn longer than ac.

6.8 Impedance

A new name is needed to refer to the current-limiting properties of this circuit as a whole. *Resistance* is appropriate to *R* and *reactance* to *C*; a combination of them is called *impedance* (symbol: *Z*). Resistance and reactance themselves are special cases of impedance. So we have still another relationship in the same form as Ohm's law; one that covers the other two, namely:

$$I = \frac{E}{Z}$$

Although *Z* combines *R* and *X*, it is not true to say *Z* = *R* + *X*. Suppose that in Figure 6.10 C is the 2 μF capacitor which we have already calculated has a reactance of 1592 Ω at 50 Hz, and that *R* is around 1000 Ω. *AB*, shall we say, is 0.1 A. Then *ac* must be 100 V and *cb* 159 V. If the phasor diagram was drawn accurately to scale it would show *ab* to be 188 V, so *Z* must be 1880 Ω, which is much less than 1592 + 1000.

As the three voltage phasors refer to the same current, to a suitable scale they also represent the corresponding three impedances: resistive, reactive and total. Also, as they make a right-angled triangle, the celebrated Theorem of Pythagoras tells us that

$$Z^2 = R^2 + X^2$$

$$\text{or } Z = \sqrt{(R^2 + X^2)}$$

In trigonometry the ratio cb/ac, which we have just seen is equal to *X/R*, is known as the tangent of the angle φ, abbreviated to tan φ. As there are tables from which the angle corresponding to any tangent can be seen and some pocket calculators include them, phase angles can be found from

$$\tan \varphi = \frac{X}{R}$$

So we have the alternative of computing Z by arithmetic instead of drawing a phasor diagram or a waveform diagram accurately to scale.

The phasor triangle also tells us, of course, that

$$ab = \sqrt{(ac^2 + cb^2)}$$

(Note that here 'ac' is really one symbol, all squared; not $a \times c^2$.) We will no doubt have already realised that 100 V and 159 V, the voltages across R and C, do not add up in the ordinary way to equal 188 V, the voltage across both. This would be very puzzling—in fact, quite impossible, and a serious breach of Kirchhoff's voltage law—if these figures were instantaneous values. But of course they are r.m.s. values, and the explanation of the apparent discrepancy is that the maximum instantaneous values do not occur all at the same time. This is most clearly shown by a waveform graph, tedious though it is to draw as a means of calculating the total voltage, etc., compared with a phasor diagram or algebra.

But this type of phasor diagram does very neatly illustrate Kirchhoff's voltage law, because the voltage phasors for any whole circuit always necessarily form a closed geometrical figure, such as our triangle.

6.9 Power in Mixed Circuits

By now we should hardly need to draw waveform diagrams such as Figure 6.8 in order to find how much power the generator delivers to the Figure 6.10 circuit. Figure 6.8, together with Figure 6.9, has shown that when phasors are at right angles, as they are when they refer to a perfect capacitor, energy flows to and fro between generator and load, but none is permanently delivered. That is because a capacitor can store electrical energy and returns it in the same form, unlike a resistor, which dissipates it all as heat just as fast as it is received. In Section 5.4 we found that the power P delivered to and dissipated by resistance R is E^2/R, where E is the r.m.s. voltage across it. As $R = E/I$, I being the r.m.s. current through R, by substituting for R in the first equation we get $P = EI$, just as with d.c. (Section 2.17). The quantities corresponding in Figure 6.10 to E and I are AB and ac, if those are being used to represent r.m.s. values. If however they are used more strictly to represent peak values, and since peak values are $\sqrt{2} \times$ r.m.s. values, P would be equal to $AB.ac/2$.

In a circuit such as this, including both R and C in series, the part of the total voltage that should be multiplied by the current to find the power dissipated is represented by the phasor that is parallel to the current phasor: ac in Figure 6.10b. In terms of the total voltage, this voltage is $ab.ac/ab$, or ER/Z. So

$$P = \frac{EIR}{Z}$$

E and I being the total voltage and the current.

In trigonometry the ratio ac/ab is called the cosine of the angle φ, abbreviated to cos φ. So another form of the same equation is

$$P = EI \cos \varphi$$

φ being the phase angle between E and I. Cos φ is known to electrical engineers as the *power factor*.

6.10 Capacitance and Resistance in Parallel

In solving this type of circuit, Figure 6.11, we revert to the practice of starting with the voltage, because that is common to both. What we do, in fact, is exactly the same as for the series circuit except that currents and voltages

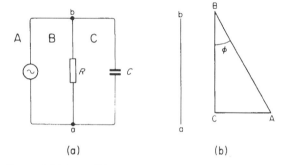

Figure 6.11 A parallel circuit such as this is similar to a series circuit such as Figure 6.10 with voltages and currents interchanged

change places. We now have two currents, CB and AC, one in phase with the e.m.f. ab and the other 90° ahead of it. By drawing the phasors accordingly, Figure 6.11b, we find the total current, AB, and the phase angle φ by which it leads ab.

By the same methods as we used in Section 6.8 we find that

$$AB = \sqrt{(AC^2 + CB^2)}$$

This is in fact the same equation, apart from the capital letters for currents instead of small ones for voltages. As with them, AC is regarded as one symbol, all squared.

Calculating the impedance is slightly more complicated, however, because the currents are *inversely* proportional to the impedances through which they flow. So $1/Z = \sqrt{(1/R^2 + 1/X^2)}$, or

$$Z = \frac{1}{\sqrt{\left(\frac{1}{R^2} + \frac{1}{X^2}\right)}}$$

which can be simplified to

$$Z = \frac{RX}{\sqrt{(R^2 + X^2)}}$$

Compare the formula for two resistances in parallel at the end of Section 2.9.

In Section 2.12 we found that calculations in parallel resistance circuits can be made as simple as those for series circuits by substituting conductance, G, for $1/R$. This idea has been extended to a.c. circuits by introducing *susceptance*, B, for $1/X$, and *admittance*, Y, for $1/Z$. The equations for parallel circuits then correspond to those for series circuits (Section 6.8):

$$Y = \sqrt{(G^2 + B^2)}$$

and $I = EY$ (compare $I = EG$ in Section 2.12)

Tan φ in Figure 6.11b is equal to CA/BC, and as these currents are proportional to susceptance and conductance respectively we have

$$\tan \varphi = \frac{B}{G}$$

Note that the method by addition of squares can be applied only to circuits in which the two currents or voltages are precisely in quadrature (90° out of phase), as when one element is a pure resistance and the other a pure reactance. Combining series impedances which are themselves combinations of resistance and reactance in parallel, or vice versa, is a rather more advanced problem than will be considered here. A circuit such as Figure 6.12, however, can be tackled by rules already given. R_2 and R_3

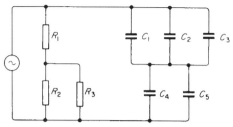

Figure 6.12 Example of a complex circuit that can be reduced to Figure 6.11 by applying the rules for resistances and capacitances in series and in parallel

are reduced to a single equivalent resistance (Section 2.10) which is then added to R_1. C_1, C_2 and C_3 are just added together, and so are C_4 and C_5; the two resulting capacitances in series are reduced to one (end of Section 6.4). The circuit has now boiled down to Figure 6.11.

Chapter 7

Inductance in A.C. Circuits

7.1 Current Flow in an Inductive Circuit

Although the phenomenon of magnetism, which gives rise to inductance, differs in many ways from electrostatics, which gives rise to capacitance, it is helpful to draw a very close parallel or analogy between them. Inductance and capacitance are, in fact, like opposite partners; and this chapter will in many ways be a repetition of the last one, but with a few basic things reversed.

We have already noticed some striking similarities as well as differences between capacitance and inductance (Section 4.10). One thing that could be seen by comparing them in Figures 3.6 and 4.10 was an exchange of roles between voltage and current.

The simple experiment of Figure 6.2 showed that, in a circuit consisting of a square-wave alternating-voltage generator in series with a capacitor, the current increased with frequency, beginning with zero current at zero frequency. This was confirmed by examining in more detail a circuit (Figure 6.1) with a sine-wave generator, in which the current was found to be exactly proportional to frequency.

Comparing Figure 4.10 with Figure 3.6, we can expect the opposite to apply to inductance. At zero frequency it has no restrictive influence on the current, which is limited only by the resistance of the circuit. But when a high-frequency alternating e.m.f. is applied the current has very little time to grow (at a rate depending on the inductance) before the second half-cycle is giving it the 'about turn'. The gradualness with which the current rises, you will remember, is due to the magnetic field created by the current, which generates an e.m.f. opposing the e.m.f. that is driving the current. It is rather like the gradualness with which a heavy truck gains speed when you push it. If you shake it rapidly to and fro it will hardly move at all.

For finding exactly how much current a given alternating voltage will drive through a given inductance (Figure 7.1) we have the basic relationship (Section 4.4):

$$\text{Volts} = \text{Amps-per-second} \times \text{Henries}$$

(Compare Section 6.3.)

Adapting the same basic method we begin with a sinusoidal current and find what voltage is required. The dashed curve in Figure 7.2 represents such a current during rather more than one cycle. At the start of a cycle (a), the current is zero, but is increasing faster than at any other phase. So the amps-per-second is at its maximum, and so therefore must be the voltage. After half a cycle (b) the current is again zero and changing at the same rate, but this time it is *decreasing*, so the voltage is a maximum *negative*. Halfway between (c) the current is at its

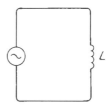

Figure 7.1 Circuit consisting of an a.c. generator with a purely inductive load

maximum, but for an instant it is neither increasing nor decreasing, so the voltage must be zero. And so on. Completing the voltage curve in the same way as we did the

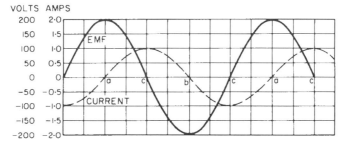

Figure 7.2 E.m.f. and current graphs for Figure 7.1, drawn to correspond with Figures 5.2 and 6.4

49

current curve in Figure 6.4, we find that it also is sinusoidal. This is very fortunate, for it allows us to say that if the sinusoidal voltage represented by this curve were applied to an inductance, the current curve would represent the resulting current, which would be sinusoidal.

7.2 The 'Ohm's Law' for Inductance

We have already found (Section 6.3) that when an alternating e.m.f. E, of frequency f, is sinusoidal, its r.m.s. volts-per-second is $2\pi fE$. The same method applies equally to current, so the r.m.s. amps-per-second is $2\pi fI$. Fitting this fact into our basic principle we get

$$E = 2\pi fIL$$

where L is the inductance in henries.

What we set out to find was the current due to a given e.m.f., so, the appropriate form of the above equation is

$$I = \frac{E}{2\pi fL}$$

Again, this is the same form as Ohm's law, $2\pi fL$ taking the place of R. This $2\pi fL$ can therefore also be reckoned in ohms, and, like $1/2\pi fC$, it is called reactance and denoted by X. Whenever it is necessary to distinguish between inductive reactance and capacitive reactance they are denoted by X_L and X_C respectively. X is the general symbol for reactance.

A question that may come to mind at this point is: can X_L and X_C in the same circuit be added in the same simple way as resistances? The answer is so important that it is reserved for the next chapter.

Note in the meantime that X_L is proportional to f, whereas X_C is proportional to $1/f$; this expresses in exact terms what our early experiments had led us to expect about the opposite ways in which frequency affected the current due to a given voltage.

An example may help to clinch the matter. What is the reactance of a 2 H inductor at 50 Hz? $X = 2\pi fL = 2\pi \times 50 \times 2 = 628\,\Omega$. So if the voltage is 240, the current will be $240/628 = 0.382$ A.

7.3 Inductances in Series and in Parallel

The previous section has shown that the reactive impedance of an inductance is directly proportional to the inductance—opposite to that of capacitance (which is proportional to $1/C$), but exactly like resistance. So inductances can be combined in the same way as resistances. The total effect of two in series, L_1 and L_2, is equal to that of a single inductance, $L_1 + L_2$. And inductances in parallel follow the reciprocal law: $1/L = 1/L_1 + 1/L_2$, or

$$L = \frac{L_1 L_2}{L_1 + L_2}$$

These principles can easily be verified by adding the reactances when they are in series, and the currents when the inductances are in parallel.

The above rules are subject to one important condition, however: that the inductors are so placed that their mutual inductance (Section 4.11) or magnetic coupling is negligible. If mutual inductance, M, does exist between two coils having separate self-inductances L_1 and L_2, the total self-inductance of the two in series is $L_1 + L_2 + 2M$ or $L_1 + L_2 - 2M$, according to whether the coils are placed so that their separate magnetic fields add to or subtract from one another. The reason M is doubled is that each coil is affected by the other, so in the combination the same effect occurs twice.

The corresponding elaboration of the formula for two inductances in parallel gives the result

$$L = \frac{(L_1 \pm M)(L_2 \pm M)}{L_1 + L_2 \pm 2M}$$

where ' $\pm$ ' signifies ' $+$ or $-$ '.

Complications arise also when the fields of two capacitors interact, but with the usual forms of construction such interaction is seldom enough to take into account.

7.4 Power in an Inductive Circuit

Comparing Figures 6.4 and 7.2 we see one difference. Whereas the capacitive current leads its terminal voltage by a quarter of a cycle (90°), the inductive current lags by that amount. If you want something to do you might care to calculate the instantaneous power at intervals throughout the cycle in Figure 7.2 and draw a power curve, as in Figure 6.8. But the same result can be achieved with less trouble simply by reversing the time scale in Figure 6.8, making it read from right to left. So except for relative phases the conclusions regarding power in a pure inductance are exactly the same as for a capacitance: the net power taken over a whole cycle is zero, because during half the time the generator is expending power in building up a magnetic field, and during the other half the energy thus built up is returning power to the generator. So in the example we took, with 240 V passing 0.382 A through 2 H, 240×0.382 does not represent watts dissipated in the circuit (as it would with resistance) but half that number of volt-amps tossed to and fro between generator and inductor at the rate of 100 times a second.

If you try an experiment similar to Figure 6.3, but using a coil of several henries in place of the capacitor, you may find that it does become perceptibly warm. This is because the coil is not purely inductive. Whereas the resistance of the plates and connections of a capacitor are usually negligible compared with its reactance, the resistance of the wire in a coil is generally appreciable. Although the resistance and inductance of a coil are physically inseparable, it is allowable to consider them theoretically as separate items in series with one another.

7.5 Phasor Diagram for Inductive Circuit

For comparison with Figure 6.9 we draw the simple circuit diagram, Figure 7.3a, of a generator feeding a load consisting of pure (i.e., resistanceless) inductance. We start with the current phasor and by applying the basic principles we know, as in Section 6.6, we find that the correct relative angle of the voltage phasor is as shown (Figure 7.3b). With the same conventions as before, it expresses the well-known rule that the current through a pure inductance lags the applied sinusoidal voltage by 90°.

7.6 Inductance and Resistance in Series

R in Figure 7.4a represents the resistance of the inductor whose inductance is represented by L, plus any other resistance there may be in series. This circuit can be tackled in exactly the same manner as Figure 6.10. As the current is common to both L and R we again begin with its phasor.

behind in phase, which is the same thing as saying cb must lead AB by that angle.

As ac is parallel to AB, φ is the angle by which AB lags the total applied voltage, ab. Again, as in Sections 6.8 and 6.9, and by similar reasoning,

$$Z = \sqrt{(R^2 + X^2)}$$

$$\tan \varphi = \frac{X}{R}$$

and $P = EI \cos \varphi$

7.7 Inductance and Resistance in Parallel

Figure 7.5 corresponds to Figure 6.11 and the phasor diagram is derived in the same way. As the current AC through L lags the applied voltage ab by 90° instead of leading it as it would with a capacitor, AC must be drawn

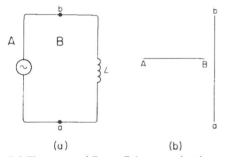

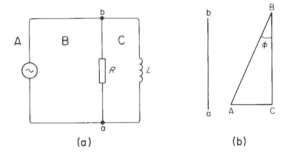

Figure 7.3 The circuit of Figure 7.1 repeated with its phasor diagrams

Figure 7.5 Parallel LR circuit and phasor diagrams

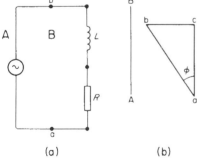

Figure 7.4 L and R in series, and corresponding phasor diagrams

The voltage phasor ac is again parallel to AB, and in fact the only real difference is that the phasor representing the voltage cb applied to L must be drawn so that AB is 90°

in the opposite direction, and φ is the angle by which the total current supplied by the generator lags its voltage.

If R is independent of L, then this circuit is somewhat unrealistic, as the resistance of the wire with which L is wound is not allowed for. If its resistance were included (in series with L) the problem could be solved by combining the methods for series and parallel impedances, as in Section 8.9. We shall see, however, in Section 8.16, that Figure 7.5a is an alternative way of taking account of the resistance of a reactive component.

The formulae for Z, Y, etc., found in Section 6.10, apply equally here, X standing for inductive instead of capacitive reactance.

The more advanced books extend the methods and scope of circuit calculation enormously; but before going on to them one should have thoroughly grasped the contents of these last two chapters. Some of the results can be summarised as follows:

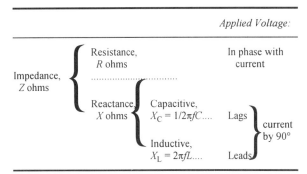

With R and X in series, $Z = \sqrt{R^2 + X^2}$ and $\tan \varphi = X/R$.

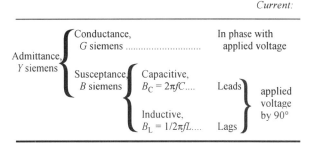

With G and B in parallel, $Y = \sqrt{G^2 + B^2}$ and $\tan \varphi = B/G$.

7.8 Transformers

In Section 4.11 we had a brief introduction to the transformer, which is an appliance very widely used for electrical purposes, including radio and electronics. We are now equipped to study it in a little more detail, beginning with Figure 7.6a. This represents a transformer connected

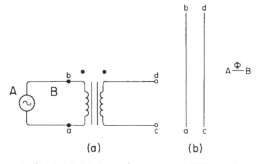

Figure 7.6 Unloaded 1:1 transformer with 100% coupling

to a generator giving the usual sinusoidal waveform. Let us assume that the primary and secondary coils have the same number of turns, and at this stage such complications as their resistance are conveniently assumed to be absent. An iron core is used to ensure that we can also assume, without serious error, that all the magnetic flux generated by the primary coil also links with the secondary, which at present is not connected to anything. The dots marking the top ends of the coils are a standard convention to exclude uncertainty about the relative direction of the windings. Without it one could not be sure about the polarity of the secondary voltage. If both coils are wound around the core in the same direction, then the dots mark both starting ends *or* both finishing ends.

With the secondary coil open-circuited, as shown, the circuit is essentially the same as Figure 7.3, so the phasor diagram is the same. We saw in Section 7.2 that the current driven by any applied voltage E through an inductance L is inversely proportional to L. And in Section 4.6 we noted that the inductance of an ideal inductor of the type we are considering is directly proportional to the permeability, μ, of the core, and to the square of N, the number of turns. The iron core of a transformer normally has a μ running into thousands, and N is usually considerable, so the inductance is likely to be very large and the current therefore quite small. That is why AB is drawn so short in Figure 7.6b. AB is called the *magnetising current* because its purpose is to generate sufficient alternating magnetic flux in the core to induce an e.m.f. in the primary coil exactly equal to that applied. As the flux is due directly to the current (not to its rate of change), it is in phase with the current, so the same phasor will do for both. This is shown by marking it with the flux symbol, Φ (Greek capital phi), in Figure 7.6b; and this also picks out AB as the magnetising current. Actually, as mentioned in Section 4.7, μ depends on the flux density in the core, varying throughout each cycle; so to induce a sinusoidal e.m.f.—necessary to match the applied one at every instant—the magnetising current has to be distorted, or non-sinusoidal, and therefore cannot strictly be represented by a phasor. This awkward complication is usually ignored, especially as the magnetising current is such a little one. With the ideal conditions assumed, it is clear that the average power required from the generator is nil, but that does not mean that there is no point in keeping the magnetising current as small as possible; the resistances of coil, generator and leads, which we are neglecting, give rise to a power loss proportional to the square of the current.

Because the secondary coil in the assumed transformer has the same number of turns as the primary and is linked by the same alternating flux, the secondary voltage *cd* must be the same as the primary, *ab*, so is represented by a phasor of the same length and angle. In fact, a single phasor would do, and this corresponds to the fact that a could be connected to c and b to d without affecting the electrical conditions. This is where the dots are helpful; if the secondary dot was at the c end it would indicate a reversal of winding and consequent polarity of secondary voltage, so paralleling the two coils as described would be disastrous!

7.9 Load Currents

The next step is to connect a load (for a full explanation of this term, see Section 11.2) to the secondary terminals, and continuing our ideal simplicity we make it a resistor, R in

Inductance in A.C. Circuits 53

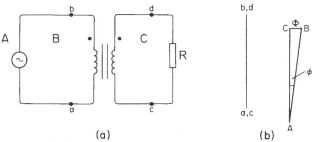

Figure 7.7 The transformer of Figure 7.6 with a resistive load

Figure 7.7a. This creates a new current mesh, C, and also raises a question about how this should be related to mesh B. Are we to suppose that the external mesh A penetrates between the windings, or should we think of them as too close together for that? According to our current notation BC is the current crossed on passing from B to C, and this is so whether we chose to see an intermediate mesh A, so that BC = BA + AC, or A is ignored and BC is seen as the same as the sum of the currents upwards in the two coils. Both ways come to the same thing. The important point is that the magnetic flux is the result of current in both coils. As the generator voltage *ab* is the same as before, the net magnetising current, which now is BC, must be the same as BA in Figure 7.6. So in Figure 7.7b we must draw BC accordingly.

Having settled that point we turn to the secondary circuit, which is identical with Figure 5.11 except for the different letters. Following the same principle we draw AC parallel to cd, which now has been made to coincide with ab as we saw it could have been in Figure 7.6b. The total current that the generator has to supply, AB, follows from this construction, and we see that it lags the generator voltage by the small phase angle φ.

The procedures for types of load other than R have been explained so need not be repeated.

We have already noted that in this example the coils could have been joined together at both ends, as we have represented by the coinciding voltage phasors. The coils themselves, indeed, could be made to coincide without affecting the working principles. Having done so, we could regard the secondary load current AC as flowing direct from the generator by the route bdca, and the combined primary and secondary coil as carrying only the magnetising current, CB. This is equivalent to going straight from C to B in the phasor current diagram, instead of via A, which takes account of the separate coil currents. These largely oppose one another in order to maintain the magnetising current CB unchanged. The very close correspondence between this type of phasor diagram and what it represents give a clear insight into the working of the circuit and makes it almost impossible to fall into such a common error as to suppose that the secondary voltage is in opposite phase to the primary.

7.10 Transformer Losses

In real transformers the primary coil has some resistance as well as inductance. Moreover, the alternating flux in the core generates a certain amount of heat in it (Section 8.14), which has to come from somewhere, and in accordance with our wider definition of resistance (Section 2.18) can be considered as equivalent to some extra resistance in the primary. The primary resistance, augmented in this way, causes the magnetising current to lag the primary voltage by less than 90°.

Nor is the power from the supply transferred to the load without loss. The primary and secondary load currents have to pass through the resistances of the windings. These resistances are represented in Figure 7.8 by R_1 and R_2. Moreover in real transformers some of the flux linking the primary coil fails to link with the secondary; similarly when secondary current is flowing it creates some flux not linked with the primary. These fluxes can be represented by small inductances L_1 and L_2, called *leakage inductances*. Note that the transformer terminals are still a-b and c-d; the internal junctions are of course not accessible as they are entirely imaginary. Figure 7.8 is an example of an *equivalent circuit* —equivalent in this case, approximately,

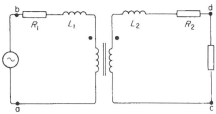

Figure 7.8 Figure 7.7a is here elaborated to include the resistances of the coils and their leakage inductances

to a real transformer, so far as any actual electrical measurements can show. It happens to be an example of a very common type of equivalent circuit, with two input and two output terminals; and because one may have to guess, from external measurements, what circuit best represents its actual behaviour, it is often called a black box.

If you feel confident about applying the phasor diagram rules, Figure 7.8 would be a good exercise. Although more complicated than anything we have tackled so far, it is quite straightforward, if one labels the four extra junctions and

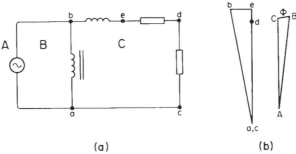

Figure 7.9 Simpler phasor diagrams (b) than those applicable to Figure 7.8 are obtained by transferring the primary resistance and leakage inductance to the secondary circuit and combining the fully coupled parts of the windings (a)

starts with the cd phasor. However, as it may be too big a jump for some, let us deal with a simplified version.

We have already seen that if the transformer turns ratio is 1:1 the two coils can be reduced to one. And as the magnetising current is normally quite small compared with the load current its contribution to the voltage drop across R_1 and L_1 can reasonably be neglected, by transferring them to the secondary circuit, so that suitably increased values of R_2 and L_2 include them. Figure 7.9a shows the simplified diagram. For easy comparison with the preceding ones, c is still shown, though it is now electrically the same as a.

AC and cd are first drawn parallel to one another. If the load had not been purely resistive they would of course have had to be drawn with the appropriate phase angle between them. de is simply an extension of cd, but eb must lead it by 90° because *eb* is across an inductance. We now see ab, the input voltage, and can therefore draw CB at right angles to it to represent an ideal magnetising current, or at a somewhat smaller angle to be more realistic. As the diagram clearly shows, one effect of coil resistance and leakage inductance is that the output voltage falls as the current taken is increased. This effect is confirmed by practical experience. And if we calculate the output power, equal to *cd.AC*, and compare it with the input power, which in Section 6.9 we found to be *ab.AB* cos φ, we can find how much power is lost in the transformer. The ratio of output to input power, which is always less than 1, is called the *efficiency* of the transformer.

These considerations are most important in mains transformers; in audio-frequency transformers (which are less used than they once were) the emphasis is rather different, and in radio-frequency transformers it is almost completely different, as will be seen later.

7.11 Impedance Transformation

So far our transformer diagrams have been related to those having the comparatively little-used 1:1 ratio. There is no reason why other ratios should not be illustrated by means of the relative lengths of the phasors, except that if the ratio were large there would be difficulty in drawing all the phasors clearly to the same scale. Of course the primary and secondary phasors of our type would need to be shown separately. On the other hand there is no strong reason why phasor diagrams should be used to show ratios other than 1:1. Transformer designers usually work in ampere-turns and volts per turn, which are the same for both primary and secondary and for any other windings embracing the whole core. So the phasors can very well be regarded as referring to ampere-turns and volts per turn, the actual amps and volts being derived by dividing or multiplying by the numbers of turns.

For we already know that when the same alternating flux links all the turns it induces the same voltage in every turn. And you have probably realised that if (say) the secondary coil has twice as many turns as the primary, so that its no-load voltage is twice that applied to the primary, the primary current due to any load current in the secondary must be twice as great, because the primary has only half the number of turns with which to offset the magnetising effect of the secondary current. It has to be so, too, to make the input watts the same (neglecting losses) as the output.

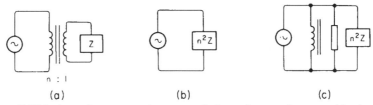

Figure 7.10 So far as the generator is concerned, the perfect transformer and load at (a) can be replaced by the impedance as at (b). An imperfect transformer can be simulated as at (c)

Now if the turns ratio of the transformer is $n:1$ and the load impedance is Z (Figure 7.10a) the primary voltage V_p is n times the secondary voltage V_s, and the primary load current I_p is I_s/n. So, while $Z = V_s/I_s$, the impedance 'looking into' the transformer from the primary side is thus n^2V_s/I_s, or n^2Z. The transformer and its load Z, then, can be replaced by a load n^2Z, as in Figure 7.10b, without making any difference from the generator's point of view, except for magnetising current and losses. If we want to be more precise and simulate these too we can do it by adding suitable parallel inductance and resistance, as in Figure 7.10c. Here is another example of an equivalent circuit. Obviously the equivalence is valid only on the primary side.

One of the uses of a transformer is for making a load of a certain impedance, say Z_s, equivalent to some other impedance, say Z_p. So it is often necessary to find the required turns ratio. Since $Z_p = n^2Z_s$, it follows that n must be $\sqrt{(Z_p/Z_s)}$.

Another use is for insulating the load from the generator. That is why 1:1 transformers are occasionally seen. But if it is not necessary to insulate the secondary winding from the primary, there is no need for two separate coils, even when the ratio is not 1:1. The winding having the smaller number of turns can be abolished, and the connections tapped across the same number of turns forming part of the other winding as in Figure 7.11. This device is called an *autotransformer*.

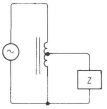

Figure 7.11 For some purposes it is practicable to obtain a voltage step-up or down without a separate secondary winding

Transformers are not limited to two windings. In practice it is quite usual to have several secondary windings delivering different voltages for various purposes, all energised by one primary coil.

A good example is the line output transformer in a television receiver. It has various secondaries used to:

(a) feed the line scanning coils
(b) feed the filament of the picture tube
(c) provide a boosted supply for the line output stage
(d) provide e.h.t. for the picture tube, usually via a voltage tripler (described in Section 28.10)
(e) provide a low-voltage supply for the transistors and integrated circuits in the receiver.

So it really is a powerhouse supplying most of the needs of the receiver.

Chapter 8 The Tuned Circuit

8.1 Inductance and Capacitance in Series

In the previous chapters we have seen what happens in a circuit containing capacitance *or* inductance (with or without resistance), and the question naturally arises: what about circuits containing both? We know that two or more reactances of either the inductive or capacitive kind in series can be combined just like resistances, by adding; and either sort of reactance can be combined with resistance by the more complicated square-root process. The outcome of combining reactances of opposite kinds is so fundamental to radio and many other electronic applications that it needs a chapter to itself.

We shall begin by considering the simplest possible series circuit, Figure 8.1, with capacitance and inductance in series. The method is exactly the same as for reactance and resistance in series; that is to say, since the current is common to both it is easiest to start from it and work

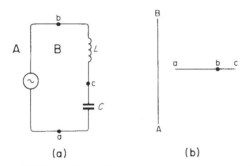

Figure 8.1 (a) circuit consisting of an a.c. generator in series with capacitance and inductance, and (b) corresponding phasor diagram

backwards to find the e.m.f. needed to drive it. As in Figure 7.4, we first draw a current phasor AB at any angle we like. The voltage applied to L (*cb*) must lead AB by 90°; *ac* on the other hand must lag AB by 90°. And so we get the complete phasor diagram. As a further check we may care to take another look at phasor and waveform diagrams for C alone (Figures 6.9 and 6.4) and for L alone (Figures 7.3 and 7.2). The conclusion must be the same: that in Figure 8.1 the voltages across C and L are in opposite phase, or 180° out of phase. So one of them must be *subtracted* from the other to give the necessary driving e.m.f. And since reactance is equal to applied voltage divided by current (the same in both cases) it follows that the total reactance of the circuit is also equal to one of the separate reactances *less* the other.

Which voltage or reactance must be subtracted from which to give the total? There is no particular reason for favouring either, but the agreed convention is to call X_L positive and X_C negative.

For example, suppose L in Figure 8.1 is 2 H, C is 2 μF, and *ab* is 240 V at 50 Hz. We have already calculated the reactances of 2 H and 2 μF at 50 Hz (Sections 7.2 and end of 6.3) and found them to be 628 Ω and 1592 Ω. So the total reactance must be 628 − 1592 = −964 Ω, which at 50 Hz is the reactance of a 3.3 μF capacitor. The generator, then, would not notice any difference (at that particular frequency) if a 3.3 μF capacitor were substituted for the circuit consisting of 2 μF in series with 2 H. The magnitude and phase of the current would be the same in both cases; namely, 240/964 = 0.294 A, leading by 90°.

To make sure, let us check it by calculating the voltage needed to cause this current. *cb* is equal to X_C times AB, = 0.249 × 1592 = 396 V. *ac* is X_L times AB, = 0.249 × 628 = 156 V. The resultant of *ac* and *cb*, namely *ab*, must be (as shown in Figure 8.1*b*) 396 − 156 = 240 V.

The fact that the voltage across the capacitor is greater than the total supplied by the generator may be surprising, but it is nothing to what we shall see soon!

8.2 L, C and R all in Series

Bringing R into the circuit introduces no new problem, because we have just found that L and C can always be replaced (for purposes of the calculation) by either L or C of suitable value, and the method of combining this with R was covered in the previous two chapters. Elaborating the equation given therein (Section 6.8) to cover the new information, we have

$$Z = \sqrt{[(X_L - X_C)^2 + R^2]}$$

If the three circuit elements are connected in the order shown in Figure 8.2*a*, the phasor diagram, drawn according to the now familiar rules, will be as at *b*. Note that *ad*, representing the voltage across C, is longer than *ab* representing the generator voltage. If L and C were replaced by a larger C to form the equivalent circuit mentioned above, *bc* would disappear and *ad* be shortened by that amount, but *ab* would still have the same length and phase angle relative to AB.

8.3 The Series Tuned Circuit

We have already seen that the reactance of a capacitor falls and that of an inductor rises as the frequency of the current supplied to them is increased (Sections 6.1 and 7.1). It is therefore going to be interesting to study the behaviour of a

56

circuit such as Figure 8.3 over a range of frequencies. For the values given on the diagram, which are typical of practical medium-wave receiver circuits, the reactances of coil and capacitor for all frequencies up to 1800 kHz are plotted as curves in Figure 8.4. The significant feature of this diagram is that at one particular frequency, about

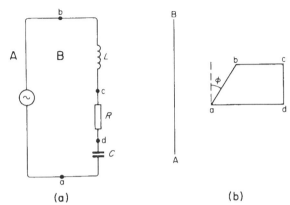

Figure 8.2 Inductance, resistance and capacitance all in series, with (b) corresponding phasor diagrams

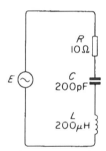

Figure 8.3 The way the reactances in this circuit vary with the frequency of E, graphed in Figure 8.4, leads to interesting conclusions regarding this type of circuit in general

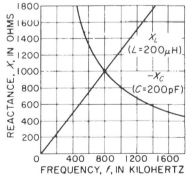

Figure 8.4 Reactances of the coil and capacitor in Figure 8.3 plotted against frequency

800 kHz (more precisely, 796 kHz), L and C have equal reactances, each amounting to 1000 Ω. So the total reactance, being the difference between the two separate reactances, is zero. Put another way, the voltage developed across the one is equal to the voltage across the other; and since they are, as always, in opposition, the two voltages cancel out exactly. The circuit would therefore be unaltered, so far as concerns its behaviour *as a whole* to a voltage of this particular frequency, by completely removing from it both L and C. This, leaving only R, would result in the flow of a current equal to E/R.

Let us assume a voltage not unlikely in radio broadcast reception, and see what happens when $E = 5$ mV. The current at 796 kHz is then 5/10 = 0.5 mA, and this current flows, not through R only, but through L and C as well. Each of these has a reactance of 1000 Ω at this frequency; the potential difference across each of them is therefore 0.5 × 1000 = 500 mV, which is just one hundred times the voltage E of the generator to which the flow of current is due.

That a simple circuit like this, without any amplifier, can enable so small a voltage to yield two such relatively large voltages, may seem incredible, but practical radio communication would not be possible without it.

8.4 Magnification

In the particular case we have discussed, the voltage across the coil (or across the capacitor) is one hundred times that of the generator. This ratio is called the *magnification* of the circuit.

We have just worked out the magnification for a particular circuit; now let us try to obtain a formula for any circuit. The magnification is equal to the voltage across the coil divided by that from the generator, which is the same as the voltage across R. If I is the current flowing through both, then the voltage across L is IX_L, and that across R is IR. So the magnification is equal to IX_L/IR, which is X_L/R or (because in the circumstances considered X_C is numerically equal to X_L) X_C/R. At any given frequency it depends solely on L/R, the ratio of the inductance of the coil to the resistance of the circuit, or on CR. (Both of these, you may have noticed, are time constants; see Sections 3.5 and 4.8.)

To obtain high magnification of a received signal (for which the generator of Figure 8.3 stands), it is thus desirable to keep the resistance of the circuit as low as possible.

The symbol generally used to denote this voltage magnification is Q (not to be confused with Q denoting quantity of electricity).

In any real tuned circuit the coil is not pure inductance as shown in Figure 8.3, but contains the whole or part of R and also has some stray capacitance in parallel with it. Consequently the magnification obtained in practice may not be exactly equal to Q (as calculated by X_L/R); but the difference is usually unimportant.

8.5 Resonance Curves

We have already seen that at 796 kHz the impedance of the series circuit consists only of the reactance R. At other frequencies the impedance of the circuit is greater, because in addition to R there is some net reactance. At 1250 kHz, for example, the individual reactances are 1570 and 637 Ω (see Figure 8.4), leaving a total reactance of 933 Ω. Compared with this, the resistance is negligible, so that the current, for the same driving voltage of 5 mV, will be 5/933 mA, or roughly 5 μA. This is approximately one hundredth of the current at 796 kHz. Passing through the reactance of L (1570 Ω) it gives rise to a voltage across it of 1570 × 5 = 7850 μV = 7.85 mV, which is only about one sixty-fourth as much as at 796 kHz.

By extending this calculation to a number of different frequencies we could plot the current in the circuit, or the voltage developed across the coil, against frequency. This has been done for two circuits in Figure 8.5. The only

The principle on which a receiver is tuned is now beginning to appear; by adjusting the values of L or C in a circuit such as that under discussion one can make it resonate at any desired frequency. Signal voltages received from the antenna at that frequency will be amplified so much more than those of substantially different frequencies that these others will be more or less tuned out.

8.6 Selectivity

This ability to pick out signals of one frequency from all others is called selectivity. It is an even more valuable feature of tuned circuits than magnification. There are alternative methods (which we shall consider later) of magnifying incoming signal voltages, but by themselves

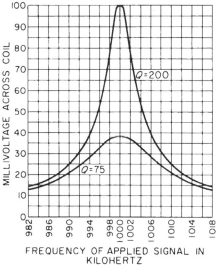

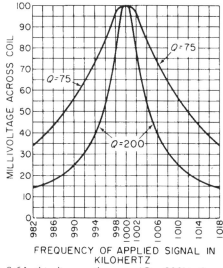

Figure 8.5 Showing how the voltage developed across the coil in a tuned circuit varies with frequency. Curves are plotted for L = 180 μH, C = 141 pF, E (injected voltage) = 0.5 mV, and R = 15 W for the Q = 75 circuit and 5.63 W for the Q = 200 circuit

Figure 8.6 In this diagram the curve 'Q = 200' is the same as in Figure 8.5, but, to enable the selectivity of the Q = 75 circuit to be more easily compared, the voltage injected into it has been raised from 0.5 mV to 1.33 mV, so as to make the output voltage at resonance equal to that across the Q = 200 circuit

difference between the circuits is that in one the resistance is 15 Ω, giving Q = 75, and in the other it is 5.63 Ω, giving Q = 200. In both, the values of L and C are such that their reactances are equal at 1000 kHz. These curves illustrate one thing we already know—that the *response* (voltage developed across the coil) when the reactances are equal is proportional to Q. They also show how it falls off at frequencies on each side, due to unbalanced reactance. At frequencies well off 1000 kHz the reactance is so much larger than the resistance that the difference between the two circuits is insignificant. The shapes of these curves show that the response of a circuit of the Figure 8.3 type is far greater to voltages of one particular frequency—in this case 1000 kHz—than to voltages of substantially different frequencies. The circuit is said to be *tuned to*, or to *resonate at*, 1000 kHz; and the curves are called *resonance* curves. This electrical resonance is very closely analogous to acoustical resonance: the way in which hollow spaces or pipes magnify sound of a particular pitch.

they would be useless, because they fail to distinguish between a programme on the desired frequency and others.

The way the curves in Figure 8.5 have been plotted focuses attention on how the value of Q affects the response at resonance. The comparative selectivity is more easily seen, however, if the curves are plotted to give the same voltage at resonance, as has been done in Figure 8.6. This necessitates raising the input to the low-Q circuit from 0.5 V to 1.33 V. Since the maximum voltage across both the coils is now 100 mV, the scale figures can also be read as percentage of maximum. Figure 8.6 brings out more clearly than Figure 8.5 the better selectivity of the high-Q circuit. For a given response to a desired station working on 1000 kHz, the 200-Q response to 990 kHz voltages is less than half that of the 75-Q circuit. In general, the rejection of frequencies well off resonance is nearly proportional to Q.

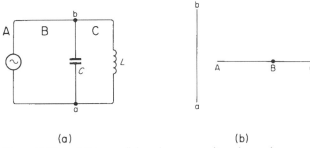

Figure 8.7 L and C in parallel, and corresponding phasor diagrams

8.7 Frequency of Resonance

It is obviously important to be able to calculate the frequency at which a circuit containing known L and C resonates, or to calculate the L and C required to tune to a given frequency. The required equation follows easily from the fact that resonance takes place at the frequency which makes the reactance of the coil equal that of the capacitor:

$$2\pi f_r L = \frac{1}{2\pi f_r C}$$

where the symbol f_r is used to denote the frequency of resonance. Rearranging this, we get

$$f_r^2 = \frac{1}{(2\pi)^2 LC}, \quad \text{so } f_r = \frac{1}{2\pi\sqrt{(LC)}}$$

If L and C are respectively in henries and farads, f_r will be in hertz; if L and C are in henries and microfarads, f_r will be in kHz. But perhaps the most convenient units for radio purposes are μH and pF, and when the value of π has been filled in the result is

$$f_r \text{ (in kHz)} = \frac{159\,155}{\sqrt{(LC)}} \text{ or } f_r \text{ (in MHz)} = \frac{159.15}{\sqrt{(LC)}}$$

If the answer is preferred in terms of wavelength, we make use of the relationship $f = 3 \times 10^8/\lambda$ (Section 1.9) to give

$$\lambda_r = 1.885\sqrt{(LC)} \quad (\lambda \text{ in m}; L \text{ in μH}; C \text{ in pF})$$

Note that f_r and λ_r depend on L multiplied by C; so in theory a coil of any inductance can be tuned to any frequency by using the appropriate capacitance. In practice the capacitance cannot be reduced indefinitely, and there are disadvantages in making it very large. A typical tuning capacitor for the medium- or low-frequency bands has a range of about 16 to 270 pF. The capacitance of the wiring and circuit components necessarily connected in parallel may add another 14 pF. This gives a ratio of maximum to minimum of 284/30 = 9.47. But the square root sign in the above equations means that if the inductance is kept fixed the ratio of maximum to minimum f_r (or λ_r) is $\sqrt{9.47}$, or 3.075. Any band of frequencies with this range of maximum to minimum can be covered with such a capacitor, the actual frequencies in the band depending on the inductance chosen for the coil.

Suppose we wished to tune from 1620 MHz to 1620/3.075 or 527 kHz, corresponding to the range of wavelengths 185 to 569 metres, approximately the limits of the medium waveband. For the highest frequency or lowest wavelength the capacitance will have its minimum value of 30 pF; by filling in this value for C and 1.620 MHz for f_r, we have $1.620 = 159.15/\sqrt{(30L)}$, from which $L = 321.5$ μH*.

If we calculate the value of L necessary to give 0.527 MHz with a capacitance of 284 pF, it will be the same.

In the same way the inductance needed to cover the short-wave band 9.8 to 30 MHz (30.6 to 10 metres) can be calculated, the result being 0.938 μH.

Notice how large and clumsy numbers are avoided by a suitable choice of units, but of course it is essential to use the equation having the appropriate numerical 'constant'. If in any doubt it is best to go back to first principles ($2\pi f_r L = 1/2\pi f_r C$) and use henries, farads, and Hz.

8.8 L and C in Parallel

Having studied circuits with L and C in series we come naturally to the simple parallel case, Figure 8.7a. It should hardly be necessary by now to point out that the voltage ab is the quantity common to both L and C, and so we start with its phasor. Drawing the current phasors in the appropriate directions and assuming that in this particular example more current goes through L than through C we get the complete diagram, Figure 8.7b.

If we go back to Figure 8.1 and compare the two phasor diagrams we see that they are identical except that current and voltage have changed places. We seem to be coming across quite a lot of this sort of thing: see Sections 4.8, 4.10 and 6.10, for example. There is a special name for this, but we shall have to wait until Section 12.2 to discover it.

Meanwhile we note the fact of current–voltage interchange sufficiently to be able to grasp, without too much repetition along the lines of Section 8.1, that in the parallel circuit the branch *currents* are in opposite phase, so

---

* For this reverse process it saves time to adapt the formula, thus:
$$f_r = 159.15/\sqrt{(LC)}$$
$$\therefore f_r^2 = 159.15^2/LC$$
$$\therefore L = 25\,333/f_r^2 C$$
The corresponding adaptation for calculating C is $C = 25\,333/f_r^2 L$ (μH, pF, MHz).

that the total current to be supplied is the *difference* between them (or their sum if one of them is taken as negative). Similarly the two susceptances, B_L and B_C, are opposite in sign, and this time it is the inductive one that is taken as negative. So the total susceptance in our example is negative and smaller numerically than either B_L or B_C.

Suppose now that the frequency is such as to make the susceptances (and therefore the reactances) equal. The currents will therefore be equal, so that the difference between them will be zero. We then have a circuit with two parallel branches, both with currents flowing in them, and yet the current supplied by the generator is nil!

It must be admitted that such a situation is impossible, the reason being that no practical circuits are entirely devoid of resistance. But if the resistance is small, when the two currents are equal the system does behave approximately as just described. It is a significant fact, which can easily be checked by calculation of the type given in Section 8.7, that the frequency which makes them equal is the same as that which in a series circuit would make L and C resonate.

8.9 The Effect of Resistance

When the resistance of a parallel LC circuit has been reduced to a minimum, the resistance in the C branch is likely to be negligible compared with that in the L branch. Assuming this to be so, let us consider the effect of resistance in the L branch (Figure 8.8), representing it by the symbol r as a reminder that it is small compared with the reactance. We have already studied this branch on its own (Figure 7.4) and noted that one effect of the resistance is to make the phase angle between current and voltage less than 90°; the greater the resistance the smaller the angle. It would be quite easy to modify Figure 8.7b by reducing the phase angle slightly, but to do the job thoroughly, so that if necessary we could draw the diagram to scale, let us start from scratch without reference to other diagrams and just follow the basic rules of phase relationships.

We start with the series branch made up of L and r, and as current is common to both we draw its phasor, AC, horizontally. Now we can draw ac in phase, small because r is small; and cb 90° ahead. That completes the voltage part. Having now got ab, not perfectly vertical this time, we can draw CB 90° ahead of it; let us assume it is equal to AC.

That shows us the supplied current, AB. Because AC = CB, the angle ABC is less than a right angle, but as we have drawn CB at right angles to ab we deduce that although the supplied current AB looks as if it might be in phase with its e.m.f. ab it is not exactly so. It could be made so by a very slight reduction in frequency, which would increase AC and reduce CB. (This sort of information would be almost impossible to obtain from waveform diagrams, even at cost of much greater time and trouble.) In practice r is likely to be less than one-hundredth as much as X_L, so even when X_L is made exactly equal to X_C the current supplied by the generator would be small and practically in phase with its e.m.f. In other words, for most purposes it can be taken as equal to the current that would flow if the two reactive branches were replaced by one consisting of a high resistance.

8.10 Dynamic Resistance

It is of considerable practical interest to know how high this resistance is (let us call it R), and how it is related to the values of the real circuit: L, C and r. We have already seen that if $r = 0$, AB is zero, so in that case R is infinitely large. Figure 8.8b shows that as r is increased, making the voltage drop across it, ac, greater, ab is swung round clockwise, and BC with it, increasing AB. This increase in current from the generator signifies a fall in R. So at least we can say that the smaller we make r in Figure 8.8, the greater is the resistance that the circuit as a whole presents to the generator.

It can be shown mathematically that, so long as r is much less than X_L, the resistance R, to which the parallel circuit *as a whole* is equivalent at a certain frequency, is practically equal to X_L^2/r. The proof by algebra is given in textbooks, but geometry-minded readers should not find it difficult to prove from Figure 8.8b.

To take an example, suppose C and L are the same as in Figure 8.3, and r is equal to 10 Ω (Figure 8.9a). At 796 kHz the reactances are each 1000 Ω (Figure 8.4). So R is $1000^2/10 = 100\,000\,\Omega$ (Figure 8.9b). If L, C and r were hidden in one box and R in another, each with a pair of terminals for connecting the generator, it would be impossible to tell, by measuring the amount and phase of the current taken, which box was which. Both would appear to be resistances of 100 kΩ, *at that particular*

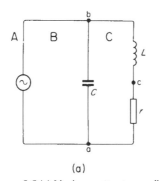
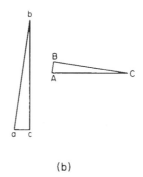

(a)   (b)

Figure 8.8 (a) Ideal capacitor in parallel with inductor having losses represented by resistance r in series, and (b) corresponding phasor diagrams

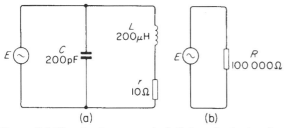

Figure 8.9 The circuit composed of C, L and r having the values marked (a) can at 796 kHz be replaced by a high resistance (b), so far as the phase and magnitude of the current taken from the generator is concerned

frequency. We shall very soon consider what happens at other frequencies, but in the meantime note that the apparent or equivalent resistance, which we have been denoting by R, has the special name *dynamic resistance*.

8.11 Parallel Resonance

Suppose E in Figure 8.9 is 1 V. Then under the conditions shown the current I must be 1/100 000 A, or 10 µA. Since the reactances of C and L are each 1000 Ω, and the impedance of the L branch is not appreciably increased by the presence of r, the branch currents are each 1000 µA. But let us now change the frequency to 1000 kHz. Figure 8.4 shows that this makes X_C and X_L respectively 800 Ω and 1250 Ω. The current taken by C is then 1/800 A = 1250 µA and the L current is 1/1250 A = 800 µA. Since r is small, these two currents are so nearly in opposite phase that the total current is practically equal to the difference between them, 1250 − 800 = 450 µA. This balance is on the capacitive side, so at 1000 kHz the two branches can be replaced by a capacitance having a reactance = E/I = 1/0.00045 = 2222 Ω, or 71.6 pF. The higher the frequency the greater is the current, and the larger this apparent capacitance; at very high frequencies it is practically 200 pF—as one would expect, because the reactance of L becomes so high that hardly any current can flow in the L branch.

Using the corresponding line of argument to explore the frequencies below 1000 kHz, we again find that the total current becomes larger, and that its phase is very nearly the same as if C were removed, leaving an inductance greater than L but not much greater at very low frequencies.

Plotting the current over a range of frequencies, we get the result shown in Figure 8.10a, which looks somewhat like a series-circuit resonance curve turned upside down.

The graph of impedance against frequency (Figure 8.10b) is still more clearly a resonance curve.

8.12 Frequency of Parallel Resonance

Before we complicated matters by bringing in r we noticed that the frequency which would make L and C resonate in a series circuit had the effect in a parallel circuit of reducing I to zero and making the impedance infinitely great. In these circumstances the frequency of parallel resonance is obviously the same as that of series resonance. But actual circuits always contain resistance, which raises some awkward questions about the frequency of resonance. In a series circuit it is the frequency that makes the two reactances equal. But in a parallel circuit this does not make the branch impedances exactly equal unless both contain the same amount of resistance, which is rarely the case. The two branch currents are not likely to be exactly equal, therefore. So are we to go on defining f_r as the frequency that makes $X_L = X_C$, or as the frequency which brings I into phase with E so that the circuit behaves as a resistance R, or as the frequency at which the current I is least?

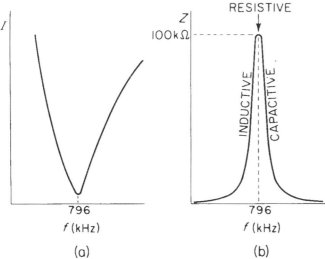

Figure 8.10 Plotting the current and impedance of the Figure 8.9a circuit as a whole at various frequencies gives the shapes shown here: (a) current, (b) impedance

Fortunately it happens that in normal radio practice this question is mere hair-splitting; the distinctions only become appreciable when r is a large-sized fraction of X_L; in other words, when Q ($= X_L/r$) is abnormally small—say less than 5. On the understanding that such unusual conditions are excluded, therefore, the following relationships are so nearly true that the inaccuracy is of no practical importance. The frequency of parallel resonance (f_r) is the same as that for series resonance, and it makes the two reactances X_L and X_C numerically equal. The impedance at resonance (R) is resistive, and equal to X_L^2/r and also to X_C^2/r (Section 8.10). And, because $X_L = X_C$,

$$R = \frac{X_L X_C}{r} = \frac{2\pi f_r L}{2\pi f_r C r} = \frac{L}{Cr}$$

from which

$$r = \frac{L}{CR}$$

In all these, r can be regarded as the whole resistance in series with L and C, irrespective of how it is distributed between them. Since $Q = X/r$,

$$R = QX, r = \frac{X}{Q}, \frac{R}{X} = \frac{X}{r} \text{ and } \frac{R}{Q} = Qr$$

All these relationships assume resonance.

8.13 Series and Parallel Resonance Compared

The resonance curve in Figure 8.10*b* showing $R = 100\,k\Omega$ applies to the circuit in Figure 8.9*a*, in which $Q = X_L/r = 2\pi fL/r = 2\pi \times 796\,000 \times 0.0002/10 = 100$. If r were reduced to $5\,\Omega$, Q would be 200, and R would be $200\,k\Omega$. But reducing r would have hardly any effect on the circuit at frequencies well off resonance; so the main result would be to sharpen the resonance peak. Increasing r would flatten the peak. This behaviour is the same as with series resonance, except that a peak of impedance takes the place of a peak of voltage across the coil or of current from the generator. Current from the generator to the parallel circuit reaches a minimum at resonance (Figure 8.10*a*), as does impedance of the series circuit.

At resonance a series circuit takes the maximum current from the generator, so is called an *acceptor circuit*. A parallel circuit, on the contrary, takes the minimum current so is called a *rejector circuit*. One should not be misled by these names into supposing that there is some inherent difference between the circuits; the only difference is in the way they are used. Comparing Figure 8.9*a* with Figure 8.3 we see the circuits are identical except where the generator is connected. To a generator connected in one of the branches, Figure 8.9*a* would be an acceptor circuit at the same time as it would be a rejector circuit to the generator shown.

If the generator in a circuit such as Figure 8.9*a* were to deliver a constant current instead of a constant e.m.f., the voltage developed across it would be proportional to the impedance of the tuned circuit and so would vary with frequency in the manner shown in Figure 8.5. So the result is the same, whether a tuned circuit is used in series with a source of voltage or in parallel with a source of current. In later chapters we shall come across examples of both.

8.14 The Resistance of the Coil

In the sense that it cannot be measured by ordinary direct-current methods—by finding what current passes through on connecting it across a 2 V cell, for example—it is fair to describe R as a fictitious resistance. Yet it can quite readily be measured by any method suitable for measuring resistances at the frequency to which the circuit is tuned; in fact, those methods by themselves would not disclose whether the thing being measured was the dynamic resistance of a tuned circuit or the resistance of a resistor.

If it comes to that, even r is not the resistance that would be indicated by any d.c. method. That is to say, it is not merely the resistance of the wire used to make the coil. Although no other cause of resistance may appear, the value of r measured at high frequency is always greater, and may be many times greater, than the d.c. value.

One possible reason for this has been mentioned in connection with transformers (Section 7.10). If a solid iron core were used, it would have currents generated in it just like any secondary winding: the laws of electromagnetism make no distinction. These currents passing through the resistance of the iron represent so much loss of energy, which, as we have seen, brings the primary current more nearly into phase with the e.m.f., just as if the coil's impedance included a larger proportion of resistance. To stop these *eddy currents*, as they are called, iron cores are usually made up of thin sheets arranged so as to break up circuits along the lines of the induced e.m.f. At radio frequencies even this is not good enough, and if a magnetic core is used at all it is either iron in the form of fine dust bound together by an insulator, or a non-conducting magnetic material called a *ferrite*.

Another source of loss in magnetic cores, called *hysteresis*, is a lag in magnetic response, which shifts the phase of the primary current still more towards that of the applied e.m.f., and therefore represents resistance, or energy lost.

The warming up of cores in which a substantial number of watts are being lost in these ways is very noticeable.

Although an iron core introduces the equivalent of resistance into the circuit, its use is worth while if it enables a larger amount of resistance to be removed as a result of fewer turns of wire being needed to give the required inductance. Another reason for using iron cores, especially in r.f. coils, is to enable the inductance to be varied, by moving the core in and out.

Even the metal composing the wire itself has e.m.f.s induced in it in such a way as to be equivalent to an increase in resistance. Their distribution is such as to confine the current increasingly to the surface of the wire as the frequency is raised. This *skin effect* occurs even in a straight wire; but when wound into a coil each turn lies in the magnetic field of other turns, and the resistance is further increased by the so-called *proximity effect*; so much so that using a thicker gauge of wire sometimes actually increases the r.f. resistance of a coil.

8.15 Dielectric Losses

In the circuits we have considered until now, r has been shown exclusively in the inductive branch. While it is true that the resistance of the capacitor plates and connections is usually negligible (except perhaps at very high frequencies), insulating material coming within the alternating electric field introduces resistance rather as an iron core does in the coil. It is as if the elasticity of the 'leashes' (Section 3.4) were accompanied by a certain amount of friction, for the extra circuit current resulting from the electron movements is not purely capacitive in phase. The result is the same as if resistance were added to the capacitance.

Such materials are described as poor dielectrics, and obviously would not be chosen for interleaving the capacitor plates. Air is an almost perfect dielectric, but even in an air capacitor a part of the field traverses solid material, especially when it is set to minimum capacitance. Apart from that, wiring and other parts of the circuit—including the tuning coil—have 'stray' capacitances; and if inefficient dielectrics are used for insulation they will be equivalent to extra resistance.

8.16 H.F. Resistance

All these causes of loss in high-frequency circuits come within the general definition of resistance (Section 2.18) as (electrical power dissipated) ÷ (current squared). In circuit diagrams and calculations it is convenient to bring it all together in a single symbol such as r in Figure 8.9a in conjunction with a perfectly pure inductance and capacitance, L and C. But we have seen that at resonance the whole tuned circuit can also be represented as a resistance $R = L/Cr$ (Figure 8.9b). We also know that a perfect L and C at resonance have an infinite impedance, so could be shown connected in parallel with R as in Figure

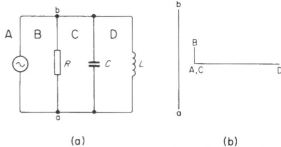

(a)   (b)

Figure 8.11 L, C and R all in parallel, and corresponding phasor diagrams

8.11a. That they are 'invisible' to the generator is shown in the phasor diagram (b) by the points A and C coinciding, meaning that the current AC is zero. At the same time the capacitor and coil currents, CD and DA, are relatively large, but equal and opposite. This current diagram is unusual in being a four-sided figure with two diagonally opposite points coinciding (to put it paradoxically!). If R had been placed between C and L the current diagram would have opened out into a slim rectangle. Figure 8.11a represents the actual circuit not only at resonance but (except in so far as R varies somewhat with frequency) at frequencies on each side of resonance.

More generally, for every resistance and reactance in series one can calculate the values of another resistance and reactance which, connected in parallel, at the same frequency, are equivalent. And if (as we are assuming) the reactance is much greater than the series resistance, it has very nearly the same value in both cases. The values of series and parallel resistance, r and R respectively, are connected by the same approximate formula $R = X^2/r$, which we have hitherto regarded as applying only at the frequency of resonance (Section 8.9). This ability to reckon resistance as either in parallel or in series is very useful. It may be, for example, that in an actual circuit there is resistance connected both ways. Then for simplifying calculations it can be represented by a single resistance, either series or parallel.

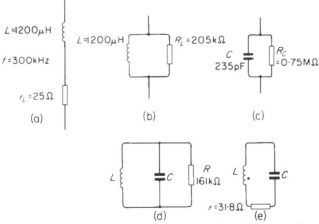

Figure 8.12 Example to illustrate series and parallel equivalents. (b) is equivalent to (a); and (d) and (e) are alternative ways of expressing the result of combining (c) and (b)

An example, worked out with the help of relationships explained in this chapter, may make this clearer. A certain coil has, say, an inductance of 1.2 mH, and at 300 kHz its resistance, reckoned as if it were in series with a perfect inductance, is 25 Ω (Figure 8.12a). At that frequency the reactance, X_L, is $2\pi fL = 2\pi \times 300 \times 1.2 = 2260\,\Omega$. Now although Q has been reckoned in Section 8.12 as the ratio of reactance to series resistance (X/r) of a coil, it has so far been considered as a property of a complete resonant circuit, in which X could be the reactance of either coil or capacitor, since at resonance they are equal. And r is the resistance of the whole tuned circuit. (Likewise in the alternative form for Q: R/X.) But this meaning of Q is often extended to cover separate reactive parts of circuits, as a measure of their 'goodness'. The Q of the coil in our example, which we could distinguish as Q_L, is therefore 2260/25 = 90.4. If tuned by a completely loss-free capacitor, that would also be the Q of the combination.

The same coil can be represented alternatively as a perfect inductance of very nearly 1.2 mH in parallel with a resistance $R_L = X_L^2/rL = 22\,602/25 = 205\,\text{k}\Omega$ (Figure 8.2b). If rL remained the same at 200 kHz, the value of R_L would be $(2\pi \times 200 \times 1.2)^2/25 = 91\,\text{k}\Omega$; so when the equivalent parallel value has been found for one frequency it must not be taken as holding good at other frequencies. As a matter of fact, neither rL nor R_L is likely to remain constant, but in most practical coils Q is fairly constant over the useful range of frequency; which is a good reason for using it as a figure of merit. So it is likely that at 200 kHz R_L would be somewhere around 135 kΩ, and r_L would be about 17 Ω.

The capacitance required to tune the coil to 300 kHz, or 0.3 MHz, is (Section 8.7, footnote) $C = 25\,333/f_r^2L = 25\,333/(0.33 \times 1200) = 235\,\text{pF}$. This, of course, includes the tuning capacitor itself plus the stray capacitance of the wiring and of the coil. The whole of this is equivalent to a perfect 235 pF in parallel with a resistance of, shall we say, 0.75 MΩ (Figure 8.12c). The reactance, X_C, is bound to be the same as X_L at resonance; but to check it we can work it out from $1/2\pi fC = 1/(2\pi \times 0.3 \times 0.000\,235) = 10^6/(2\pi \times 0.3 \times 235) = 2260\,\Omega$. Q_C is therefore $R_C/X_C = 750\,000/1260 = 332$.

If the two components are united as a tuned circuit, the total loss can be expressed as a parallel resistance, R, by reckoning R_L and R_C in parallel: $R_L R_C/(R_L + R_C) = 0.205 \times 0.75/0.955 = 0.161\,\text{M}\Omega$ (Figure 8.12d). Although this is less than R_L, it indicates a greater loss. The Q of the tuned circuit as a whole is $R/X = 161\,000/2260 = 71$. Incidentally, it may be seen that this is $Q_L Q_C/(Q_L + Q_C)$. The total loss of the circuit can also be expressed as a series resistance, $r = X^2/R = 2260^2/161\,000 = 31.8\,\Omega$ (Figure 8.12e). Calculating Q from this, $X/r = 2260/31.8$, gives 71 as before.

Before going on, the reader would do well to continue calculations of this kind with various assumed values, taking note of any general conclusions that arise. For instance, a circuit to tune to the same frequency could be made with less inductance; assuming the same Q, would the dynamic resistance be lower or higher? Would an added resistance in series with the lower-L coil have less or more effect on its Q? And what about a resistance in parallel?

8.17 Cavity Resonators

As radio communication has developed, carrier frequencies have become higher. Although the carriers of less than 1 MHz that were used in the early days of broadcasting are still in use, they are now supplemented by u.h.f. television broadcasts in the 470–960 MHz band and satellite broadcasts in the 11.7–12.0 GHz band.

The inductance and capacitance used in tuned circuits have correspondingly decreased; for example, a long-wave tuning coil might have 250 turns, a medium-wave one 80 turns and a coil for 100 MHz about five turns. A coil with less than one turn is difficult to envisage and, to resonate at hundreds of MHz, a different type of tuning system is adopted using transmission lines (Section 16.10). To achieve resonance at frequencies above approximately 2000 MHz even this technique has to be abandoned and the following system is used.

Imagine a conventional LC tuned circuit in which, to achieve a high resonance frequency, the inductor has been reduced to a single loop of wire and the capacitor to two small parallel plates. Such a tuned circuit, illustrated in Figure 8.13a, resembles that used by Hertz in his early experiments on radio, and might achieve a frequency in the low hundreds of MHz. To raise the frequency we must reduce inductance and capacitance. Inductance can be reduced by connecting several loops of wire in parallel and capacitance by making the plates smaller and increasing their separation as shown at b. In search of yet higher frequencies we continue this process, increasing the number of loops until they touch each other along the whole of their length and so become a continuous conducting surface. By reducing the area of the capacitor plates they become, in the limit, the edges of an opening in the spherical surface as shown at c. The LC circuit has thus become a conducting vessel with an aperture, and its resonance frequency depends primarily on the dimensions of the vessel and the aperture. This type of tuned circuit, known as a *cavity resonator*, is used in klystron valves (Section 14.14). A rather old-fashioned name for a cavity resonator is a *rhumbatron*.

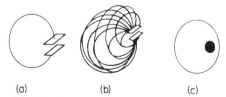

(a)   (b)   (c)

Figure 8.13 Development of the cavity resonator (a) shows a single-loop inductor, (b) multiple loops generating a continuous surface and (c) a spherical vessel with an aperture.

Chapter 9 Diodes

9.1 Electronic Devices

We now have all the essential basic principles of circuits, and it is time to strike out along a new line—electronics. Although all the electric currents we have been discussing in the last seven chapters consisted of electrons in motion there has been no real need to consider them as such. It would have been just about as easy to deal with the subject using one of the theories of the electric current in vogue before electrons were discovered. But now we are going to study electronic devices, which could not be satisfactorily explained without reference to electrons. It is these devices that are responsible for the rapid and multifarious developments included under the broad name of electronics.

There are two main divisions of electronic devices: those in which the currents pass through vacuum or gas, and those in which they pass through semiconducting solids. The first includes those called valves in Britain and tubes in America, and the second includes transistors.

Although some simple semiconductor devices were used, without knowing how they worked, for several decades before 1948, it was not until that year, when the transistor was invented as a result of research into the electronic nature of semiconductors, that they began to overtake thermionic valves (as they are named more fully), which are reckoned to have started in 1904.

Electronic devices can be, and often are, classified according to the number of their *electrodes*. An electrode is not very easy to define precisely, but roughly it is any part of the device where a current leaves or enters, or by means of which it is electrically controlled. Often each electrode has a terminal or wire by which connection is made to a circuit, but sometimes electrodes are connected together inside the device.

9.2 Diodes

The least number of electrodes is two, and devices so constituted are called *diodes*. Although they are relatively limited in what they can do—their chief capability is allowing currents to flow only one way—they embody most of the basic principles needed for understanding all electronic devices. So as not to be too theoretically minded, however, we may at once note two most important practical uses of diodes. Working at low (or power) frequencies such as 50 Hz, they enable a.c. to be converted to d.c. such as is required for running nearly all electronic equipment. This function is called *rectification*. And by rectifying radio-frequency currents, diodes enable radio signals to be made evident, as mentioned in Section 1.11. This function, although basically rectification, is often distinguished as *detection* and is the subject of Chapter 18.

Although thermionic diodes have been superseded almost entirely by semiconductor (or solid-state) types, their principles are easier to understand and are a necessary introduction to other valves and to cathode-ray tubes. So we begin with them.

9.3 Thermionic Emission of Electrons

So far we have considered an electric current as a stream of electrons along a conductor. The conductor is needed to provide a supply of 'loose' or mobile electrons, ready to be set in motion by an e.m.f. An e.m.f. applied to an insulator causes no appreciable current because the number of mobile electrons in it is negligible.

To control the current it is necessary to move the conductor, as is done in a switch or rheostat (a variable resistor). This is all right occasionally, but quite out of the question when (as in radio) we want to vary currents millions of times a second. The difficulty is the massiveness of the conductor; it is mechanically impracticable to move it at such a rate.

This difficulty can be overcome by releasing electrons, which are inconceivably light, from the relatively heavy metal conductor. Although in constant random agitation in the metal, they have not enough energy at ordinary temperature to escape beyond its surface. But if the metal is heated sufficiently they 'boil off' from it. A source of free electrons such as this is called a *cathode*. To prevent the electronic current from being hindered by the surrounding air, the space in which it is to flow is enclosed in a glass bulb and as much as possible of the air pumped out, giving a vacuum. We are now well on the way to manufacturing a thermionic valve.

Cathodes are of two main types: *directly heated* and *indirectly heated*. The first, known as a *filament* and now seldom used, consists of a wire heated by passing a current through it. To minimise the current needed, the wire is usually coated with a special material that emits electrons freely at the lowest possible temperature. The indirectly heated cathode is a very narrow tube, usually of nickel, coated with the emitting material and heated by a separate filament, called the *heater*, threaded through it. Since the cathode is insulated from the heater, three connections are necessary compared with the two that suffice when the filament serves also as the source or electrons, unless two are joined together internally. In either case only one connection counts as an electrode—the cathode; that it is heated electrically is only for convenience. So except in Figure 9.1 the heating arrangement will be omitted.

One advantage of indirectly heated cathodes is that they can be heated by a.c., as explained in Section 27.12; another is that the whole cathode is at the same potential.

The electrons released by the hot cathode do not tend to move in any particular direction unless urged by an electric field. Without such a field they accumulate in the space around the cathode. Because this accumulation consists of electrons, it is a negative charge, known as the *space charge*, which repels new arrivals back again to the cathode, so preventing any further build-up of electrons in the vacuous space.

9.4 The Vacuum Diode Valve

To overcome the barrier caused by the negative space charge it is necessary to apply a positive potential. This is introduced into the valve by means of the second electrode, a metal plate or cylinder called the *anode*.

When the anode is made positive relative to the cathode, it attracts electrons from the space charge, causing its repulsion to diminish so that more electrons come forward from the cathode. In this way current can flow through the valve and round a circuit such as in Figure 9.1. But if the battery is reversed, so that the anode is more negative than

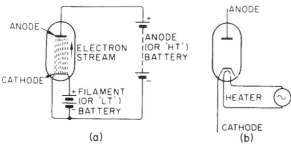

Figure 9.1 (a) Electron flow from cathode to anode in a directly heated diode. The alternative indirect method of heating is shown at (b)

the cathode, the electrons are repelled towards their source, and no current flows. The valve therefore permits current to flow through it in one direction only, like the air valve attached to a tyre, and this is why it was so named. The anode battery or other supply is often called the h.t. (signifying 'high tension') and the filament or heater source the l.t. ('low tension').

If the anode of a diode is slowly made more and more positive with respect to the cathode, as for example by

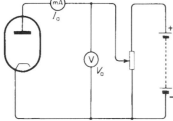

Figure 9.2 Circuit for finding the relationship between the anode voltage (V_a) and the anode current (I_a) in a diode

moving the slider of the potentiometer in Figure 9.2 upwards, the attraction of the anode for the electrons is gradually intensified and the current increases. To each value of anode voltage V_a there corresponds some value of anode current I_a, and if each pair of readings is recorded on a graph a curve like that of Figure 9.3 (called a valve I_a/V_a characteristic curve) is obtained.

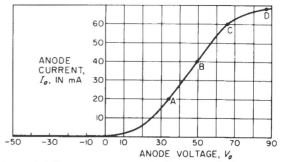

Figure 9.3 Characteristic curve of a thermionic diode

The shape of the curve shows that at low voltages the anode collects few electrons, being unable because of its greater distance from the cathode to overcome much of the repelling effect of the space charge. The greater the positive anode voltage the greater the negative space charge it is able to neutralise; that is to say, the greater the number of electrons that can be on their way between cathode and anode; in other words, the greater the anode current. By the time the point C is reached the voltage is so high that electrons are reaching the anode practically as fast as the cathode can emit them; a further rise in voltage collects only a few more strays, the current remaining almost constant from D onwards. This condition is called *saturation*. Because the saturation current is limited by the emission of the cathode, and that depends on the temperature of the cathode, it is often called a *temperature-limited* current, to distinguish it from the *space charge-limited* current lower down the curve. Normally the saturation current is far above the working range.

At B an anode voltage of 50 V drives through the valve a current of 40 mA; the valve could therefore be replaced by a resistance of $50/40 = 1.25$ kΩ without altering the current flowing at this voltage. This value of resistance is therefore the equivalent d.c. resistance of the valve at this point. The curve shows that although the valve is in this sense equivalent to a resistor, it does not conform to Ohm's law (compare Figure 2.3); its resistance depends on the voltage applied. To drive 10 mA, for example, needs 25 V: $V/I = 25/10 = 2.5$ kΩ.

This is because the valve, considered as a conductor, is quite different from the material conductors with which Ohm experimented. Its so-called resistance—really the action of the space charge—is like true resistance in that it restricts the current that flows when a given voltage is applied, but it does not cause the current to be exactly proportional to the voltage as it was in Figure 2.3. In other words, it is non-linear.

The kind of non-linearity just examined is particularly important in an undesirable way in valves having more than two electrodes, as we shall see later. The most significant non-linearity of diodes is that because the anode emits no electrons there is no possibility of current when the anode voltage is negative—compare Figure 9.3 again with Figure 2.3. It is this that makes diodes so useful. Figure 9.3 makes clear that if a source of alternating voltage of, say, 50 V peak were applied instead of the battery in Figure 9.1 to the anode the positive half-cycles would cause a peak current of 40 mA, but the negative half-cycles would cause none. So the anode current would consist of d.c. pulses.

For this process of rectification the thermionic diode has two disadvantages, both of them wasteful of power: it requires an auxiliary source of current to heat the cathode, and the space charge necessitates quite a number of volts to neutralise it. In high-power applications the latter disadvantage can be mitigated by admitting a controlled amount of gas or vapour into the tube. Directly electrons start moving across to the anode with sufficient velocity—which can be imparted by a fairly low anode voltage—they collide with the gas atoms (or small groups of atoms called molecules) violently enough to knock out some of their electrons. These electrons join the stream and augment the anode current, but the more important result is that the gas molecules with electrons missing are positively charged ions (Section 2.2). (The process just described is *ionisation*, which is important in many connections.) These ions are therefore attracted to the cathode, but being far heavier than the electrons they move comparatively slowly. So each one lingers in the space charge long enough to neutralise the negative charge not only of the electron torn from it but also of many of those emitted from the cathode. In fact, the negative space charge is practically eliminated, and with it nearly all the resistance of the diode.

Even this interesting improvement succeeded only in delaying for a time the replacement of thermionic diodes by semiconductor ones, both for power and signals (rectifiers and detectors). So now we go on to study these.

9.5 Semiconductors

In Section 2.3 solid materials were grouped into conductors and insulators. The difference was explained by supposing that conductors are substances in which the electrons belonging to their atoms are free to move, under the influence of an electric field, in one direction (towards the positive pole) and so to create an electric current, whereas the electrons of insulators are, as it were, held on elastic leashes so that an electric field (unless extremely strong) is unable to move them far from the home atoms. This is a very much over-simplified view of the matter, but it will do as a first approach. For a more advanced but still not highly mathematical treatment, see M. G. Scroggie's *The Electron in Electronics*.

Not all the electrons in atoms are involved, but only the outermost ones, known as *valence* electrons. This term first arose in chemistry, because it happens that the chemical behaviour of substances depends mainly on the number and arrangement of these outer electrons. There are never more than seven of them per atom, though the normal total number of electrons per atom, called the *atomic number*, can be anything up to about 100.

Substances in which all the atoms have the same atomic number are called *elements*. The semiconductor materials most important in electronics are the elements silicon and germanium, which are tetravalent (meaning that they have four valence electrons per atom). Their atomic numbers are 14 and 32 respectively, but for our purpose all except the four valence electrons may be lumped together with the remainder of the atoms; that is to say, the positively charged nucleus. Around this relatively heavy main body of the atom the valence electrons can be visualised as satellites (Figure 9.4). The positive charge on any atom is exactly

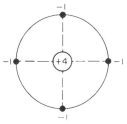

Figure 9.4 The four valence electrons in a tetravalent atom can be visualised as satellites around the rest of the atom. These parts are electrically charged as shown

neutralised by the whole normal complement of electrons, so if four electrons are subtracted there must be a positive charge of four electron units, as shown.

Now it happens that the full capacity of the outermost or valence orbit is eight electrons. Atoms show a general tendency to fill up vacancies in such orbits in some way or other, and one of these ways is the creation of what are called *covalent bonds*. If each atom shares each of its four valence electrons with another atom, so that both atoms have two electrons in common, every atom has eight electrons in its valence orbit and is attached to four other atoms. These are arranged symmetrically in three dimensions, so a model would be needed to represent the system properly; but Figure 9.5 shows the result of squashing a bit of such a model flat so as to get it on a sheet of paper. Because every atom is identical with every other, the structure so formed is made up of atoms in perfectly regular array, like a case of tennis balls packed for export. But we must remember that the atoms are inconceivably small and numerous. Broken-off pieces of such an atomic structure naturally tend to have flat faces at certain definite angles to one another, and this description fits what we call crystals.

9.6 Holes

A crystal of this kind, if perfect, would have no electrons free to roam around and would therefore be a perfect insulator. And in fact one of the tetravalent elements, carbon, crystallises in this way (though not very readily!) as diamond, which is a very good but rather expensive insulator. But the atoms of silicon—and even more those of

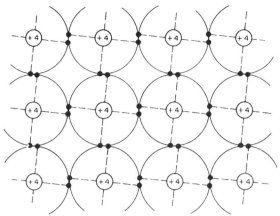

Figure 9.5 Diagrammatic representation of the four-fold linkages of tetravalent atoms such as silicon

germanium—hold on to their electrons less firmly, and at ordinary room temperatures there is enough heat energy to 'shake out' a valence electron here and there, making it available for electrical conduction. The atom from which it came is then an electron short.

Now 'an electron short' is, in effect, a positive charge. We have already noted that the inner part of the atom, taken by itself, is positively charged, as shown in Figure 9.4, because its four valence electrons are being considered separately. Taken as a whole, with all its electrons, the atom is neutral. But when one of the valence electrons moves off, the atom as a whole becomes a positive ion (Section 2.2). The electron vacancy thus created is called a *hole*. The curious and interesting thing is that this hole can wander about and cause conduction, almost like the loose electron. Although the exact way in which this happens is difficult to understand, the following analogy or working model should give sufficient insight for practical purposes into the significance of electronic holes.

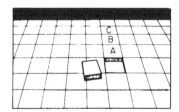

Figure 9.6 Tile analogy of the creation of a 'hole' by the displacement of an electron from an atom

Figure 9.6 shows part of a floor laid with small square tiles. One tile has been removed and laid aside. It represents an electron, and the lump it makes on the floor represents its negative charge. The hole it has left behind, being the reverse of a lump, represents the positive charge created by an electronic hole. Now introduce a tile-motive force towards yourself. That draws the displaced tile towards you and represents a tiny electric current away from you. Most of the other tiles (representing electrons fixed in the crystal structure) are unable to move in response to the e.m.f., but the one marked A can do so because there is an adjacent hole into which it can slide. That leaves tile B free to move into the hole left by A, and so on. The net effect is that the hole moves away from you, and as it represents a positive charge it too means an electric current away from you.

An electronic hole can for most purposes be regarded as a mobile electron, except for its being electrically positive instead of negative. When a meter reads 1 mA it could be due to 6.24×10^{15} electrons per second moving through it one way, or an equal number of holes the other way—or a mixture of both to the same total. Conduction caused by heat disturbance is therefore twofold: each electron released creates a hole, and when there is an electric field (due, very likely, to an e.m.f.) the resulting current is made up of electrons moving towards the positive end and holes towards the negative end. Owing to the more complicated way in which holes 'move', their mobility is less than that of electrons. This means that they move more slowly.

Just as the only things that really moved in Figure 9.6 were tiles, so the only things that move in conduction by holes are electrons. Teachers are sometimes so anxious to emphasise that holes are not real positive charges that there may be difficulty later on when certain experiments show that holes really do behave like positive charges. Although not strictly scientifically, Figure 9.6 covers this point too. We can all agree that both kinds of tile current are brought about solely by movements of tiles. That corresponds to the statement that a hole movement one way is really an electron movement the opposite way. But looking at the two 'charge carriers' (often called 'current carriers') we must also admit that one is a hump and the other a hollow. As these irregularities stand for negative and positive polarities respectively, one kind of current is a negative charge moving one way, and the other kind is a positive charge moving the other way. So although the causes are the same the results are different. That corresponds to the statement that a hole behaves physically as a positive charge.

9.7 Intrinsic Conduction

Having grasped the significance of holes, we may perhaps gain further insight into this kind of conduction by considering a vast array of small balls on a flat tray. They are kept in their places by resting in shallow depressions ('holes') in the tray. A difference of potential between the ends of the tray is represented by raising one end slightly, the raised end being 'negative'. Such a tilt would cause no movement of balls ('electrons'), except perhaps a slight lurch to the side of each hole nearest the 'positive' end, representing a displacement or capacitive current (Section 3.8). This is like a crystal at a very low temperature, when it is an insulator.

The effect of raising the temperature could be shown by making the tray vibrate with increasing vigour. At a certain intensity a ball here and there would be shaken loose, and on a perfectly level tray (no e.m.f.) would wander around at random. In doing so it would sooner or later fall into a hole—not necessarily the one from which it came. There would therefore be some random 'movement' of holes. At any given degree of vibration (temperature) a balance would eventually be reached between the rate of shaking loose and the rate of reoccupation (called in electronics recombination), but the greater the vibration the greater the number of free electrons and holes at any one time, and the greater the drift of balls towards the 'positive' end of the tray when tilted. One result of this movement would be a tendency for more holes to be reoccupied at the lower end than the upper; in effect, there would be a movement of holes up the tray towards the negative end.

To represent a sustained e.m.f. it would be necessary for the balls that reached the bottom of the tray to be conveyed by the source of the e.m.f. to the top end, where their filling of the holes would be equivalent to a withdrawal of the holes there.

If you have been able to visualise this model sufficiently you should by now have gathered that at very low temperatures a semiconductor crystal is an insulator, but that as the temperature rises it increasingly conducts. In other words its resistance falls, in contrast to that of metals, which rises with temperature. This sort of conduction is called *intrinsic*, because it is a property of the material itself.

A similar effect is caused by light; a fact that is turned to advantage in the use of semiconductors as light detectors and measurers.

Unless a substance crystallises under ideal conditions, it tends to be a jumble of small crystals, at the boundaries of which there are numerous breaks in the regularity of the lattice, releasing space electrons; this is yet another cause of conduction.

Even although conduction due to these causes is contributed to by both electrons and holes, the combined number falls far short of the free electrons in metals, so the result is only 'semiconduction'. In fact, as we shall see later, to be suitable material for most electronic devices the intrinsic conduction at working temperatures should be as nearly as possible negligible. It is because silicon fulfils this condition better than germanium that it has largely superseded germanium, especially for use at high temperatures.

9.8 Effects of Impurities

So far we have assumed that our crystal, however it may be disturbed by heat, light, or its original formation, is at least all of one material. In practice, however, nothing is ever perfectly pure. Now if the impurities include any element having five valence electrons (such as phosphorus, arsenic or antimony) its atoms will be misfits. When they take their places in the crystal lattice—as they readily do—one electron in each atom will be at a loose end, more or less free to wander and therefore to conduct. Such an incident is represented in Figure 9.7. Intruding atoms that yield spare electrons are called donors. It might be supposed that because each surplus electron leaves behind it an equal positive charge it thereby creates a hole, but it should be understood that the technical term 'hole' applies only to a mobile positive charge. In this case all four valence electrons needed for the lattice are in fact present, so there is no vacancy into which a strolling electron can drop.

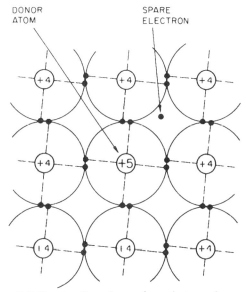

Figure 9.7 Showing the release of an electron wherever a pentavalent (donor impurity) atom finds it way into a tetravalent crystal lattice structure

Therefore the positive charges due to donor impurities are fixed, and conduction is by electrons only. This is a most important thing to remember. Because electrons are *n*egative charges, a semiconductor with do*n*or impurity is classified as *n*-type.

An incredibly small amount of impurity is enough to raise the conductivity appreciably; far less than can be detected by the most sensitive chemical analysis. As little as one part in a hundred thousand million (1 in 10^{11}) can be significant. This is the ratio of about one twenty-fifth of a

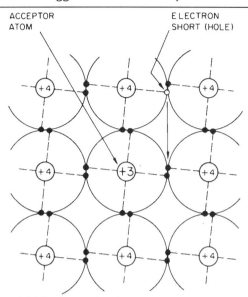

Figure 9.8 Where a trivalent (acceptor impurity) atom takes a place in the tetravalent lattice a hole is created by its capturing an electron

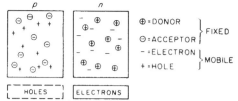

Figure 9.9 Simplified representation of pieces of impure semiconductor, showing only the electric charges of broken impurity atoms. In p-type material the mobile charges are predominantly positive (holes), and in n-type material negative (electrons). Note the uniform distribution of the mobile charges

person to the entire population of the world! So the manufacture of semiconductor devices is a tricky business.

Trivalent elements, such as aluminium, gallium and indium, have only three valence electrons per atom; and when any of these substances is present the situation is the opposite of that just described. The vacancy in such an atom's valence orbit is a hole, and because trivalent atoms attract electrons more strongly than the surrounding tetravalent atoms hold on to theirs, the hole is quickly filled (Figure 9.8). The impurity atom in this case is an *acceptor*. Note that when the hole has migrated from the acceptor—in other words, an electron from elsewhere has filled up or cancelled the hole in the acceptor atom—that atom constitutes a negative charge, because its main body has only three net positive charges to offset what are now four valence electrons; but such negative charges are fixed and take no direct part in current flow. Because conduction is by holes only, which are *positive* charges, a semiconductor with acce*p*tor impurity is called *p*-type.

So far as impurity conduction is concerned, we see that the semiconductor itself plays only a passive part, as a framework for the impurity atoms. In fact, it corresponds to the vacuum in a thermionic valve, and ideally would be non-conducting. Intrinsic conduction is just a nuisance. In the diagrams that follow, therefore, all except the relatively few impurity atoms will be omitted in order to focus attention on them. Figure 9.9 shows the symbols used for this purpose. The mobile charges—electrons and holes—are represented by simple minus and plus signs respectively; and the fixed charges—donor and acceptor atoms—are distinguished by encircled signs. Although these atoms are all members of the crystalline lattice, they are such a small minority that when the tetravalent atoms are omitted, as here, the pattern can no longer be discerned. In each type of material the mobile charges are distributed as uniformly as the fixed, and clearly they must be equal in number, so the material is electrically neutral throughout.

One might ask what happens when both donor and acceptor impurities are present at the same time. The answer is that they tend to cancel one another out by electrons filling holes, and when present in equal quantities the semiconductor behaves as though it had no impurity at all. But although this *compensation* as it is called is used for converting n-type into p-type (or vice versa) by adding more than enough opposite impurity to neutralise the impurity already present, the balance would be too fine to enable heavily contaminated material to be made apparently pure by the same method.

We must, however, take care not to concentrate so exclusively on the important impurity charge carriers that we forget all about the background intrinsic conduction, for it is always present in semiconductor devices and in fact sets an upper limit to the temperatures they can be allowed to reach. Do the two sorts of conduction go on independently, so that the total numbers of electrons and holes are the sum of those due to impurity atoms and those due to heat and other disturbances? The answer is no, because there is a law that the product of the densities of electrons and holes (i.e., number per unit volume of material) is constant. Suppose that in a certain very small volume of pure material there are 14 free electrons and 14 holes. The product, 14^2, is 196. Now add n-type impurity to the extent of 96 electrons. The total numbers are now 110 and 14, and the product is $110 \times 14 = 1540$, which is much too large. What happens is that 12 of the impurity electrons fill 12 of the intrinsic holes, reducing the numbers to 98 and 2, to restore the product to 196. The electrons in this example are called *majority carriers* and the holes are *minority carriers*. In p-type material the roles are reversed, of course.

More hole-and-electron pairs are continually being generated by heat, but these are balanced by recombinations, so at a fixed temperature the numbers remain practically constant. The numbers taken in our example were extremely small, to make the calculation clear; in such a small sample they would fluctuate considerably about the averages quoted. But we should realise that in as little as one cubic millimetre (about the size of a pin head) of typical material at ordinary room temperature the number of carriers is of the order of 10^7 for silicon and 10^{10} for germanium. Even so, they are very rare compared with the atoms.

9.9 P-N Junctions

We can have, then, p-type material, in which conduction is by holes (positive carriers); n-type, in which it is by electrons (negative carriers); and what is sometimes called i-type (for 'intrinsic'), in which it is by both in equal numbers. None of these by itself is particularly useful; it is when more than one kind are in contact that interesting things begin to happen. Just bringing two pieces together is not good enough, however; the combination has to be in one crystalline piece, divided into two or more regions having different impurities. Various manufacturing methods have been devised for achieving this. The process of introducing a desired proportion of p or n impurity is called *doping*.

Figure 9.10 shows with standard symbols a small section of such a combination at its boundary between p and n

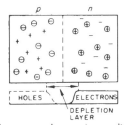

Figure 9.10 Where p and n regions adjoin in the same crystal, diffusion across the junction and reciprocal cancellation of mobile charges create a depletion layer. Here the fixed charges set up a potential difference that stops further diffusion

materials, both equally doped let us assume. The boundary is called a p-n junction. Compare it with Figure 9.9 where the two materials are shown separately. Let us suppose first that there is no electric field to influence matters. But at room temperature, which is nearly 300°C above absolute zero (−273°C), the spare electrons on the n side and holes on the p side are in continuous random motion, like the intrinsic ones illustrated by the tray experiment. This motion tends to make any concentration of them gradually spread out into a uniform distribution—a process known as *diffusion*. It is by this process that when a bottle of ether is unstoppered in a room the smell pervades the whole room within seconds. Unlike intrinsic mobile charges, which for the moment we are ignoring, and impurity mobile charges within their own separate zones in Figure 9.9, all of which from the start are distributed uniformly throughout the material because that is how they are generated, the holes have the entirely holeless n region in front of them, so they begin to diffuse across the junction. Directly they do so they encounter electrons diffusing in the opposite direction and combine with them, eliminating both so far as our diagram (restricted to electric charges) is concerned. The same story applies in reverse to the electrons. Because of this recombination process the mobile charges are prevented from diffusing far into the opposite territory. It is like a frontier between two equally matched warring nations. As a result of the casualties there is a thin layer on each side of the junction in which there are hardly any mobile charges; a sort of no man's land. It is called a *depletion layer* (Figure 9.10). And it really is thin; typically one thousandth as thick as a sheet of paper.

The fixed charges are still there, of course; positive on the n side and negative on the p. The cancelling out of the mobile charges therefore builds up a difference of potential between the two regions, of such a polarity as to repel mobile charges moving towards the junction, until it is strong enough—a few tenths of a volt—to bring further traffic across the junction to a halt. The array of unneutralised negative fixed charges on the p side behaves rather like the space charge in a thermionic diode towards electrons seeking to follow up from the n side, but this time there is a corresponding setup of opposite polarity on the other side, holding back the holes.

If you connect a sensitive voltmeter to the two sides of a p-n junction to measure this p.d. you will not get a reading. This does not prove that the p.d. does not exist. In the complete circuit there are inevitably other inter-material p.d.s that bring the total to zero, provided all is at the same temperature.

Next, consider what happens when an external e.m.f. is applied, as in Figure 9.11. Electrons in the n region are

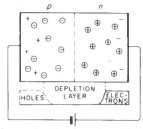

Figure 9.11 If an external voltage is applied, positive to n, the p.d. in Figure 9.10 is augmented and the depletion layer widened

attracted by its positive pole and tend to move to the right, while holes are attracted to the left. There are no charge carriers of the correct polarity to be drawn across the junction. Instead, more fixed charges on each side of it are 'uncovered' by retreating carriers, and the barrier p.d. is thereby increased. When this process has gone on enough for the p.d. between the regions to be equal and opposite to the applied e.m.f. another no-current balance is established. In other words, to an e.m.f. of this polarity (called a reverse bias) the junction behaves as a non-conductor. Because it develops a back voltage equal to that applied, it is like a small capacitor.

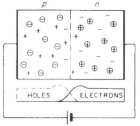

Figure 9.12 If sufficient external voltage is applied, positive to p, the internal p.d. is overcome and current flows

Suppose, however, that the e.m.f. is applied the other way round, as in Figure 9.12, and is increased gradually from zero. The polarity of this forward bias, as it is called, is now such as to repel electrons and holes *towards* the junction, and the first effect is to neutralise some of the fixed charges there and reduce the potential barrier. This movement constitutes only a small temporary 'capacitive' current. But when the applied e.m.f. is sufficient to exceed the initial barrier p.d. the carriers are able to flow across the junction. Because of recombinations when the electrons and holes meet, the flow of holes to the right begins to die out even before the junction is reached, and has done so completely a short distance beyond it. So current through the semiconductor, except close to the junction, is by holes on the p side and by electrons on the n side, as indicated at the foot of Figure 9.12. The fact that near the junction there are some carriers of the 'wrong' polarity will be found later to be very important. Their number can be greatly increased if one region is doped more heavily than the other, instead of equally as we have been assuming.

If we remember that there are no positive charge carriers in metals, we may wonder where the supply of holes comes from to replenish the p zone as they are drawn to the right and are neutralised by electrons. They are in fact created by withdrawal of electrons at the positive terminal of the semiconductor.

9.10 The Semiconductor Diode

Summing up, a p-n junction allows current to flow one way (provided that the e.m.f. applied is not too small) but not the other.

In short, it is a rectifier. The foregoing explanation of why it rectifies may seem complicated in comparison with the action of the vacuum diode (Section 9.4), so let us look at them side by side, as in Figure 9.13. Here diodes of both vacuum and pn junction types are connected both ways across a battery. At *a* we are reminded that current can consist either of negative charges flowing to + or positive charges to −. Or, of course, both at once. When a vacuum diode is connected as at *b*, current can flow, because the heated cathode is an emitter of electrons, which are negative charges. But when reversed, as at *c*, there is no current, because the anode does not emit electrons. And neither electrode emits positive charges. In a pn diode connected p to + as at *d*, current flows, because the n region emits electrons, and because the p region emits holes, which are positive charges. When reversed (*e*) both regions emit the wrong charges, so there is no current. The BS-recommended general symbol for a diode is shown at *g* and the two alternatives shown at *f* are also still used. Any of the three could replace both *b* and *d* if there was no need to specify which kind.

The semiconductor diode rectifier has some important advantages over the vacuum type: it needs no cathode heating; it is less fragile; and the emitted electrons do not form a space charge tending to oppose the anode voltage, because their charge is neutralised in the crystal by the fixed positive charges, so less voltage is lost in the diode when current flows and as a rectifier it is more efficient. And because the working region of a semiconductor diode is contained in a microscopically thin layer the whole thing can be made exceedingly small.

On the other hand there are one or two complications. So far we have ignored the intrinsic conduction which, as we have seen, results from the releasing of electrons and holes in equal numbers throughout the material, so does not depend on the direction of the e.m.f. As a small extra current in the forward direction it is unimportant, but there should be no current in the reverse direction. So to the extent that intrinsic conduction results in a reverse or leakage current it reduces the effectiveness of the diode as a rectifier. One cause of intrinsic or minority current—light—can easily be excluded by an opaque covering. But the amount of heat at ordinary temperatures is sufficient for the current of a germanium diode to be appreciable, and it approximately doubles with every 9°C rise in temperature. This is one of the most serious limitations of germanium. In this respect silicon is usually preferred because its intrinsic conductivity is about a thousand times less. On the other hand silicon has nearly three times the barrier potential to be overcome, so the forward current does not 'get off the ground' so easily.

This imperfection of a semiconductor diode cannot be fully represented in an equivalent circuit diagram by a resistor in parallel with a perfect diode, because the current through an ordinary ohmic resistor is proportional to the applied voltage, whereas intrinsic semiconductor current is limited by the number of carriers released at the temperature of the material and cannot be increased by applying more volts—short of a limit to be mentioned later. With no bias, there is an exact balance between intrinsic electrons released on the p side (where they are minority carriers) moving to the n side because of the attraction of the positive space charge there, and electrons released on the n side with sufficient force to propel them across the barrier. Hole flow is balanced in the same way. So there is no net current. But once enough reverse bias has been

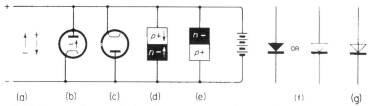

Figure 9.13 Showing how in diodes the presence or absence of current depends on the presence or absence of the right kind of charge carriers. (g) is the general symbol for a diode

applied to raise the barrier too high for any majority carriers to get over it, the only current flow consists of minority carriers only, and is in the direction of the bias.

The behaviour of this leakage current, in reaching a saturation value unaffected by further increase in voltage but steeply increasing with rise in temperature, may remind us of the temperature-limited vacuum diode (Section 9.4) but the mechanism of the effect is quite different.

9.11 Diode Characteristics

We can now account for the features of the current/voltage characteristics of a typical silicon junction diode, Figure 9.14. To positive voltages below about 0.6 V the response

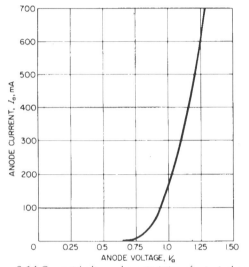

Figure 9.14 Current/voltage characteristics of a typical small-signal silicon junction diode

is very small, as they are occupied mainly in dismantling the potential barrier at the junction. Above that level of applied voltage the barrier is down and the forward current increases steeply; soon it is limited chiefly by the resistance of the semiconductor material and external circuit. Compared with such a large impurity current the intrinsic current at ordinary temperatures is small, so the effect of a moderate change in temperature on the total current is also small and is not shown in the diagram.

A germanium diode has a similar characteristic but current starts at a lower forward voltage—about 0.2 V. Moreover a germanium junction diode has a significant reverse current which increases rapidly with rise in temperature. In silicon diodes the reverse current is much smaller and, at normal temperatures, can be ignored. The first junction diodes were introduced in the mid-1950s and were germanium, but a few years later silicon diodes were produced and these have now superseded germanium. Although current does not start until the forward voltage of a silicon diode reaches 0.6 V, it gets away at a much lower anode voltage than does a thermionic diode (compare Figure 9.3).

If an attempt were made to extend the curve in the forward direction, the power dissipated in the diode would sooner or later overheat it so much that the intrinsic current would no longer be a minor effect but the major one. And because it can flow equally in both directions the diode would cease to be an effective rectifier. It could hardly even be called a semiconductor, since the material would be more like a metal. The junction would be destroyed and one would say that the diode was burnt out.

If the voltage is increased in the reverse direction, the current remains nearly constant at a very low value until a point is reached at which the intrinsic carriers are driven so fast that they knock out more electrons from the crystal atoms and start a chain reaction, called *avalanche effect*, that increases the current until either it is limited by external means such as resistance or the diode is destroyed, rather as a capacitor is destroyed by excessive voltage (Section 3.4). On the other hand the avalanche effect is put to good use in the zener diode (Section 27.9).

And indeed a semiconductor diode is also a capacitor, as we have already noted. Because its dielectric—the depletion layer—is so extremely thin it has quite a large capacitance per square millimetre. It provides a bypass for a.c., which may be a serious disadvantage. High-power rectifiers need large junction areas to pass heavy current without getting too hot, but fortunately the frequencies for which such diodes are needed are usually low—50 Hz, for example—and the capacitance is then not too serious.

However the capacitance can be so reduced in modern methods of construction of junction diodes that they can be used successfully at television and higher frequencies. Figure 9.15a illustrates one method of construction used for small-signal diodes. The diode element is mounted on a

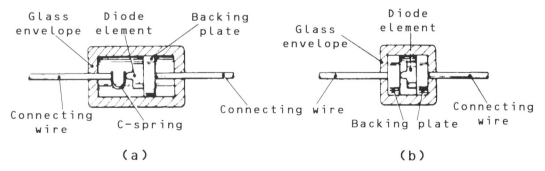

Figure 9.15 Construction of a small-signal silicon junction diode. (*a*) spring-contact, (*b*) 'whiskerless'

metal backing plate and contact with the other face of the diode is via a C-shaped spring, the whole assembly being sealed in a glass envelope. Diagram *b* shows an alternative method of construction, known as the 'whiskerless' method, which gives better protection against vibration. Here the diode element is sandwiched between two plates, held rigidly by the glass envelope.

A wide variety of characteristics are obtainable by varying the impurity in the material, and other details. For example, see Section 14.8. The capacitance of a silicon diode, undesirable for signal rectification, can be made use of in voltage-varied capacitors, or *varactors*. The capacitance of a diode, being dependent on the thickness of the depletion layer, and that thickness varying with the reverse voltage, as shown by comparing Figure 9.11 with Figure 9.10, the capacitance must vary with reverse voltage. Since increasing the voltage increases the thickness of the depletion layer, it reduces the capacitance (Section 3.3). This property has uses such as tuning receivers (Section 21.6).

One quite essential thing we have been taking for granted is the possibility of connecting the p and n regions of a diode to an external circuit. This routine necessity is actually too difficult a subject to discuss in this book, and the science of it is still not fully understood, so manufacturers have to rely mainly on practical experience. The problem is to make a joint between the semiconductor crystal and metal wire, that does not rectify on its own. The ideal is a connection that has no resistance at all; that being impossible, the aim is a resistance that is low and linear (ohmic).

9.12 Recapitulation

Even an elementary introduction to semiconductors is not easy to grasp at once, and as it is important to have at least the main points clear before going any further, let us recapitulate. To avoid a lot of 'ifs' and 'buts' let it be assumed that the basic semiconductor material is silicon ('Si') or germanium ('Ge') and that exceptional conditions are excluded.

At low temperatures, perfectly pure material would be an insulator, because the active or valence electrons would all be occupied in holding the atoms together in regular crystalline array and there would be none to spare as charge carriers. At slightly below room temperatures (Si) or well below (Ge) the heat energy is sufficient to ionise enough atoms (i.e., release electrons from them) to cause perceptible conduction. Each displaced electron leaves behind a hole, which behaves like a positive charge. So negative and positive charge carriers are created in equal numbers throughout the material. At any given temperature this population increases until balanced by the rate at which pairs recombine. Their density (number per unit volume) then remains steady at that temperature. But it increases so fast with temperature that the conductivity doubles with about every 9°C rise. The ultimate conductivity can be very high—comparable with that of metals—because *all* the atoms are involved. But at room temperatures only perhaps 1 in 10^{10} happens to be ionised at any moment, so the material behaves like a poor insulator. So much for intrinsic conduction.

When trivalent impurities are present they provide one hole per impurity atom for conduction, so there is a surplus of holes. Some of them mop up some of the intrinsic electrons, so the net effect is to reduce electrons and increase holes. Electrons are then minority carriers and holes are majority carriers, and the material is p-type. At anything above an extremely low temperature all the impurity atoms are ionised, so even if they are outnumbered perhaps $1:10^{11}$ by the basic material they are still likely to contribute more charge carriers at room temperature. But as their contribution is already at its maximum it is rapidly rendered insignificant at high temperatures by the huge intrinsic reserves. At the same time, of course, the disparity in numbers between electrons and holes practically vanishes too. As semiconductor devices rely on the disparity they must be worked below the temperature at which intrinsic conduction begins to take over.

Pentavalent impurity has the same results except for introducing electrons instead of holes, so material with it is n-type. If both kinds of impurity are present, the carriers created by the smaller quantity of impurity are neutralised by those from the larger quantity, so only the surplus impurity of one kind counts.

All the impurity atoms in p-type material are fixed negative charges, and all those in n-type are fixed positive charges. Unless the mobile charge carriers are moved by an electric field they mill about at random, and because they are equal and opposite to the fixed charges any volume of the material as a whole is neutral.

This random motion tends to make the carriers fill any available volume uniformly, so at a p-n junction the electrons on the n side begin to diffuse into the p region, and when they meet the holes doing the same thing in the opposite direction they fill them, so creating a depletion layer, in which there is a shortage of both kinds. So the fixed charges of opposite polarity on each side of the junction establish a potential barrier that stops further diffusion. Applying an external voltage, negative to p and positive to n, heightens the barrier and no majority-carrier current can flow. Intrinsic (minority) carriers can flow, because they are of the right polarity to do so, but the resulting current is limited by their number, which is increased by a rise in temperature but not by a rise in voltage—short of breakdown voltage. Applying an external voltage of the opposite (forward) polarity overcomes the barrier with roughly its first 0.2 V (Ge) or 0.6 V (Si), and all above that causes a current limited mainly by incidental resistance.

At the present time silicon is by far the most-used semiconductor, germanium having largely fallen out of favour. But for special purposes there are other options, such as compounds of trivalent and pentavalent elements, e.g., gallium arsenide.

A reverse-biased diode has a capacitance that decreases with increase in bias voltage.

Chapter 10 Triodes

10.1 The Vacuum Triode Valve

Valuable though diodes are, radio and electronics would not have got very far on only two electrodes. The important date is 1906, when De Forest tried adding a third electrode as a sort of tap to control the flow of current between cathode and anode. It is called the *grid*, and usually takes the form of an open spiral of thin wire wound closely around the cathode so that in order to reach the anode the electrons from the cathode have to pass between the turns of wire.

If the potential of the grid is made positive with respect to the cathode it will assist the anode to neutralise the space charge of electrons around the cathode, so increasing the anode current; and, being nearer to the cathode, one grid volt is more effective than one anode volt. If, on the other hand, it is made negative it will assist the space charge in repelling electrons back towards the cathode.

Figure 10.1 shows the apparatus needed to take characteristic curves of this three-electrode valve, or triode. And in Figure 10.2 are some such curves, showing the

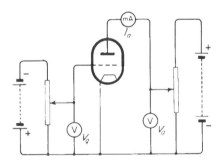

Figure 10.1 Circuit for taking characteristic curves of a thermionic triode

effects of both anode and grid voltages. (These curves are somewhat idealised, being straighter and more parallel than with real valves.) Each of them was taken with the fixed grid voltage indicated alongside. Notice that this voltage, like all others relating to a valve, is reckoned from the cathode as zero. If, therefore, the cathode of a valve is made two volts positive with respect to earth, while the grid is connected to earth, it is correct to describe the grid as 'two volts negative', the words 'with respect to the cathode' being understood. In a valve directly heated by d.c., voltages are reckoned from the negative end of the filament.

10.2 Amplification Factor

One would expect that if the grid were made neither positive nor negative the triode would be much the same as if the grid were not there; in other words, as if it were a diode. This guess is confirmed by the curve in Figure 10.2 marked '$V_g = 0$', for it might easily be a diode curve.

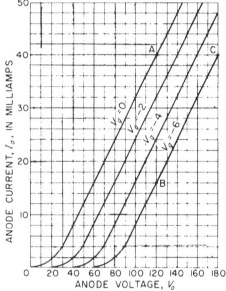

Figure 10.2 Characteristic curves of triode valve, at different grid voltages

Except for a progressive shift towards the right as the grid is made more negative, the others are almost identical. This means that while a negative grid voltage reduces the anode current in the way described, this reduction can be counterbalanced by a suitable increase in anode voltage. In the valve for which curves are shown, an anode current of 40 mA can be produced by an anode voltage of 120 if the grid is held at zero potential. This is indicated by the point A. If the grid is now made 6 V negative the current drops to 16 mA (point B), but can be brought up again to its original value by increasing the anode voltage to 180 V (point C).

Looked at another way, the distance A to C represents 60 V on the V_a scale and –6 V on the V_g scale, with no change in I_a. So we can say that a change of 6 V at the grid can be compensated for by a change of 60 V, or ten times as much, at the anode. For reasons that will soon appear, this ratio of 10 to 1 is called the *amplification factor* of the valve. It is yet another thing to be denoted by μ.

10.3 Mutual Conductance

Another measure of the control that the grid has over the electron stream in a valve is the change of anode current that results from a given change in grid voltage, the anode voltage meanwhile remaining constant. Comparing points A and B in Figure 10.2 we note that increasing the grid voltage by 6 V increases the anode current by 24 mA. In Section 2.12 the number of amps made to flow, per volt, was the definition of the siemens, the basic unit of conductance. Here now we have a current change brought about in a circuit (the anode circuit) by a voltage somewhere else (between cathode and grid). To distinguish this conductance from the ordinary sort it is called *mutual conductance* and given the special symbol g_m. Although its value in this example would correctly be stated as 4 mS (4 millisiemens) one is more likely to come across it in milliamps per volt. Mutual conductance is known as *transconductance* in the USA and is commonly called 'slope', as later in this chapter.

To find the value of g_m from an I_a/V_a graph, as we have just done, it must have curves for at least two different grid voltages. A more suitable form of characteristic curve for this purpose is obtained by plotting I_a against V_g, taking care to keep V_a constant, as in Figure 10.3.

A single curve of this kind is sufficient to show g_m. For instance BC represents the increase in I_a caused by an increase in V_g represented by AB; so g_m is given by BC/AB—in this case 4 mA/V, as before. Although they can be drawn by direct measurement, using the Figure 10.1 set-up, the curves in Figure 10.3 were actually derived from those in Figure 10.2, by noting (for example) the anode currents corresponding to the four grid voltages where they cut the vertical $V_a = 100$ line; and so on for the other three values of V_a.

The usefulness of identifying g_m with the slope of such curves is that one can see at once that it is not everywhere the same. Near the foot of each curve the slope, and therefore g_m, is less. And if we examine genuine valve curves we find that they are not so straight and parallel as these, so that g_m varies to some extent even along the upper stretches. Thus, although g_m used sometimes to be called a valve constant, in actual fact it is not at all constant, so is better referred to as a *parameter* of the valve. The other parameter we have encountered, µ, also varies to some extent with V_a and V_g, but much less so than g_m does.

If a characteristic curve really is appreciably curved (technically, non-linear) then not only does its slope vary from point to point but even at a given point it depends on the amplitude of the signal. In fact, slope hardly has a meaning unless the signal amplitude is small enough for the part of the curve used by it to be practically straight (linear), so a.c. parameters are often called small-signal parameters.

10.4 Anode Resistance

By this time you may be wondering if the slopes of the curves in Figure 10.2 have any special significance and name. Clearly it is a ratio of change in I_a caused by a change in V_a, V_g being kept constant. So it too is a conductance, and this time a straightforward one, because both current and voltage refer to the same thing: the cathode-to-anode path through the valve. This slope is therefore called anode conductance, and its symbol is g_a. More often, however, it is turned upside down, $1/g_a$, and this reciprocal of g_a is, as one would expect, called *anode resistance* and denoted by r_a.

When considering the diode curve, Figure 9.3, we calculated a kind of resistance in the same way as we did with Ohm's law, as V_a/I_a, but there was not very much point in doing so, because, as we saw then, a valve is not a proper resistance and does not conform to Ohm's law. We shall see later that in most uses of valves one is more interested in *changes* in I_a, V_g, etc., than in their values relative to zero, so that is why in regard to triodes we have been working in terms of changes or differences. It represents what a valve does with an alternating voltage or current superimposed on the d.v. or d.c. corresponding to a selected point on the valve curve. So r_a is more precisely termed *anode a.c. resistance*. In Figure 10.2 it would be found as AC/BA = 60/24 = 2.5 kΩ (kΩ instead of Ω because BA is in mA). And of course g_a, being $1/r_a$, is $1/1.25 = 0.4$ mA/V or mS.

If we have been keeping the figures for this example in mind we will remember that g_m is 4 mA/V, which is now seen to be 10 times g_a. And µ is 10. This is not just a coincidence. Because V_g has µ times the effect on I_a that V_a has, and µ is the ratio of their effectivenesses in this respect, it must always be true (provided all three parameters refer to the same valve under the same conditions) that

$$g_m = \mu g_a$$

In terms of r_a, therefore

$$g_m = \frac{\mu}{r_a}, \text{ or } \mu = g_m r_a, \text{ or } r_a = \frac{\mu}{g_m}$$

We see, then, that if any two of these three parameters are known they are all known.

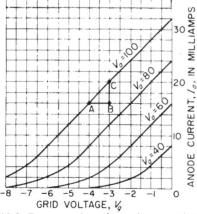

Figure 10.3 Four samples of anode-current/grid-voltage characteristic curve, each taken at a different anode voltage

10.5 Alternating Voltage at the Grid

In Figure 10.4 we have an I_a/V_g curve for a typical triode. Suppose that, as suggested in the circuit diagram, we apply a small alternating voltage, v_g, to the grid of the valve, what will the anode current do? If the batteries supplying anode

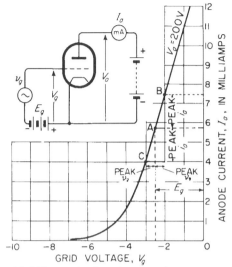

Figure 10.4 The result of applying an alternating e.m.f. (v_g) to the grid of a triode can be worked out from the I_a/V_g curve, as here. The fixed starting voltage, E_g, in this case is −2.5 V

and grid give 200 and −½ V respectively, the anode current will set itself at about 5¾ mA; point A on the curve.

If v_g has a peak value of 0.5 V, the total voltage on the grid will swing between −2 and −3 V, alternate half-cycles adding to or subtracting from the steady voltage E_g. The anode current will swing correspondingly with the changes in grid voltage, the points B and C marking the limits of the swing of both. The current, swinging between 7½ and 4 mA, is reduced by 1¾ mA on the negative half-cycle and increased by the same amount on the positive one. The whole is therefore equivalent to the original steady current with an alternating current of 1¾ mA peak superposed on it. Incidentally this, divided by the peak grid voltage, 0.5, gives 3.5 mA/V, the g_m.

There are two ways in which this a.c. in the anode circuit can be usefully employed. It can be used for producing an alternating voltage, by passing it through an impedance. If the voltage so obtained is larger than the alternating grid voltage that caused it, we have a voltage amplifier. Alternatively, if the alternating current is strong enough it can be used to operate a loudspeaker or other device, in which case the valve is described as a power amplifier. It is this ability to amplify that has made modern radio and the many kindred developments comprised in electronics possible. The subject is so important that several chapters will be devoted to it, beginning with the next one.

10.6 Grid Bias

You may have been wondering why the grid of the valve has never been shown as positive. The reason is that a positive grid attracts electrons to itself. It is, in effect, an anode. The objection to this is not so much that it robs the official anode of electrons—for up to a point it continues to increase the flow there as well as to itself—but because current flowing in the grid circuit calls for expenditure of power there, thus depriving the valve of one of its most attractive features—its ability to release or control power without demanding any from the source of input signal. For this reason valves are often called *voltage-operated* or *voltage-controlled* devices. Another way of expressing the same idea is to say that the valve, when operating normally, has an infinite input resistance. If, however, the grid is driven positive, the input resistance falls to a low value.

To prevent the flow of grid current during use, a fixed negative voltage, known as *grid bias*, is applied. The amount of bias required is equal to—or preferably a volt or so more than—the positive peak of the signal input voltage which the valve is expected to accept. The reason for making the bias greater than the signal peak is that some electrons are emitted from the cathode with sufficient energy to force a landing on the grid against a negative bias of anything up to about one volt.

10.7 The Transistor

Considering how enormously the usefulness of the thermionic valve was increased by adding a third electrode to the original diode, one is hardly surprised that attempts were made to do the same with the semiconductor diode. Owing to the greater difficulty of the basic theory, success was not achieved until 1948. But the result, the *transistor*, in spite of its late start has supplanted thermionic valves.

Figure 10.5 reproduces parts of Figure 9.13a, being a thermionic diode connected in the forward direction. Current flows because the cathode emits electrons, which

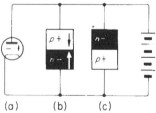

Figure 10.5 To remind us of Figure 9.13

are attracted across the vacuum by the positive anode. This was converted into a triode by interposing an electrode (the grid) to control the current—normally by reducing it. A junction diode, also connected in the forward direction, is shown at b. Besides electron flow to the anode there is hole flow to the cathode. So the valve control method would not work here; its effect on the electrons would be counteracted by its opposite effect on the holes.

Coming now to c, a reverse-biased diode, we recall that the potential barrier at its junction prevents both electrons and holes from crossing it. However, the junction p.d. that is a barrier to those charge carriers has the opposite effect on carriers of the opposite polarities, such as those responsible for intrinsic conduction (Section 9.10). In rectifiers these are a nuisance, to be kept to a minimum, because they are not under control. What we want is a supply of (say) electrons that can be introduced into the p region in numbers that can be controlled. But this is what we already have in diode *b*. The number of electrons introduced into the p region there depends on the positive bias applied to it. So all we have to do to diode *c* is to add on an n region below it and apply a positive bias to p, relative to the new n region, as in Figure 10.6. The upper pn junction is the reverse-biased one of Figure 10.5*c* and

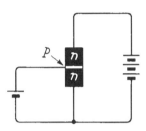

Figure 10.6 Applying bias of the right polarity to the middle layer causes increased current between the outers

the lower junction is the forward-biased one of Figure 10.5*b*. This device, which is called an n-p-n transistor, does indeed work, but if it is to do so to a useful degree at least two important structural features must be included.

In considering pn junctions with the aid of Figures 9.10–12, we assumed that the p and n regions were doped with opposite impurities equally. As Figure 9.12 indicates, half the electrons have combined with oncoming holes before they have even crossed the frontier, and the survivors are nearly all mopped up before they get very far. So the chances of reaching the second junction and crossing it would seem poor. In practice, therefore, the lower n region is doped much more heavily, say a hundred times more; thus the electrons, by so greatly outnumbering the p-region holes, nearly all manage to cross the first junction. And when they have arrived in the p layer the rate of recombination is slower, so more of them get further. This result is represented in Figure 10.7.

The other feature must now be obvious: make the p layer as thin as possible, so that the electrons have hardly got into it before they come under the influence of the relatively high positive potential of the upper n region and are pulled into it. For this reason the p layer is usually made less than 0.004 cm thick and in modern transistors intended for operation at high radio frequencies is less than one-hundredth of this.

Making the journey to the positive n region so short and easy has the effect of making the alternative path to the source of p bias relatively less easy, so that the proportion of electrons finding that route is small. This is very much to the good because it reduces the power needed for control. There would not be much point in being able to control a current if the power needed to do so was not less—and preferably very much less—than the power controlled.

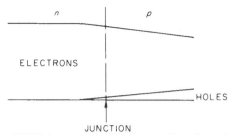

Figure 10.7 If the n region of a diode is 'doped' much more heavily than the p region, so that electrons greatly outnumber holes, they are able to penetrate an appreciable distance into the p region, as shown here and as is required in the middle layer of a transistor

The purpose of the lower n region is to provide a large source of free electrons for emission into the p layer, which without it is virtually an electron vacuum. So it corresponds to the cathode of a vacuum valve, and is called the *emitter* (a name that could equally well be applied to the valve cathode). The upper n region is there to collect the electrons so emitted, so is called the *collector* (a name that would equally fit the anode of a valve). The amount of collector current depends on the amount of bias given to the p layer, which is therefore analogous to the valve grid, and is called the *base*. Unlike the other two names, this one seems strangely inappropriate. Its origin was historical rather than functional. And the emitter is usually said to *inject* rather than emit carriers.

Obviously there is an alternative way of making a transistor, by sandwiching an n layer between two p regions, giving a p-n-p type. To make it work, the base and collector polarities have to be negative, instead of positive as in Figure 10.6.

Although in some ways a transistor is remarkably like a vacuum triode valve, there are important differences. The valve is a rectifier connected to a supply voltage in the right direction to make current flow through it. This current can be either increased or decreased by applying grid bias, according to whether it is positive or negative; but to avoid the flow of grid current it is usual to make it negative, so that in practice the anode current is controlled downwards. The transistor, on the contrary, faces the supply with a rectifier in reverse, so except for a small leakage the normal state is no current. Biasing the base negative would merely add another reversed rectifier to the score, so to obtain any useful result it is necessary for the bias to be positive; and although under these conditions base current is inevitable it enables a much larger collector current to be created and controlled. In electronic jargon a valve is said to be usually operated in the depletion mode and a transistor in the enhancement mode.

As with the junction diode, the transistor does not contain an unneutralised space charge to impede current flow, so only a few volts are needed at the collector—certainly not enough to be dignified by the description 'h.t.'. The base bias is small, too; usually only a fraction of a volt. And the emitter provides vast numbers of electrons without any heater. So transistors clearly save much power and some circuit complication compared with valves.

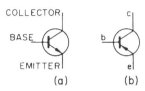

Figure 10.8 (a and b) Usual symbol for a transistor, n-p-n and p-n-p respectively

Figure 10.8 gives the graphical symbols recommended by BSI for p-n-p and n-p-n transistors. In both symbols the emitter is distinguished from the collector by carrying an arrow head which also distinguishes p-n-p from n-p-n types by pointing in the direction of (positive) current.

These aptly suggest the point-contact transistor (long obsolete and not described in this book) but are poor representations of the type of transistor we have been considering. However, in the interests of national and international standardisation we will use these symbols throughout the book.

10.8 Transistor Characteristic Curves

To plot transistor curves we can use much the same set-up as for the triode valve (Figure 10.1) except that there is no need for cathode heating, the supply voltages should be lower and both of the same polarity relative to the emitter, and a low-reading milliammeter is needed for base current; see Figure 10.9. We must remember that what in Figure 9.14 was the 'working' (forward) part of the graph has no place in connection with the collector-to-emitter circuit of a transistor, which always includes one reverse-biased junction no matter what the polarity of the voltage applied. The only place where the forward curve has any relevance is in the control (base) circuit by itself. First, though, let us deal with the collector circuit.

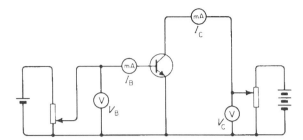

Figure 10.9 Arrangement, similar to Figure 10.1, for obtaining characteristic curves of a transistor

With the base disconnected, what we have is a reverse-biased p-n junction in series with a forward one. The only current that can flow through the reversed junction is the intrinsic or leakage current, which should be small enough to neglect for most practical purposes. The bias needed to make this small current flow through the forward emitter-to-base junction is also negligible. So the first curve, whether we label it '$I_B = 0$' or '$V_B = 0$', will show a small and nearly constant (over a wide range of V_C) collector current I_C.

As our circuit provides for measuring both I_B and V_B, we have a choice of two methods. We could follow the same procedure as with valves and plot I_C/V_C curves at regular intervals of positive V_B (instead of negative V_g). If we did we would find that instead of the regular spacing of the valve curves as in Figure 10.2 the transistor curves would be crowded together at first, with progressive thinning out, as shown in Figure 10.10a. This shows that the increases in I_C are not in constant ratio with the increases in V_B causing them. However, it is more usual to draw the I_C/V_C curves for transistors in terms of base current I_B. Figure 10.10b is an example, the curves applying to the same transistor as characteristics *a*. These curves, known as collector or output characteristics, are spaced comparatively regularly, depending on the type of transistor. This can be seen more clearly by plotting I_C against I_B at a constant V_C as in Figure 10.11.

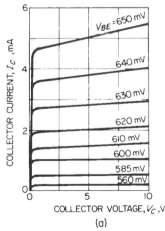
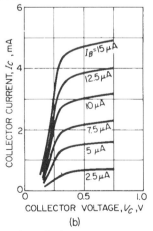

Figure 10.10 I_C/V_C characteristics for a typical small silicon n-p-n transistor plotted in terms of (a) base voltage and (b) base current

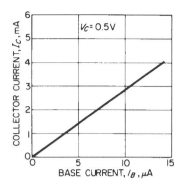

Figure 10.11 I_C/V_C for a fixed V_C (0.5 V) derived from Figure 10.10b showing how directly proportional I_C is to I_B

The corresponding I_C/V_B curve can be deduced from Figure 10.10a and this is given in Figure 10.12. It corresponds to Figure 10.3 for a valve but it is scarcely

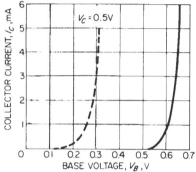

Figure 10.12 I_C/V_B characteristic for a fixed V_C (0.5 V) derived from Figure 10.10a showing the non-linearity of the I_C/V_B relationship. The dashed curve is the corresponding characteristic for a germanium transistor

necessary to draw more than one such curve for the transistor because it is so little affected by changing the collector voltage. The curve illustrates the non-linearity mentioned above, and it also shows that the base bias voltage needed to give appreciable collector current (known as the offset voltage) is around 0.6 V. This applies to all silicon transistors. For comparison Figure 10.12 also gives, as a dashed line, the I_C/V_B characteristic for a germanium transistor. This is situated well to the left of the curve for the silicon transistor and shows that the offset voltage for the germanium transistor is only about 0.2 V.

For germanium and silicon transistors, if the temperature is raised sufficiently, all the I_C curves move upwards, i.e., to higher collector currents.

Another transistor characteristic which is of interest is the base or input characteristic, i.e., the relationship between I_B and V_B. This can be measured using the circuit of Figure 10.9 if allowance is made for the voltage drop in the I_B milliammeter, which must be deducted from the reading of the V_B voltmeter—necessarily a low-range one

to suit the low working bias voltages of transistors. If, on the other hand, the voltmeter is connected directly to the base, the current it takes must be deducted from the reading of the I_B meter. The I_B/V_B characteristic for our particular n-p-n transistor is given in Figure 10.13. Because I_C is so directly proportional to I_B it has a shape very similar to the I_C/V_B characteristic. It is, of course, the characteristic of a forward-biased p-n junction diode.

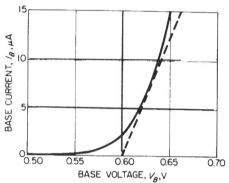

Figure 10.13 I_B/V_B characteristic for a fixed value of V_C (0.5 V)

Curves for p-n-p transistors are essentially the same except for all voltage and current polarities being reversed. Strictly, their curves should be drawn in the lower left-hand quarter where the scales are negative, but more often they are drawn in the upper right-hand quarter like valve and n-p-n transistor curves, sometimes without even minus signs to the scales to show that they refer to a p-n-p type.

We did not bother to measure emitter current in our Figure 10.9 experiment, because Kirchhoff's current law (Section 2.13) assures us that it is equal to $I_C + I_B$. As both flow into the n-p-n transistor, I_E must flow out of it. And as I_B is normally very much less than I_C, I_E is approximately equal to I_C but a little larger.

10.9 Transistor Parameters

The most obvious differences between the I_C/V_C curves in Figure 10.10 and the corresponding ones for a valve (Figure 10.2) are the very rapid rise of I_C followed by a nearly flat top. In a triode valve there is a plentiful supply of electrons emitted by the cathode, and the amount of anode current depends on how much anode voltage there is to counteract the combined throttling-back effect of the negative space charge and (usually) negative grid voltage. An n-p-n transistor, on the other hand, relies on a positive base voltage to make available a supply of electrons, and a fraction of a volt at the collector is enough to get nearly all of them across the junction. Little can be done by further increasing V_C.

Although transistor characteristics differ so much from those of the triode valves seen so far, they resemble quite closely those of tetrode and pentode valves.

When considering valve curves we saw that the steeper the slope of a current/voltage characteristic the higher the conductance and the lower the resistance to current changes at that part of the curve. So Figure 10.10a shows a very low resistance for the first fraction of a volt, and a very high resistance thereafter. Consider the resistance between emitter and collector terminals with 5 V V_C and 620 mV V_B, so 2 mA I_C. To a.c. it can be found by noting that a change in V_C of 2.5 V brings about a change in I_C of 0.05 mA. So the a.c. resistance is 2.5/0.000 05 = 50 kΩ. To d.c., however, the resistance at the 5 V/2 mA point is only 5/0.002 or 2 kΩ. The a.c. resistance, which could be denoted by r_C, is in practice seldom used, for reasons that will be explained in Chapter 12.

The mutual conductance of a transistor, as we would expect from Section 10.3, is given by the slope of the I_C/V_B curve, Figure 10.12. We see that its value depends very much on the part of the curve chosen, so this parameter too is less used than with valves.

Because the voltage amplification factor of a transistor would be equal to the above two parameters multiplied together (Section 10.4) it likewise would depend very much on working conditions, in contrast to a valve's μ which is fairly constant. On the other hand, Figure 10.11 shows that the ratio of collector current to base current is remarkably constant. The slope, being practically the same throughout, is easily calculated as the ratio of any corresponding pair of currents, say 3.0 mA and 10 μA, so in this case it is 300. If we consider how a transistor works we shall realise that this figure means that out of 3.01 mA crossing from emitter to base 3.00 mA is bagged by the collector and only 0.01 mA is lost by flowing into the base circuit. This current ratio parameter is called the *current amplification factor*. Several symbols have been used for it, such as α and β. The official one is h_F, again for reasons that will appear in Chapter 12.

It should be admitted that not all types of transistor show such a constant h_F as our example, but the tendency is there.

If h_F is truly constant the I_B/V_B curve, Figure 10.13, is just as non linear as Figure 10.12. In some types, however, the I_C/I_B characteristic is appreciably curved, and in such a way that I_C/V_B is less curved. In any case the input a.c. resistance, which can be found from the I_B/V_B curve, is important, even if we have to realise that it varies considerably from point to point. Let us try the one marked in Figure 10.13. Drawing the dashed line to show the slope there, we calculate the resistance to a.c. by dividing the voltage step (0.66 – 0.60 = 0.06) by the corresponding current step (15 μA = 0.015 mA) and get 0.06/0.015 or 4 kΩ. So, in this example at least, it is less than the collector resistance. This is very different from a valve, which when suitably biased has almost infinite resistance between grid and cathode.

For broadly comparing one triode valve's characteristics with another's it is, as we have seen, sufficient to know any two of the three parameters we have studied: r_a, μ and g_m. The third can always be very simply calculated from the other two (Section 10.4). For a transistor, because there is base voltage *and* current and another complication that will emerge in Chapter 12, we must know at least four, selected from a bewilderingly large list of possible parameters. Unfortunately at the present time there is no general agreement as to which four. The choice depends largely on why they are needed, or perhaps on what the manufacturer publishes on his data sheets. In the next chapter we shall begin to look at the significance of triode characteristics under working conditions.

10.10 Field-effect Transistors

The general class of triode transistors considered so far is sometimes called *bipolar*, because it employs charge carriers of both polarities: electrons and holes. There is another class, distinguished as *unipolar*, working on such a different principle that its right to be called a transistor at all is open to doubt, but it is nevertheless known as the field-effect transistor, usually abbreviated to f.e.t. To some extent it combines the most useful features of valves and bipolar transistors; in particular, the power economy and compactness of the transistor and the almost infinite input resistance of the valve. On the other hand, it is much more easily destroyed by misuse than valves.

An f.e.t. consists essentially of a channel of semiconductor material, either p- or n-type, on the side of which is a control electrode of the opposite type, as shown in Figure 10.14. Unlike the bipolar transistor, the p-n

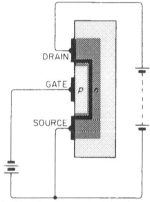

Figure 10.14 Showing in principle how a field-effect transistor is constructed

junction so formed is biased in the reverse polarity; hence its high resistance. Again, we have to learn a new set of names for the electrodes. In place of the valve cathode or transistor emitter we have the *source*. In place of the anode or collector there is the *drain*. There is, however, a tendency to use the terms emitter and collector for f.e.ts too. But the third electrode in an f.e.t. is always the *gate*. Increasing its reverse bias increases the depletion layer, as explained in Section 9.9, and thus narrows the conducting channel between source and drain, reducing the amount of current the applied voltage can pass, just like a water or gas tap. If sufficient bias is applied the channel is closed and the drain current cut off.

With an n-type channel, as in Figure 10.14, the reverse gate bias is normally negative relative to the source, and the drain voltage is positive. In both these respects, and in starting with maximum current at zero bias voltage, this f.e.t. resembles a valve. But drain-current/drain-voltage (I_D/V_D) curves, shown in Figure 10.15, are more like transistor curves; for quite a different reason, however.

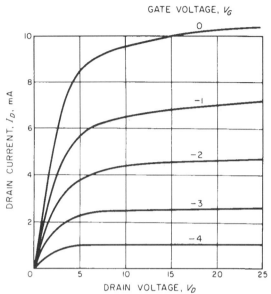

Figure 10.15 Typical I_D/V_D characteristic curves for junction field-effect transistor. Compare Figures 10.1 and 10.10

Increasing the drain voltage accelerates the electrons and so tends to increase the current, as with ordinary resistance, but at the same time it makes all parts of the channel progressively more positive towards the drain end, thereby increasing the reverse bias across the junction and narrowing the channel. Beyond a certain V_D, called the *pinch-off voltage*, the two effects almost balance, so the curve flattens out as shown. The a.c. resistance is therefore fairly high. Note, however, that the curves level out more gradually than in bipolar types, and in general the working drain voltages tend to be higher than collector voltages. F.e.t. mutual conductances are of the same order as in valves; typical I_D/V_G curves are so similar to valve I_a/V_g curves such as those in Figure 10.3 that there is really no need to show any. In fact, f.e.t.s can sometimes be substituted for valves with little or no circuit modification.

The f.e.t. illustrated so far is an n-channel type, corresponding to an n-p-n bipolar transistor. Just as there is the alternative p-n-p type, so there is the p-channel f.e.t., requiring negative V_D and positive V_G bias. Both of them work in what is called the depletion mode, the effect of applying bias being to reduce the drain current.

10.11 Insulated-gate F.E.T.s

There is another group of f.e.t.s which offer not only a choice of polarities (n-channel or p-channel) but also a choice between the depletion mode and the enhancement mode, in which the effect of applying bias is to increase the drain current from zero or thereabouts. Instead of the

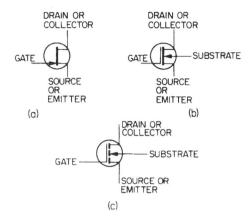

Figure 10.16 Symbols for n-channel (a) junction-gate f.e.t., (b) depletion insulated-gate f.e.t., (c) enhancement insulated-gate f.e.t. For p-channel types the arrow head is reversed

resistance between gate and drain being in megohms as in the f.e.ts we have been considering, it is in thousands of megohms, or gigohms (GΩ), at least.

The reason for this remarkably high control resistance, greater even than that of valves, is that the gate electrode is very highly insulated. So in contrast to the junction-gate f.e.ts (jugfets) already described they are known as insulated-gate f.e.t.s (igfets) or, alternatively, because of a particular construction favoured, metal-oxide-silicon or metal-oxide-semiconductor f.e.t.s (mosfets or mosts). This basic difference is indicated in the symbols for igfets (Figure 10.16b and c compared with that for a jugfet, a). All these symbols refer to n-channel types, with positive drain voltages; for p-channel the arrow heads are reversed. Both depletion and enhancement n-channel igfets are laid down on a block of p-type material (the substrate), making a fourth terminal, which may be connected straight to the source, or not, as required. Below the source and drain terminals are n-type regions, heavily doped to keep their resistance low. Between them is the gate electrode, insulated from the body of the device by a very thin layer of silicon dioxide.

Except for the insulation, a depletion igfet resembles in principle a jugfet, for there is an n-type channel between drain and source, as shown in Figure 10.17, so that the device normally conducts. Applying a negative bias to the

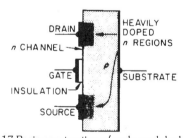

Figure 10.17 Basic construction of n-channel depletion igfet

gate repels electrons present in the channel, reducing its conductivity, and ultimately cutting off drain current entirely. The characteristic curves are therefore similar to those for jugfets (Figure 10.15) except that the zero-V_G curve is usually much lower, but because the gate is insulated it can be used to increase as well as reduce the drain current. Doing this with a jugfet would drastically reduce the input resistance.

In an enhancement igfet the channel is missing, so with no gate bias there is no conduction between source and drain. This feature is indicated in its symbol (Figure 10.16c) by the gaps between source and drain. To turn the device on, positive gate bias is needed. In this respect it resembles bipolar transistors, but in all other respects it is similar to the depletion igfet. Because the bias has to make the p-type material that lies under the gate intrinsic before it can begin to convert it into an n-type channel, several volts of bias are needed to bring the device to a suitable working point.

The gate insulation in an igfet being so extremely thin, of the order of 0.001 mm, it can be permanently broken down by quite a low voltage—25, say—and because the capacitance is no more than perhaps 4 pF the quantity of electric charge needed to destroy the device instantly is extremely small. Substituting the above figures in $Q = VC$ (Section 3.2), $25 \times 4 \times 10^{-12} = 10^{-10}$ C. This can easily be generated by taking a step or two on a dry floor in rubber soled shoes. So precautions have to be taken, such as wrapping the device in foil until installed, and shunting it by protective diodes.

10.12 Light-sensitive Diodes and Triodes

We have already seen that the reverse (leakage) current of a p-n diode depends very much on the temperature, because heat energy splits up atoms into free electrons and holes, both of which are charge carriers. Light has the same effect, and for this reason ordinary semiconductor diodes and transistors have opaque containers. But for many purposes small light-sensitive devices are useful. So what are called *photodiodes* are provided with little windows through which light can fall on the junction in such a way as to reach as many as possible of the semiconductor atoms. The resulting free electrons and holes are driven in opposite directions by the reverse bias voltage, creating a current through the diode. The brighter the light, the more the current. The very small leakage that flows when there is no light is called the *dark current*.

Often the changes in light that are to be detected or measured cause insufficient change in current to do what is required, so the current changes have to be amplified. A separate amplifier can be attached of course, but it may be more convenient to use instead of the photodiode a *phototransistor*, in which the light is applied to the base-to-collector junction (which of course is reverse-biased). The current changes here cause changes about h_F times greater in the emitter circuit (Sections 10.9, 13.3). Changes due to temperature, which are comparatively slow, or due to slow changes in light, can be filtered out by a blocking capacitor (Section 14.6). Phototransistor characteristic curves are like Figure 10.10b but with brightness of illumination in place of base current values.

In both of the above types of device, light controls a current driven by a voltage source. There are some applications (e.g., photographic exposure meters) where providing this source is inconvenient. Fortunately there are such things as *photovoltaic diodes*, in which the light itself provides it. We saw in Section 9.9 that the free electrons on the n side of a p-n junction and the holes on the p side diffuse across the junction and set up a difference of potential between the two sides. As mentioned there, a voltmeter connected to the two sides of the junction would give no reading, because if both sides are at the same temperature and receive the same light intensity there is a balance. However, by suitable design of the junction and applying the light to one side there develops between the terminals a net voltage, which can drive a current around an external circuit. The strength of the current is proportional to the illumination, and can be used as a measure of it.

On a larger scale, photovoltaic diodes are used to generate power from sunlight—particularly useful in space vehicles and in remote regions on earth.

When a semiconductor diode is passing *forward* current, electrons and holes recombine in large numbers, as described at the end of Section 9.9. Consequently energy is released, usually as heat. But there are certain semiconductors other than germanium and silicon (such as gallium arsenide) which emit some of the energy as light. The colour of the light is determined by the materials used to make these *light-emitting diodes* (l.e.d.s), and the intensity of the light is controlled by the current, so they have many uses, some described in Section 23.14.

An *opto-isolator* consists of an l.e.d. in an opaque encapsulation with a phototransistor. Current in the l.e.d. stimulates current in the phototransistor, providing a means of transferring a signal between isolated circuits.

Thermionic diodes are now practically obsolete, but if we recall from Section 9.3 that certain substances readily emit electrons when heated, and that we have just noted that light has an effect similar to heat as a stimulator of electron movement, it is hardly surprising that light-sensitive devices analogous to thermionic diodes are possible. In fact, they were used extensively until superseded by the semiconductor devices just described. They are still used as 'targets' in television cameras of the *photoemissive* type (Section 23.6).

There are other substances which are sensitive to light in that their resistance is reduced by it. Among other applications these too are used in television cameras of the *photoconductive* type.

Photoelectric devices, which are in general those that emit or are stimulated by light, are distinguished in circuit diagrams by a pair of diagonal arrows, representing the light, placed above the basic symbol (Figure 10.18).

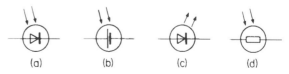

Figure 10.18 Symbols for (a) photodiode, (b) photovoltaic diode, (c) light-emitting diode and (d) photoconductive cell

Chapter 11
The Triode at Work

11.1 Input and Output

Although in the last chapter attention has been focused on triode valves and transistors themselves, various hints along the line (especially Section 10.5) must have conveyed some ideas about how these devices are employed. Now is the time to turn our attention to this aspect.

We have seen that basically triodes are devices by which very small amounts of electrical power—sometimes negligibly small, because they require voltage without appreciable current—are able to control relatively large amounts, somewhat as the power of a car engine can be controlled by slight variations in pressure on a pedal. Two circuits are thus involved: the controlling or *input* circuit and the controlled or *output* circuit. Each of these must be connected to the amplifying device (such as a triode) at not less than two points. As a triode has by definition only three terminals, one of them must be common to both input and output circuits. The choice of common electrode is a very important question which we shall have to look into in the next chapter. So as not to complicate things too much right at the start, however, one particular choice has been made for each device considered: cathode, emitter and source, which in principle are all the same. This is the choice we have been assuming, because it is by far the most usual in practice, so will continue to be assumed unless one of the others is indicated.

11.2 Source and Load

Running a car engine in neutral gives it no scope for making use of the power it can develop, and is likely to be harmful if the accelerator is pressed hard down. To fulfil its purpose it needs a *load*. This term is the one used also in electronics and other branches of electrical engineering, but we must beware because it is also used to mean several things that are not quite the same—as indeed it is in general conversation (e.g., 'he's loaded'). When we say something is overloaded we usually mean that too much power is going into it and it is in danger of burning out. But here the word 'load' will be used primarily to mean whatever receives output power; for example, a loudspeaker.

Sometimes a load is fed with power direct from a generator. In this respect a television receiver as a whole is a very small part of the load connected to a generator in a power station. Simpler still, the lamp in a torch is the load fed by the battery. This book is more concerned with situations in which there is an important link between the source of power and the load; for example, a transformer, which does not itself amplify but has to be suitably adapted to its load. Or the output part of a triode, which is a link between the source of anode or collector voltage and the load. But more often the triode and any power supplies it needs are regarded as one unit (an amplifier, for example) which is a link between the controller and the load.

This controller, typified by the driver of a car, is usually called the source of the voltage or current needed to control the *device**, and is the essential item in its input circuit, just as the load is in the output circuit.

Figure 11.1 Triodes, like many other devices and complete units for such purposes as amplification, have an input and an output, as shown here in general

The substance of these first two sections can be shown quite simply as Figure 11.1. Later on, when we see the importance of the impedances between the four pairs of terminals, we shall have to elaborate this diagram somewhat.

11.3 Feeds and Signals

In the meantime there are a few practical details and terms that need to be made clear before we go on.

You have gathered, no doubt, that any triode (and the same is true of electronic devices with more than three electrodes) can only work properly if it is supplied with certain suitably chosen currents and voltages. What is needed can usually be ascertained with the aid of characteristic curves—together with some understanding of how the device works, which by now we are supposed to have. For instance, our study of triodes would have made clear that trying to work a valve with its anode-to-cathode supply voltage reversed would be futile, but that this might not necessarily be true of the drain-to-source voltage of an f.e.t.

We may have gathered that certain combinations of these feeds would be harmful to the device. For example, too high a collector voltage would puncture the junction. A lower voltage might be all right within a certain range of base voltage, but above that the collector current multiplied

---

* Components in electronics are often termed devices. Those, such as valves and transistors, which can generate signals and amplify are called *active devices*, and components which cannot (typically resistors, inductors and capacitors) are known as *passive devices*.

by the collector voltage would be a larger power than the transistor could stand without overheating and finally failing. The manufacturers supply information on these maximum ratings, which put certain areas of the curve sheets 'out of bounds'.

With all this in mind we find a suitable *working point* for the purpose in view, and provide power supplies to maintain the device in the conditions represented by that point. It is then in a fit state to receive controlling currents or voltages from the source. These are referred to as input *signals*, whether they are in fact what most people would understand by that word, or are electrical alternations corresponding to sound waves (Section 1.3) or to pictures, or even a current we want to measure and which is too small for any meter we have. By signal currents and voltages, then, current and voltage *changes* are to be understood.

All this was illustrated in Section 10.5 by a simple example, but it is worth refreshing the memory about it because the apparatus we have been using so far (for characteristic curves) has been concerned mainly with feed currents and voltages. While these parts of a circuit are absolutely essential, and arranging for them is a large part of practical circuit design, for discussing signal handling in circuits it is clearer, and saves a lot of time, if we take the feed arrangements for granted and do not trouble to show them. It is quite usual for a designer to deal with the two matters separately, each with a circuit diagram showing only one, before combining them in a diagram of the complete circuit. It is also customary, as in Figure 10.4, to use capital letters for feeds and small ones for signals.

In most (but not all) applications of electronic devices the signals are alternating at a sufficiently high frequency (which actually can be quite low) to be easily separated from the constant feeds in the circuits themselves as well as in the diagrams. Much use is made of the fact that a capacitor is a complete circuit break (open circuit) to d.c., but allows a.c. to pass. And inductance, on its own, offers no impedance to d.c., so is a short circuit, yet may offer such a high impedance to a.c. as to be almost an open circuit. In Figure 10.4, however, the input feed source of R_g and the input signal source of V_g were connected simply in series, and no load was shown at all. We must now consider how an amplifier is affected by its load.

11.4 Load Lines

At this early stage in the study of electronic circuits we must take care not to get so involved in various incidental complications that the basic principles are not clearly seen. For instance, amplification of high (radio) frequencies is beset with considerable complications, which will be kept back until Chapter 22. So for a start we shall assume signals of low (audio) frequency. They are not only simpler to deal with but are of practical interest in radio, television, tape recorders, record players and much else.

Triode valves are now almost never used for a.f. amplification, but they are freer from sidetracking issues than transistors, so the basic principles of amplification can be demonstrated more clearly at the start using a triode valve as the active device; once grasped, these principles provide a good foundation for studying other types.

Figure 11.2, then, shows a common-cathode triode valve as the simplest possible amplifier. It is in fact Figure 10.4 with the addition of a 20 kΩ load resistance (R) in the

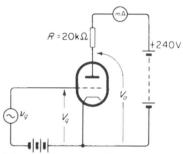

Figure 11.2 Simple voltage-amplifier circuit

anode (output) circuit. This immediately faces us with a difficulty. It is easy enough to calculate the output signal current (a.c.) when R is not there; it is equal to the milliamps of anode signal current per volt of grid signal (i.e., g_m) multiplied by the grid signal voltage (v_g). In

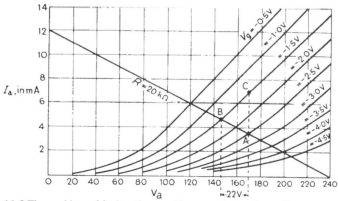

Figure 11.3 The problem of finding the amplification of the valve in Figure 11.2 is here worked out on an I_a/V_a curve sheet

Figure 10.4 we had $g_m = 3\frac{1}{2}$ and the peak v_g was $\frac{1}{2}$ V, so the peak I_a should be $\frac{1}{2} \times 3\frac{1}{2} = 1\frac{3}{4}$ mA, and this agrees with what we found from the I_a/V_g graph. But whenever the anode current varies in Figure 11.2 the voltage drop across R must vary correspondingly, and as the total voltage provided in this anode circuit is fixed at 240 V, the voltage across the valve (V_a) must vary oppositely to that across R. But the definition of g_m says V_a must be constant. So that method of calculation fails. And the alternative method, by Ohm's law, using r_a as the resistance of the valve (between anode and cathode) also breaks down, because the definition of r_a specifies constant grid voltage. What are we to do?

The answer is to combine the anode characteristic curves of the valve with a 'characteristic curve' of R in such a way as to make the total voltage across both of them equal to that provided, in this example 240 V. Suppose the curves in Figure 11.3 are the ones relating to our valve. They tell us, of course, all the possible combinations of I_a at any V_g for which a curve is provided. When $V_g = -0.5$ V, for example, the combinations include 4 mA and 100 V and 10 mA and 160 V, and any number of others. But in our loaded circuit the V_a corresponding to any particular I_a is fixed by the fact that V_a must be 240 V less the amount dropped in R, namely $I_a R$. When $I_a = 0$, then $V_a = 240 - 0$, $= 240$ V; when $I_a = 1$ mA, $I_a R = 20$ V, so $V_a = 220$ V; when $I_a = 2$ mA, $V_a = 200$ V; and so on. Plotting all these points on Figure 11.3 we get the straight line '$R = 20$ kΩ'. Since R is a load resistance, this line is called a load line. It would be equally easy to draw a load line for any other value of R.

The only possible values of I_a are those that are on the appropriate load line. So now for any given load line each value of V_g has only one possible corresponding value of I_a. To check this, think of a number for I_a, say $3\frac{1}{2}$ mA. The only point corresponding to this on the 20 kΩ load line is marked A. It tells us that V_a is 170 V and V_g is -2.5 V. $3\frac{1}{2}$ mA flowing through 20 kΩ drops $3\frac{1}{2} \times 20 = 70$ V. Deducting this from the anode battery voltage, 240, leaves 170 V, which checks.

11.5 Voltage Amplification

Suppose now we alter the grid voltage to -1.5 V. The working point must move to B, because that is the only point on both the '$V_g = -1.5$ V' curve and the load line. The anode current rises from 3.5 mA to 4.6 mA: an increase of 1.1 mA. The voltage drop due to R therefore increases by $1.1 \times 20 = 22$ V. So the voltage at the anode (V_a) falls by that amount.

Note that a grid voltage change equal to 1 V has caused an anode voltage change of 22 V, so we have achieved a voltage multiplication—called amplification or *gain*—of 22. If the anode had been connected direct to a 170 V battery there could of course have been no change in V_a at all, and the working point would have moved to C, representing an I_a increase of 3.3 mA. The reason why the increase with R in circuit was only 1.1 mA was the drop of 22 V in V_a, which partly offset the rise in V_g.

Another thing to note is that making the grid less negative caused the anode to become less positive. So an amplifier of this kind reverses the sign of the signal being amplified, sometimes described as inverting the signal, because if the input and output signals are depicted, one is upside down with respect to the other.

11.6 An Equivalent Generator

It would be very convenient to be able to calculate the voltage amplification when a set of curves was not available and only the valve parameters were known. To understand how this can be done, let us go back to Figure 10.1 with its controls for varying anode and grid voltages. First, leaving the V_g control untouched, let us work the V_a control. The result, indicated on the milliammeter, is a variation in anode current. The same current variation for a given variation in V_a would be obtained if an ordinary resistance equal to r_a were substituted for the valve (Section 10.4). Considering only the variations (signals), and ignoring the initial V_a and I_a (feeds) needed to make the valve work over an approximately linear part of its characteristics, we can say that from the viewpoint of the anode voltage supply the valve looks like a resistance r_a.

Now keep V_a steady and vary V_g. To the surprise of the V_a supply (which does not understand valves!), I_a again starts varying. If the V_a supply could think, it would deduce that one (or both) of two things was happening: either the resistance r_a was varying, or the valve contained a source of varying e.m.f. To help it to decide between these, we could vary V_a slightly—enough to check the value of r_a by noting the resulting change in I_a—at various settings of the V_g control. Provided that we took care to keep within the most linear working condition, the value of r_a measured in this way would be at least approximately the same at all settings of V_g. So that leaves only the internal e.m.f. theory in the running. We (who do understand valves) know that varying V_g has μ times as much effect on I_a as varying a voltage directly in the anode circuit (namely V_a).

So we can now draw a diagram, Figure 11.4 (which should be compared with Figure 11.2), to show what the valve looks like from the point of view of the anode circuit. Its behaviour can be accounted for by supposing that it contains a source of e.m.f., μv_g, in series with a resistance, r_a. We have already come across examples of substituting

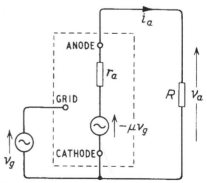

Figure 11.4 The 'equivalent generator' circuit of a valve, which can be substituted for Figure 11.2 for purposes of calculating the performance of the valve

(on paper or in the imagination) something which, within limits, behaves in the same way as the real circuit, but is easier for calculation. One of them was a dynamic resistance in place of a parallel resonant circuit (Figure 8.9). And now this trick of the *equivalent generator*, which is one of the most important and useful of all equivalent circuits.

But 'within limits' must be remembered. In this case, for a valve to behave like an ohmic resistance it must be working at feed voltages corresponding to linear parts of its characteristic curves, and they are never perfectly linear, so at best r_a is no more than a fair approximation to an ohmic resistance. And we must remember Section 11.3 and realise that the equivalent generator takes account only of signals and ignores feeds, which are merely incidental conditions necessary for achieving reasonable linearity.

It is failure to separate in one's mind these two things, feeds and signals, that leads to confusion about the next idea—the minus sign in front of μv_g in Figure 11.4. Some people argue that because a positive change in v_g (say from -2.5 V to -1.5 V) increases I_a, the imaginary signal voltage μv_g must be positive. And then they are stuck to account for v_a, the amplified signal voltage, being negative as we noticed at the end of the last section. But if they realised that the feed voltage and current do not come into Figure 11.4 they would avoid this dilemma. For it just happens that the feed current flows anticlockwise around the circuit, but as we have replaced the real valve and its feeds by an imaginary signal generator there is no reason for departing from the usual convention of reckoning voltages with reference to the common electrode, here the cathode. If the generator voltage were $+\mu v_g$, a positive signal on the grid would make the anode go positive, contrary to fact. To put this right by turning the generator upside down and reckoning the cathode voltage with respect to the anode is contrary to established convention and common sense. The logical course is to reckon the generator voltage as $-\mu v_g$.

A further point about equivalent generators is that such equivalence as they have applies only to the external circuit. In general we would be wrong to draw any conclusion from them about conditions within the valve or other device simulated.

Figure 11.5 is an alternative version of the diagram, using the notation explained in the latter part of Chapter 5. As Figure 11.4 is the kind of diagram commonly used, and Figure 11.5 has a number of advantages over it which grow with increasing circuit complexity, it will be as well to be able to interpret both. One of the advantages of Figure 11.5 is that while we can refer to the outflowing i_a as AB, the people who insist on reckoning the anode signal current as inwards can call it BA, and they can have their positive generator voltage bk, $= \mu k_g$. Just now when we are going to discuss valves in general, i_a etc. will be a little more convenient than AB etc., because one does not have to refer to a particular diagram to know what they mean, and in such a simple case we are not likely to get the signs wrong. And without any reactances there is little justification for a phasor diagram. But in preparation for less simple circuits, and as a further check on our understanding of this basic circuit, Figure 11.5b is its phasor diagram according to the conventions explained in Chapter 5. The absence of reactances is indicated by all the phasors being parallel to one another—a condition which the usual arrowed type of diagram tends to confuse rather than elucidate. In accordance with the equation $kb = -\mu k_g$, kb is drawn opposite to kg and μ times as long. It represents the generator voltage, which is divided between r_a and R. Point b is of course inaccessible in a real valve; ab is drawn dashed to emphasise this.

See Section 12.1 for a fuller account of equivalent generators.

11.7 Calculating Amplification

Now let us apply this equivalent generator technique to the problem we set out to solve—finding the amplification of a valve. The signal current i_a in Figure 11.4 is (by Ohm's law) the signal e.m.f. divided by the total circuit resistance: $-\mu v_g/(R + r_a)$. The signal output voltage, v_a, is caused by this current flowing through R, so is $= \mu v_g R/(R + r_a)$. The voltage amplification (which is often denoted by A) is v_a/v_g, so

$$A = \frac{-\mu R}{R + r_a}$$

This minus sign agrees with our finding that the signal output voltage is inverted with respect to the input. Since it is generally understood that v_a is in opposite phase to v_g, the minus is often omitted. Let us apply this equation to the case we considered with the help of Figure 11.3. Using the methods already described in Sections 10.2 and 10.4 we find that in the region of point A the valve in question has an r_a of about 14.5 kΩ and μ about 39. Substituting these, and $R = 20$ kΩ, in the formula: $A = -39 \times 20/(20 + 14.5) = -22.6$, which agrees pretty well with the figure obtained graphically. (It must be remembered that with electronic devices we are in the realm of the approximate. Performance depends so much on conditions, and even when conditions are the same, different samples of the same type of device give different results. So there is no point in specifying them with high precision.)

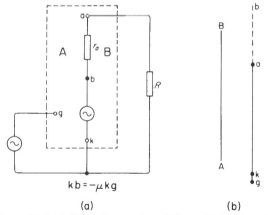

Figure 11.5 (a) Alternative version of Figure 11.4, adapted for phasor diagrams (b)

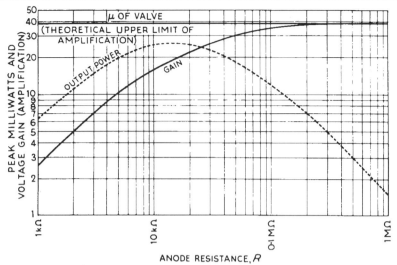

Figure 11.6 The full line shows how the voltage amplification depends on the value of load resistance R. It is calculated for a valve with μ = 39 and r_a = 14.5 kΩ. The dashed line shows the peak milliwatts delivered to R for a grid input of 1 V peak

The amplification formula that we have just used shows at once that making $R = 0$ would make $A = 0$. It also shows that making R so large that r_a was negligible in comparison with it would make A almost equal to μ, which is the reason for calling this parameter the amplification factor. What happens at intermediate values of R can best be seen by using the formula to plot a graph of A against R—the full-line curve in Figure 11.6, in which, incidentally, logarithmic scales (Appendix B, Section B6) have been used to cover wide ranges of values without squeezing most of them into a corner. According to this, it looks as if the higher the more, so far as R and A are concerned. But although feeds do not come directly into the equivalent generator (from which this curve was derived) the assumption that they are adequate to maintain the parameters at the values assumed must never be forgotten. Looking at Figure 11.3 we see that if R were increased to a much larger value the load line would have to swing round until it was nearly horizontal, cutting the valve curves at points where their decreased slopes would mean much larger values of r_a too. So although very high load resistances are sometimes used, it is more likely to be for some other reason, such as keeping the feed current small, than for extracting the maximum possible voltage gain. The alternative solution, increasing the supply voltage to compensate for the loss in R, is usually uneconomic with valves, and exposes transistors to serious risk of breakdown when current is cut off. In practice there is often little or no choice of load resistance, and things may have to be designed to fit it, rather than vice versa.

11.8 The Maximum-power Law

For some purposes (such as working loudspeakers) we are not interested in the voltage output so much as the power output. This, of course, is equal to $i_a v_a$ or $i_a^2 R$ (Section 2.17), and can therefore be calculated by filling in the value of i_a we found; namely $-\mu v_g/(R + r_a)$. Denoting the power by P, we therefore have

$$P = \left(\frac{\mu v_g}{R + r_a}\right)^2 R$$

A graph of this expression, for $v_g = 1$ V, and μ = 39 and r_a = 14.5 kΩ as before, is shown dashed in Figure 11.6. The interesting thing about it is that it has a maximum value when R is about 15 kΩ. The only thing in the circuit that seems to give any clue to this is r_a, 14.5 kΩ. Could the maximum power result when R is made equal to r_a?

It could, and does, as can be proved mathematically. This fact is not confined to this particular valve, or even to valves in general, but (since, you remember, it was based on Figure 11.4) it applies to all circuits which consist basically of a generator having internal resistance and working into a load resistance. So the fact that it has been demonstrated for a device not now commonly used does not detract from its importance. Making the load resistance equal to the generator resistance (which in this case simulates an amplifying device's output resistance) is called *load matching*.

It does not follow that it is always desirable to make the load resistance equal to the generator resistance. Attempting to do so with a power-station generator would cause so much current to flow that it would be disastrous! But this law, that the maximum power from a generator (whether 'equivalent' or real) giving a *fixed* voltage, a.c. or d.c., is obtained when the load resistance is equal to the generator resistance, is a very important one. In the form just stated, both generator and load are assumed to be purely resistive; if there is reactance too it should be cancelled out by reactance of the opposite kind (Section 8.3).

11.9 Transistor Load Lines

We have seen that the output current/voltage curves of transistors, both bipolar and field-effect, are shaped very differently from those of vacuum triodes. It happens that valves of the types that were most used, such as pentodes, also had the same sort of shape, rising very steeply at first and then flattening out. So we ought to see how the foregoing principles apply to curves of this general shape.

Figure 11.7 is a set of output curves for a typical low-power silicon transistor. Note that this time (like Figure 10.10a) they have been drawn for equal intervals of base

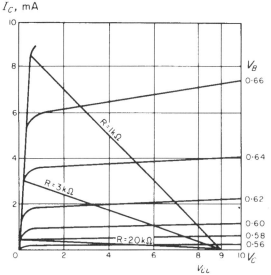

Figure 11.7 Various load lines applied to a set of I_C/V_C curves for a small silicon transistor

voltage. As we would expect from Figure 10.13, the corresponding steps in collector current are not at all equal. If, mindful of the maximum power law, we were to try drawing a load line having the same slope (but the other way) as the transistor lines we would see that whatever our choice the load line would have to be extended a very long way to the right before it hit the zero-current line, so that the point where it did would almost certainly represent a collector voltage above the rated maximum for the transistor. That being so, if the collector current were cut off for any reason (a large negative input signal, or a lack of bias), the transistor's life would probably come to an abrupt end. In practice one is seldom vitally interested in getting the maximum power output *per input volt or milliamp* regardless of load resistance. A commoner requirement is maximum power output for any reasonable input, and this will come up for discussion in Chapter 19.

Meanwhile, an interesting thing to notice about devices with curves of this shape is that owing to the very steep initial rise of current the load line can be extended nearly to zero V_C regardless of the slope at which it is drawn. For the same reason the maximum output voltage obtainable is practically the same for all load resistances. The three sample load lines in Figure 11.7, based on 9 V as the collector supply voltage (V_{CC}), bring out these facts clearly. Comparing this diagram with Figure 11.3 we see that something like twice as much of the supply voltage can be utilised.

It remains true, as with valve triodes, that the higher the load resistance the higher the voltage gain, again with the practical restriction that if we are too greedy we get into trouble with the closing up of the curves at what is called the *bottom bend*—more clearly seen as such when the relationship between I_C and V_B is expressed directly as in Figure 10.13.

While the voltage output is limited at the lower end of the load line by the bottom bend, it is stopped abruptly at the top end by coming up against the steep rise of the curves, an action that is commonly called *bottoming*. This term may appear rather an odd one for something that happens at the top end of the line, furthest from the bottom bend. The explanation is not an unseemly anatomical obsession but a reference to the output voltage, which is here at the bottom of its range—only a fraction of a volt above zero V_C. An alternative name for this state is *saturation*.

More appropriate to bipolar transistors than voltage gain is current gain. By looking at Figure 10.10b and mentally drawing load lines across it we should be able to see that maximum current amplification is obtained with $R = 0$. Just as maximum voltage gain with no current is seldom useful, so is maximum current gain with no voltage. Both conditions give zero power output. However, by turning the vertical line representing zero R through a moderate anticlockwise angle we shall find that a useful voltage output is obtained without losing very much current. Try something like 500 Ω.

11.10 Class A, B and C Operation

The bias conditions so far described yield what is known as Class A amplification or operation, in which the input and output waveforms are similar, though one may be inverted. Another feature of Class A is that the mean current drawn from the power supply is constant, being unaffected by changes in input-signal amplitude, even by its complete removal. Clearly such an amplifier does not use the power taken from the supply very economically!

There is, however, another mode of amplifier operation in which the active device is biased almost to the point of cut-off of output current. Only half of each input sine wave can be reproduced by such an amplifier, but, by using two similar active devices, and applying the input signal to them in such a way that one reproduces positive half-cycles and the other negative (a widely used circuit arrangement known as push-pull), distortionless amplification can be achieved. This is known as Class B amplification. Because it is more economical of supply current it is widely used in amplifiers where that is important; e.g., in equipment run off batteries.

By further increasing the bias, the output can be made to flow only at the peaks of the input signals. In such an amplifier there is no proportionality between input and output signals. Nevertheless there are circumstances, as we shall see later, where this mode, called Class C, is useful, and of course it is even more economical than Class B.

Chapter 12

Transistor Equivalent Circuits

Don't be perturbed if you find this chapter tougher than any of the others, especially Section 12.7 onwards. It is tougher! The first half should be persevered with, and if you then feel the need for some relief you may be glad to know that a complete grasp of the second half is not essential to understanding the rest of the book. At first it should be enough to study Figures 12.14, 12.16, 12.18 and 12.19.

12.1 The Equivalent Current Generator

When, in Section 10.9, we attempted to get to grips with the leading transistor parameters on the lines that worked so well and simply with valves, we were put off by statements that what looked like the obvious equivalents for transistors were actually not very popular, and by promises that the reasons would emerge more clearly in due course. That moment has now arrived.

Two of the three main valve parameters relate output to input and are therefore called *transfer* parameters: g_m is the ratio of output signal current (i_a) to input signal voltage (v_g), output signal voltage being zero; μ is the ratio of the voltage of the imaginary output generator to the input voltage. Knowing them both, then for a given input voltage we know the maximum values of both output voltage and current. As g_m is defined for zero load resistance, the whole of that voltage is engaged in driving that current through the internal resistance of the valve, which can therefore be found as the ratio of voltage to current, or μ/g_m, and is denoted by r_a. (It can also be found directly and independently, from I_a/V_a characteristic curves.)

Both of the transfer parameters for valves refer to the input *voltage*. That is natural enough, for any input current is incidental. It is the grid voltage that does the controlling. Because the input resistance of a transistor is comparatively low, both voltage and current are needed, but our study of how a transistor worked showed that it was the base *current* that decides how much collector *current* is released. So it is the ratio of these two, the current amplification factor, h_f, that best expresses the control capability of a transistor. Like μ for a valve, h_f has the merit that its value depends less on working conditions than do most other parameters. It seems, then, that in bipolar transistors (which are the kind to be assumed unless indicated otherwise) current plays a role corresponding to voltage in valves. So, as our valve equipment was a voltage generator, should a transistor equivalent be a current generator? Is there such a thing?

The answer to the second question is definitely yes; but to the first it is maybe, for in truth either kind of generator can be used to represent either a transistor or a valve, though it is also true that for transistors the current form is usually chosen.

In Section 11.7 we used the equivalent voltage generator circuit to calculate the usable output voltage from the valve. The generator voltage, μv_g, was in series with the valve's own internal resistance and the load resistance, r_a and R, as shown again in Figure 12.1a. So the fraction of it that comes across R is $R/(R + r_a)$, and therefore

$$v_a = \mu v_g \frac{R}{R + r_a}$$

If we remember that $\mu = g_m r_a$ we can rewrite this equation as

$$v_a = g_m r_a v_g \frac{R}{R + r_a} = g_m v_g \frac{R r_a}{R + r_a}$$

At the end of Section 2.9 we saw that the product of two resistances, divided by their sum, was the combined resistance of the two in parallel. So the second form of the equation says that the output voltage of a valve is equal to

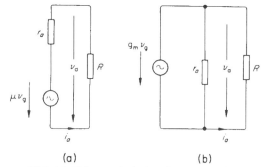

Figure 12.1 (a) Valve equivalent voltage generator as in Figure 11.5, except that as equivalent generators in general are now being led up to there is no need to retain the minus sign appropriate to valve equivalents in particular, so the current direction is shown reversed. (b) Alternative equivalent current generator

$g_m v_g$ multiplied by the combined resistance of valve and load in *parallel*. Now $g_m v_g$ is a current—the output signal current from a valve with zero load resistance—so if we connect a generator of that amount of current in parallel with r_a and R, as in Figure 12.1b, this diagram also conforms to the equation and is interchangeable with a so far as the external circuit is concerned. For example, if $R = 0$ then all the current from the generator will take that path of least resistance—and that checks with the way g_m is defined. At the opposite extreme, if R is infinitely large (an open circuit) all the current has to go through r_a, setting up a voltage $g_m v_g r_a$ between the terminals. This is the same thing as μv_g, which is the voltage we would get between the open terminals in Figure 12.1a. If you are still not convinced, try working out some more results for each equivalent circuit in turn, using any numerical values you like for r_a, g_m and R.

The same principles hold good for ideal voltage and current generators in general: a voltage generator in series with a resistance (Figure 12.2a) can be replaced by a current generator in parallel with the same resistance, the

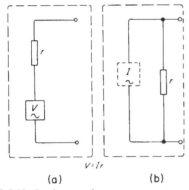

(a)   (b)

Figure 12.2 Ideal voltage and current generator symbols, (a) and (b) respectively, favoured by BSI in place of the circular general generator used exclusively so far, as in Figure 12.1

amount of current being whatever is needed to set up the same voltage by flowing through that resistance (*b*). Similarly, if you have a current generator giving a certain voltage across its parallel resistance, with no load connected, that is the voltage of the equivalent voltage generator, which should be in series with the same resistance.

This ideal voltage generator, which we have been using in valve equivalent circuits in the general generator circular form, is often called a constant voltage generator, because (since it has no resistance of its own) it maintains its voltage regardless of the amount of current drawn from it. If it was 'dead-shorted' it would deliver an infinitely large current. Although no real generator can do this (for all have some internal resistance) the idea is easy to accept because close approximations to it do exist, such as large accumulator batteries for d.c. and the mains electricity supply for a.c. Real generators can be simulated by an ideal voltage generator in series with the appropriate resistance, as we have seen. Note that the current supplied by a constant voltage generator is whatever is necessary to develop a back voltage equal to its own, across whatever resistance is connected to it.

An ideal or constant current generator is one that delivers a certain current regardless of what is connected to it. The voltage between its terminals is whatever is needed to make that current flow. So if it is open-circuited the voltage becomes infinitely large. This kind of generator is even more difficult to imagine because nothing very like it exists. The nearest is a very high voltage in series with a very large resistance, so that inserting any reasonable load resistance makes very little difference to the amount of current. But we shall see in Chapter 27 that a constant current almost regardless of resistance (within certain limits) can be provided by special electronic circuits. Any real generator can be simulated by an ideal current generator in parallel with the appropriate resistance, or conductance.

The symbol shown in Figure 12.3 is often used in place of that in Figure 12.2b but is undesirable for this purpose as it is an international standard symbol for a transformer.

Figure 12.3 This symbol for a current generator is deprecated

12.2 Duality

Although it will interrupt our quest for a transistor equivalent circuit, this may be as good a moment as any to take note of an important general circuit concept—*duality*. As far back as Section 2.12 we saw that an alternative to Ohm's law in the form $E = IR$ is $I = EG$. This is because G, the conductance, is defined as $1/R$, and substituting it in the first form of Ohm's law, after having divided both sides by R, we get the second form. Comparing the two forms, we see that current and voltage have changed places, and so have resistance and conductance. This is not just a coincidence. We can extend the list:

$$
\begin{array}{rcl}
\text{Voltage, } V \text{ or } E & \leftrightarrow & \text{Current, } I \\
\text{Resistance, } R & \leftrightarrow & \text{Conductance, } C \\
\text{Capacitance, } C & \leftrightarrow & \text{Inductance, } L \\
\text{Reactance, } X & \leftrightarrow & \text{Susceptance, } B \\
\text{Impedance, } Z & \leftrightarrow & \text{Admittance, } Y
\end{array}
$$

From any equation in which any of these quantities appear, another equation can be derived by changing each of them for the other in the pair. In Section 7.2 we found that the 'Ohm's law' for inductance was

$$I = \frac{E}{2\pi f L}$$

If we do the exchanging trick we get

$$E = \frac{I}{2\pi f C}$$

which we went to some trouble to prove in Section 6.3.

These are simple examples, but the principle holds good for the most complicated ones. Having established one relationship, we can prove another in this easy way; two formulae for the price of one, as it were. They are called *duals* of one another.

More interesting still, if an equation applies to a certain circuit its dual equation applies to its dual circuit, which is constructed by exchanging series for parallel and short circuits (zero resistance, infinite conductance) for open circuits (zero conductance, infinite resistance), capacitance for inductance, etc. We do not have to look far for an example of this, for Figure 12.1 shows two circuits that externally are equivalent to one another; the first comprises a voltage generator in series with two resistances; the second, a current generator in parallel with what could be expressed as two conductances—more conveniently, in fact, for they can be simply added to give the total conductance, just as the resistances in the first circuit are added to give the total resistance.

The equation derived from Figure 12.1a for finding v_a (the part of the generated voltage received by the load) was

$$v_a = \frac{\mu v_g R}{R + r_a}$$

We can find the part of the total generated current in Figure 12.1b reaching the load by applying duality to the above equation, substituting the generator current $g_m v_g$ for the generator voltage μv_g:

$$i_a = \frac{g_m v_g G}{G + g_a}$$

where of course $G = 1/R$ and $g_a = 1/r_a$. If you still do not trust duality by itself you can work it out from the equation for v_a, for $i_a = v_a G$. And if you still prefer resistances to conductances you should easily be able to find that

$$i_a = \frac{g_m v_g r_a}{R + r_a}$$

Lastly, dual circuits have dual phasor diagrams. We have already noted examples of this, such as Figures 8.1 and 8.7. Because the circuits concerned are duals, the phasor diagrams are identical except for interchange of current and voltage.

12.3 Voltage or Current Generator?

How do we decide whether to use a voltage or current generator in an equivalent circuit? Remember that although they are exactly equivalent externally they are not equivalent inside. Our domestic electricity supply could be regarded as a voltage generator of 240 V r.m.s. in series with perhaps 0.1 Ω. With nothing switched on, there would be no power consumption in the equivalent generator either. The equivalent current generator would have to be 2400 A in parallel with 0.1 Ω, which would be consuming $2400^2 \times 0.1 = 576\,000$ W all the time!

More important for us are the terms in which the two components are expressed. For a valve the generator part can be in terms of μv_g or $g_m v_g$—voltage or current. The other part is the same in both, though in one it is normally expressed as r_a and in the other as g_a. In a triode valve particularly, r_a is usually moderate, and we could (if we used such a valve at all) make R relatively large. So for a rough approximation we could neglect r_a in comparison with R in $R + r_a$ in the equation we found in Section 11.7 for a valve's voltage gain:

$$A = \frac{\mu R}{R + r_a}$$

It would then reduce to the simple approximation $A \simeq \mu$. ($\simeq$ means 'is approximately equal to'.) See also Figure 11.6. The parameter thus gives a better idea of the valve's capability than g_m, for the current generator circuit shows that when R is much greater than r_a only a small part of the generated current gets into it. When $R > r_a$, then, the voltage generator circuit is usually preferred. We have seen that both bipolar and field-effect transistors have such large a.c. output resistances that R is usually small by comparison. The same is even more true of many valves other than triodes. If then in the equation for A we neglect R in the denominator (lower part) we get

$$A \simeq \frac{\mu R}{r_a}$$

and as $\mu/r_a = g_m$ this further simplifies to

$$A \simeq g_m R$$

which is a very useful approximation. So where the load resistance is less than the device's output resistance the current generator, with g_m, is usually preferred. This preference is strengthened by the fact that in practical circuits there are a number of resistances and reactances in parallel at the output, and if they are expressed as conductances and susceptances they can be added in quadrature to find the total admittance.

12.4 Transistor Equivalent Output Circuit

Our conjecture in Section 12.1 that a current generator might be the one to use for representing a transistor is thus justified. The output current with $R = 0$ is, as we know, $h_f i_b$, so that is the correct output of the imaginary generator. In parallel with it is the internal resistance of the transistor, preferably expressed as a conductance. We might make its symbol g_c, but for a reason to appear later h_o has been adopted.

So now we have the two components needed for constructing the output part of a transistor equivalent circuit. But a question arises again about the sign of the generator's output. In Section 11.6 the reason was given for preferring a minus sign for the voltage, μv_g. It seems logical that when a current generator is substituted the sign of the current should be the same. However, there is a very widely used convention that instead of a minus sign for $h_f i_b$ the direction of the current should be shown reversed, so that the positive direction of current through the load is into the collector, as in Figure 12.4. The effect of a positive i_e is to

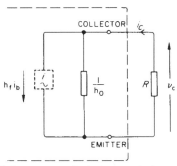

Figure 12.4 Output part of transistor equivalent circuit using the conventions for h parameters

make the emitter end of R positive, so for v_C to be positive it would have to point downwards. As the convention is to make it point upwards, for positive inputs $+v_C$ will always have negative values. This seems confusing, but fortunately our arrowless notation does not tie one down to either convention but covers both. So Figure 12.5a shows the same thing as Figure 12.4, the convention therein being expressed by $DA = h_f i_b$. The other, $AD = -h_f i_b$, means exactly the same thing, but because e_C is in any case equal to R multiplied by the current flowing down through it (CA) and up through the generator (AD) a negative value of it comes naturally as a result of the minus sign in front of $h_f i_b$, corresponding to positive values of i_b. Figure 12.5b is the corresponding phasor diagram, and by comparing it with Figure 11.5b we see that it is essentially the same except that voltage and current have changed places. These are therefore dual phasor diagrams. Again, a dashed line has been used for what is inaccessible in the real circuit. If you choose to regard the collector current as inwards, it is represented by CA (instead of AC), and ec shows that the voltage e_C is negative relative to it.

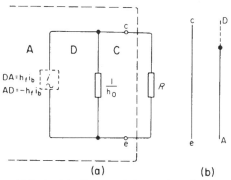

Figure 12.5 At (a), Figure 12.4 is adapted for phasor diagrams (b)

So far an n-p-n transistor has been assumed, but no polarity changes in equivalent circuit diagrams are needed for p-n-p types, because when they are *all* reversed they are still relatively the same. That is why it is possible to use one-way arrows for a.c. signals, which are reversing all the time. But in circuits with reactances, where phase relationships are not just simple positive or negative, arrows can be confusing, whereas the notation explained in the latter part of Chapter 5 is always unambiguous.

We next have to consider the input part of the transistor, which is more complicated than that of a valve, because both voltage and current are involved. The ratio of the two is the input resistance, which we found from the slope of the I_B/V_B characteristic as in Section 10.9. It has been given the symbol h_i, and its value depends very much on V_B, as we can tell from the variableness of the slope in Figure 10.13. And that is not the end of the matter, for a further complication arises which involves a principle of such importance that we shall have to devote some time to it.

12.5 Some Box Tricks

Suppose we are given a 'black box' with two terminals and are told to measure the resistance between them. We do this in a simple straightforward way by applying a known voltage and measuring the current going through the box. Suppose we try 6 V and the current meter reads 50 mA. We conclude that the box contains a resistance of 6/0.05 = 120 Ω. However, just to make sure we try 12 V, and expecting the current to be 100 mA we are disconcerted to find it is 200 mA. After having just been thinking of a non-linear characteristic curve, we may put forward the theory that the box contains a non-linear resistor; a transistor, perhaps. Determined to find out, we plot more points in order to trace the curve, and find it is a perfectly straight

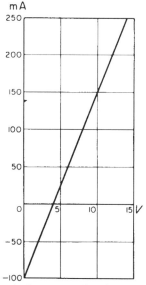

Figure 12.6 Graph showing results of an experiment with a sealed box

line, Figure 12.6. Odder still, when we apply 4 V we get no current at all, and when we join the terminals by the milliammeter alone we get −100 mA!

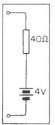

Figure 12.7 Contents of the box revealed

The straight line tells us that the box contains a linear resistance. Its slope, which being constant can of course be measured anywhere to give the same result, tells us that the resistance is 40 Ω. But obviously there must be a source of e.m.f. too. If we assume it is a voltage source, its voltage is best measured when there is no current to cause a voltage drop in the resistance. The point where the line crosses the zero-current line tells us that it is 4 V. So we deduce that the contents of the box could be as in Figure 12.7. (Alternatively they could be—in theory at least—a current source of 0.1 A in parallel with 40 Ω.) Actually the box could contain a whole lot of batteries and resistors connected in an elaborate network, in which case Figure 12.7 would be the simplest possible equivalent circuit.

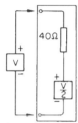

Figure 12.8 In this box the internal voltage is directly proportional to the voltage applied

Next, imagine that instead of the e.m.f. in the box being constant it is directly proportional to the voltage applied between the terminals; say, always half as much, and in the reverse direction, Figure 12.8. The effect of applying any voltage would now be the same as applying half that voltage to a plain 40 Ω resistor. There would be only half the expected current, so the box would behave exactly like an 80 Ω resistor. There would be no means, without opening the box or X-raying it, of telling whether it contained just an 80 Ω resistor or some other resistance and a source of e.m.f. proportional to that applied. And we can easily see that if the internal source in Figure 12.8 was reversed, so that it increased the flow of current, it would make the resistance appear less than it really was.

Or imagine again that instead of being in series with a proportional voltage source our resistor was in parallel with a proportional current source, as in Figure 12.9. For every milliamp we put into the resistor from outside, the unseen current generator puts another through it in the same direction. So the voltage drop across the resistor, which we

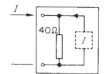

Figure 12.9 Box that includes a current source equal to that applied

measure between the terminals, is twice what it should be with 40 Ω. It would be reasonable to deduce, therefore, that the box contained an 80 Ω resistor. If the internal source sent current through the resistor in the reverse direction it would make the resistance seem less than it actually was.

Here we have another example of duality: voltage in opposition simulates higher resistance; current in opposition simulates higher conductance. And conversely for assisting direction.

Let us call the apparent resistance, as inferred from the externally applied voltage V and current I, R_{app}, and the actual resistance in the box, R_{act}. Let v be the internal voltage aiding V (negative value if it is opposing it). Then

(1) $R_{app} = \dfrac{V}{I}$ and (2) $R_{act} = \dfrac{V + v}{I}$

From (2), $I = \dfrac{V + v}{R_{act}}$, and substituting this in (1) we get

$$R_{app} = R_{act} \dfrac{V}{V + v}, \text{ or } \dfrac{R_{act}}{1 + v/V}$$

Check this with our example. Then use duality and get

$$G_{app} = \dfrac{G_{act}}{1 + i/I}, \text{ or } R_{app} = R_{act}(1 + i/I)$$

where i is an unseen current augmenting I, in Figure 12.9 for example.

12.6 Input Resistance

The previous section with its imaginary voltage and current sources adjusting themselves as if by magic to fixed proportions of whatever we cared to try at the input terminals, deceiving us into believing that the box contained only an ordinary resistor of a different value, was not just a vain flight of fancy. It demonstrates one of the most important basic principles of practical electronic circuits, known as *feedback*. We need it now in order to continue our study of transistors.

So far we have taken for granted that we could connect our test circuit, such as Figure 10.9, directly to the p-n junction. Inevitably, however, there is some resistance between them and each of the connecting wires. These resistances are represented in Figure 12.10 by resistors. The important one just now is r_e because it is in the common lead so has both input and output currents flowing through it, and the output current is usually much the larger, so is likely to have a drastic effect on the value of r_e as seen by the input signal.

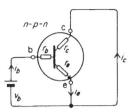

Figure 12.10 The conventional transistor symbol is here elaborated to show the internal resistances

Feeds are again omitted and v_b is a small d.v. signal applied between emitter and base (input). It causes an input current i_b, and as we are assuming small signals we take v_b/i_b as the input resistance, for which we have noted the symbol is h_i. The main result of this input signal is to cause an output current h_f times as large, if there is zero load resistance, as is normally so when making measurements on transistors or valves. The positive direction of this current, i_c, is as shown; check with Figure 12.4. We now have exactly the sort of situation imagined in Figure 12.9. For every microamp of i_b flowing through r_e via r_b there are h_f extra microamps flowing through it via r_c. So to the input circuit r_e behaves as if it were $r_e(1 + i_c/i_b)$, $= r_e(1 + h_f)$; and $h_i = r_b + (1 + h_f)r_e$. r_c includes a reversed junction, so its resistance is high enough for its shunting effect to be neglected.

Typically for a small transistor, r_b might be $1000\,\Omega$, r_e $10\,\Omega$ and h_f 100. So h_i would in this case be just over $2000\,\Omega$.

For the purpose of deriving characteristic curves under the simple conditions provided by the apparatus used (Figure 10.8) this internal complication can be ignored. It makes no difference whether we attribute the measured results to the signal voltage applied or to the somewhat smaller voltage actually reaching the base-emitter junction plus the fed-back voltage. But under working conditions there is normally a non-zero load resistance, so i_c is less than $h_f i_b$, the difference (in a current-generator equivalent circuit) being diverted into the h_o by-pass. So the feedback is that much less, and the apparent extra resistance due to it is less, and the input resistance is less than h_i (which assumes $R = 0$). In short, *the input resistance of a transistor depends to some extent on the load resistance.*

Without the feedback effect the input resistance would be $r_b + r_e$. The extra resistance introduced by full feedback is $h_f r_e$. This could be represented by a voltage generator opposing the input signal. However, because R, if not exactly zero, is normally much less than $1/h_o$, it is much more convenient to start from $R = 0$ as the reference condition, allowing in h_i for full feedback, and to correct for the reduction in feedback due to non-zero R by imagining a voltage generator assisting the signal and thereby reducing the apparent input resistance, as in Figure 12.11.

The problem now is to find what this voltage is, in terms of R or something related to it. It must be equal to r_e multiplied by the reduction in current through it due to R not being zero. That reduction is equal to the current by-passed through h_o, which must be equal to $v_c h_o$. So the output of the voltage generator is $v_c h_o r_e$, in the direction shown. Now $h_o r_e$ is really a ratio between two resistances, r_e/r_c, and typically is a few ten-thousandths. It has been allocated the symbol h_r and named the reverse voltage feedback ratio.

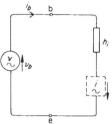

Figure 12.11 Consideration of Figure 12.10 shows that the input circuit of a transistor can be represented by a fixed resistance (h_i) in series with a voltage generator to simulate the feedback effect

As shown in Figure 12.11 it is not reverse, since it seems to be assisting the input signal to drive i_b. That is because we have departed from our practice of reckoning all voltages with respect to the common terminal, the emitter in this case. If we reverse the arrow the voltage of the generator must be negative with respect to the input, just as with the output voltage generator in a valve equivalent. The conventions shown in Figure 12.4 bring this right automatically, because v_c is always negative for positive input, so $h_r v_c$ must be negative, since h_r like all the other three h parameters is taken as positive. Again, either direction we care to give the output current is catered for in the arrowless notation. So in Figure 12.12 both systems are shown, as they were for valves (Figures 11.4, 11.5).

As $DA = h_f AB$, in the phasor diagram AB is in the opposite direction to AD (compare kg and kb in Figure 11.5b), and, as $ea = h_r ec$, ea is in the same direction as ec; and ab, representing the voltage applied to h_i to cause AB, is in phase with it.

12.7 Complete Transistor Equivalent Circuit

We now have a complete equivalent circuit; one of many, but worthy of a special place because its four h parameters are measurable practically. By the way, the h stands for hybrid, because they are a mixed bag: h_i is a resistance, h_o a conductance, h_f a current ratio and h_r a voltage ratio.

We have seen that when measuring h_i the load resistance must be kept at zero, or as near to it as makes little difference. This obviously applies also to measuring h_f so that all the current passes through the external measuring instrument. When measuring h_r, by applying a signal voltage to the output terminals and observing the resulting voltage (or more likely microvoltage) at the input terminals, no signal current must be allowed to flow in that circuit, or some of the voltage would be lost in r_b. Lastly, h_o can also be measured by applying a voltage to the output terminals and noting the current that goes in. We would again have to make sure that no feedback signal current flows in the input

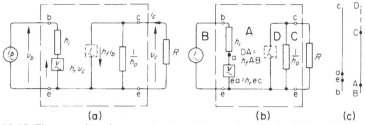

Figure 12.12 The complete *h*-parameter transistor equivalent circuit is shown at (a) in conventional form and at (b) adapted for phasor diagrams (c)

circuit, because an amplified current due for it would pass through h_o and confuse the measurement. (For the same reason, when plotting I_C/V_C curves for fixed values of I_B it may be necessary to adjust V_B slightly to keep I_B constant.)

Just as the input resistance is exactly equal to h_i only when there is zero output voltage and therefore nothing from the internal voltage generator, so the output conductance is exactly equal to h_o only when there is zero input current and therefore nothing from the internal current generator. (Note duality again.) And just as the effective input resistance is in general not h_i but depends on the load resistance (for that is what gives rise to the v_c or ea factor in the reverse feedback voltage), so the effective output conductance is in general not h_o but depends on the signal source conductance (for that is what gives rise to the i_b or AD factor in the forward transfer current).

Suppose we try to measure h_o and the condition of zero input source conductance is not fulfilled. Then the measuring current applied to the output, say by making ec positive so that it flows into c, develops a positive $h_i ec$ which drives a positive current BA round the input circuit. This is a negative AB, so AD is negative twice and therefore positive. Some of this current goes through h_o, in the same direction as that supplied. So (see end of Section 12.5) it makes the output conductance appear less than h_o. So the greater the load resistance the less the input resistance, and the greater the input source conductance the less the output conductance.

12.8 A Simpler Equivalent

It is because bipolar transistor input and output circuits react on one another in this way that calculating their circuit performance is more complicated than for valves and f.e.t.s. But if the load resistance, which we should now distinguish as R_L, is considerably less than $1/h_o$, as it very often is, and the input signal source or generator resistance (R_G) is considerably greater than h_i, as it often is or can be made, then the effect of h_r is small and its voltage generator can be omitted, as in Figure 12.13, and calculations are quite easy. Even if this simplification gave rise to errors of the order of 30% these could be acceptable if we kept in mind that the parameters of individual transistors of the same type are likely to differ even more widely. We shall see later (Section 13.1, for example) that one of the objects of good circuit design is to minimise the influence of transistor parameters on the results.

To get the feel of the thing let us try a few figures. Suppose h_i is $3\,\text{k}\Omega$, h_f is 80 and R_L is $4\,\text{k}\Omega$. The approximate current gain, A_i, is simply h_f, so is 80. The voltage gain is

$$A_v = \frac{v_c}{v_b} \simeq -\frac{i_c R_L}{i_b h_i} = -\frac{h_f R_L}{h_i} = -\frac{80 \times 4}{3} = -107$$

The input resistance, r_{in}, $\simeq h_i$, and the output conductance, g_{out}, would on our assumption be zero, but if h_o is known it might as well be used as an approximate g_{out}. And we get better approximations for A_i and A_v, without adding much to the work, by taking account of it. Suppose it is $25\,\mu\text{S}$ ($= 25 \times 10^{-6}\,\text{S}$). The load conductance, G_L, $= 1/R_L$, is 1/4000, or $250\,\mu\text{S}$. So the total conductance into which the output current goes is $275\,\mu\text{S}$, and the load's share being 250 our figure for A_i is reduced to $80 \times 250/275$, or 73. A_v is reduced in proportion to 97.

Compare these approximations with the results using the full equivalent circuit. At this stage the derivation of the

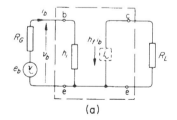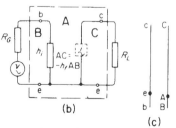

Figure 12.13 Simplified transistor equivalent circuit, good enough in many cases. The three diagrams correspond with those in Figure 12.12

full formulae would be rather too much, so here they are just stated:

$$A_i = \frac{h_f}{h_o R_L + 1} \quad A_v = \frac{-h_f}{h_i(h_o + G_L) - h_r h_f}$$

$$r_{in} = h_i - \frac{h_r h_f}{h_o + G_L} \quad g_{out} = h_o - \frac{h_r h_f}{h_i + R_G}$$

Suppose h_r is 5×10^{-4} and R_G is $5\,k\Omega$. The results of filling in the values in the above, with the previous approximate results for comparison, are:

	Without h_r or h_o	Without h_r	Full working
A_i	80	73	73
A_v	−107	−97	−102
r_{in}	3 kΩ	3 kΩ	2.855 kΩ
g_{out}	0	25 μS	20 μS

So the simplest diagram gives very good approximation (for this kind of work) to A_v, and with h_o gives correct A_i. r_{in} is not much less than h_i; but g_{out} is rather more affected by the finite value of R_G, though still not enough to worry about. Even without allowing for h_o the answers, obtainable in a few seconds, are good enough for a preliminary estimate.

Sometimes the values of r_b and r_e (Figure 12.10) are given instead of h_i and h_r. The working r_{in} is then $r_b + (h_f + 1) r_e$ as explained in Section 12.6, and as 1 is usually negligible compared with h_f the formula can be simplified to $r_b + h_f r_e$. Given h_f (and perhaps h_o) the current and voltage gains are calculable as before.

12.9 Other Circuit Configurations: Common Collector

So far we have considered only the circuit configuration or mode in which the emitter (or cathode, or source) is the electrode common to both input and output. As a triode has two other electrodes there are two other possibilities—unless we make three more by interchanging output and input, but as these do not amplify they are not usually counted.

Figure 12.14 shows the three useful ones as applied to a bipolar n-p-n transistor. With the feed batteries reversed the same diagrams would apply to a p-n-p transistor. And there are corresponding varieties for f.e.t.s and valves. Although in b the junction between input and output circuits is not actually at the collector (since the collector feed source comes in between) so far as signals are concerned an ideal source is non-existent and is omitted from our theoretical diagrams.

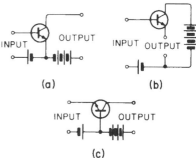

Figure 12.14 The three useful transistor transfigurations, or modes of use: (a) common-emitter, (b) common-collector, and (c) common-base

One of the advantages of the h parameters, already noted, is that they are suited to practical measurement. Another advantage is that, being related to input and output terminals and not to parts of the transistor itself, they are not affected in principle by the way the transistor is connected between those terminals. For the same reason, neither are the formulae for A_i, etc., given in the previous section. But because the h parameters are measured between different points of the transistor in three different modes their values are different and so therefore are A_i and the other derived properties. To prevent confusion all the h parameters are given an additional subscript letter to denote the common electrode. So what we have been calling h_f is more explicitly h_{fe}, and similarly for the others.

The common-emitter mode (Figure 12.14a), having been assumed in all we have done on transistors so far, should already be fairly familiar. It is also the most used. So perhaps the best line to take with the other two is to find how the h values, and consequently the circuit properties, compare.

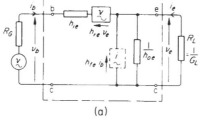 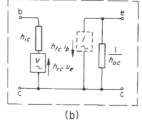

Figure 12.15 (a) Common-emitter transistor equivalent circuit (Figure 12.12a) redrawn to fit the common-collector terminal arrangement. The standard h-parameter circuit (b) can be made equivalent to this by changing the values of the parameters where necessary

Beginning with the common-collector mode we redraw our common-emitter equivalent circuit (Figure 12.12) to fit the common-collector input-output terminal scheme, Figure 12.15a. Our object is to find the values of h_c parameters that make the standard h form of circuit (b) equivalent to this.

The simplest way of regarding the change from common-emitter to common-collector is as merely a transfer of the load from collector to emitter. So whenever the output terminals are short-circuited (as they are for measuring h_f and h_i) the two circuits are exactly the same. As v_e is then zero, $h_{re}v_e$ is likewise, and to the input Figure 12.15 consists only of h_{ie}. So

$$h_{ic} = h_{ie}$$

It would be falling into a trap to assume without further thought that $h_{fc} = h_{fe}$, for if we look at the diagrams again we see that when the output terminals are shorted not only does the whole of $h_{fe}i_b$ flow that way but i_b as well. Moreover they flow towards the common terminal (c) instead of away from it as in Figure 12.12a. So

$$h_{fc} = -(h_{fe} + 1)$$

Remembering that the positive h_{fe} went along with the fact that the output was reversed relative to the input, would we be correct in interpreting the negative h_{fc} as giving an output of the same sign as the input? We would indeed, as we can check by putting a positive input signal on Figure 12.14b; it increases the emitter current and so makes the emitter more positive.

For measuring h_o, R_L is removed and the input terminals are open circuited (to signals) so that there is no i_b and the v_e and i_e used in the test see only h_{oe}. So

$$h_{oc} = h_{oe}$$

Where the most drastic difference comes is in regard to h_{rc}, the reverse voltage transfer ratio. Multiplied by the output voltage across the load, it tells us the voltage fed back to the input. If we look again at Figs. 12.14b and 12.15 we see that the whole of the output voltage is fed back to the input. So instead of being usually a few parts in 10 000 like h_{re},

$$h_{rc} = 1$$

We must note again, however, that because the output voltage for a positive input is positive instead of negative the fed-back voltage, $h_{rc}v_e$, is positive, so opposes v_b around the input circuit, and instead of slightly reducing the zero-R_L input resistance it greatly increases it. We can quite easily find how much it does so. If only i_b were flowing through $1/h_{oe}$ and R_L in parallel, their combined resistance would seem to the input to be $1/(h_{oe} + G_L)$, where R_L is transformed into a conductance for easy addition. But we know that $h_{fe} + 1$ (or $-h_{fc}$) times as much current is passing through in the same direction, so its resistance seems like $(h_{fe} + 1)/(h_{oe} + G_L)$, or $-h_{fc}/(h_{oe} + G_L)$, besides which h_{ie} can

usually be neglected. With $h_{rc} = 1$, this agrees with the formula for r_{in} given in the previous section.

To compare the configurations let us use the formulae on the same data as before, namely

$h_{ie} = 3$ kΩ   so h_{ic} = 3 kΩ
$h_{oe} = 25$ μS h_{oc} = 25 μS
$h_{fe} = 80$ h_{fc} = −81
$h_{re} = 5 \times 10^{-4}$ h_{rc} = 1
$R_L = 4$ kΩ and R_G = 5 kΩ

The results are

$$A_i = \frac{-81}{0.1+1} = -73.6$$

$$A_v = \frac{81}{0.82 + 81} = 0.99$$

$$r_{in} = 3000 + \frac{81}{(25+250) \times 10^{-6}} = 298\ \text{k}\Omega$$

$$g_{out} = 25\ \mu S + \frac{81 \times 10^6}{8000}\ \mu S$$

$$= 25 + 10\ 125\ \mu S = 10.15\ \text{mS}$$

or $r_{out} = 98.6\ \Omega$

Some intermediate steps are shown here, to give an idea of how much the various data contribute to the results. Provided that R_L is not abnormally small, h_{ic} is a negligible part of r_{in} compared with the amount due to the 100% voltage feedback. So where a far higher input resistance is needed than a common-emitter connection can provide, the common-collector is indicated. This goes along with a far lower output resistance than $1/h_{oe}$ alone. The current gain is practically the same, but the voltage 'gain' is less than 1, which means that there is a loss. Admittedly the loss is very small in any practical case, but one would not choose common-collector for voltage gain! Nor is it any better than common-emitter for current gain, but it is used quite a lot for matching high to low resistance.

The reason for A_v being always less than 1 is very clear in the phasor diagram. We could draw phasor diagrams corresponding to Figures 12.12 and 12.15b but they will be clearer if they are referred to the actual transistor circuits, as in Figure 12.16. (Helpful though equivalent circuits are as a logical basis for calculations, we must never become so at-home in their world of make-believe that we forget that they are not real circuits, any more than the algebraical symbols.) The phasor diagrams for common-emitter and common-collector turn out to be identical. The only distinction we can make is to show which is the common terminal by encircling its point. We must remember that so far as the transistor is concerned the working input terminals are always e and b. In Figure 12.16a these also happen to be the external input terminals, but in Figure 12.16b the input terminals embrace the output voltage too, so the gross input must always be much more than the output needed to provide the working input. This is a result of 100% voltage feedback. The voltage phasors in Figure

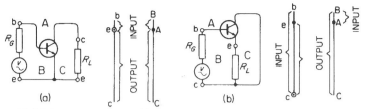

Figure 12.16 Basic transistor amplifier circuits and their phasor diagrams: (a) common emitter: (b) common-collector. Note that the phasor diagrams themselves are identical

12.16 show that whereas in *a* the input and output are in opposite directions and very different in magnitude, in *b* the emitter output potential is nearly the same in both magnitude and direction as the input, so a transistor in this mode is often called an *emitter follower*. The emitter (output) potential follows that of the base (input). Similarly the common-anode valve is called a cathode follower.

Substituting various values of R_L and R_G in the formulae shows that the input and output resistances depend very much on them. When we have dealt with the common-base mode we shall look at some comparative curves for all three modes.

12.10 Common Base

Although now perhaps the least used of the three modes, historically this was the first. Whereas the common collector has 100% voltage feedback, common base has 100% current feedback. The consequences can be seen if we follow the same plan as in the last section.

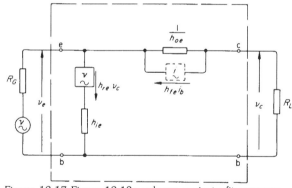

Figure 12.17 Figure 12.12a redrawn again to fit common-base terminals

Figure 12.17 shows the common-emitter equivalent circuit redrawn to fit the common-base terminal scheme. We begin again by short circuiting R_L. Then, the basic physical behaviour of a transistor is that any input current i_e going towards the emitter divides inside the transistor, a small part going direct to the base, and h_{fe} times as much going via the collector. So for every $(h_{fe} + 1)$ microamps of input, the output is h_{fe} microamps, slightly less. This mode therefore gives a small current loss. This current 'gain' is also negative in phase because the output current flows in the opposite direction to that from the h_{fe} generator in Figure 12.17. So

$$h_{fb} = -\frac{h_{fe}}{h_{fe} + 1}$$

The historical symbol for h_{fb}, still sometimes used, is α, corresponding to β for h_{fe}.

As the shorted R_L makes v_c zero, we can ignore the internal voltage generator, but the effect of the current generator, as we have just seen, is to cause $(h_{fe} + 1)$ times as much input current to flow as could be accounted for by the resistance h_{ie} alone. So

$$h_{ib} = \frac{h_{ie}}{h_{fe} + 1}$$

Next, we leave the input open, remove R_L and substitute a source to make current flow into the collector. As $i_e = 0$, the same current must flow out from the base, via h_{ie}. Positive i_b is reckoned into the base, so $i_b = -i_c$. So for every microamp of i_c we sent through $1/h_{oe}$ from outside, the internal current generator sends $(h_{fe} + 1)$ times as much v_c as can be accounted for by i_c alone, and therefore its resistance appears to be $(h_{fe} + 1)/h_{oe}$. Compared with this, the effects of the other two items in circuit are negligible, and

$$h_{ob} \simeq \frac{h_{oe}}{h_{fe} + 1}$$

While we are applying v_c we should notice that the equivalent circuit forms a potential divider. Using double-subscript notation,

$$v_c = v_{bc} = v_{be} + v_{ec}$$

The purpose of the parameter h_{rb} is to indicate what proportion of v_{bc} appears as v_{be}:

$$h_{rb} = \frac{v_{be}}{v_{bc}}$$

We have just found that the resistance between c and e is $(h_{fe} + 1)/h_{oe}$, and the resistance between e and b is h_{ie}. So the part of h_{rb} due to these resistances is

$$\frac{h_{ie}}{(h_{fe} + 1)/h_{oe}} = \frac{h_{ie}h_{oe}}{h_{fe} + 1}$$

Figure 12.17 shows that there is another voltage contributing to v_{be}; it is $-h_{re}v_c$. Divided by v_c it is $-h_{re}$. So the total h_{rb} is

$$h_{rb} = \frac{h_{ie}h_{oe}}{h_{fe}+1} = -h_{re}$$

With the same data as before for an example, the results of using the above formulae are

$$h_{ib} = 37\,\Omega \qquad h_{ob} \approx 0.31\,\mu S$$

$$h_{fb} = -0.9875 \qquad h_{rb} = 4.26 \times 10^{-4}$$

$$A_i = \frac{-0.9875}{0.00124 + 1} = -0.986$$

$$A_v = \frac{0.9875}{(9.26 \times 10^{-3}) + (0.42 \times 10^{-3})} = 102$$

$$r_{in} = 37 + 1.68 = 38.7\,\Omega$$

$$g_{out} = 0.31\,\mu S + 0.084\,\mu S = 0.394\,\mu S$$

or $r_{out} = 2.54\,M\Omega$

We see that unless R_L is exceptionally large A_i is practically the same as h_{fb}, and with the same stipulation we can neglect h_{ob} compared with G_L, and can also neglect $h_{rb} h_{fb}$, so that

$$A_v \approx \frac{R_L}{h_{ib}}$$

that is to say, the ratio of load resistance to zero-R_L input resistance. And if R_G is large, g_{out} is not much more than h_{ob}, but in practice there could be an appreciable difference. Note the much lower r_{in} and g_{out} than with common-emitter. The contrast with common-collector is even greater. So choice of mode (and to some extent of R_G and R_L) gives one a wide range of control over input and output resistances of a transistor.

We do not really need another diagram to compare with Figures 12.12a and 12.15b, for the only differences would be the terminal labels and the second subscripts to the h parameters. But Figure 12.18 is for comparison with Figure 12.16, and again we see the phasor diagram is the same except for the encircled point marking the common

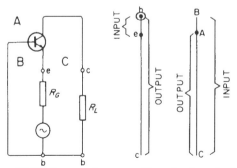

Figure 12.18 Common-base amplifier circuit and phasor diagrams for comparison with those for the other modes (Figure 12.16)

electrode. The voltage diagram shows at once that only common-emitter gives a signal inversion. Study of the three phasor diagrams helps one to see how the common-emitter gives both voltage and current gain, whereas common-collector gives only current gain and common-base only voltage gain. The obtainable power gain of the common-emitter mode is therefore obviously much greater than that of the other two, which accounts for its popularity. But it must be re-emphasised that we are assuming low-frequency signals.

Note that A_i and A_v are in terms of the input current and voltage actually reaching the input terminals, not those supplied by the input current or voltage signal generator. So although in our numerical example the common-base mode had the same A_v as the common-emitter, the higher input current with common-base would mean far greater loss in the generator resistance, R_G.

12.11 A Summary of Results

Perhaps the best way of taking in the results of the last three sections is to look at Figure 12.19. This shows how current, voltage and power gain, and input resistance, of our typical transistor vary with load resistance over the range $100\,\Omega$ to $1\,M\Omega$, and also how output resistance depends on the resistance of the input signal source over the same range. To cover this wide range, and the ranges of gain, etc., the scales are logarithmic.

The curves tell the story without further comment, but three qualifications should be underlined: these curves are derived from one particular set of transistor parameters; they are small-signal parameters; and they are restricted to frequencies low enough for reactances to be neglected. But the general shapes of the curves are affected hardly at all by substituting other values of the parameters typical of transistors intended for small-signal amplification. We shall see later (in Chapter 25) that transistors serve other purposes, for which h parameters are not the most helpful approach.

To complete this summary, here follow the formulae and values used for plotting the curves in Figure 12.19. You may notice that for values of R_L within the range $100\,\Omega$ to $10\,k\Omega$ the curves are either almost flat or sloping at 45°, for which simpler approximate formulae given at the end are good enough.

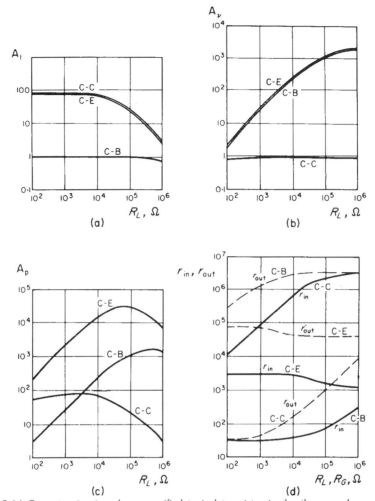

Figure 12.19 (a) Current gain given by a specified typical transistor in the three modes, related to load resistance; (b) voltage gain related to load resistance; (c) power gain related to load resistance; (d) input resistance related to load resistance and output resistance related to signal generator resistance

h_i = input resistance with output shorted (i.e., $R_L = 0$)
h_o = output conductance with input open (i.e., $G_G = 0$)
h_f = forward current transfer ratio
h_r = reverse voltage transfer ratio
R_L = load resistance. G_L = load conductance = $1/R_L$
R_G = external resistance between input terminals
$G_G = 1/R_G$

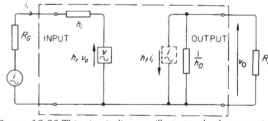

Figure 12.20 This circuit diagram illustrates the h parameters and associated symbols

Transistor Equivalent Circuits

	TYPICAL VALUES					
	common-emitter		*common-collector*		*common-base*	
h_{ie}	3 kΩ	$h_{ic} = h_{ie}$	3 kΩ	$h_{ib} = \dfrac{h_{ie}}{h_{fe}+1}$	37 Ω
h_{oe}	25 μS	$h_{oc} = h_{oe}$	25 μS	$h_{ob} \approx \dfrac{h_{oe}}{h_{fe}+1}$	0.31 μS
h_{fe}	80	$h_{fc} = -(h_{fe}+1)$	−81	$h_{fb} = -\dfrac{h_{fe}}{h_{fe}+1}$	0.9875
h_{re}	5×10^{-4}	$h_{rc} = 1$	1	$h_{rb} = \dfrac{h_{ie}h_{oe}}{h_{fe}+1} - h_{re}$	4.26×10^{-4}
$h_{re}h_{fe}$	0.04	$h_{rc}h_{fc}$	−81	$h_{rb}h_{fb}$	-4.21×10^{-4}

Current gain, $A_i = \dfrac{h_f}{h_o R_L + 1}$

Voltage gain, $A_v = \dfrac{-h_f}{h_i(h_o + G_L) - h_r h_f}$

Power gain, $A_p = A_i A_v$

Input resistance, $r_{in} = h_i - \dfrac{h_r h_f}{h_o + G_L}$

Output conductance, $g_{out} = h_o - \dfrac{h_r h_f}{h_i + R_G}$

Output resistance, $r_{out} = \dfrac{1}{g_{out}}$

	APPROXIMATIONS (R_L or $R_G = 10^2 - 10^4$ unless stated)		
	common-emitter	*common-collector*	*common-base*
A_i	h_{fe}	h_{fc}	1 (RL up to 10^5)
A_v	$h_{fe}\dfrac{R_L}{h_{ie}}$	1 (R_L above 10)	$h_{fe}\dfrac{R_L}{h_{ie}}$
A_p	$h_{fe}^2\dfrac{R_L}{h_{ie}}$	A_i ($R_L\ 10^3 - 10^4$)	A_v (R_L up to 10^5)
r_{in}	h_{ie}	$-h_{rc}h_{fc}R_L$	h_{ib}
g_{out}	h_o (R_G above 3×10^3)	$-\dfrac{h_{rc}h_{fc}}{h_{ic}+R_G}$	h_{ob} (R_L above 3×10^3)

12.12 F.E.T.s and Valves

With their near-infinite low-frequency input resistance, field-effect transistors resemble valves much more than they resemble the bipolar transistors we have been considering, so that the same simple equivalent circuits and gain calculations studied for valves in Chapter 11 will do for f.e.t.s. If the valves are of the pentode type the output-current/output-voltage curves are very similar too and in this respect they both differ little from bipolar transistors except for the greater voltage before the nearly flat (high-resistance) parts are reached. Unless the contrary is mentioned, operation is assumed to be confined to these parts. Because of the normally high output resistance, the equivalent current generator is usually preferred to the voltage generator.

Under the simple low-frequency small-signal conditions assumed, then, the bipolar parameters h_r and h_i are not applicable, nor (except in common-gate or common-grid) are A_i and A_p, nor h_{in}. And g_{out} is simply g_a for valves (= $1/r_a$) and g_d for f.e.t.s. These correspond to h_o for bipolar transistors. The most common parameter is g_m, the mutual conductance or ratio of short-circuit output signal current to input voltage (Section 10.3). To bring it more into line with bipolar transistor nomenclature, it is alternatively called g_{fs}, the subscripts indicating 'forward' and '(common) source'.

This is a reminder that so far we have considered valves and f.e.t.s only in their common-cathode or common-source mode. Although it is the most used, the other two also have their uses. Their circuit diagrams are of course the same as those in Figures 12.14, 12.16 and 12.18 except for device symbols and electrode initials. So too are the phasor diagrams if the current BA is reduced to zero.

Figure 12.21 compares the equivalent circuits for the three configurations; all three are actually the same circuit internally, only the positions of the terminals and the ways the source and load are connected being different. In b and c the current phasors are reversed, but this is not a discrepancy; it arises because the mesh labels A, B and C cannot be arranged so that the current directions they indicate are the same in both actual and equivalent sets of diagrams. In Figure 12.21b and c, AB is opposite to AB in a, so to agree with physical behaviour the minus sign is omitted from b and e.

Comparing now the common-drain and common-anode modes, b (also known as source follower and cathode follower for the reason given in Section 12.9) there are two main effects, both similar to those already studied in bipolar transistors: the voltage gain is reduced to a little less than 1, and the effective output resistance is also greatly reduced. The reason for the first effect is exactly the same as before: the whole of the output voltage is fed back in opposition to the input, which must therefore be augmented equally to overcome it. The reason for the low output resistance can be seen by using the same method as before: substituting for R_L a source of signal voltage, which in this case would be called ds, and noting the resulting current. To ensure again that no current comes from the equivalent generator the input sg must be zero, and as this time it is a voltage instead of a current the no-signal condition is obtained by *short*-circuiting the input terminals. In diagram a under this condition we would encounter only the resistance $1/g_d$. But in b the test voltage ds comes reversed between s and d, so $sg = sd = -ds$, and the generator consequently gives a current $AB = sd.g_m$. This means the same as $BA = sd.g_m$, and this current flows in the same direction as the current $ds.g_d$ directly due to ds. So the total current is $ds.g_d + ds.g_m$, and the total conductance is therefore

$$g_{out} = g_m + g_d$$

Usually g_d is negligible compared with g_m, so it is near enough to say that the output conductance is g_m. A typical value is 3 mS, which is a resistance of 333 Ω.

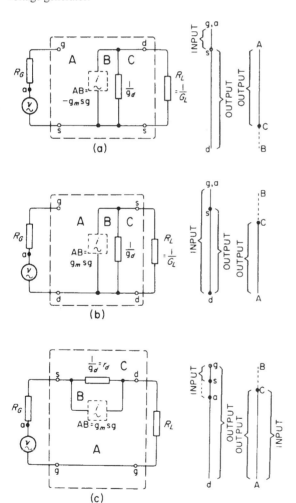

Figure 12.21 Field-effect transistor equivalent current-generator circuits and corresponding phasor diagrams (compare Figures 11.5 and 12.5): (*a*) common-source, (*b*) common-drain, and (*c*) common-gate

The voltage 'gain' can quite easily be worked out; it can be put in various forms, of which perhaps the simplest and clearest is

$$A_v = \frac{g_m}{g_m + g_d + G_L}$$

in which g_d is usually small enough to neglect, and G_L can be made fairly small, so that A_v is only slightly less than 1.

Both the above results apply to valves if g_a is substituted for g_d.

Lastly, Figure 12.21c shows very clearly that input and output are in series, so the same current passes through both and $A_i = 1$. Input and output both affect one another, R_G comes into the account and the situation is nearly as complicated as with a bipolar transistor. We can see from the generator equation that current is injected into the circuit in the same direction as that due directly to the signal and g_m times as much. A little of it is lost in g_d, but most of it flows right round the circuit. So the apparent input resistance is much less than $(r_d + R_L)$. It is in fact

$$r_{in} = \frac{r_d + R_L}{g_m r_d + 1} = \frac{r_d + R_L}{\mu + 1}$$

The output resistance is measured with no input voltage g_a, but this does not mean that the internal current generator is doing nothing, for the circuit current flowing through R_G sets up an input voltage. The internal generator current due to this sets up a voltage across r_d opposing the applied voltage g_d, so the effect of R_G is to increase r_{out}, which comes to

$$r_{out} = r_d + R_G(g_m r_d + 1) = r_d + R_G(\mu + 1)$$

The safest way of arriving at these results is to apply Kirchhoff's laws to the circuit and solve the equations so obtained. If we continue this work to find the voltage gain, as before in terms of the input voltage reaching the input terminals, we get

$$A_v = \frac{R_L(g_m r_d + 1)}{R_L + r_d} = \frac{R_L(\mu + 1)}{R_L + r_d}$$

If we allow for the loss in R_G and reckon the input as g_a, we get a smaller gain; call it A_v:

$$A_v = \frac{R_L}{R_G + \frac{R_L + r_d}{g_m r_d + 1}} = \frac{R_L}{R_G + \frac{R_L + r_d}{\mu + 1}}$$

All the above results are given alternatively in terms of μ, which is often known, for valves at least, in which cases of course r_a would take the place of r_d.

As an example let us assume $g_m = 3$ mA/V, $r_d = 30$ kΩ, $R_L = 10$ kΩ and $R_G = 0.5$ kΩ. Then

$$r_{in} = 0.44 \text{ k}\Omega \ (440 \ \Omega);$$

$$r_{out} = 75.5 \text{ k}\Omega; \ A_V = 22.75;$$

$$A_v = 10.65$$

We can review this section by looking again at the phasor diagrams in Figure 12.21. The chief advantage of f.e.t.s and valves over bipolar transistors is the absence of input current in a and b, but in c this advantage is sacrificed, so common-gate and common-grid are less attractive than the other modes.

Having now achieved complete transistor and valve equivalent circuits, at least for frequencies at which reactances can be neglected, we may find this a good moment for pausing to realise that within the limits of the assumptions made we can represent all actual circuits by equivalent circuits made up of only five different elements: resistance, capacitance, inductance, voltage generators and current generators. Even this short list could be reduced, because voltage and current generators are alternatives, and at any one frequency inductances and capacitances can be expressed as positive and negative reactances.

So much attention having been given to the rather artificial concepts of small signal equivalent circuits, before turning to applications of transistors and valves it will be as well to consider their working requirements; in a word, feeds.

Chapter 13 The Working Point

13.1 Feed Requirements

We have been studying at some length the parameters and performance of transistors and valves, but of course these qualities cannot be obtained unless the devices are fed with d.c. power adjusted to achieve a suitable working point. What is suitable depends on the kind of work they will be required to do, and for a start we shall continue to assume low-frequency low-power amplification. Modifications for other kinds of work will appear when we come to them.

First of all let us deal with n-p-n transistors. The same information can be applied to p-n-p types if all polarities are reversed. F.e.t.s and valves will follow.

As we have seen, the emitter-to-base junction is a diode which under working conditions is forward biased. The easily remembered rule for this is that positive bias supply goes to the *p* side of the junction. The collector-to-base junction is always reverse biased. (A transistor as a whole is more than just two diodes connected together, of course.) With reference to the emitter, the collector and base biases therefore have the same polarity. In this respect they differ from valves and most f.e.t.s, in which the control electrode is reverse biased (negative to gate or grid). The valve diode, which has the cathode as one of its electrodes, is forward biased, with positive to anode. (With valves at least, the term 'bias' is usually restricted to the grid supply.)

Figure 13.1 summarises these requirements. It also shows the symbols used for feed voltages (and currents, where applicable): a capital V (or I) with double capital subscripts to indicate the electrode for which it is intended, followed if necessary by a third letter to indicate the reference electrode. (Note, this is opposite to the convention used for phasor diagrams.) K rather than C is used for cathode, as C is reserved for collector or capacitance. As anything to do with signals is omitted, there is no need to show common-collector, etc., arrangements separately; so far as feeds are concerned they are the same as common-emitter, though the actual voltages required may differ. For the same reason the feed voltages look the same as those actually applied to the electrodes, but where loads, etc., intervene, the electrode voltages are distinguished by single subscripts as used in previous chapters.

Although two batteries are shown for each device, arrangements are usually made for both electrodes to be fed from one source. The procedure for obtaining device data, plotting characteristic curves and drawing load lines, and thereby deciding on a suitable working point, by which the feed voltages and currents are specified, has already been

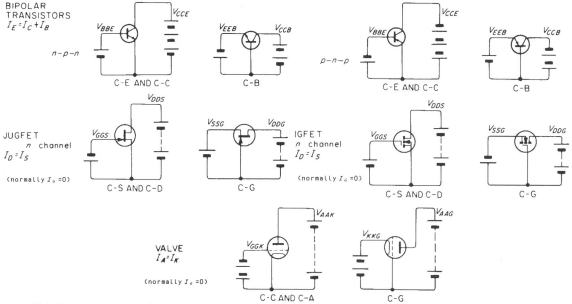

Figure 13.1 Comparison with polarities and comparative magnitudes of feed voltages needed by various amplifying devices in their three configurations

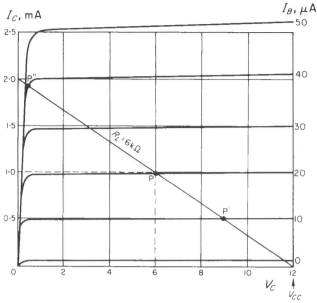

Figure 13.2 Typical I_C/V_C curves and load line for low-power transistor. The designed working point (P) may be drastically altered (to P' or P'', for example) by changes in h_{fe}

described in Sections 10.1 and 10.8. What we now have to realise is that this procedure takes far too much time to put into practice for every individual transistor. Yet individual devices, even when they bear the same type number, differ so widely that their feeds seem to need individual tailoring. Even if they got it, satisfaction could still not be guaranteed, for temperature changes can upset the most carefully adjusted working points.

The answer to this problem is to design the circuit so that any changes tending to displace the working point are automatically counteracted to a sufficient extent to maintain performance reasonably close to that intended. To do this we must know about the causes of working point displacement.

13.2 Effect of Amplification Factor Variations

Figure 13.2 shows a typical I_C/V_C curve sheet with load line. P has been selected as the working point because it is in the middle of the usable part of the load line, so that positive and negative input half-cycles, each of 20 μA peak value, could be accepted and would yield apparently equal output half-cycles, each of about 5½ V peak.

The curves tell us that for each increase of 10 μA in I_B the collector current increases by about 500 μA, so h_{fe} is about 50. If this was classed as 'typical', for others of the same type it could range from 25 to 100. With a 25 h_{fe} the 20 μA I_B line would be about where the 10 μA line is in Figure 13.2, so the working point would be displaced to P', halving the maximum signal that could be handled without cutting off its peaks. With a 100 h_{fe} sample the 20 μA line would be about where the 40 μA line is in Figure 13.2, so the working point would be P'' and while negative half-cycles could be handled splendidly the positive halves would be cut off almost completely, since V_C is only a fraction of a volt. It looks as if our aim should be to maintain the designed working collector current rather than base current.

13.3 Influence of Leakage Current

Although leakage current came into Chapter 10 it has been ignored since, because it has no place in equivalent circuits that are concerned only with signals. Even now it may not appear to be important, for it is represented in graphs like Figure 13.2 by the height of the zero-I_B line above the zero-I_C line, and unless it is exaggerated it is often too small to show clearly, even for a germanium transistor, let alone silicon. All it seems to do is shorten to an insignificant extent the useful part of the load line at the high-V_C end. But that is leaving temperature out of account. And we must remember that the temperature concerned is that of the junction itself, which is affected not only by the external ('ambient') conditions but by the electrical power dissipated in the transistor. Temperature is what limits the power a transistor can handle, and the reason is the rapid increase in leakage current that it causes.

There are actually three leakage currents, depending on which of the three electrodes is open-circuited, or at least reduced to zero current. Of these, I_{EBO}, the current from emitter to base with collector unconnected, is unimportant, because this diode is normally forward biased. I_{CBO} is the reverse current of the collector-base diode, as shown in Figure 13.3a, and is the one most often given on data sheets. The leakage current represented by the $I_B = 0$ line in Figure 13.2 is I_{CEO}, which passes through both diodes (Figure 13.3b) and is the most important in practice. Note that the reverse voltage used for measuring leakage current is not specified. Provided it is not less than say 1 V and is not high enough to threaten breakdown, its value hardly matters, as Figure 13.2 shows.

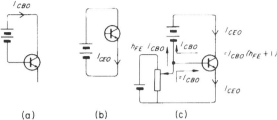

(a) (b) (c)

Figure 13.3 For the elucidation of leakage currents in transistors

Thoughtful readers looking at Figure 13.3b may be puzzled. It doesn't seem to conform to the rule that $I_C = h_{FE}I_B$, for here I_B is zero but I_C is not. Observant readers will have noticed that although the voltage between collector and base in Figure 13.3a has the normal working polarity for an n-p-n transistor the base current is flowing the wrong way for normal operation. If an adjustable bias were added between base and emitter, as at c, there would be an adjustment at which the currents into and out from the base were equal, leaving zero net current and so reproducing the conditions of diagram b. The current going into the base, which we have made equal and opposite to I_{CBO}, by transistor action causes h_{FE} times as much current to flow to the collector. The total collector current, which in these circumstances is by definition I_{CEO}, is thus equal to $I_{CBO}(h_{FE} + 1)$. It is therefore many times larger than I_{CBO}. Note the capital letter subscripts in h_{FE}, signifying that it is the d.c. no-signal ratio I_C/I_B, not the signal ratio i_c/i_b as in the last chapter. h_{FE} and h_{fe} are usually of the same order, at least; but both depend somewhat on the working point and temperature.

A typical value of I_{CBO} for a low-power germanium transistor at 25°C is 3 μA. I_{CEO} could therefore be something like 200 μA or 0.2 mA. As the leakage doubles for about every 9°C, at 70°C this would have risen to almost 6 mA. In a case such as that shown in Figure 13.2, lifting the whole array of I_B lines so that the zero one was at about 6 mA I_C would result in the working point being buried in the nearly vertical saturation part of the curves, putting the transistor virtually out of action. Negative base bias would be needed.

Again, it seems clear that stabilising collector current, rather than base current, should be the aim.

13.4 Methods of Base Biasing

The simplest method of providing base bias current is from the collector supply through a resistor, R_B in Figure 13.4. Suppose, for example, that the required bias is 30 μA and V_{CCE} is 9 V; then R_B should be 9/30 = 0.3 MΩ. In this calculation the difference between the potential of the base and that of the emitter has been neglected. As we noted in Section 9.12, the forward voltage bias for a germanium diode is about 0.2 V and for silicon 0.6 V, and because the current rises steeply (see Figure 10.13, for example), which means that the g_m of bipolar transistors is high, the working range of bias voltage is quite small and never far from the above values. For d.v. calculations the base-to-emitter voltage can therefore be assumed to be approximately these values. Anyone who wanted to calculate R_B more accurately by deducting the appropriate p.d. from V_{CCE} could do so, but he would be wasting his time because nothing about transistors is as precise as that. In any case in practice the choice of resistance of resistors is confined to 'preferred values'. Furthermore, this method of biasing is unsuitable for most practical circuits, because the required R_B is generally much larger than any variations of the input resistance of the transistor, so in conjunction with V_{CCE} is virtually a constant-current generator (Section 12.1). The last two sections have shown that constant bias current is unlikely to maintain a satisfactory working point against changes in temperature and transistor parameters.

Figure 13.4 does, however, show two capacitors, which are usually necessary to prevent the base and collector potentials from upsetting or being upset by whatever is connected to the input and output terminals. At the same

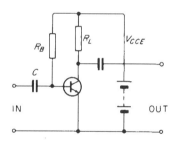

Figure 13.4 A simple but not entirely satisfactory method of providing base bias for an amplifying transistor

time, if of sufficient capacitance in relation to the resistances, they allow signals to pass freely (Section 14.6).

Because the full working range of collector current results from varying the base voltage within a range which typically is only a fraction of a volt (Figure 10.12), establishing the working point by using a constant-voltage source might seem to be even more unlikely to succeed. And so it would be if such bias were introduced directly between emitter and base. Adjustment of the bias voltage would be extremely critical, and when carried out for one transistor might well be quite unsuitable for others of the same type. But the very steep control of collector current by base voltage—the reason for dismissing this method—is actually the basis for a modification which is the most popular of all base biasing methods.

Figure 13.5 shows an experimental circuit in which a resistor R_E has been introduced in the emitter lead. Suppose R_E is of the order of 1 kΩ. With the base voltage control slider at the bottom, $V_B = 0$ and the transistor is 'cut off'.

Now raise the slider gradually. At about 0.6 V (for a silicon transistor) a base current begins to flow, causing a much larger emitter current. Flowing through R_E, this I_E causes the emitter to become more positive, reducing the working bias of the base relative to the emitter. So I_E is not as much as it would have been without R_E. As V_B is increased, V_E

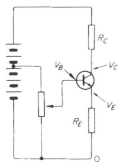

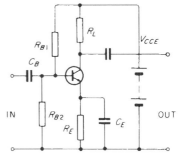

Figure 13.6 This base-biasing scheme is more stable than the one shown in Figure 13.4 and is usually preferred

The extra capacitor C_E (which is not invariably used) is to short-circuit R_E to signal currents.

There are other methods of biasing, such as keeping the base at zero and biasing the emitter negatively, but this calls for an extra current-supplying source and is not always convenient. The only other one that perhaps deserves showing now is outlined in Figure 13.7. The object of connecting R_B to the collector instead of as in Figure 13.4 is to compensate in a simpler way for changes in I_C. If I_C is too large, it increases the voltage drop in R_L

Figure 13.5 Experimental circuit for studying base biasing

follows close behind all the way, for this is a form of emitter-follower (Section 12.9). Even for quite a large I_E the increase of V_{BE} above the initial 0.6 V or so is quite small, so we are not far out in saying that V_E is $V_B - 0.6$ over the whole range of adjustment, at least until saturation is reached (V_B and V_C nearly equal). Note that the value of R_E does not enter into this. So long as it is not too small relative to R_C it is not very important. At a fixed setting of V_B, V_E remains almost constant regardless of R_E or the amount of current drawn. The arrangement is a form of nearly constant voltage generator. It is also a basic circuit that has other applications. But just now we are interested in it because the amount of emitter current is fixed within fairly close limits by the setting of V_B and the value of R_E. From the foregoing

$$I = \frac{V_E}{R_E} \approx \frac{V_B - 0.6}{R_E}$$

for silicon; for germanium substitute 0.2 for 0.6. I_C is of course nearly the same as I_E.

Substituting a transistor having a different value of h_{FE}, and therefore demanding a different value of I_B to maintain the desired I_C, merely causes I_E to change to the small extent needed to adjust V_{BE} sufficiently to change to the new I_B. The larger V_E is, the smaller the necessary changes in I_E and the less any differences in actual V_{BE} required by different transistors matter.

Figure 13.6 shows how this biasing scheme is usually arranged in practice. Success depends on V_B being reasonably constant, so the values of the potential dividing resistors R_{B1} and R_{B2} should not be so high that the potential of the base is affected to a substantial extent by changes in base current. In other words, the current flowing through them when the base is disconnected should be at least several times I_B. But we do not want the resistances to be so low that they waste a lot of power from the collector supply, or reduce the input resistance enough to cause serious loss of signal.

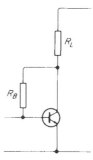

Figure 13.7 This base-biasing circuit is much simpler than Figure 13.6 and does have stabilising action, but owing to certain disadvantages is less used

and so reduces the voltage applied to R_B, reducing I_B and therefore I_C; and oppositely if I_C is too small. There are several disadvantages, however, one of them being that signal as well as feed current is fed back from collector to base. For these reasons the circuit is not used in this simple form, but in more sophisticated circuits using pairs of transistors (described in a later chapter) these drawbacks are avoided. So successful are these circuits that i.c.s using them can maintain a satisfactory performance at temperatures as low as $-15°$ C and as high as $+70°$ C.

Common-collector transistors call for no separate attention as they automatically provide compensated bias of the Figure 13.6 type to a maximum degree.

There is no difference in principle in biasing transistors in the common-base mode; for example, Figure 13.6 can be converted by short-circuiting the input terminals (so that C_B earths the base to signals), the input signal source being connected between emitter and R_E.

13.5 Biasing the F.E.T.

The shape of the f.e.t. output characteristic curves (Figure 10.15, for example) is similar to those for bipolar transistors, and so is the drawing of load lines and finding a working point, allowance being made for the higher saturation or bottoming voltage; about 5 V in the example referred to. (Drain voltages are usually higher than collector voltages, too.) But when biasing a depletion-mode device we must remember that instead of beginning at the bottom with base current rising from zero and driven by a voltage of the same polarity as the collector we begin at the top with gate voltage falling from zero and opposite in polarity to the drain.

The oppositeness of polarity does not mean that a special gate voltage source must be provided. We have seen in Figure 13.6 how an emitter can be held at a positive potential by making use of the voltage drop in a resistor connected in the emitter lead. The base, because it must be positive to the emitter, has to be provided with a slightly larger positive voltage. Things are simpler with an f.e.t. because the required gate bias is negative, so the gate can be left at zero potential and the resistance R_S in Figure 13.8 adjusted to bias the source positively to the required extent. This is the same thing as negative gate bias, except that it is obtained at the expense of V_{DS}, the effective drain voltage. And it has the advantage, as in Figure 13.6, of automatically adjusting the amount of bias to counteract any tendency for the source current (= drain current) to depart from the designed value—but less effectively, because f.e.t. mutual conductances are lower.

If the device is permanently connected to a signal source which provides a d.c. path but no d.v., then C_G and R_G can be omitted. The purpose of R_G is to ensure that the gate is held at the common terminal potential (reckoned as zero). Because of the nearly zero gate current with negative bias, R_G can be high; typically 10 MΩ.

Because f.e.t. characteristics are especially subject to temperature and manufacturing variations, the degree of

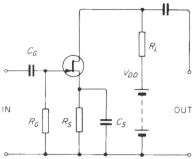

Figure 13.8 This method of providing gate bias for f.e.t.s is applicable also to valves

compensation provided by simple source bias as in Figure 13.8 may not always be enough. More compensation is obtained by increasing R_S, but if this alone were done it would develop excessive bias, so it must be accompanied by positive gate bias, which can conveniently be provided by connecting an appropriate high resistance from V_{DD} to gate, making a potential divider similar in form to that in Figure 13.6 but easier to design because of the absence of gate current. This new resistor would normally have to be several times higher in resistance than R_G, and such values are not usually very satisfactory in situations such as this. To enable conveniently lower values to be used without losing the potentially high input resistance of the f.e.t., a single high resistance can be connected between the junction of the potential divider and the gate.

13.6 Valve Biasing

Hardly anything need be said under this heading at the present stage, because the methods for f.e.t.s can be used for valves, the simpler form (Figure 13.8) usually giving adequate compensation. Other methods for both valves and f.e.t.s are commonly used, however, but they will be easier to understand in connection with the applications considered later (Section 27.8).

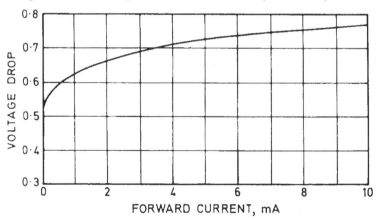

Figure 13.9 Showing how the voltage across a typical silicon junction diode changes little with the current through it, over a wide range

13.7 Biasing by Diodes

In many circuits there is a need for a fairly small constant bias voltage. Provided one can make do with about 0.2 V or 0.65 V or a combination of these, a convenient, reliable and cheap method is to pass a current through a germanium or silicon junction diode or through two or more in series, in the forward or low-resistance direction, and use the voltage drop across it or them. As Figure 13.9 shows, the precise current used does not matter, nor does it have to be anywhere near constant. Figure 19.15 is an example of this technique, but see Section 27.9 for the production of higher constant voltages by special reverse-connected diodes. Figure 13.9 is, of course, the diode forward characteristic (e.g. Figure 9.14) with the axes interchanged.

Chapter 14 Oscillation

14.1 Generators

Having now a fundamental knowledge of the few basic electrical and electronic circuit elements, we can begin to look at their applications. Elementary amplifiers have already appeared, and frequent references have been made to signals. Generators of signals have so far been represented by conventional symbols. The only sort of a.c. generator actually shown (Figure 6.2) is incapable of working at the very high frequencies needed in radio and other systems, and produces a square waveform which for many purposes is less desirable than the sine shape (Figure 5.2).

The alternating currents of power frequency—usually 50 Hz in Great Britain and many other countries—are generated by rotating machinery which moves conductors through magnetic fields produced by electromagnets (Section 4.3). Such a method is quite impracticable for frequencies of millions of cycles per second. This is where the curious properties of inductance and capacitance come to the rescue.

14.2 The Oscillatory Circuit

Figure 14.1 shows a simple circuit comprising an inductor and a capacitor in parallel, connected as the load of an f.e.t. The f.e.t. is not essential in the first stage of the experiment, which could be performed with a battery and switch, but it will be needed later on. A valve would do just as well, or (with minor modifications) a bipolar transistor. With the switch as shown, there is no negative bias, and we can assume that the current flowing through the coil L is fairly

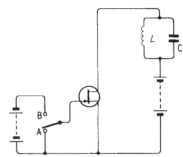

Figure 14.1 Apparatus for setting up oscillations in the circuit LC

large (Figure 10.15). It is a steady current, and therefore any voltage drop across L is due to the resistance of the coil only. As none is shown, we assume that it is too small to take into account, and therefore there is no appreciable voltage across L and the capacitor is completely uncharged. The current through L has set up a magnetic flux which by now has reached a steady state, and represents a certain amount of energy stored up, depending on the strength of current and on the inductance L (Section 4.9). This condition is represented in Figure 14.2 by a, so far as the circuit LC is concerned.

Now suppose the switch is moved from A to B, putting such a large negative bias on the gate that drain current is completely cut off. If the current in L were cut off instantaneously, an infinitely high voltage would be self-induced across it (Section 4.4), but this is impossible owing to C. What actually happens is that directly the current through L starts to decrease, the collapse of magnetic flux induces an e.m.f. in such a direction as to oppose the

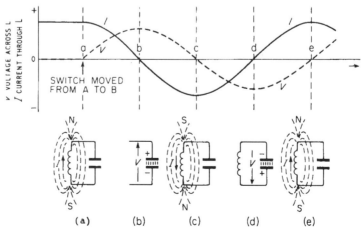

Figure 14.2 Sequence of events in Figure 14.1 after the switch has been moved to B. Stages (a) to (e) cover one complete oscillation

decrease, that is to say, it tends to make the current continue to flow. It can no longer flow through the f.e.t., but there is nothing to stop it from flowing into the capacitor C, charging it up. As it becomes charged, the voltage across it rises (Section 3.2), demanding an increased charging voltage from L. The only way in which the voltage can rise is for the current through it to fall at an increasing rate, as shown in Figure 14.2 between a and b.

After a while the current is bound to drop to zero, by which time C has become charged to a certain voltage. This state is shown by the curves of I and V and by the circuit sketch b. Because the current is zero there can no longer be any magnetic field; its energy has been transferred to C as an electric field.

The voltage across C must be balanced by an equal voltage across L, which can result only from the current through L continuing to fall; that means it must become negative, reversing its direction. The capacitor must therefore be discharging through L. Directly it begins to do so it inevitably loses volts, just as a punctured tyre loses pressure. So the rate of change of current (which causes the voltage across L) gradually slackens until, when C is completely discharged and there is zero voltage, the current is momentarily steady at a large negative value (point c). The whole of the energy stored in the electric field between the plates of C has been returned to L, but in the opposite polarity. Assuming no loss in the to-and-fro movement of current (such as would be caused by resistance in the circuit), it has now reached a value equal and opposite to the positive value with which it started.

This current now starts to charge C, and the whole process just described is repeated (but in the opposite polarity) through d to e. This second reversal of polarity has brought us to exactly the same condition as at the starting point a. And so the whole thing begins all over again and continues to do so indefinitely, the original store of energy drawn from the battery being forever exchanged between coil and capacitor; and the current will never cease oscillating. Note that from the moment the switch was moved the f.e.t. played no part whatever. The combination of L and C is called an *oscillatory circuit*.

It can be shown mathematically that voltage and current vary sinusoidally.

14.3 Frequency of Oscillation

From the moment the f.e.t. current was switched off in Figure 14.1 the current in L was bound to be always equal to the current in C because, of course, it was the same current flowing through both. Obviously, too, the voltage across them must always be equal. For a sinusoidal current, reactance is the ratio of voltage to current (Sections 6.2, 7.2), and it follows that the reactances of L and C must be equal, that is to say $2\pi fL = 1/2\pi fC$. Rearranging this we get

$$f_o = 1/2\pi\sqrt{(LC)}$$

where f_o denotes frequency of oscillation. This is the same as the formula we have already had (Section 8.7) for the frequency of resonance. It means that if energy is once imparted to an oscillatory circuit (which is now seen to be the same thing as a tuned circuit) it goes on oscillating at a frequency called the natural frequency of oscillation, which depends entirely on the values of inductance and capacitance. By making these very small—using only a few small turns of wire and small widely spaced plates—very high frequencies can be generated, running if necessary into hundreds of millions of cycles per second, or even more with combined structures (Sections 8.17, 16.10).

14.4 Damping

The non-stop oscillator just described is as imaginary as perpetual motion. It is impossible to construct an oscillatory circuit entirely without resistance of any kind. Even if it were possible it would be merely a curiosity, without much practical value as a generator, because it would be unable to yield more than the limited amount of energy with which it was originally furnished. After that was exhausted it would stop.

The inevitable resistance in all circuits—using the term resistance in its widest sense (Section 2.18)—dissipates a proportion of the original energy during each cycle, so each is less than its predecessor, as in Figure 14.3. This effect of resistance is called *damping*; a highly damped oscillatory circuit is one in which the oscillations die out quickly.

In Section 8.4 the symbol Q was given to the ratio of reactance (of either kind) to resistance in a tuned circuit. So a highly damped circuit is a low-Q one. If the Q is as low as 0.5 the circuit is said to be critically damped; below that figure it is non-oscillatory and the voltage and current do not reverse after the initial kick due to an impulse but tend towards the shape shown in Figures 3.6 and 4.10. A high-Q circuit continues oscillating for many cycles. Oscillation is created in any tuned circuit by any sudden electrical impulse, and the circuit is said to be shock-excited. The same kind of thing happens to a car when it encounters a

Figure 14.3 The first few cycles of a train of damped oscillations

hole or bump in the road, unless sufficient mechanical resistance is imparted to the springs in the form of shock absorbers (or dampers).

14.5 Maintaining Oscillation

What we want is a method of keeping an oscillator going by supplying it periodically with energy to make good what has been usefully drawn from it as well as what has unavoidably been lost in its own resistance. We have plenty of energy available from batteries, etc., in a continuous d.c. form. The problem is to apply it at the right moments.

The corresponding mechanical problem arises in a clock. The pendulum or the balance wheel is an oscillatory system which on its own comes to a standstill owing to frictional resistance. The driving energy of the mainspring is 'd.c.' and would be of no use if applied direct to the pendulum. The problem is solved by the escapement, which is a device for switching on the pressure from the spring once in each oscillation, the 'switch' being controlled by the pendulum itself.

In Figure 14.1 oscillation was started by moving the switch from A to B at the stage marked a in Figure 14.2. By the time marked c a certain amount of energy would have been lost in resistance, so that the negative current would be a little less than the positive current with which it started, and by e the loss would be greater. But by switching positive current on again at c until e the upsweep of the current I would be augmented. The amount of current required would not be as much as the original (before a) but only enough to make good the loss. In a high-Q circuit this would be comparatively little.

Operating an actual switch by hand to reduce the bias during half of every cycle is not practicable, but it can easily be arranged to be done automatically by the oscillation itself. From a to c in Figure 14.2, V is positive. This is the period during which we want the downward swing of I to be speeded. Reducing the f.e.t. drain current by applying or increasing negative bias does just this. During the negative half-cycle of V we want to increase the current again by reducing bias. If we inductively couple a small coil to L as in Figure 14.4 we generate in it a voltage that alternately augments and diminishes the fixed bias provided by the bias battery. This voltage is exactly synchronised with V; its amount can be adjusted by the number of turns in the gate-circuit coil and by how closely

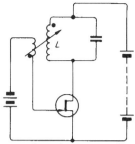

Figure 14.4 Showing how the circuit in Figure 14.1 can be modified to enable oscillations to be self-maintained continuously

it is coupled to L, and the relative polarity of the voltage applied to the gate is decided by which way the coil is turned. One of these ways, shown by the dots, is right for maintaining oscillation. This system is often called positive feedback (or regeneration), and the coil is a feedback or reaction (or, more precisely, retroaction) coil.

When the coupling is at least sufficiently close and in the right direction to maintain oscillation there is no need to do anything to start it; the ceaseless random movement of electrons in conductors (Section 2.3) provides impulses to start a tiny oscillation which soon builds up.

There is another way of looking at a self-oscillator circuit. Suppose the f.e.t. or other active device is arranged in a way with which we are quite familiar: Figure 14.5a. v_g is a small alternating input signal, graphed at b. We have seen in Section 8.10 that at the frequency of oscillation a tuned circuit is equivalent to a high resistance. If L and C are tuned to the frequency of v_g they are such a resistance, so voltage amplification takes place and an enlarged and inverted copy of v_g is obtained across LC. This is v_d, also shown at b.

If then we take a fraction v_g/v_d of the output voltage and invert it, as in Figure 11.2, for example, we can substitute it for the generator in Figure 14.5a. The amplifier supplies its own input by positive feedback and has become an oscillator.

In one respect this amplifier approach can be misleading. An amplifier has input and output pairs of points, and its behaviour depends on which electrode of the amplifying device is made common to both. Chapter 12 was largely about this. It is perfectly allowable to consider oscillator circuits in this way, and it is often done, but doing so tends

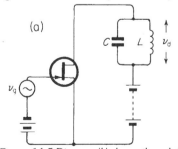

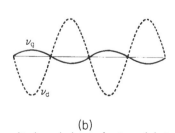

Figure 14.5 Diagram (b) shows the relative amplitude and phase of gate and drain voltages when their frequency is the same as the natural frequency of the circuit LC

to obscure the basic unity of oscillator circuits. Figure 14.6 shows an oscillatory circuit *LC* with a potential divider across it. If a triode of any kind that can be used for amplification is connected as shown (where, in order to focus attention on signals, feeds have been omitted) and its gain is sufficient to cancel losses, oscillation is maintained *and the question of which electrode is common to input and output does not arise*. If we choose to regard c as the common point (as in Figure 14.4, for example) then the input *ca* and the output *cb* need to be opposite in polarity, as they in fact are with common source (or emitter or cathode). But anyone is entitled to say that a is the common point and that this is a common-gate circuit, in which the amplified output *ab* has the same sign as the input (*ac*). A third party could equally maintain that this was a common-drain circuit, in which the output voltage *bc* is always not only of the same sign as the input (*ba*) but slightly smaller, and oscillation is equally well maintained on this basis, for different points of view of observers make no difference to the circuit!

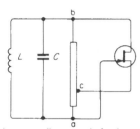

Figure 14.6 A basic oscillator, with feeds omitted for clarity. Note that the question of which electrode of the amplifying device is the common one does not arise

14.6 Practical Oscillator Circuits

Basically, then, an oscillator consists of an oscillatory circuit (*L* and *C* in Figure 14.6) and a maintaining amplifier (the rest of Figure 14.6). Owing largely to the necessity for feeding the amplifier with d.c., practical oscillator circuits usually look very different from Figure 14.6, and it is often quite difficult to see that in principle they are the same.

A resistive potential divider is rarely used, for it adds to the losses. And it is quite unnecessary; the oscillatory circuit can be made to act as its own potential divider. One method is to 'tap' the coil, as in Figure 14.7. (To show no bias in favour of any particular amplifying device, here a

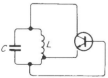

Figure 14.7 This variety of Figure 14.6 is the basis of the important Hartley oscillator circuit

bipolar transistor is used.) This is the basic form of what is called the Hartley circuit. Alternatively we can divide the oscillatory voltage in the *C* side. This could be done by inserting a third plate between the two, but usually it is more convenient to use two separate capacitors as in Figure 14.8, of such capacitances that the two in series are equal to *C*. This is the Colpitts oscillator circuit.

Figure 14.8 If the oscillatory circuit is tapped on the capacitive side instead of on the inductive side as in Figure 14.7 it is a Colpitts circuit

The coil in Figure 14.7 can be regarded as an auto-transformer (Figure 7.11). In Section 7.11 we saw that a resistance connected across one winding of a transformer affects the other similarly to connecting across it a resistance of a different value, depending on the turns ratio. The same principle applies to an auto-transformer, and if a capacitor of a suitably larger capacitance than *C* is connected across part of the coil it may still be tuned to the same frequency.

Figure 14.9 is therefore essentially the same as Figure 14.7, but we can more easily recognise it as the retroaction

Figure 14.9 The capacitor in the oscillatory circuit is often connected to the tapping point; when provided with feeds this circuit is seen to be the same as in Figure 14.4

coil circuit with which we began, Figure 14.4. In general, the Hartley and Colpitts types enable the amplifier to be connected more effectively to the tuned circuit, so (especially the Colpitts) are particularly useful in 'difficult' cases, such as very high frequencies, at which oscillation is less easily maintained. A practical advantage of the separate retroaction coil type is that no special devices are needed to prevent feeds from being short-circuited, but in the circumstances just mentioned there might be difficulty in getting close enough coupling.

Figure 14.10 is a practical Hartley circuit, in which the biasing components have the same labels as in Figure 13.6 for easy identification. C_B, which completes the oscillation-frequency connection between one end of *L* and the base without upsetting the bias by short-circuiting R_{B1} to d.c., is called a *blocking* capacitor. The resistance R_{B1} must be high enough not to load the left-hand part of *L* excessively. The tapping connection for the emitter goes via the battery and C_E (for which reason this is called a series-fed Hartley circuit), and to provide an easier path for the a.c. a capacitor is often connected from tap to emitter. The same object is achieved in a different manner in what is called

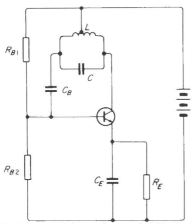

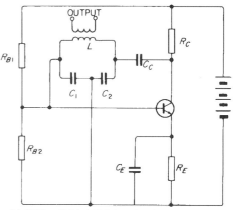

Figure 14.10 Practical form of series-fed Hartley oscillator circuit

the parallel-fed Hartley circuit, Figure 14.11, where for another change a valve is the active device. L_A is a coil of relatively high inductance, called a choke coil, which keeps

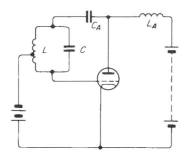

Figure 14.11 Parallel-fed Hartley circuit, with series grid bias

a.c. out of the anode current supply while allowing the d.c. feed to flow freely. C_A serves the same purpose as C_B in Figure 14.10. In this particular circuit the grid bias is fed in series with part of L, but often the oscillatory circuit is tapped straight to the common or earth terminal and bias is parallel-fed too. Figure 14.12 shows a parallel-fed Colpitts transistor circuit in which R_C and C_C take the place of L_A

Figure 14.12 Parallel-fed Colpitts circuit

and C_A as the parallel-feed components. This diagram also shows the output of the oscillator taken from a coil coupled to the inductor of the resonant circuit. As explained in Section 14.11 this gives a better output waveform than taking the output from the collector current, which can be distorted.

Practice in 'reading' circuit diagrams is needed before one can tell quickly and certainly which components are for blocking d.c. or a.c. and which are parts of the oscillatory circuit. One must bear in mind, too, that commercial designers like to make one component do the jobs of two or more. Then the intentional tuning capacitance is always augmented by stray capacitances due to other components and wiring. Diagrams of very-high-frequency oscillators are likely to be especially confusing because the tuning and coupling capacitances required are often so small that inter-electrode capacitances of the active device, and other 'strays', are sufficient.

As an example of this, at v.h.f. the inter-electrode capacitances of a transistor, though small, can be used to form the capacitance tapping in a Colpitts oscillator. Figure 14.13a gives the circuit diagram of such an oscillator, and here the collector-emitter and the base-emitter capacitances are shown in dashed lines to facilitate comparison with the basic Colpitts oscillator circuit shown in Figure 14.8. This is an interesting circuit because it requires only two

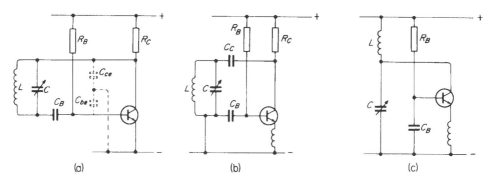

Figure 14.13 A v.h.f. Colpitts oscillator circuit using inter-electrode capacitances is shown in (a). A modification to permit one side of the tuning capacitor to be earthed is shown at (b) and a simplification of this circuit at (c).

connections to the *LC* circuit in order to obtain oscillation compared with the three (at least) required by the amplifier maintaining circuits discussed previously.

A disadvantage of circuit *a* is that both ends of the *LC* circuit are at r.f. potential with respect to earth so that it is impossible to earth the moving vanes of the tuning capacitor *C*, often a practical requirement. This difficulty can be overcome, as shown at *b*, by including in the emitter circuit an inductor with a high reactance at the oscillation frequency. This permits the emitter to be at r.f. potential and the base (connected to the moving vanes of *C*) to be earthed. The capacitance C_C is included to isolate L (one end of which is now earthed) from the positive potential of the collector. Because the lower end of L is at earth potential it is possible to introduce the positive supply voltage for the collector at this point and to eliminate the parallel-feed components R_C and C_C. The circuit diagram now appears as at *c*, and without showing the two fundamental inter-electrode capacitances this is very difficult to recognise as an oscillator circuit, let alone a Colpitts oscillator.

14.7 Resistance-Capacitance Oscillators

We have approached the subject of oscillators by considering an *LC* circuit with its natural frequency of oscillation, and then connecting an amplifier to feed it with regular periodical doses of energy to prevent those natural oscillations from dying out owing to circuit resistance or loading. We can alternatively regard the *LC* circuit simply as a positive-feedback path from output to input of the amplifier, so designed that only a.c. of one particular frequency is able to reach the input at the strength and phase relationship capable of maintaining itself via the amplifier. A circuit that selects or rejects particular frequencies or bands of frequencies is known as a *filter*. Filters are not necessarily oscillatory circuits, or even combinations of *L* and *C*. Figure 6.10 shows how, when a voltage such as *ab* is applied to a capacitance and resistance in series, the voltages across *C* and *R* separately lead or lag *ab* by a phase angle determined by *C* and *R* and the frequency of the voltage. Now, as Figure 14.5 showed, the input to a common-source (or cathode, or emitter) amplifier required to maintain oscillation must be inverted with respect to the output, and at least $1/A$ times the output in strength, A being the gain of the amplifier. Now for a sine wave inversion gives the same waveform as 180° phase shift. So a network which gives 180° phase shift at the required frequency of oscillation will also be satisfactory. Study of the situation illustrated by Figure 6.10 shows that

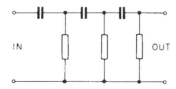

Figure 14.14 Three-stage CR filter circuit for obtaining the 180° phase shift required by a self-oscillator

by the time the phase difference is 90° the 'output' voltage is nil. But by using three stages of this elementary form of filter, as in Figure 14.14, a total phase difference of 180° can be achieved with an overall loss which can be matched by the gain of a transistor or valve. The output of this filter is connected to the input of the amplifier, whose output is connected to the input of the filter, as in Figure 14.15.

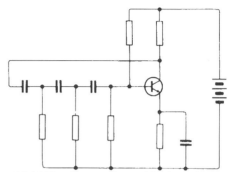

Figure 14.15 The filter in Figure 14.14 can be developed into an oscillator by connecting a simple amplifier between OUT and IN

This type of oscillator has a rather limited use, because the component values and the active device have to be well chosen to ensure oscillation; varying the frequency of oscillation is not very convenient because several component values have to be varied in order to do so; because of stray capacitances it is not really suitable for high frequencies; and the waveform is prone to depart from the sine shape. But it does illustrate the possibility of a sinusoidal oscillator without the inductance that is essential in a natural oscillatory circuit, and *RC* oscillators in more elaborate forms are used as audio-frequency and low-frequency signal generators.

14.8 Negative Resistance

Going back to the *LC* oscillatory circuit, we recall that the reason for the dying out of its oscillation was resistance. In Chapter 8 we studied the total effective resistance of such circuits and came to the conclusion (Section 8.16) that it could be represented as a (normally low) resistance in series with L and C, or as a (normally high) resistance in parallel with them. Figure 14.16 may bring it to mind.

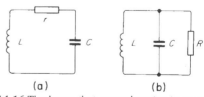

Figure 14.16 The losses that cause damping in an oscillatory circuit can be represented by either series (*a*) or parallel (*b*) resistance

Since then we have come across duality (Section 12.2) and can now see *a* and *b* as duals of one another and be inclined to refer to the high resistance as a low conductance. Now if we had such a thing as negative resistance we could insert $-r$ ohms in series in Figure 14.16a and thereby reduce the total resistance to zero, so that oscillations once started would continue unabated. Or if we had negative conductance we would connect $-1/R$ siemens of it in parallel with Figure 14.16b and get the same result.

Although there is no such thing as negative resistance or conductance as a passive component, an active device can be arranged to simulate it, and this is how we can regard the maintaining amplifier in each of our oscillator circuits.

We have seen that in a graph of current against voltage applied to a conductance the value of the conductance is represented by the *slope* of the graph. The slope of a graph of voltage against current represents the resistance concerned. Our first graphs of this kind, such as Figure 2.3, referred to ordinary resistors and were straight lines passing through the origin. When we came to active devices we found that graphs of their voltages and currents were not by any means straight; their varying slopes implied resistances and conductances that varied with the values of current and voltage. But whatever the slopes, an increase in voltage always went with an increase in current, so that the slopes, which are the ratios of the increases, were always positive. Load lines are not an exception, because their voltage scales (not shown) run from right to left. However, there are a few active devices which have current/voltage characteristics which over parts of their ranges slope the 'wrong' way. An increase in applied voltage, say, causes a fall in current.

One such device is a type of semiconductor diode in which the impurity content is vastly greater than in the usual types. It is called a tunnel or Esaki diode, and for reasons which would take too long to go into here its I/V characteristic takes the form shown in Figure 14.17. We

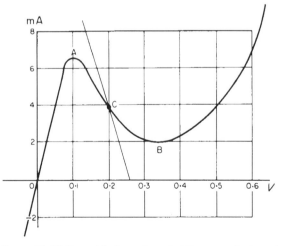

Figure 14.17 Current/voltage characteristic curve of a tunnel diode, showing negative resistance range, A to B

can say that over the range A to B its conductance is negative. Connecting it in parallel with an oscillatory circuit having a positive conductance not greater than the negative conductance at the selected working point results in sustained oscillations. If however the positive conductance is the greater, as shown by the greater slope of the load line cutting the curve at C, the net conductance is positive, the working point remains stable at C, and any oscillations are damped—though not so heavily as without the diode.

Devices are known in which there is a negative slope in part of the V/I characteristic, representing negative resistance, which is capable of cancelling positive resistance in series.

Negative conductances and resistances can be represented in an extension of the equivalent generator method explained in Section 12.5. Suppose, as in Figure 14.18, an internal voltage generator provides a reverse

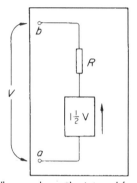

Figure 14.18 When, as here, the internal feedback voltage in Figure 12.8 exceeds the applied voltage, the box appears to contain a negative resistance

voltage *exceeding* that applied. The applied voltage V then apparently makes current flow out from instead of into the resistor. In the example shown, the total circuit voltage corresponding to $+V$ applied would be $V - 1\frac{1}{2}V = -\frac{1}{2}V$, and the current would be $\frac{1}{2}V/R$, so the apparent resistance would be $V \div \frac{1}{2}V/R, = -2R$. This could of course be expressed as a conductance $-\frac{1}{2}R$, and could alternatively be simulated by the equivalent current generator in parallel with R (Section 12.1).

The tunnel diode oscillator is another example of an oscillator circuit which requires only two connections to the resonant circuit.

14.9 Amplitude of Oscillation

So far it has been assumed that the amplifier or other negative-resistance device used for maintaining oscillation was applied only just sufficiently to make good the losses. This is quite an impractical condition, because the adjustment would be impossibly delicate in the first place, and even if it could be achieved it would be upset by the slightest change in temperature, component or device parameters, etc., so that oscillation would either start dying out or building up. In the latter event, what stops it growing so that it levels out? The answer is that the range of voltage

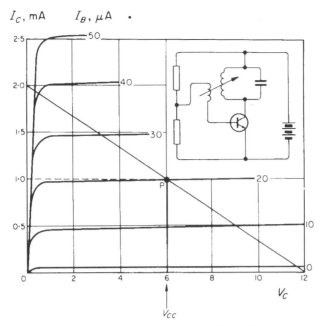

Figure 14.19 A simple oscillator circuit investigated by the load-line technique

or current over which the negative-resistance or amplifying characteristics exist is limited. This is very easily seen in Figure 14.17 in which the extent of the negative-resistance region of the characteristic is limited to approximately 0.25 V. The peak-to-peak amplitude of the oscillation is limited to this value by the damping imposed on the tuned circuit by the positive-resistance regions below A and above B. Similarly we have already noted that the amplifying parameters we studied in Chapter 12 are valid only so far as the slopes of the characteristics at the working point continue linearly in both directions. This is why they were called small-signal parameters. Sooner or later those slopes come to an end, and with them the properties that made the amplitude of oscillation increase.

This can be illustrated by an example, Figure 14.19. Although not identical with any of the previous oscillator circuits, all its features have already been explained. The I_C/V_C characteristics and load line are the same as in Figure 13.2, and so is the working point P, but there is an interesting difference in the conditions represented thereby. In Figure 13.2 the load was a real resistance, so of the supply voltage V_{CC}, which was 12 V, half was lost in the resistance and only the remaining half reached the transistor as V_C at point P. In Figure 14.19, however, the resistance of the coil L to d.c. is likely to be negligible, so only a 6 V battery is needed to maintain the working V_C at 6 V.

The load line represents the dynamic resistance of the LC circuit, at the frequency of oscillation only (Section 8.10) and indicates by its slope that this is 6 kΩ. It is the parallel equivalent of the oscillation-frequency losses in L and C and also the resistance across the feedback coil multiplied by the square of the effective transformer step-up ratio from it to L (Section 7.11). Part of the resistance across the feedback coil is the input resistance of the transistor, which we know depends very much on the working point—see Figure 10.12, for instance—but let us suppose that its variableness is reduced to a sufficiently small proportion by the constant resistance of the biasing potential divider for it not to have much effect on the total load resistance of the transistor. Suppose also that the amplitude of oscillation across L is 3 V peak, which means that V_C varies between 3 V and 9 V. Tracing this along the load line shows that in order to maintain this amplitude L needs to induce in the feedback coil sufficient voltage to make the base current oscillate between 30 μA and 10 μA; i.e., 10 μA peak reckoned from the working point level, 20 μA. If in fact it induces a shade more, the amplitude of oscillation grows, but by the time it has doubled, swinging between 0 and 40 μA, there is practically no scope for further increase.

In this case there is a sudden limiting at both ends, but sometimes the working point and load line are so placed that cut-off occurs first at one end, or more gradually—especially in the I_B/V_B characteristic, the effect of which we have been neglecting. The amplitude at one end may then continue to increase, until the reduction of h_{fe} at the other end, perhaps at zero, over an increasingly large proportion of each cycle, automatically limits the feedback power to a point where a balance is reached.

Sometimes a diode is used to provide a sharp amplitude-limiting effect that is more easily controlled in this respect than the amplifying device itself.

Up to this point we have assumed that f.e.t.s and valves are biased to a point that allows a range of oscillation between cut-off at one end and the start of grid or gate current at the other. Within this range practically no input current flows and the oscillatory circuit is not loaded thereby. But beyond it the amplitude of oscillation tends to be limited by output current being cut off, or the LC circuit being loaded by input current, or both.

14.10 Automatic Biasing of Oscillators

In many oscillator circuits, however, input current is allowed to flow and is used to generate bias for the active device. Figure 14.20 shows the circuit diagram of a Hartley

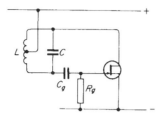

Figure 14.20 A Hartley oscillator circuit using an f.e.t. and incorporating automatic gate bias

oscillator using this technique. The f.e.t. has zero gate bias, R_g being returned to the source terminal, and the transistor (assumed a depletion type) amplifies, so initiating oscillation in LC. As the amplitude of oscillation builds up, each positive peak at the gate causes the transistor to take a burst of gate current. This has two effects. Firstly it applies damping to the tuned circuit. Secondly it charges the capacitor C_g and the polarity of the charge biases the gate negatively. After the pulse of gate current has ceased, C_g begins to discharge through R_g and the voltage across C_g falls. The extent of the fall depends on the time constant $R_g C_g$ and if this is long the fall is negligible before the next pulse of gate current restores the charge on C_g. In this way an almost steady voltage is maintained across C_g, and this provides the transistor with negative gate bias. Its value may be such that the transistor is cut off during negative peaks of oscillation at the gate, and takes a burst of gate current and drain current on positive peaks. The pulses of drain current provide the energy necessary to maintain the LC circuit in oscillation and the bursts of gate current abstract energy from it. The amplitude of oscillation thus stabilises at a value at which the energy received from the drain circuit just balances the losses due to gate current and the inherent resistance in the LC circuit.

This biasing method can be used only with active devices such as valves, bipolar transistors and jugfets which take input current. It cannot be used with igfets which do not take input current.

14.11 Distortion of Oscillation

Section 14.9 can be summarised by saying that the amplitude of oscillation grows until it is limited by non-linearity of the characteristics of the maintaining device. As we shall be seeing more fully in Chapter 19, non-linearity causes distortion of the oscillation waveform, which in its natural state is sinusoidal. One cycle appears in full line in Figure 14.21. Under the conditions shown in Figure 14.19

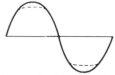

Figure 14.21 If the amplitude of oscillation goes past the ends of the load line in Figure 14.19 the peaks of collector current will be flattened as shown

both peaks of collector current would be cut off fairly sharply, as shown by the dashed line, representing severe distortion of the current waveform. But it would be wrong to suppose that this represents distortion of the oscillation waveform. To repeat: this is the waveform of the collector current. Its job is merely to make good the loss of energy in the oscillatory circuit through which it flows. Our study of this basic circuit in Chapter 8 showed that the oscillatory circulating current is Q times as much as needs to be supplied from outside. So no matter how badly this supplied current is distorted, it cannot affect the waveform of the oscillatory current much if Q is large. To go back to our clock analogy, the pendulum maintains its steady oscillation 'undistorted' although the periodical pushes administered by the escapement are anything but sinusoidal in 'waveform'. And it is quite usual for electronic oscillators to be kept going by square waves, pulses or whatever, as we have already seen in the last section.

But to obtain a very pure waveform—and this is often required of an oscillator—it is sensible to make sure that the oscillatory circuit itself has as high a Q as practicable, and that this is not spoilt by resistance transferred into it by connecting the maintainer. In other words, the maintainer should not be coupled more closely to the LC circuit than is necessary to ensure adequate oscillation in all conditions of use. The higher the Q the less maintenance is needed. The loose-coupling policy also tends to prevent even the small maintaining part of the current from being badly distorted.

For many purposes non-sinusoidal waveforms are required; methods of generating them will be deferred until Chapter 25.

14.12 Constancy of Frequency

The frequency of an LC oscillator, as we have seen, is adjusted by varying the inductance and/or capacitance forming the oscillatory circuit. For most purposes it is desirable that when the adjustment has been made the frequency shall remain perfectly constant. This means that the inductance and capacitance (and to a less extent the resistance) shall remain constant. For this purpose inductance includes not only that which is intentionally provided by the tuning coil, but also the inductance of its leads and the effects of any coils or pieces of metal (equivalent to single short-circuited turns) inductively coupled to it. Similarly capacitance includes that of the coil, the wiring and the maintaining device. Resistance depends on many things, such as loading and feed voltage variations. So some of the conditions for constant frequency are the same as those for undistorted waveform: high Q and minimum coupling with the maintaining device.

Inductance and capacitance depend mainly on dimensions, which expand with rising temperature. If the tuning components are shut up in a box along with the maintaining device, the temperature may rise considerably, and the frequency will drift for some time after switching on. Frequency stability therefore is aided by placing any components which develop heat well away from the tuning components; by arranging effective ventilation; and by designing tuning components so that expansion in one dimension offsets the effect of that in another. Fixed capacitors are obtainable employing a special ceramic dielectric material whose variations with temperature oppose those of other capacitors in the circuit; this reminds one of temperature compensation in watches and clocks.

An alternative method of achieving constant frequency is explained in the next section.

14.13 Crystal Oscillators

Oscillators for certain purposes require very good frequency stability. An example is the oscillator acting as the source of carrier frequency in a radio transmitter. The function of this, usually called the master oscillator, is to define the frequency of the station at least as accurately and constantly as the now very narrow international regulations require. Many such oscillators rely on an important aid to constant frequency which has not so far been mentioned. It exploits the remarkable properties of certain crystals, notably quartz. Such crystals are capable of vibrating mechanically up to high radio frequencies and in doing so they develop an alternating voltage between two opposite faces. Conversely, an a.v. applied between the faces causes vibration. This is known as the *piezo-electric* effect.

The subject is a large and specialised one, but for our purpose the crystal can be regarded as a tuned circuit of remarkably high inductance and Q. Unlike the coil-and-capacitor circuit, a crystal once it has been cut to the size and shape needed to vibrate at the required frequency has hardly anything about it that can vary. There is a small change in frequency with temperature, so where requirements are stringent the crystal is kept in a temperature-controlled box often called an oven.

The crystal is mounted between two metal plates, which are sometimes coated on to it, and for best results is enclosed in a vacuum. Figure 14.22 shows an oscillator circuit in which the crystal itself is the only tuned circuit. The nature of this circuit is not clear unless the capacitances of the valve's anode and grid (shown dashed) are taken into account; it then turns out to be a Colpitts oscillator (Figure 14.8). A disadvantage of this circuit is that most crystals are capable of oscillating at what are called spurious frequencies. To prevent this the anode coupling resistor is often replaced by an LC tuning circuit, which has insufficient impedance to sustain oscillations at any frequency other than the one to which it is tuned. Many other crystal-controlled oscillator circuits, both valve and transistor, have been devised.

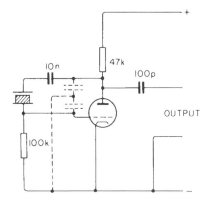

Figure 14.22 One of the many types of crystal oscillator circuit

Although the frequency of a crystal-controlled oscillator can be adjusted within very narrow limits (a few parts in 10 000) by means of a variable capacitor in parallel, it is virtually a fixed-frequency oscillator. Crystals being relatively inexpensive, they can be used in a transmitter working on a number of spot frequencies. But if it is necessary to be able to vary the oscillator over a band of frequencies an LC-tuned variable frequency oscillator (v.f.o.) must be used, with appropriate precautions.

14.14 Microwave Oscillators

We saw in Section 8.17 that at microwave frequencies cavity resonators take the place of LC circuits, and we shall now see how such resonators are used in oscillators. One type of microwave oscillator consists essentially of an amplifier which provides its own input, but the mechanism of amplification is new and is illustrated in Figure 14.23, which shows a valve with a cathode at one end and an

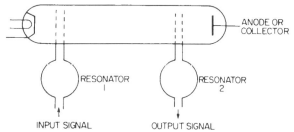

Figure 14.23 Essential features of a two-cavity klystron amplifier. By interconnecting the two resonators a microwave oscillator can be produced

anode at the other. Between these electrodes are two pairs of grids, each pair connected to a resonator. Suppose that resonator 1 is given an input signal at its resonance frequency. The electron stream is attracted to the anode's positive charge but, in passing through the first pair of grids, is alternately accelerated and retarded according to the instantaneous signal voltage between the grids, a process known as *velocity modulation*. As a result, in the space between the pairs of grids, the swifter electrons overtake slower ones, and the design is such that the electron stream reaching resonator 2 has well-defined bunches of electrons. These act as pulses and stimulate resonator 2 into oscillation, enabling a small-amplitude signal applied to resonator 1 to generate a larger signal in resonator 2. This type of microwave amplifier is known as a two-cavity klystron and, by coupling the resonators, a microwave oscillator can be obtained.

It is not necessary, however, to have two resonators. If, having passed through the grids of a resonator, the electron stream is reversed by a negatively charged electrode (the repeller) so that it passes through the same two grids again (this time in the form of bunches) oscillation will result and the oscillation frequency can be controlled by mechanical deformation of the resonator or by adjusting the repeller voltage. Such a valve is known as a reflex klystron.

Chapter 15 Modulation

15.1 The Need for Modulation

We have now reached the point—in theory at any rate—at which we can generate a continuous stream of waves of constant frequency. This, however, is completely useless for communication of messages, let alone speech, music or pictures. It merely establishes a link between transmitter and receiver, and to effect communication via the link some kind of organised change is needed. The technical word for it is *modulation*. The raw material that we have provided ready for modulation is called the carrier wave, although really it is a continuous succession of waves.

15.2 Amplitude Modulation

As a simple example of modulation suppose the 'programme' to be sent is a 1 kHz tuning note, which is heard as a high-pitched whistle. If it is to be a 'pure' sound its waveform must be sinusoidal, as in Figure 15.1a. It can be generated by a suitable a.f. oscillator but, for the reasons given in Chapter 1, waves of such a low frequency cannot be radiated effectively through space. We therefore choose an r.f. carrier wave, say 1000 kHz. As 1000 of its cycles occur in the time of every one of the a.f. cycles, they cannot all be drawn to the same scale, so the cycles in Figure 15.1b are representative of the much greater number that have to be imagined. What we want is something which, like b, is entirely radio frequency (so that it can be radiated), and yet carries the exact imprint of a (so that the a.f. waveform can be extracted at the receiving end). One way of obtaining it is to vary the amplitude of b in exact proportion to a, increasing the amplitude when a is positive and decreasing it when a is negative, the result being c. As we can see, this modulated wave is still exclusively r.f. and so can be radiated in its entirety. But although it contains no a.f. current, the tips of the waves trace out the shape of the a.f. waveform, as indicated by the dashed line, called the *envelope* of the r.f. wavetrain.

The same principle applies to the far more complicated waveforms of music and speech, or of light and shade in a televised scene—their complexity can be faithfully represented by the envelope of a modulated carrier wave.

Since receivers are designed so that an unmodulated carrier wave (between items in the programme) causes no sound in the loudspeaker, it is evident that a graph such as Figure 15.1c represents a note of some definite loudness, the loudness depending on the amount by which the r.f. peaks rise and fall above and below their mean value. The amount of this rise and fall, relative to the normal amplitude of the carrier wave, is the *depth of modulation* or, to use the BS preferred term, the *modulation factor*.

For distortionless transmission the increase and decrease in carrier amplitude, corresponding to positive and negative half-cycles of the modulating voltage, must be in proportion. The limit to this occurs when the negative half-cycles reduce the carrier-wave amplitude exactly to zero, as

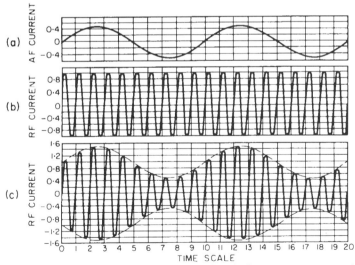

Figure 15.1 Curve (a) represents two cycles of audio-frequency current, such as might be flowing in a microphone circuit; (b) represents the much larger number of radio-frequency cycles generated by the transmitter. When these have been amplitude modulated by (a), the result is as shown at (c)

shown in Figure 15.2. At its maximum it will then rise to double its steady value. Any attempt to make the

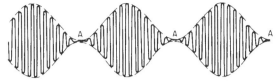

Figure 15.2 Carrier wave modulated to a depth of 100%. At its minima (points A) the r.f. current just drops to zero

maximum higher than this will result in the carrier actually ceasing for an appreciable period at each minimum; over this interval the envelope of the carrier amplitude can no longer represent the modulating voltage and there will be distortion.

When the carrier has its maximum swing, from zero to double its mean value, it is said to have a modulation factor of 100%. In general, half the variation in amplitude, maximum to minimum, expressed as a percentage of the mean amplitude, is taken as the measure of modulation factor. Thus, when a 100 V carrier wave is made to vary between 40 V and 160 V, the modulation factor is 60%.

In transmitting a musical programme, variations in loudness of the received music are produced by variations in modulation factor, these producing corresponding changes in the amount of a.f. output from the receiver.

Similarly in transmitting a television programme the variations in light and shade occurring along each of the scanning lines are represented by the amplitude of the carrier wave, but the necessity for including synchronising signals and colour information complicates the modulation process as explained in Sections 23.6 and 23.11.

15.3 Methods of Amplitude Modulation

In the early days, modulation was usually effected by varying the voltage of one of the electrodes of the oscillator valve by means of an amplified version of the sound, etc., to be communicated. This inevitably varied the frequency of oscillation to a small but no longer acceptable extent, so the practice now is to modulate the carrier wave in one of the r.f. amplifier stages. This is preferably not done in the first stage, in case it reacts on the oscillator, but the earlier in the chain it is done the less is the a.f. power needed for modulation. This advantage of what is called low-level modulation is offset by the need to avoid any distortion of the envelope from there on. That rules out Class C for a start (Section 11.10). Class B r.f. amplification of a modulated carrier within tolerable limits of distortion is possible but difficult. So in spite of the very large amount of a.f. power needed, modulation is often applied at the final power amplifier stage.

Figure 15.3 shows in outline the classical method of modulation, often called the Heising method, which varied the anode voltage. This common-cathode amplifier was fed with input signal from a tuned circuit $L_1 C_1$, the output being developed in the tuned circuit $L_0 C_0$. Both LC circuits were assumed tuned to the carrier frequency and, for sake of simplicity, various arrangements necessary in a practical version of this circuit have been omitted.

C had sufficient capacitance to provide a direct path for r.f. current to cathode but not enough to shunt the a.f. transformer T appreciably. This transformer coupled the output of an a.f. amplifier to the r.f. power amplifier. The amount of a.f. power needed was anything up to half as much as that supplied by the d.c. anode feed source, and of the same order as the r.f. output, so for a high-power transmitter the transformer had to be a very substantial one, and so had the a.f. amplifier feeding it. At 100% modulation its peak output voltage was almost equal to the feed d.v., and it therefore swung the total anode supply voltage from nearly zero to almost double the normal, thereby making the tuned circuit current vary as shown in Figure 15.2.

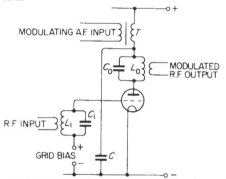

Figure 15.3 Outline of Heising method of amplitude modulation

It used to be common practice to use a single inductor (a 'choke coil') in place of the transformer, and to connect the final a.f. amplifying valve across C, so that it was fed from the same source as the r.f. valve. This form of anode modulation was known as choke control.

Because of the large amount of power and heavy equipment needed for anode modulation, the a.f. voltage was sometimes applied to the grid. This was not so simple or satisfactory as might have been expected; it tended to increase distortion and certainly reduced the r.f. efficiency, so was not much favoured. Other methods of modulation have been devised, some of them very complicated, in attempts to reduce power consumption without increasing distortion.

15.4 Frequency Modulation

Instead of keeping the frequency constant and varying the amplitude, one can do vice versa. Figure 15.4 shows the difference; *a* represents a small sample—two cycles—of the a.f. programme, as in Figure 15.1*a*, and *b* is the corresponding sample of r.f. carrier wave after it has been amplitude-modulated by *a*. If, instead, it were frequency-modulated it would be as at *c*, in which the amplitude remains constant but the frequency of the waves increases and decreases as the a.f. voltage rises and falls. In fact, with a suitable vertical scale of hertz, *a* would represent the frequency of the carrier wave.

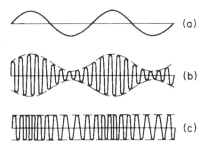

Figure 15.4 Curve (a) represents two cycles of audio frequency used to modulate an r.f. carrier wave. If amplitude modulation is used, the result is represented by (b); if frequency modulation, by (c)

The chief advantage of frequency modulation ('f.m.') is that under certain conditions reception is less likely to be disturbed by interference or noise. Since with f.m. the transmitter can radiate at its full power all the time, it is more likely to outdo interference than with amplitude modulation ('a.m.'), which has to use a carrier wave of half maximum voltage and current and therefore only one quarter maximum power. At the receiving end, variations in amplitude caused by noise, etc., coming in along with the wanted signals can be cut off without any risk of distorting the modulation.

Another contribution to noise reduction results from the fact that modulation can be made to swing the carrier-wave frequency up and down over a relatively wide range or 'band' as it is called. Exactly how this advantage follows is too involved to explain here, but since the change of frequency imparted by the modulator is proportional to the a.f. voltage it is obvious that the greater the change of frequency the greater the audible result at the receiving end.

In a.m., 100% modulation varies the amplitude between zero and double. If an f.m. station varied its frequency between zero and double it would (apart from other impracticabilities!) leave no frequencies clear for any other stations, so some arbitrary limit must be imposed, to be regarded as 100% modulation. This limit is called the *rated frequency deviation*, f_D, and for sound broadcasting is commonly 75 kHz above and below the frequency of the unmodulated carrier wave. Stations working on the 'medium' broadcast frequencies (525 to 1605 kHz) are spaced only 9 or 10 kHz apart, and if they used f.m. they would spread right across one another. For this and other reasons it is necessary for f.m. stations to use much higher frequencies, usually, in the v.h.f. band, 30 to 300 MHz.

A silent background to reception is especially desirable in broadcasting, and for this reason f.m. is much used for sound programmes, with or without accompanying vision. It is also widely used in radio telegraphy and still-picture transmission.

15.5 Telegraphy and Keying

To produce telegraph signals such as Morse, a simple but extreme form of modulation is used: its depth is always either 0 or 100%. It used to be done by a Morse key—a form of switch that makes a contact when pressed and breaks it when let go—in such a way that it turned the output of the transmitter on and off. Nowadays most of the remaining radiotelegraph communication is done at high speed by punched tape controlling contactors, which are electrically operated switches. There are a great many alternative positions in the transmitter circuits where a key or contactor would start and stop the output, but in practice quite a number of problems arise. So a.m. has been largely displaced by f.m., or frequency-shift keying, in which the radiation is going all the time but alternates between two closely spaced frequencies.

15.6 Methods of Frequency Modulation

The frequency of oscillation of an *LC* circuit can be altered by changing the value of the inductance or of the capacitance, and frequency modulation could therefore be achieved by using for the latter a component of which the capacitance can be controlled by the modulating signal. Such a component is a capacitor microphone because its capacitance varies in accordance with the sound waves striking it. Another possibility is the varactor diode, a current-controlled capacitor (Section 9.11). However, these devices are unlikely to be used in practice because of the difficulty of achieving the required degree of linearity between frequency deviation and modulating signal, of controlling the extent of the deviation and of ensuring constancy in the unmodulated frequency (known as the centre frequency). These factors are important in f.m. transmitters and to achieve them circuits known as reactance modulators (Section 19.21) are used.

A simple method of frequency modulation is possible using a reflex klystron as the oscillator (Section 14.14) by applying the modulating signal to the repeller electrode. Microwave links sometimes use such a modulator.

15.7 Sources of Modulating Signal

So far we have learnt something about two of the boxes at the sending end of Figure 1.11: the r.f. generator and the modulator.

A device by which sound, consisting of air waves, is made to set up electric currents or voltages of identical waveform, which in turn are used to control the modulator, is one that most people handle more or less frequently, in telephones. It is the *microphone*. There are a number of ways in which the variations in air pressure constituting a sound wave can be made to produce corresponding variations in an electrical signal which can be applied to the modulator of a transmitter. Most microphones have a small diaphragm which responds to sound waves by vibrating in accordance with them.

In the capacitor microphone the diaphragm is usually of metal, with a metal plate very close behind. When vibrated by sound, the capacitance varies, and the resulting charging and discharging currents from a battery, passed through a high resistance, set up a.f. voltages. Its disadvantages are very low output and high impedance.

The crystal or piezo-electric microphone depends on the same properties as are applied in crystal control; the varying pressures directly give rise to a.f. voltages. Crystals of Rochelle salt are particularly effective.

The type most commonly used for broadcasting and the better class of 'public address' is the electromagnetic. There are many varieties: in some a diaphragm bears a moving coil in which the voltages are generated by flux from a stationary magnet; while in the most favoured types a very light metal ribbon is used instead of the coil. Electromagnetic and crystal microphones generate e.m.f.s directly, so need no battery; but electromagnetic types need a transformer to step the outputs up.

The microphone is not the only source of audio signal used in a broadcasting system. Record players, CD players and tape reproducers, described in Chapter 27, are also widely used, as are audio signals from the sound track of cinema film.

The counterpart of the microphone in a vision system is the television camera. In a plain television camera an image of the scene is focused as in a photographic camera; but in place of the film the television camera contains an electrode (called the target) inside a special form of cathode-ray tube (Section 23.6). This electrode is sensitive to light, not chemically as in photography, but electrically (Section 10.12). There are different types in use, but the general result is a charge pattern over the inner face, corresponding to the light pattern of the image projected on to the outer face of the target. The electron beam is made to scan the target as described in Section 1.8 and picks off the charges; their voltages are amplified and used to amplitude-modulate the carrier wave of the transmitter.

A colour television camera is required to produce three output signals corresponding to the red, green and blue components in the image of the scene and from these signals the modulating signal is derived. The image of the scene is split into the three colour components by an optical filter system, and the resulting three images are focused on the targets of three camera tubes in which the scanning beams move in exact synchronism. The plain television picture signal (which is transmitted in a colour system in addition to the colour information) can be obtained by combining the colour signals in the correct proportions but it is now common to incorporate a fourth tube in a colour camera specifically to generate the plain television signal; its light input is, of course, derived from the scene image before this enters the colour-splitting filter.

The television camera is not the only source of video signal in a television broadcasting system. Video tape reproducers, described in Chapter 27, are also used and video signals can also be obtained from cinema film using equipment known as telecine. Computers and video games also generate video signals.

So far we have confined our discussion to transmission of sound and vision signals as in broadcasting, but much of the information applies to the equipment used in radio telephones and by amateur broadcasters (radio hams). Transmitters have a host of other applications—military, in radio astronomy, in space travel, in Citizen's Band radio, by police, hospital services, fire brigades to name but a few. They are also used for remote metering purposes—e.g., to give information about the depth of water in a reservoir. This is an example of telemetry, and the modulating signal is likely to be in pulse form.

15.8 Theory of Sidebands

Let us now return to the waveform of an amplitude-modulated signal (Figure 15.1c) and examine it in detail; in particular, to consider its frequency. Strictly speaking, it is only an exactly recurring signal that can be said to have one definite frequency. The continuous change in amplitude of a carrier wave during amplitude modulation makes the r.f. cycle of the modulated wave non-recurring, so that in acquiring its amplitude variations it has lost its constancy of frequency.

A mathematical analysis shows that if a carrier of f_1 Hz is modulated at a frequency f_2 Hz the modulated wave is exactly the same as the result of adding together three separate waves of frequencies f_1, $(f_1 + f_2)$, and $(f_1 - f_2)$. It is not easy to perform the analysis of the modulated wave into its three components by a graphical process, but the corresponding synthesis, adding together three separate waves, demands little more than patience.

Figure 15.5 shows at a, b, and c three separate sine-wave trains, there being 35, 30 and 25 complete cycles, respectively, within the width of the diagram. By adding the heights of these curves above and below their own zero lines point by point, the composite waveform at d is obtained. It has 30 peaks of varying amplitude, and the amplitude rises and falls five times in the period of time represented. If this is a thousandth part of a second, curve d represents what we have come to know as a 30 kHz carrier amplitude-modulated at 5 kHz. Figure 15.5 shows it to be identical with three constant-amplitude wavetrains having frequencies of 30, 30 + 5 and 30 − 5 kHz.

Thus a carrier modulated at a single frequency is equivalent to three simultaneous signals: the unmodulated carrier itself and two other steady frequencies spaced from the carrier on each side by an amount equal to the frequency of modulation.

The same facts can be illustrated by a phasor diagram. In such a diagram, as we know (Section 5.7), a single unmodulated carrier wave of frequency f is represented by a phasor such as OP in Figure 5.5, pivoted at one end (O) and rotating f times per second. The length of the phasor represents (to some convenient scale) the peak voltage or current, and the vertical 'projection' PQ represents the instantaneous value, which of course becomes alternately positive and negative once in each cycle.

Modulation 127

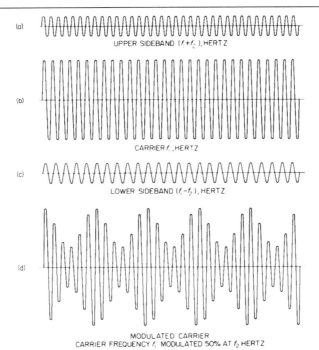

Figure 15.5 Showing that a modulated carrier wave (d) is identical with three unmodulated waves ((a), (b) and (c)) added together

If this were related to a waveform diagram such as Figure 15.5, it should be clear that amplitude modulation would be represented by a gradual alternate lengthening and shortening of the phasor as it rotated. Thus, modulation of a 600 kHz carrier to a depth of 50% at a frequency of 1 kHz would be represented by a phasor revolving 600 000 times per second, and at the same time increasing and decreasing in length by 50% once during each 600 revolutions.

The attempt to visualise this rapid rotation makes one quite dizzy, so once we have granted the correctness of the phasor representation we may forget about the rotation and focus our attention on how the alternate growing and dwindling can be brought about by phasorial means.

OP in Figure 15.6a, then, is the original unmodulated-carrier phasor, now stationary. Add to it two other phasors, PA and AB, each one-quarter the length of OP. If they are both in the same direction as OP, they increase its length by 50%; if in the opposite direction they decrease it by 50%. The intermediate stages, corresponding to sinusoidal modulation, are represented by the two small phasors rotating in opposite directions at modulation frequency. Various stages in one cycle of modulation are shown at Figure 15.6b–j, the total length of all three vectors added together being OB in every case. Notice that in spite of the small phasors going through complete 360° cycles of phase relative to OP, the triple combination OB is always exactly in phase with OP. It therefore gives the true lengthening and shortening effect we wanted.

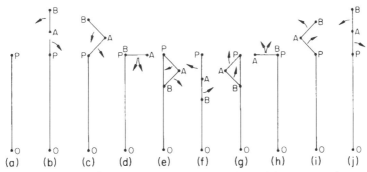

Figure 15.6 Progressive series of phasor diagrams, showing how adding a pair of oppositely rotating phasors (PA and AB) alternately lengthens and shortens the original phasor (OP) without affecting its phase

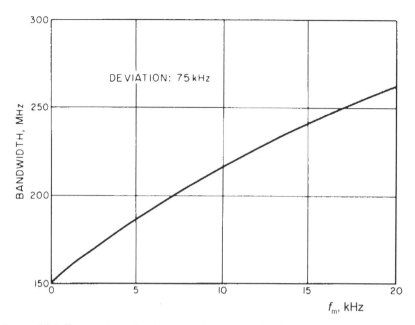

Figure 15.7 Showing how the frequency bandwidth of a frequency-modulated carrier wave depends on the maximum modulation frequency, f_m

Having agreed about this, we can now set OP rotating at 600 000 rev/s again. In order to maintain the cycle of changes shown in Figure 15.6 the little phasor AB will therefore have to rotate at 600 000 rev/s plus once in every 600; that is, 601 000: while PA must rotate at 600 000 less once in every 600; 599 000.

Translated back into electrical terms, this confirms the statement that amplitude-modulating a carrier wave of f_1 Hz at a frequency of f_2 Hz creates side frequencies, $f_1 + f_2$ and $f_1 - f_2$. And Figure 15.6 also shows that the amplitude of each of these side waves relative to that of the carrier is *half* the modulation depth (Section 15.2).

In a musical programme, in which a number of modulation frequencies are simultaneously present, the carrier is surrounded by a whole array of extra frequencies. Those representing the lowest musical notes are close to the carrier on each side, those bringing the middle notes are further out, and the highest notes are the farthest removed from the carrier frequency. This spectrum of associated frequencies on each side of the carrier is called a *sideband*, and as a result of their presence a musical programme, nominally transmitted on a frequency of, say, 1000 kHz, spreads over a band of frequencies extending from about 990 to 1010 kHz. Expressed generally, the band of frequencies occupied by an amplitude-modulated r.f. signal is double the highest modulation frequency.

With frequency modulation we might expect the band of frequencies to be double the rated deviation (Section 15.4), f_D. In fact, however, the matter is very involved mathematically, and the pattern of sidebands is affected by the modulation frequency (f_m) as well as by its depth. Only at low modulation frequencies is the bandwidth not appreciably higher than $2f_D$. In Figure 15.7 the bandwidth is plotted against f_m for f_D = 75 kHz, which is the usual figure for broadcasting.

Many people have had the bright idea that by making f_D small—say 1 kHz or even less—the bandwidth required could also be made small. These attempts to defeat the course of nature all come to grief because the sidebands always extend across at least $2f_m$, not less with f.m. than with a.m. but rather more.

A most important point to remember is that with either system of modulation the frequency band occupied depends entirely on the modulation and not at all on the frequency of the carrier wave.

15.9 Channel Separation

It is now clear that in order to receive a modulated signal completely the receiver must accept not one frequency but a whole band, preferably with equal effectiveness throughout. Outside this frequency band the receiver should ideally have no response, so that the sidebands of transmissions on nearby carriers are rejected. This is illustrated for a.m. in Figure 15.8 in which the dashed line shows an ideal overall receiver response curve embracing the whole of transmission A and completely rejecting transmission B. This curve illustrates the selectivity of the receiver; i.e., its ability to accept a wanted transmission and to reject those on neighbouring frequencies.

For sound, the modulation frequencies are audio frequencies, which go up to at least 15 kHz. The

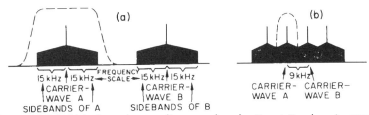

Figure 15.8 At (a) is shown how medium-wave broadcasting stations' carrier waves ought to be spaced on the frequency scale in order to permit full reception of each without interference from others. How they actually are spaced (b), shows that interference is almost inevitable if full reception of sidebands is attempted

corresponding sidebands would occupy at least 30 kHz. Ideally, then, the carrier waves of a.m. sound broadcasting stations should be spaced at intervals of not less than 30 kHz, plus a margin for selectivity, as suggested in Figure 15.8a.

Unfortunately the pressure of national demands for broadcasting 'channels' has squeezed this ideal severely; the standard interval in Europe on the low and medium frequencies is only 9 kHz. Obviously this faces receivers with an impossible task, for it brings 9 kHz modulation of one station on to the same frequency as the carrier wave of the next and the 9 kHz modulation of the next but one. Therefore few, if any, broadcasting stations on these frequencies actually radiate sidebands up to 15 kHz; 8 kHz is a more realistic maximum. There is therefore some loss at the start. And even what is transmitted cannot be fully reproduced without interference, unless the wanted transmission is so much stronger than those on adjacent frequencies as to drown them. So receivers for other than local-station reception must be designed to cut off at about 4.5 kHz, sacrificing quality of reproduction in the interests of selectivity (Figure 15.8b), and even then liable to some interference from adjacent channel sidebands if they are strong enough. This is heard as what is called monkey chatter.

Because the spacing of carrier waves is related to modulation it is on a frequency basis, not on a wavelength basis. If a station transmits on 200 kHz, the two nearest in frequency will be on 191 and 209 kHz. The corresponding wavelengths (Section 1.9) are 1571, 1500 and 1435 m, so the average wavelength spacing is 68 m. Three consecutive carrier-wave frequencies higher in the scale, with the same 9 kHz spacing, are 1502, 1511 and 1520 kHz. The corresponding wavelengths are 199.7, 198.6 and 197.3 metres, so the wavelength spacing here is only 1.2 metres.

Turning now to f.m. sound broadcasting, we saw that the bandwidth occupied by a transmission was slightly greater than twice the rated deviation f_D so that, if $f_D = 75$ kHz, the bandwidth is around 150 kHz and, to allow for the needs of selectivity, f.m. channels are generally spaced at intervals of 200 kHz. At this spacing the long waveband would scarcely accommodate one channel and even the medium waveband would hold only five! So f.m. transmissions are confined to much higher carrier frequencies—to the v.h.f. band, in fact, where the frequency range is great enough to accommodate a useful number of channels. The disadvantage of such high carrier frequencies is their comparatively short range (Section 17.12).

Since in amplitude modulation the carrier itself conveys no information, and the second sideband duplicates the first, by omitting one of them one can get twice as many channels of communication into a given frequency band, and there is a large saving in power. So in point-to-point radio, in which the necessary complications in transmitter and receiver equipment can be accepted, this single-sideband system is usually adopted. But to equip millions of broadcast radio receivers to work on this system seems impracticable. However, in television it is possible to eliminate most of one of the sidebands without causing difficulties at the receiving end. Because only a small part of one sideband is transmitted this is called a vestigial sideband (v.s.b.) system. Even with this economy the bandwidth occupied by a television transmission is enormous. In a 625-line system the modulation frequencies (video frequencies) extend to 5.5 MHz, so an a.m. transmission would occupy 5.5 MHz on a single-sideband system and, say, 6.75 MHz on a v.s.b. system. To allow for selectivity and the accompanying sound carrier the channels are spaced at 8 MHz in the UK. This spacing exceeds the combined width of all the bands available for broadcasting on low, medium and high frequencies! Television is therefore accommodated on the u.h.f. band.

15.10 Multiplex

We have so far discussed methods of modulating a carrier by a single modulating signal. But in stereo radio and colour television broadcasting, a second modulating signal is needed (as explained in later chapters) and it would be very convenient if the same carrier could be used to carry both, provided that the signals could be recovered without mutual interference at the receiver.

This can be achieved, and the same method is used in both radio and television. The second signal is used to amplitude-modulate a second carrier, known as a subcarrier, which is added to the first signal, the frequency of the subcarrier being so chosen that its sidebands do not interfere with those of the first signal. In the Zenith-GE stereo system used in the UK, the subcarrier is at 38 kHz and its sidebands occupy the range 23–53 kHz. The sidebands of the first signal (known as a baseband signal because it begins near z.f.) extend to 15 kHz so that there is a guard band of 8 kHz between the two sets of sidebands. The composite signal is used to modulate the main carrier. Receivers intended for monophonic sound reproduction or plain television ignore the subcarrier and its modulation.

This technique can be extended to embrace any number of signals. By choosing a different carrier frequency for each signal, the sidebands can be made to occupy a band without interference in the same way as radio channels are arranged within a waveband. This is known as frequency-division multiplex (f.d.m.) and is exploited in telephone systems to enable up to 900 telephone calls to be transmitted simultaneously over a single link.

15.11 Pulse Modulation Systems

Extensive use is made in modern electronics of pulses, which are waves of approximately rectangular shape. A train of similar pulses can be used to modulate a sinusoidal carrier, and can themselves be modulated to transmit information. For example, the amplitude or the frequency of the pulses can be modulated as in the a.m. and f.m. systems described above, but a pulse signal permits other modulation methods not possible with a sinusoidal carrier such as pulse duration and pulse position modulation. These systems, apart from pulse amplitude modulation, employ trains of constant-amplitude pulses and these have a great advantage over sinusoidal systems. If a received pulse signal is weak it is likely to be almost obscured by the noise inevitably generated by the active and passive components in any transmission system. But, provided their amplitude exceeds that of the noise, the pulses can be used to trigger a pulse generator (such as a multivibrator, Section 25.5) which will produce much stronger constant-amplitude noise-free pulses with exactly the same timing and hence carrying the same modulation as the received pulses.

A number of modulating signals can be transmitted independently using a pulse train. To do this the pulses are divided into groups, and the first pulse in each group is used to transmit modulating signal A, the second for modulating signal B, and so on. The modulating signals share the transmission path on a time division basis; this is accordingly known as a time-division multiplex system (t.d.m.).

In another pulse modulation system the pulses are again divided into groups but the pulses in each group are arranged to represent in binary notation (Section 26.4) the amplitude of the modulating signal. This system, known as pulse code modulation (p.c.m.), is used by the BBC to distribute its sound programmes (including television sound) to the broadcasting transmitters throughout the country.

It is also used to record sound on compact discs (Section 27.5) and is largely responsible for the superb sound quality and absence of background noise on reproduction from such discs.

Chapter 16 Transmission Lines

16.1 Feeders

In dealing with circuit make-up we have for the most part assumed that resistance, inductance and capacitance are concentrated separately in particular components, such as resistors, inductors and capacitors. True, any actual inductor (for example) has some of all three of these features, but it can be imitated fairly accurately by an imaginary circuit consisting of separate L, C and R (Section 8.16). In the next chapter, however, we shall turn attention to antennas, in which L, C and R are all mixed up and distributed over a considerable length. Antenna calculations are therefore generally much more difficult than anything we have attempted, especially as antennas can take so many different forms.

But one example of distributed-circuit impedance does lend itself to rather simpler treatment, and makes a good introduction to the subject, as well as being quite important from a practical point of view. It is the *transmission line*. The most familiar example is the cable connecting the television antenna on the roof to the set indoors. For reasons just explained (Section 15.9) television uses signals of very high frequency, which are best received on antennas of a definite length (Section 17.7). It is desirable that the wires connecting antenna to receiver should not themselves radiate or respond to radiation, nor should they weaken the v.h.f. or u.h.f. signals. The corresponding problem at the sending end is even more important, because a large amount of power has to be conveyed from the transmitter on the ground to an antenna hundreds of feet up in the air. A transmission line for such purposes is usually called a *feeder*.

To reduce radiation and pick-up to a minimum, its two conductors must be run very close together, so that effects on or from one cancel out those of the other. One form is the parallel-wire feeder (Figure 16.1a); still better is the coaxial feeder (b), in which one of the conductors totally encloses the other.

16.2 Circuit Equivalent of a Line

With either type, and especially the coaxial, the closeness of the two leads causes a large capacitance between them; and at very high frequencies that means a low impedance,

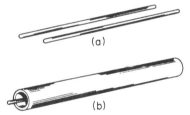

Figure 16.1 The two main types of r.f. transmission line or feeder cable: (a) parallel-wire, and (b) coaxial

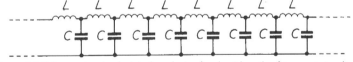

Figure 16.2 Approximate electrical analysis of a short length of transmission line

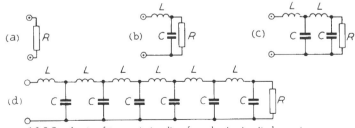

Figure 16.3 Synthesis of transmission line from basic circuit elements

which, it might be supposed, would more or less short-circuit a long feeder, so that very little of the power put in at one end would reach the other. But it must be remembered that every centimetre of the feeder not only has a certain amount of capacitance in parallel, but also, by virtue of the magnetic flux set up by any current flowing through it (Figure 4.2), a certain amount of inductance in series. Electrically, therefore, a piece of transmission line can be represented by standard circuit symbols as in Figure 16.2, in which L and C are respectively the inductance and capacitance of an extremely short length, and so are extremely small quantities. Each inductance is, of course, contributed to by both wires, but it makes no difference to our argument if for simplicity the symbol for the total is shown in one wire; it is in series either way. As we shall only be considering lines that are uniform throughout their length, every L is accompanied by the same amount of C; in other words, the ratio L/C is constant.

Having analysed the line in this way, let us consider the load at the receiving end; assuming first that its impedance is a simple resistance, R (Figure 16.3a). If we were to measure the voltage across its terminals and the amperage going into it, the ratio of the two readings would indicate the value R in ohms.

Next, suppose this load to be elaborated by adding a small inductance in series and a small capacitance in parallel, as at b. Provided that the capacitance really is small, its reactance, X_C, will be much greater than R. By an adaptation of the method used in Section 8.16 a relatively large reactance in parallel with a resistance can, with negligible error, be replaced by an equivalent circuit consisting of the same resistance in series with a relatively small reactance. Calling this equivalent series capacitive reactance X_C', we have $X_C'/R = R/X_C$, or $X_C' = R^2/X_C$. And if we make this X_C' equal to X_L, it will cancel out the effect of L (Section 8.3), so that the whole of Figure 16.3b will give the same readings on our measuring instruments as R alone. X_L is, of course, $2\pi fL$, and X_C is $1/2\pi fC$; so the condition for X_C' and X_L cancelling one another out is

$$2\pi fL = R^2 \times 2\pi fC$$

or,

$$L = R^2 C$$

or,

$$R = \sqrt{\left(\frac{L}{C}\right)}$$

Notice that frequency does not come into this at all, except that if it is very high then L and C may have to be very small indeed in order to fulfil the condition that X_C is much larger than R.

Assuming, then, that in our case $\sqrt{(L/C)}$ does happen to be equal to R, so that R in Figure 16.3a can be replaced by the combination LCR in b, it will still make no difference to the instrument readings if R in b is in turn replaced by an identical LCR combination, as at c. R in this load can be replaced by another LCR unit, and so on indefinitely (d). Every time, meters connected at the terminals will show the same readings, just as if the load were R only.

The smaller L and C are, the more exactly the above argument is true at all practical frequencies. This is very interesting, because by making each L and C smaller and smaller and increasing their number we can approximate as closely as we like to the electrical equivalent of a parallel or coaxial transmission line in which inductance and capacitance are uniformly distributed along its length. In such a line the ratio of inductance to capacitance is the same for a metre or a mile as for each one of the infinitesimal 'units', so we reach the conclusion that provided the far end of the line is terminated by a resistance R equal to $\sqrt{(L/C)}$ ohms the length of the line makes no difference; to the signal-source or generator it is just the same as if the load R were connected directly to its terminals.

That, of course, is exactly what we want as a feeder for a load which has to be located at a distance from the generator. But there is admittedly one difference between a real feeder and the synthetic one shown in Figure 16.3d; its conductors are bound to have a certain amount of resistance. Assuming that this resistance is uniformly distributed, every 10 metres of line will absorb a certain percentage of the power put into it. Obviously, a line made of a very thin wire, of comparatively high resistance, will cause greater loss than one made of low-resistance wire. So, too, will a pair separated by poor insulating material (Section 8.15). Although the loss due to resistance is an important property of a line, it is generally allowable to neglect the resistance itself in calculations such as we have just been making.

16.3 Characteristic Resistance

The next thing to consider is how to make the feeder fit the load, so as to fulfil the necessary condition, $\sqrt{(L/C)} = R$. $\sqrt{(L/C)}$ is obviously a characteristic of the line itself, and to instruments connected to one end the line appears to be a resistance; so for any particular line it is called its *characteristic resistance*, denoted by R_0. (If loss resistance is taken into account the expression is slightly more complicated than $\sqrt{(L/C)}$, and includes reactance; so the more comprehensive and strictly accurate term is characteristic (or surge) impedance, Z_0. With reasonably low-loss lines there is not much difference.)

If you feel that the resistance in question really belongs to the terminating load and not the line, then you should remember that the load resistance can always be replaced by another length of line, so that ultimately R_0 can be defined as the input resistance of an infinitely long line. Alternatively (and more practically) it is a resistance equal to whatever load resistance can be fed through any length of the line without making any difference to the generator.

In any case, L and C depend entirely on the spacing and diameters of the wires or tubes of which the line consists, and on the permittivity and permeability of the spacing materials. So far as possible, to avoid losses, feeders are air-spaced. The closer the spacing, the higher is C (as one would expect) and the smaller is L (because the magnetic field set up by the current in one wire is nearly cancelled by the field of the returning current in the other). So a closely

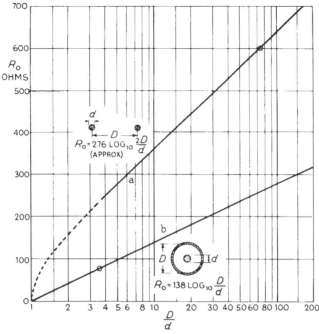

Figure 16.4 Graphs showing characteristic resistance (R_0) of (a) parallel-wire and (b) coaxial transmission lines in terms of dimensions

spaced line has a small R_0, suitable for feeding low-resistance loads. Other things being equal, one would expect a coaxial line to have a greater C, and therefore lower R_0, than a parallel-wire line. Formulae have been worked out for calculating R_0; for a parallel-wire line (Figure 16.4a) it is practically $276 \log_{10}(2D/d)$ so long as D is at least 4 or 5 times d, and for a coaxial line (Figure 16.4b) $138 \log_{10}(D/d)$. Graphs such as these can be used to find the correct dimensions for a feeder to fit a load of any resistance, within the limits of practical construction. It has been found that the feeder dimensions causing least loss make R_0 equal to about 600 Ω for parallel-wire and 80 Ω for coaxial types; but a reasonably efficient feeder of practical dimensions can be made to have an R_0 of anything from, say 200 to 650 Ω and 40 to 140 Ω respectively. We shall see later what to do if the load resistance cannot be matched by any available feeder.

In the meantime, what if the load is not a pure resistance? Whatever it is, it can be reckoned as a resistance in parallel with a reactance. And this reactance can always be neutralised or tuned out by connecting in parallel with it an equal reactance of the opposite kind (Section 8.8). So that little problem is soon solved for any desired frequency.

16.4 Waves along a Line

Although interposing a loss-free line (having the correct R_0) between a generator and its load does not affect the voltage and current at the load terminals—they are still the same as at the generator—it does affect their phase. The mere fact that the voltage across R in Figure 16.3b is equal to the voltage across the terminals makes that inevitable, as one can soon see by drawing phasor diagrams. Figure 16.5a shows a few sections as in Figure 16.3, and the phasor diagrams (b) are drawn by beginning at the load R and working backwards. They take the form of continuously opening fans. We can easily judge from this that for an actual line, in which the sections are infinitesimal, the diagrams would be circles. If we consider what happens in a line when the generator starts generating (Figure 16.6) we can see that what the gradual phase lag along it really means is that the current takes time to travel from generator to load.

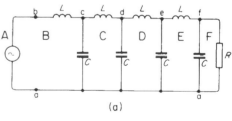

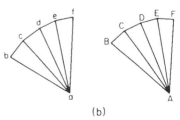

Figure 16.5 (a) Figure 16.3d labelled for phasor diagrams (b)

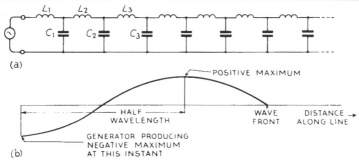

Figure 16.6 (a) Electrical representation of the generator end of a transmission line, and (b) instantaneous voltage (and current) diagram, of a cycle from the start of generating a sine wave

During the first half-cycle (positive, say) current starts flowing into C_1, charging it up. The inductance L_1 prevents the voltage across C_1 rising to its maximum until a little later than the generator voltage maximum. The inertia of L_2 to the growth of current through it allows the charge to build up in C_1 a little before C_2, and so on. Gradually the charge builds up on each bit of the line in turn. Meanwhile, the generator has gone on to the negative half-cycle, and this follows its predecessor down the line. At its first negative maximum, the voltage distribution will be as shown at *b*. With a little imagination one can picture the voltage wave flashing down the line, rather like the wave motion of a long rope waggled up and down at one end.

Our theory showed that wherever the line is connected to measuring instruments it appears as a resistance, which means that the current is everywhere in phase with the voltage, so the wave diagram implies power flowing along the line.

The speed of the wave clearly depends on its capacitance and inductance; the larger they are the longer the time each bit of line takes to charge and the longer the time for current to build up in it. We have already noted that C is increased by spacing the two conductors more closely, but that the L is reduced thereby. As it happens, the two effects exactly cancel one another out, so the speed of the wave is unaffected; the only result is that the ratio L/C—and hence R_0—is decreased. The only way of increasing C without reducing L is to increase the permittivity (ε) in the space between the conductors (Section 3.3); similarly L is affected by the permeability (μ) there (Section 4.6). It has been discovered that the speed of the waves is equal to $1/\sqrt{(\varepsilon\mu)}$. The lowest possible permittivity and permeability are in empty space (though air is practically the same); in SI units they are 8.854×10^{-12} and 1.257×10^{-6} respectively, so the speed of the waves works out at $1/\sqrt{(8.854 \times 1.257 \times 10^{-18})}$ = very nearly 300 million metres per second. Significantly, this is the same as the speed of light and electric waves in space (Section 1.9).

There have to be solid supports for the conductors, and sometimes the space between is completely filled. In practice the permeability of such material would not differ appreciably from that of empty space, but the permittivity would be higher. Consequently the speed of the waves would be lower; for example, if the relative permittivity were 4 the speed would be reduced to a half. But it would still be pretty high! So the time needed to reach the far end of any actual line is bound to be very short—a small fraction of a second. But however short, there must be some interval between the power first going into the line from the generator and its coming out of the line into the load. During this interval the generator is not in touch with the load at all; the current pushed into the line by a given generator voltage is determined by R_0 alone, no matter what may be connected at the far end; which is further evidence that R_0 is a characteristic of the line and not of the load.

During this brief interval, when the generator does not 'know' the resistance of the load it will soon be required to feed, the R_0 of the line controls the rate of power flow tentatively. In accordance with the usual law (Section 11.8), the power delivered to the line by a constant-voltage generator during this transient state is a maximum if R_0 is equal to r, the generator resistance.

Neglecting line loss, the voltage and current reaching the load end will be the same as at the generator. So if the load turns out to be a resistance equal to R_0, it will satisfy Ohm's law, and the whole of the power will be absorbed just as fast as it arrives. For example, if a piece of line having an R_0 of 500 Ω is connected to a 1000 V generator having an r of 500 Ω, the terminal voltage is bound to be 500 V and the current 1 A until the wave reaches the far end, no matter what may be there. If there is a load resistance of 500 Ω, the current will, of course, continue at 1 A everywhere.

16.5 Wave Reflection

But suppose that the load resistance is, say, 2000 Ω. According to Ohm's law it is impossible for 500 V applied across 2000 Ω to cause a current of 1 A to flow. Yet 1 A is arriving. What does it do? Part of it, having nowhere to go, starts back for home. More scientifically, it is *reflected* by the mismatch. The reflected current, travelling in the opposite direction, can be regarded as opposite in phase to that arriving, giving a total which is less than 1 A. The comparatively high resistance causes the voltage across it to build up above 500; this increase can be regarded as an in-phase reflected voltage driving the reflected current. If half the current were reflected, leaving 0.5 A to go into the load resistance, the voltage would be increased by a half, making it 750. A voltage of 750 and current 0.5 A would fit a 1500 Ω load, but not 2000 Ω, so the reflected proportion has to be greater—actually 60%, giving 800 V and 0.4 A at the load. In general terms, if the load resistance is R the fraction of current and voltage reflected, called the *reflection coefficient* or *return loss* and denoted by ρ, is $(R - R_0)/(R + R_0)$.

We now have 1 A, driven by 500 V, travelling from generator to load, and 0.6 A, driven by 300 V, returning to the generator. (The ratio of the reflected voltage to the reflected current must, of course, equal R_0.) The combination of these two at the terminals of the load gives, as we have seen, $1 - 0.6 = 0.4$ A, at $500 + 300 = 800$ V. But at other points on the line we have to take account of the phase lag. At a distance from the load end equal to quarter of a wavelength ($\lambda/4$) the arriving and returning waves differ in phase by half a wavelength ($\lambda/2$) or 180° as compared with their relative phases at the load (because a return journey has to be made over the $\lambda/4$ distance). So at this point the current is $1 + 0.6 = 1.6$ A at $500 - 300 = 200$ V. At a point $\lambda/8$ from the load end the phase separation is $\lambda/4$ or 90°, giving 1.16 A at 582 V calculated as explained in Section 5.8. At intervals of $\lambda/2$, the two waves come into step again.

16.6 Standing Waves

Calculating the current and voltage point by point in this way and plotting them, we get the curves in Figure 16.7. It is important to realise that these are not, as it were, flashlight photographs of the waves travelling along the line; these are r.m.s. values set up continuously at the points shown, and would be indicated by meters connected in or across the lines at those points (assuming the meters did not appreciably affect the R_0 of the line where connected). Because this wavelike distribution of current and voltage, resulting from the addition of the arriving and reflected waves travelling in opposite directions, is stationary, it is called a *standing wave*. For comparison, the uniform distribution of current and voltage when the load resistance is equal to R_0 is shown dashed.

The ratio of maximum to minimum current or voltage is called the *standing wave ratio* (s.w.r.); in our example it is $800/200 = 4$. (Because the ratio is the same for voltage and current the common practice of referring to voltage standing wave ratio (v.s.w.r.) is unnecessary and liable to mislead one into thinking that the ratio must be different for current.) Applying simple algebra to the way we have found the s.w.r. in this example we can easily find that in general it is equal to R/R_0. Its usual symbol is S.

In due course the reflected wave reaches the generator. It is in this indirect way that the load makes itself felt by the generator. If the 2000 Ω load had been directly connected to the generator terminals, the current would have been $1000/(500 + 2000) = 0.4$ A, and the terminal voltage $0.4 \times 2000 = 800$ V, and the power $800 \times 0.4 = 320$ W. This, as we have seen, is exactly what the load at the end of the line is actually getting. But the power that originally went out from the generator, being determined by R_0, was $500 \times 1 = 500$ W. The reflected power is $300 \times 0.6 = 180$ W, so the net outgoing power is $500 - 180 = 320$ W, just as it would have been with the load directly connected.

16.7 Line Impedance Variations

So the power adjustment is (in this case) quite simple. But the current and voltage situation at the generator is complicated by the time lag, and is not necessarily the same as at the load. Unless the length of the line is an exact multiple of $\lambda/2$, the phase relationships are different. Suppose, for example, it is an odd multiple of $\lambda/4$; say, $5\lambda/4$, as in Figure 16.7. Then the current and voltage at the generator end will be 1.6 A and 200 V respectively. That makes 320 W all right—but compare the power loss in the generator. 1.6 A flowing through $r = 500$ Ω is $1.6^2 \times 500 = 1280$ W, whereas at an 800 V point the current would be only 0.4 A and the loss in the generator only $0.4^2 \times 500 = 80$ W! So when there are standing waves, the exact length of the line (in wavelengths) is obviously very important. If there are no standing waves, it does not matter if the line is a little longer or shorter or the wavelength is altered slightly; and that is one very good reason for matching the load to the line. The usual way of making sure that the load is right is by running a voltmeter or other indicator along the line and seeing that the reading is the same everywhere, except perhaps for a slight gradual change due to line loss.

Connecting the generator to a point on the line where the voltage and current are 200 V and 1.6 A respectively is equivalent to connecting it to a load of $200/1.6 = 125$ Ω. The impedance of the line at all points can easily be derived from the voltage and current curves, as at the foot of Figure

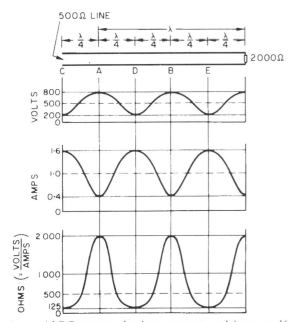

Figure 16.7 Diagrams of voltage, current, and (as a result) impedance at the load end of an unmatched transmission line, showing standing waves

16.7. This curve indicates the impedance measured at any point when all the line to the left of that point is removed. The impedance depends, of course, not only on the distance along the line to the load but also on R_0 and the load impedance. From a consideration of the travelling waves it can be seen that at $\lambda/4$ intervals the phases of the currents and voltages are exactly the same as at the load, or exactly opposite; so that if the load is resistive the input resistance of the line is also resistive. At all other points the phases are such as to be equivalent to introducing reactance.

In our example we made the generator resistance, r, equal to R_0. If it were not, the situation would be more complicated still, because the reflected wave would not be completely absorbed by the generator, so part would be reflected back to the load, and so on, the reflected power being smaller on each successive journey, until finally becoming negligible. The standing waves would be the resultant of all these travelling waves. Even quite a long line settles down to a steady state in a fraction of a second, but it is a state that is generally undesirable because it means that much of the power is being dissipated in the line instead of being delivered to the load.

It should be noted that mismatching at the generator end does not affect the standing wave ratio, but does affect the values of current and voltage attained.

Another result of making $r = R_0$ is that any reflection from the load or elsewhere inevitably impairs the generator-to-line matching. That is so even if the generator is connected to a point where the impedance curve coincides with the 500-Ω line in Figure 16.7, because then the impedance is reactive. But if r were, say, 2000 Ω, so that it would be mismatched to the 500-Ω line during the brief moment following the start, there would be a chance that when the reflected wave arrived it would actually improve the matching, even to making it perfect (e.g., if connected at A or B in Figure 16.7); but on the other hand it might make it worse still (if connected at C, D or E).

It should be remembered that 'perfect' matching is that which enables maximum power to be transferred; but in practice there may be good reasons for deliberately mismatching at the generator. To take figures we have already had, it may be considered better to deliver 320 W with a loss of 80 W than the maximum (500 W) with a loss of 500 W.

16.8 The Quarter-wave Transformer

An interesting result of the principles exemplified in Figure 16.7 is that a generator of one impedance can be perfectly matched to a load of another by suitably choosing the points of connection. For instance, a generator with an internal resistance of 125 Ω would be matched to the 2000 Ω load if connected at C, D or E. The line then behaves as a 1 : 4 transformer. Points can be selected giving (in this case) any ratio between 1 : 4 and 4 : 1, but except at the lettered points there is reactance to be tuned out.

It is not necessary to use the whole of a long line as a matching transformer; in fact, owing to the standing waves by which it operates it is generally undesirable to do so. We can see from Figure 16.7 that the maximum ratio of transformation, combined with non-reactive impedance at both ends, is given by a section of line only quarter of a wavelength long. We also see that the mismatch ratio to the line is the same at both ends; in this particular case it is 1 : 4 (125 Ω to 500 Ω at the generator end and 500 Ω to 2000 Ω at the load end), which, incidentally, is equal to the voltage ratio of the whole transformer. In general terms, if R_1 denotes the input resistance of the line (with load connected), R_1 is to R_0 as R_0 is to R; which means $R_1/R_0 = R_0/R$, or $R_0 = \sqrt{(R_1 R)}$.

This formula enables us to find the characteristic resistance of the quarter-wave line needed to match two unequal impedances, R_1 and R. Suppose we wished to connect a 70 Ω load to a 280 Ω parallel-wire feeder without reflection at the junction. They could be matched by interposing a section of line $\lambda/4$ in length, spaced to give a R_0 of $\sqrt{(70 \times 280)} = 140 \Omega$. Parallel wires would have to be excessively close (Figure 16.4a), but the problem could be solved by using a length of 70-Ω coaxial cable for each limb and joining the metal sheaths together as in Figure 16.8, putting the 70 Ω impedances in series across the ends of the 280-Ω line.

16.9 Fully Resonant Lines

Going now to extremes of mismatch, we inquire what happens when the 'load' resistance is either infinite or zero; in other words, when the end of the line is open-circuited or short-circuited. Take the open circuit first. If this were done to our original 500 Ω example, the current at the end would obviously be nil, and the voltage would rise to 1000—double its amount across the matched load. This condition would be duplicated by the standing waves at points A and B in Figure 16.7; while at C, D, and E the voltage would be nil and the current 2 A. The impedance curve would consequently fluctuate between zero and infinity.

With a short-circuited line, there could be no volts across the end, but the current would be 2 A; in fact, exactly as at E with the open line. A shorted line, then, is the same as an open line shifted quarter of a wavelength along. Reflection

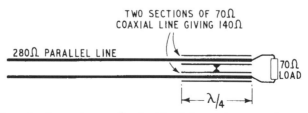

Figure 16.8 Example of a quarter-wavelength of line being used as a matching transformer

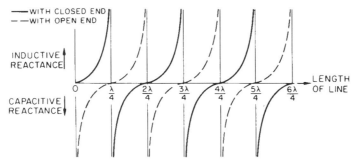

Figure 16.9 Showing how the reactances of short-circuited and open-circuited lines vary with their length

in both cases is complete, because there is no load resistance to absorb any of the power.

If the generator resistance is very large or very small, *nearly* all the reflected wave will itself be reflected back, and so on, so that if the line is of such a length that the voltage and current maximum points coincide with every reflection, the voltages and currents will build up to high values at those maximum points—dangerously high with a powerful transmitter. This reminds us of the behaviour of a high-Q resonant circuit (Section 8.3). The maximum current or voltage points are called *antinodes*, and the points where there is *no* current or voltage are *nodes*.

When the length of a short-circuited or open-circuited line is a whole number of quarter-wavelengths, the input impedance is approximately zero or infinity. An *odd* number of quarter-wavelengths gives opposites at the ends—infinite resistance if the other end is shorted, and vice versa. An *even* number of quarter-wavelengths gives the same at each end.

In between, as there is now no load resistance, the impedance is a pure reactance. At each side of a node or antinode there are opposite reactances—inductive and capacitive. If it is a current node, the reactance at a short distance each side is very large; if a voltage node, very low. Figure 16.9 shows how it varies. It is clear from this that a short length of line—less than quarter of a wavelength—can be used to provide any value of inductance or capacitance. For very short wavelengths, this form is generally more convenient than the usual inductor or capacitor, and under the name of a *stub* if often used for balancing out undesired reactance.

16.10 Lines as Tuned Circuits

A quarter-wave line shorted at one end is, as we have just seen, a high—almost infinite—resistance. But only at the frequency that makes it quarter of a wavelength (or an odd multiple of $\lambda/4$). This resistance is closely analogous to the dynamic resistance of a parallel resonant circuit; and in fact a line can be used as a tuned circuit. At wavelengths less than about 1 metre it is generally more efficient and easily constructed than the conventional sort.

The lower the loss resistance of the wire the higher the dynamic resistance across the open ends; and to match lower impedances all that is necessary is to tap it down, just as if it were any other sort of tuned circuit.

If the line is made slightly shorter than $\lambda/4$ the open end behaves as an inductor and can be tuned by connecting a variable capacitor across it. Figure 16.10a is an example of a conventional tuning circuit, such as might be used in a receiver, and *b* is the coaxial equivalent. A coaxial feeder is also used to connect the antenna, and being normally about 80 Ω it is tapped low down, near the earthed end. The impedance at the top end is normally thousands of ohms, and is too high for the input of a transistor so this too is tapped down.

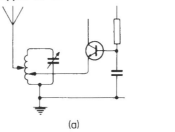 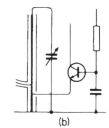

(a)   (b)

Figure 16.10 (a) Conventional aerial input circuit using 'lumped' components, and (b) distributed equivalent using a quarter-wave coaxial line

This is in essence the type of circuit used in most u.h.f. tuners in television receivers, but the variable capacitor is usually replaced by a varactor diode. Sometimes it is more convenient to make the short-circuited line slightly less than $3\lambda/4$ long, or alternatively an open-circuited line of just under $\lambda/2$ may be used. Both give the required inductive impedance but the $\lambda/2$ line has the advantage that the centre conductor—being open-circuited—can be used to supply the tuning control current to the varactor diode. See Section 22.8.

16.11 Waveguides or Radio-wave Plumbing

Above approximately 200 MHz the loss by radiation from a twin-wire transmission line becomes excessive, but coaxial lines can be successfully used up to 3000 MHz. At still higher frequencies a different mode of propagation of radio

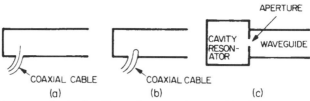

Figure 16.11 Three methods of putting radio energy into or taking it from a waveguide: (a) by a probe, (b) by a loop, (c) by an aperture coupling the guide to a cavity resonator

waves is possible in a tube without an inner conductor, in which the wave progresses along the tube in a series of internal reflections at the tube walls. In this way a radio wave can be guided along a tube as water passes along a pipe. Such tubes, often of rectangular cross-section, are termed waveguides and are extensively used in radar. For successful transmission the wavelength must be less than twice the diameter of the tube (or width of the rectangular guide); in other words, the frequency must be above a particular critical value which depends on the cross-sectional dimensions of the guide. For a rectangular guide there are two critical frequencies, one applying to waves which zigzag between the smaller sides and the other for waves reflected at the larger sides of the guide.

Radio waves can be reflected at any boundary between two different materials and thus a waveguide can be constructed of a non-metallic and non-conducting material such a polystyrene; in fact a solid rod of insulating material could be used as a waveguide, the wave being confined within the rod by reflections at its boundary with the outside air.

The use of waveguides to transfer energy from one point to another has some unusual features and poses some interesting questions. Where are the two conducting legs of the circuit? Without them, how is a signal put into or taken from a waveguide? There are, in fact, a number of ways. One is by a probe or loop projecting into the guide via a hole as suggested in Figures 16.11a and b. Or a signal can be transferred to or from a guide via an aperture as at c, which shows a guide coupled to a cavity resonator.

It was mentioned above that a rod of insulating material could be used as a waveguide. Suppose the rod diameter is reduced until we have a very fine flexible strand of a transparent material such as glass or plastic. This can convey electromagnetic waves of very short wavelength, i.e., of very high frequency. In fact light waves can be transmitted successfully within such strands, which are usually called optical fibres. These have a number of applications: they can be used as waveguides to communicate over considerable distances; a bunch of fibres can transmit light over short distances, e.g., in car instrument panels; and an assembly of fibres can convey an optical image from one point to another, even round corners, permitting, for example, seeing inside the human body.

This section is a very brief introduction to waveguides. It would take the whole of a book of this size to do justice to the subject.

Chapter 17 Radiation and Antennas

17.1 Bridging Space

In the first chapter of this book the processes of radio communication were traced very briefly from start to finish. Before going on to the receiver we should know something about how the space in between is bridged.

We have just seen how electrical power can be transmitted from one place to another at the not inconsiderable speed of nearly 300 000 km per second, in the form of waves along a line consisting of two closely spaced conductors. A Southerner who was asked to explain how it was possible to signal by telegraph from New York to New Orleans replied by pointing out that when you tread on the tail of a dog the bark comes out at its head. A telegraph was like a very long dog, with the tail in New York and the head in New Orleans. 'But what about wireless?' he was asked. 'Same thing, 'cept the dawg am imagin'ry'. This reply, though ingenious, is not perhaps altogether satisfying. How is it possible to transmit the waves without any conductors to carry the currents to and fro?

The significant thing about the section on how waves are transmitted along a line (Section 16.4) is that conduction was hardly mentioned; the important properties of a line are its distributed capacitance and inductance. In other words, transmission of waves along a line is primarily a matter of oscillating electric and magnetic fields, located in the space between and around the two conductors. These conductors are not required to carry current from one end to the other, any more than waves on a pond carry the water from one side to the other; all that the electrons in the conductors do is oscillate a very short distance locally.

17.2 The Quarter-wave Resonator Again

Near the end of the same chapter we found that a line only quarter of a wavelength long and open at the end, as in Figure 17.1a, is equivalent from the generator's point of view to a series resonant circuit b; that is to say, the only impedance the generator faces is the incidental resistance of the circuit. We saw in Section 8.4 that if this resistance is small a very little generator voltage can cause a very large current to flow and build up large voltages across C and L. This remarkable behaviour of C and L is due to the electric field concentrated mainly between the plates of C and the magnetic field concentrated mainly inside and close around the turns of L. The main difference between a and b is that in a the inductance and capacitance—and therefore the fields—are mixed up together and uniformly distributed along the length. The distance of the dashed lines from the conductors in Figure 17.1a indicates the distribution of current and voltage along them; compare Figure 16.7. The

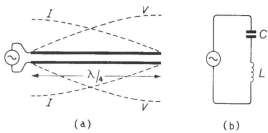

Figure 17.1 (a) A quarter-wavelength line, showing the distribution of current and voltage, and (b) its circuit equivalent

dashed lines can be regarded as marking the limits of the alternations of current and voltage, just as the dashed lines in Figure 1.3 marked the limits of vibration of a stretched string.

An essential feature of a transmission line, even such a short one as this, is that the distance between the conductors is small compared with a wavelength—or even quarter of a wavelength. The reason is that the explanation of its action is based on neglecting the time taken for the fields to establish themselves when a current flows or a p.d. is created. Provided nearly all of each field is close to the conductors, this assumption is fair enough. But when a current is switched on, the resulting magnetic field cannot come into existence instantaneously everywhere; just as a current takes time to travel along the line, so its effect in the form of a magnetic field takes time to spread out. And when the current is cut off, the effect of the remoter part of the magnetic field collapsing takes time to get back to the conductor. At points in the field one-eighth of a wavelength from the conductor of the current, the combined delay is quarter of a wavelength, or quarter of a cycle, which upsets the whole idea of self-inductance. In fact, this phase delay between a change of current and the self-induced e.m.f. calls for a driving e.m.f. in phase with the current, just as if there were a resistance in the circuit—so far as that piece of field is concerned.

Suppose that in order to study this we open out the two conductors of our quarter-wave parallel line to the fullest possible extent, as in Figure 17.2. Corresponding to our lumped L and C approximation to the original parallel line we could draw Figure 17.3. The opening out greatly increases the area through which the magnetic field can pass, so increasing L. The lines of electric field are on the average much longer, so C is decreased. Again, these two effects almost exactly cancel out, so the line is still nearly resonant to the same generator frequency. It can be brought exactly into tune by shortening it a very few per cent, which will be assumed to be done even though (as in Figure 17.2) the total length is for simplicity marked λ/2. So the

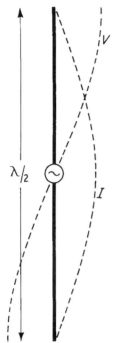

Figure 17.2 The quarter-wave line of Figure 17.1 opened fully out

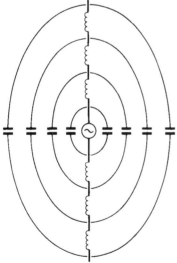

Figure 17.3 Approximate representation of the distributed inductance and capacitance in the half-wavelength system of Figure 17.2

distribution of current and voltage along the conductors is much the same as before. Looking at Figure 17.3 we would say that the current was greatest at the middle because it has to charge all the capacitance in parallel, whereas near the ends there is hardly any to charge.

Figure 17.4 shows some typical electric and magnetic lines of force. At any one radius from the generator they would trace out a sort of terrestrial globe, on which the magnetic lines were the parallels of latitude and the electric lines the meridians of longitude. This is helpful as a picture, but we must remember once again that field lines are as imaginary as lines of latitude and longitude.

17.3 A Rope Trick

It is clear that quite an appreciable portion of each field is now far enough away to be subject to a phase delay in its reaction on the conductors. To the generator the effect is as if some resistance had been inserted in the circuit. If so, that means the generator is now supplying energy. Where does it go?

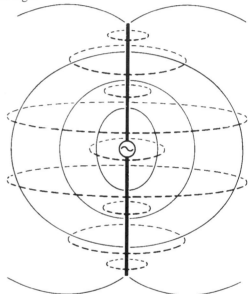

Figure 17.4 The distributed inductance and capacitance in Figure 17.2 are better indicated by drawing a few lines of force. The full lines are electric and the broken lines magnetic

When we were young we must all have taken hold of one end of a rope and waved it up and down. If this is done slowly, nothing much happens except that the part of the rope nearest us moves up and down in time with our hand, as in Figure 17.5a. But if we shake it rapidly, the rope forms itself into waves which run along it to the far end (b). This is quite an instructive analogy. The up-and-down movement of the hand represents the alternation of charge in the λ/2 rod caused by the generator. When the current is upwards it charges the upper half positively and the lower half negatively, resulting in an electric field downwards between the two. Half a cycle later the pattern is reversed. The up-and-down displacement of the rope from its middle position represents this alternating electric field. The displacement travels outward from the hand as each bit of

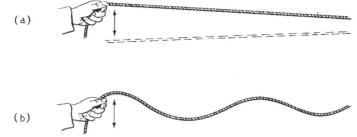

Figure 17.5 Well-known results of waving a rope up and down, (a) slowly, (b) quickly

rope passes it on to the next. If the time of each cycle of hand movement is long compared with the time for the movement to pass along the rope, all the rope for at least quite a distance appears to move as a whole 'in phase'. Because the electric field has to spread out in all directions, and not just in one direction like the rope, its amplitude falls off much more rapidly with distance from the source than does that of the rope.

When the hand is moved so fast that the rope displacement has not gone far before the hand reverses, the displacements become progressively out of phase with one another, and this is why travelling waves are formed, as in Figure 17.5b. The same applies to the alternations of electric field; they form a wavelike pattern which travels outwards, just like the electric field between the conductors of a long line in that direction, except for a falling off in amplitude because it is not going out only along one radius but all around the 'equator'.

17.4 Electromagnetic Waves

A good question at this stage would be: what keeps it going? The rope wave would not keep going unless the rope had not only displacement but *inertia*. The fact that we have already noted a resemblance between inertia and inductance (Section 4.9) may remind us that we seem to have forgotten about the magnetic field. But of course that too is spreading out, and in fact is essential to the wave motion. When we did the experiments with iron filings around a wire carrying current (Section 4.1) we said that the cause of the magnetic field was the electric current in the wire. But we know that an electric current consists of electric charges in motion, and charges are surrounded by an electric field. So the electric current, which generates the magnetic field, is really a moving electric field. We know, too, that a moving magnetic field generates an e.m.f. We tend to think, perhaps, of an e.m.f. only in a circuit, but it is really an electric field, whether there happen to be any circuits or conductors or not. Putting all this concisely: moving electric field causes magnetic field; moving magnetic field causes electric field.

Without going into rather advanced mathematics, one might guess the possibility of the two fields in motion keeping one another going. Such a guess would be a good one, for they can and do. The speed they have to keep up in order to be mutually sustaining is the same as when they are guided by the transmission line. The combination of fast-moving fields is known as an *electromagnetic wave*. The union of the two fields is so complete that each depends entirely on the other and would disappear if it were destroyed. The speed of the wave is the same as that of light—a significant fact, which led to the discovery that light waves are the same as those used for radio communication, except for their much higher frequency.

17.5 Radiation

An important thing to note about the rope trick is that the rope itself does not travel away from us—only a wavy pattern does. But there is another thing: some of the energy we put into it travels to the far end, for our waves are capable of making the post supporting the end of the rope vibrate. Electromagnetic waves are similar in this respect too; they convey energy away from the source. The outward movement of the waves from the source is called *radiation*. All variations in current cause radiation, but when the field energy is mainly localised, say between the conductors of a transmission line or the plates of a capacitor, the amount is much less than when it is spread out as in Figure 17.4. Radiation also increases with the rate of variation; the 50 Hz variations of ordinary a.c. power systems cause negligible radiation, but 50 MHz is quite a different matter. So although we have spoken of the waveforms of low-frequency a.c., this should not be taken as suggesting that they are waves in the sense used in this chapter—electromagnetic waves.

We saw in connection with the lumped tuned circuit of Figure 14.2 that the electric and magnetic fields, responsible for the capacitance and inductance, are quarter of a cycle out of phase with each other. Even in the most spread-out circuit, such as our half-wavelength rod, the majority of the energy of the surrounding fields is able to return to the circuit twice each cycle, to be manifested there as its distributed capacitance and inductance. The majority portion of each field, mainly near the source, is similarly out of phase with the other. It is distinguished by the term *induction field*, and it falls off inversely as the square of the distance, which means that most of it is quite close to the source.

In order to convey energy right away from the source, however, there have to be components of fields in phase with one another, and it is this portion, called the *radiation field*, in which we are interested now. It is in inverse proportion to the distance, which means that although it starts as a minority it spreads much more, and far away from the source it is very much in the majority.

17.6 Polarisation

Figure 17.4 shows that the electric and magnetic fields are at right angles to one another and to the direction in which they radiate outwards as electromagnetic waves. Imagine we are looking towards the vertical source and the oncoming waves. Then the electric field alternates up and down and the magnetic field side to side, as shown in Figure 17.6. This pattern can be imagined as due to a

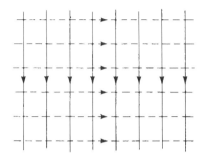

Figure 17.6 Relative direction of electric (full line) and magnetic (broken line) fields in an electromagnetic wave advancing towards the viewer

transmission line consisting of parallel strips top and bottom. At the instant depicted, the potential is positive at the top, so the electric field is downwards from it. This p.d. would be driving current towards us through the top conductor and away from us through the bottom one. The 'corkscrew rule' (Figure 4.2) indicates a clockwise magnetic field round the bottom current and anticlockwise round the top one; both combine to give a magnetic field from left to right, as shown.

If for any reason we prefer the directions of the two fields to be interchanged—electric horizontal and magnetic vertical—that can easily be arranged, by turning the radiator horizontal. Similarly for any other angle. The direction of the field in an electromagnetic wave is known as its *polarisation*. Since there are two fields, mutually at right angles, there would be ambiguity if it had not been agreed to specify the polarisation of a wave as that of its electric field. This custom has the advantage that the angle of polarisation is the same as that of the radiator.

Because radiation starts off with a certain polarisation it does not follow that it will arrive with the same. If it has travelled far, and especially if on its way it has been reflected from earth or sky, it is almost certain to have become disarranged. Another thing is that polarisation is not necessarily confined to a single angle. After reflection, radiation often has both vertical and horizontal components, and the maximum may be at some odd angle. If it is equal at all angles, the wave is said to be circularly polarised, because at any point it consists of fields of constant strength rotating continuously at the rate of one revolution per cycle. Sometimes for special purposes the radiator is designed to make the waves circularly polarised from the start.

17.7 Antennas

Having considered the process of radiation, we now turn our attention to the radiator. If intended for that purpose it is called an aerial or *antenna*. The form we developed from a quarter-wavelength transmission line, known as a half-wave *dipole*, is perhaps the most important both from the theoretical and (since the arrival of f.m. broadcasting and television) practical point of view.

We have already noted that this straight metal wire or rod is a tuned circuit, and that its sound-wave analogue is a stretched cord as in Figure 1.3. The fundamental frequency at which it resonates is the one that makes half a wavelength equal to (or, to be precise, a few per cent greater than) the length of the rod.

There are certain higher frequencies at which the same length of antenna can be set into vibration at harmonic frequencies; but because of complications due to currents flowing in opposite directions, tending to cancel out radiation, this condition is seldom used. It is not essential for an antenna to be in resonance at all, but as the radiation is proportional to the current, which is a maximum at resonance, antennas are practically always worked in that condition.

Just as with the tuned circuits in Chapter 8, the originator of the current has been shown as a generator in series; actually at the centre of the dipole. But in practice a tuned circuit is often 'excited' by an external field, as, for example, a magnetic field in inductive coupling. The same applies to our metal rod. If it is exposed to electromagnetic waves, the magnetic field pattern sweeping across it induces an e.m.f. in it. This statement must not be taken to mean that if there were no metal there would be no e.m.f. It is the movement of the magnetic field that induces the electric field, as we have seen; and that electric field constitutes an alternating e.m.f. in space.

It is unimportant whether the e.m.f. acting on the antenna is credited to the electric field or the magnetic field—they are simply two different descriptions of a single action—but it is usually convenient to work in terms of the electric component of the waves. Near a radiator it is reckoned in volts per metre, or microvolts per metre if far off. If a vertical wire two metres long is exposed to vertically polarised radiation of $100\,\mu V$ per metre, then an e.m.f. of $200\,\mu V$ will be induced in it. But if either the wire or the radiation is horizontal, no e.m.f. is induced. In general, e.m.f. is proportional to the cosine of the angle between the wire and the direction of polarisation.

It is in this way that an antenna is used at the receiving end. The only difference in principle between a sending and a receiving antenna is that one is excited by a r.f. generator connected to it, and the other by the electromagnetic waves created by the former. The factors that make for most effective radiation in any particular direction are identical with those for the strongest reception from that direction, so whatever is said about antennas as radiators can readily be adapted from the reception standpoint.

17.8 Radiation Resistance

We have noted that the dipole antenna, like the quarter-wave line from which we developed it, has inductance and capacitance. What about resistance? There is of course the resistance of the metal of which it is made, together with the various sources of loss resistance described in Sections 8.14 and 8.15. But we have seen that the waves radiated from a sending antenna carry away energy from it, otherwise they would not be able to make currents flow in receiving antennas. It is convenient to express any departure of energy from a circuit in terms of resistance (Section 2.18). But whereas the other kinds of antenna resistance are wasteful, this kind, called *radiation resistance*, measures the ability of an antenna to do its job.

The radiation resistance of an ordinary tuning circuit is generally very small compared with its other resistance (radiation, after all, is not its job!); at the other extreme the loss resistance of a dipole, made perhaps of copper tubing, is small compared with its radiation resistance, which at resonance is about 70 Ω if it is measured at the centre where the current is greatest. If therefore a current of, say, 2 A can be generated there, the power radiated, calculated in the usual way as I^2R, is $2^2 \times 70 = 280$ W. If the radiation resistance were measured at some other point along the dipole it would be greater, but this would not mean greater radiation, because the current there would be less.

A receiving antenna too has radiation resistance, and some of the energy imparted to it by the waves is re-radiated—a fact of great importance in the design of receiving antennas, as we shall see.

17.9 Directional Characteristics

We based our study of radiation from a dipole on the production of waves along a rope. But of course there is no reason for electromagnetic radiation to be confined along a single line like this; it goes out equally in all directions around the 'equator'. So the wave pattern in this plane can be likened to the ripples created on the surface of a pond by dropping in a stone. A skeleton of it could be produced by having ropes radiating from one's hand like the spokes of a horizontal wheel. But what about other angles, upwards and downwards?

Is it too much to imagine oneself in outer space, where ropes have no weight, at the centre of great numbers of them stretching in every direction? Those immediately above and below receive no wavy pattern; only lengthwise stretching and relaxing. At intermediate angles there would be a combination of the two effects, as illustrated by the few samples in Figure 17.7. By filling in as many others as possible, we can get an idea of the whole pattern of radiation: most strongly around the equator, and none at all along the axis. The pattern would be still more completely represented by the effect on a vast light elastic jelly of making a point at its centre vibrate up and down. The relative strength of radiation in different directions is, however, usually represented by what is called a *polar diagram*. This is a curve drawn around a centre, such that the distance from the centre to the curve in any direction represents the radiation in that direction. Ideally, a polar diagram would be a three-dimensional surface, but on paper one usually has to be content with the curves, which are sections of the surface at particular planes.

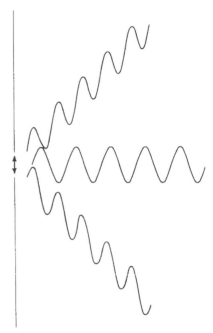

Figure 17.7 If the rope trick of Figure 17.5(b) were performed with ropes radiating in all directions, the results would look something like this

Figure 17.8a is the horizontal polar diagram of a vertical dipole, showing that radiation (or reception) is equal in all

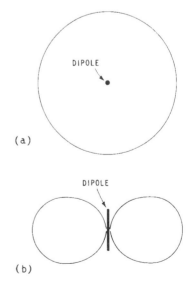

Figure 17.8 Polar diagrams of radiation from a simple half-wavelength dipole, seen (a) end-on, and (b) side view

horizontal directions. At *b* is the vertical polar diagram, showing how the radiation varies from maximum at right angles to the dipole to nil end-on. So if a dipole is placed horizontally it can be used to cut out undesired reception from two opposite directions. As regards reception, these diagrams assume, of course, that the radiation arriving has the correct polarisation. A horizontal dipole does not respond to vertically polarised waves from any direction.

17.10 Reflectors and Directors

Any metal rod is a tuned circuit which will have maximum currents set going in it by waves twice its own length. These currents radiate waves, just as if they were caused by a directly connected generator. If such a rod is placed close to a radiating dipole, the re-radiation from it is strong enough to modify the primary radiation very considerably, subtracting from it in some directions and adding to it in others. When the parasitic antenna (as it is called) is about the same length as the driven one, and quarter of a wavelength away from it, the phase of the re-radiation is such as to reverse much of the radiation beyond the rod, which is therefore called a *reflector*—see Figure 17.9. The same principle applies to a receiving antenna, and reflectors

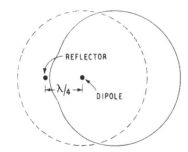

Figure 17.9 Showing how the equal-all-round radiation of Figure 17.8a (repeated here dashed) is modified by an unconnected dipole placed quarter of a wavelength away from the first, and of such a length as to have a 45° phase angle

are frequently used; not so much to strengthen the wanted signal (though that is usually helpful) but to cut out interference from the opposite direction.

The effect just described depends on the phase relationship between the primary and secondary radiation. When considering tuned circuits we may have noticed (from Figure 8.8, for example) that a small departure from exact tuning causes a large phase change between voltage and current. It is the same with dipoles; a small increase or decrease in length alters the phase and therefore the polar diagrams. The phase can also be varied by the spacing between the rods. If the parasitic rod is shorter than the driven one, and only about 0.1 λ away from it, the effect is opposite to that shown in Figure 17.9, so such a rod is called a *director*. The use of reflectors or directors affects the resistance of the driven dipole as a load, usually lowering it.

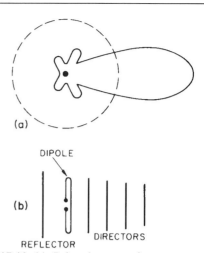

Figure 17.10 (a) Polar diagram of an antenna system consisting of two parallel units each made up of a dipole, one reflector and four directors (b). The diagram for a single dipole is again shown dashed for comparison

The one-way tendency can be increased by using both a reflector and a director, one on each side; or even several directors, as in the array shown in Figure 17.10*b*, which is called a *Yagi* antenna. Its polar diagram (*a*) shows a very considerable advantage over that of the single dipole. For the rather peculiar main dipole see Section 17.15.

But this is only the beginning of what can be done. The number of driven (or receiver-connected) dipoles can be increased, by placing them side by side, or in stacks one above the other, or both, and they also can be provided with parasitic elements. The underlying principle of all is the same: to arrange them so that the separate radiations add up in phase in the desired directions and cancel out in the undesired directions. It is necessary to consider not only the distribution horizontally but also the angle of elevation above the horizon. By using a sufficient number of dipoles, suitably arranged, the radiation can be concentrated into a narrow beam in any desired direction.

Section 16.9 mentions the high impedance of a short-circuited λ/4 length of transmission line. Suppose we have two such lengths of twin-wire line connected in parallel as in Figure 17.11*a*. The input impedance is still very high and the arrangement is useless as an antenna because radiation from one vertical conductor is cancelled by that from the other. But let us now increase the width of the conductors in the vertical plane to such an extent that the arrangement becomes a vertical slot in a conducting sheet as at *b*. This change greatly modifies the shape of the electric and magnetic fields and, as a result, the impedance is lowered to about 500 ohms and the slot becomes an efficient radiator.

Surprisingly a vertical slot radiates horizontally polarised waves. In fact the slot can be regarded as a dual of the equivalent dipole. The dipole is a λ/2 conductor in non-conducting space; the slot is a λ/2 non-conductor in a conducting sheet. The electric field from the slot corresponds to the magnetic field of the dipole; the magnetic field from the slot corresponds to the electric field of the dipole.

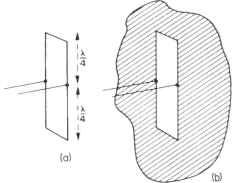

Figure 17.11 Two λ/4 short-circuited lengths of twin-wire transmission line (a) become a slot antenna when conductor width is indefinitely extended (b)

Slot antennas are used at v.h.f./f.m. transmitters. The mast top is a vertical conducting cylinder with vertical slot radiators at 90° intervals around it to give unidirectional radiations.

17.11 Antenna Gain

Concentrating the radiation in the direction or directions where it will be most useful increases the apparent power of the transmitter. The number of times the intensity of radiated power in any particular direction is greater than it would be if the same amount of power were radiated equally in all directions is called the *power gain* (or just 'gain') of the antenna in that direction. For example, if the radiated power were concentrated uniformly into one-tenth of the whole surrounding sphere the gain would be 10. Within that beam, the radiation would be the same as if ten times the power were being radiated equally in all directions. The effective power of a transmitter is therefore equal to the actual radiated power multiplied by the antenna gain; this figure is known as the *effective radiated power* (e.r.p.).

In broadcasting to the area around the transmitter, the aim is to increase the e.r.p. by directing as much of the radiation as possible horizontally, rather than upwards where it would be wasted. The apparent power of a transmitter, so far as any particular receiver is concerned, can, of course, be further increased by using a receiving antenna with gain in the direction of the transmitter.

A simple dipole, by favouring the equatorial as compared with polar directions, has a gain of 1.64, and the gains of v.h.f. and u.h.f. receiving antennas are sometimes quoted relative to this rather than to the theoretical spherical radiator.

17.12 Choice of Frequency

A feature of the dipole antenna and its derivatives which we have considered so far is that their dimensions are strictly related to the wavelength and therefore to the frequency (Section 1.9). What decides the choice of frequency: a convenient aerial length, or something else?

The shorter the wavelength and higher the frequency, the smaller and cheaper the antenna and the more practical it is to direct its radiation and increase its gain. For example, as long ago as 1931 a radio link was established across the English Channel on a wavelength of 17 cm, a half-wave antenna for which is only about 8.89 cm long! Why, then, erect at vast expense antennas hundreds of metres high?

The main reason is that such waves behave very much like the still shorter waves of light, in not being able to reach anywhere beyond the direct line of 'view' from the

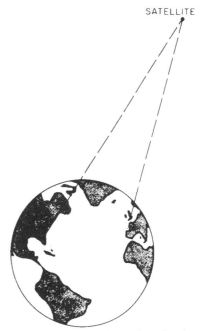

Figure 17.12 'Geostationary' satellite for long-distance microwave communication

transmitting antenna. So if the cheapness is to be maintained the range is usually restricted to a few kilometres. It can be increased to one or even two hundred kilometres by setting up the antennas on high towers or mountains, and to thousands of kilometres by using man-made satellites, for re-transmission, as in Figure 17.12. The most useful type of satellite for this purpose is in orbit at such a distance (35 860 kilometres) that its period of revolution around the earth is 24 hours, so that if established in the same direction as the earth's daily rotation the satellite maintains a constant station relative to places on earth. It is therefore suitable for both television broadcasting and point-to-point communication.

Wavelengths shorter than 10 m are used for television, not because the shortness of wavelength in itself has any particular merit for that purpose, but because it corresponds to a very high frequency, which, as we saw in Chapter 15, is necessary for a carrier wave that has to carry modulation frequencies of up to several MHz. They are also used for short-distance communication such as in police cars, radiotelephone links between islands and mainland, and other specialised short-range purposes.

17.13 Influence of the Atmosphere

At various heights between 96 and 480 kilometres above the earth, where the atmosphere is very rarefied, there exist conditions which cause radio waves to bend, rather as light waves bend when they pass from air into water. The very short (high-frequency) waves we have just been considering normally pass right through these layers and off into space, as shown in Figure 17.13a. They are therefore suitable for communication with and the control of spacecraft. Where they are close to the ground, they are rapidly absorbed, as indicated by the shortness of the ground-level arrow. So their effective range there is limited.

As the frequency is reduced below about 30 MHz—the exact dividing line depends on time of day and year, sunspot activity, and other conditions—the sky wave is bent so much that it returns to earth at a very great distance, generally several thousands of kilometres. Meanwhile the range of the wavefront travelling along the surface of the earth (and therefore termed the ground wave) increases slowly (Figure 17.13b). Between the maximum range of the ground wave and the minimum range of the reflected wave there is an extensive gap, called the *skip distance*, which no appreciable radiation reaches.

As the frequency is further reduced this gap narrows, and earth reflections may cause the journey to be done in several hops (c). Since the distances at which the sky waves return to earth vary according to time and other conditions as mentioned, it is rather a complicated business deciding which frequency to adopt at any given moment to reach a certain distance. But a vast amount of data has been accumulated on this and a judicious choice enables reliable communication to be maintained at nearly all times. As waves usually arrive at the receiver by more than one path simultaneously and tend to interfere with one another, fading and distortion are usual unless elaborate methods are adopted for sorting them out. At a certain frequency, of the order of 2 MHz, the ground wave and reflected wave begin to overlap at night (e), while during daylight the reflected wave is more or less absent. Over the ranges at which there is overlap the two waves tend to interfere and cause fading and distortion, as they do with more than one reflected wave.

Finally, the range of the ground wave increases and becomes less affected by daylight or darkness, so that frequencies of 50 kHz–20 kHz have a range of thousands of kilometres and are not at the mercy of various effects that make long-distance short-wave communication unreliable. For this reason they were originally selected as the only wavelengths suitable for long ranges, but now only a small fraction of long-distance communication is borne by very long waves. The disadvantages of the latter are (1) the enormous size of antenna needed to radiate them; (2) the low efficiency of radiation even with a large and costly antenna system; (3) the higher power needed to cover long ranges, due largely to (4) the great intensity of atmospherics—interference caused by thunderstorms and other atmospheric electrical phenomena; and (5) the very limited number of stations that can work without interfering with one another, because the waveband is so narrow in terms of frequency—which is what matters; see Chapter 15.

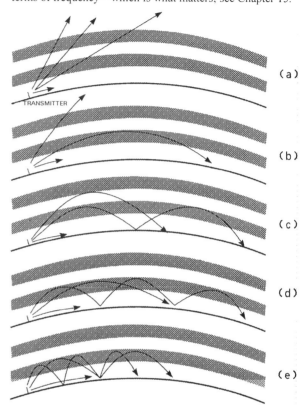

Fig 17.13 Showing (not to scale) the relative ranges of ground wave and reflected wave from very high frequencies (a) to medium (e)

17.14 Earthed Antennas

We see, then, that although for convenience of dipole antenna size one would choose frequencies of a few hundred MHz, these are not suitable for all purposes. For distances of more than about 80 kilometres it is usually necessary to use longer waves (lower frequencies). Until now we have assumed that the antenna is suspended well clear of the earth, but this is not convenient with a dipole for wavelengths of many metres, and the length of the antenna itself may be unwieldy. Marconi used the ground itself as the lower half of a vertical dipole; the remaining half sticking up out of it is therefore often called a quarter-wave Marconi antenna. The current and voltage distribution are shown in Figure 17.14, which is the same as the top half of Figure 17.2.

For the earth to be a perfect substitute for the lower half of a dipole it must be a perfect conductor, which of course it never is. Sea water is a better approximation. To overcome the loss due to imperfectly conducting earth, it is a common practice to connect the lower end of the antenna to a radial system of copper wires. An insulated set of wires stretched just above the ground is known as a *counterpoise*. Alternatively, to avoid tripping up people or animals, the wires are often buried just below the surface, as shown dashed in Figure 17.14.

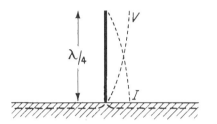

Figure 17.14 The earth can be used as one-half of a radiator, leaving a quarter-wavelength wire above earth

17.15 Feeding the Antenna

One advantage of the Marconi antenna is that it places the middle of the dipole at ground level, which is helpful for coupling it to the transmitter or receiver. Even so, it is not always convenient to have the output circuit of a transmitter close enough to the foot of the antenna to couple it directly, and it is seldom convenient to have either transmitter or receiver at the centre of a suspended dipole, so there is a need for some means of linking the antenna with the ground equipment. Chapter 16 showed how the problem can be conveniently and efficiently solved by means of a transmission line designed to match the impedance of the antenna. The object of matching, we recall, is to avoid reflections, which prevent the maximum transfer of power to or from the antenna and cause excessive line currents and voltages at a transmitter, or signal echo effects at a receiver.

We have noted that an ordinary centre-fed dipole has a radiation resistance of about $70\,\Omega$. At resonance, which is the normal working condition, inductance and capacitance cancel one another out. Other kinds of resistance are usually negligible, so the dipole can be treated as a $70\,\Omega$ load. The simplest scheme is to use a line having a characteristic resistance, R_0, of about $70\,\Omega$, as in Figure 17.15. Here the dipole can be regarded as a radiating extension of the line. At the other end, the line should be matched to the transmitter output circuit.

Figure 16.4 shows that $70\,\Omega$ is an awkward R_0 for a parallel-wire line, for the wires have to be almost touching; it is also a long way from the lowest-loss value. On the other hand it is about optimum for coaxial cable. Moreover the coaxial type radiates less and picks up less interference. But it has the disadvantage of being unsymmetrical; that is to say, unbalanced with respect to earth. The outer conductor is in fact itself normally earthed, so although the line can easily be coupled to the ground equipment (which

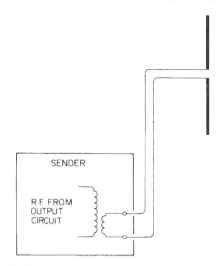

Figure 17.15 A simple dipole connected to a transmitter (or receiver) by means of a matched-impedance transmission line

is usually earthed) it is not strictly correct to connect it to a dipole, which is symmetrical. Where maximum efficiency and optimum radiation pattern are important, as in a transmitter, a balance-to-unbalance transformer (commonly called a balun) is used; for receiving, the unbalance is often tolerated. Alternatively, parallel-wire line of relatively high R_0 can be matched to the dipole by means of a transformer such as the one shown in Figure 16.8.

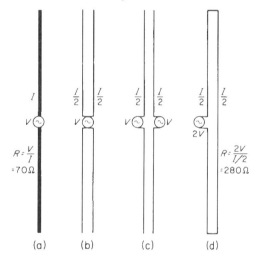

Figure 17.16 Development of folded dipole

In some cases it may be more convenient to alter the input resistance of the antenna itself. For example, in a large array where many dipoles have to be fed in parallel the total resistance would be very low if they were fed at their centres, which is analogous to feeding tuned circuits in series. We found that a tuned circuit in parallel was a relatively high resistance, and the equivalent of this is

Figure 17.17 A common method of tuning an antenna to a desired wavelength

feeding a dipole at its end. With one dipole this resistance (about $3600\,\Omega$) would be too high to match any kind of feeder, but a large number in parallel can be so matched.

It is often convenient to feed a dipole from a parallel-wire line having an R_0 of the order of $300\,\Omega$. There is a method of adapting the dipole to match this: its theory can be developed as in Figure 17.16, where *a* shows the ordinary dipole with its input resistance, at its centre, of about $70\,\Omega$. Imagine now that it is split down the middle, as at *b*. Both halves are still connected to the generator, which is maintaining the same voltage, V, so half the original current (I) goes into one half dipole and the other half into the other. Next, suppose the energising of the two half antennas is entrusted to separate generators (*c*) each supplying $I/2$. If they are in phase with one another, the ends of the two dipoles will be at the same potential so can be joined together without making any difference. Finally (*d*) the two voltages V are supplied by one generator, which must therefore provide twice the voltage of the original one (*a*) to make half the original current flow into the modified antenna, called a *folded dipole*. Its resistance is therefore four times that of an ordinary dipole, so is matched by a $280\,\Omega$ feeder.

A folded dipole is shown as the main element in the Yagi antenna in Figure 17.10, because with many directors and reflectors the impedance of a dipole is much reduced and a folded type may be a better match than an ordinary one for a coaxial line. Moreover, the centre point of the non-fed half is at zero potential so needs no insulation if the antenna is supported there.

17.16 Tuning

A dipole antenna is self-tuned, by virtue of its chosen length, to a particular frequency. But an antenna is often required to work over a very wide range of frequency, especially for receiving. Instead of varying the length of the antenna wire, it is usual to augment its distributed inductance and capacitance by means of variable capacitors or inductors, as, for example, in Figure 17.17. These tuning components reduce the resonant length of the antenna for a given frequency, which may be very desirable; for example, even a quarter-wave self-tuned antenna for 30 kHz would have to be over 12 000 metres high! It is quite usual for a considerable proportion of the reactance of an antenna system, especially for medium and long waves, to be in concentrated form; but it must be understood that this is at the expense of radiation or reception.

Where height is limited, by restrictions imposed by aviation or by the resources of the owner, it is possible to increase radiation moderately by adding a horizontal portion, as in the T or inverted L antennas, Figure 17.18. The effect of the horizontal extension is to localise the bulk of the capacitance in itself so that the current in the vertical part, instead of tailing off to zero, remains at nearly the

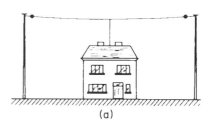

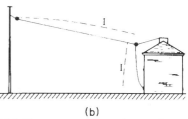

Figure 17.18 To increase the radiation from a vertical antenna without increasing its height, a horizontal top is commonly added, forming (a) the T or (b) the inverted L type. Note the more uniform current distribution in the vertical part (b)

maximum and therefore radiates more. The top portion radiates too, but with a different polarisation, so the addition due to it may not be very noticeable in a vertical receiving antenna.

17.17 Effective Height

In general, the greater the radiation resistance of an antenna the more effective it will be for sending. But a more useful piece of information about an antenna when it is used for receiving is what is called its *effective height*. It is the height which, when multiplied by the electric field strength, gives the e.m.f. generated in it. For example, if the wave from a transmitter has a strength of 2 mV per metre at a receiving station, and the effective height of the antenna is 5 metres, the signal voltage generated in the antenna will be 10 mV. The effective height is less than the physical height. It can be calculated for a few simple shapes, such as a vertical wire rising from flat open ground, but usually has to be measured, or roughly estimated by experience.

17.18 Microwave Antennas

What happens if, instead of making the wavelength longer than is most handy for dipoles, it is made much shorter—say a few centimetres? The radiation from a single dipole is of course limited by its size, and to bring the e.r.p. (Section 17.11) up to a useful figure a relatively large reflector is needed. Instead of an enormous number of small dipoles, it is made of continuous metal, or, to reduce weight and wind resistance, wire netting. The currents induced in this by the driven dipole cause reflection, but in a more controllable pattern, for the metal can be formed into a parabolic reflector, like a large headlight or bowl heater, usually called a dish. In this way the radiation or reception can be confined very closely in the required direction.

Just as at these frequencies the two-conductor transmission line gives place to a waveguide (Section 16.10), the dipole gives place to a horn, which is an expanding end to the waveguide, arranged to direct the radio power on to the reflector.

Such frequencies and antennas are commonly used for radar, where narrow beaming is essential, and for point-to-point transmission of television signals.

17.19 Inductor Antennas

In all the antennas considered, the circuit between the terminals is completed by the capacitance from one arm to the other or to earth. Although they have inductance, they can be regarded roughly as opened-out capacitors. It is possible instead to use an opened-out inductor, in which of course there is a wire path all the way. This type is known as a frame or loop antenna, and because its loss resistance is usually larger than the radiation resistance it is seldom used as a radiator, but mainly for receiving, especially for direction finding.

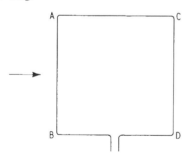

Figure 17.19 The effectiveness of a receiving antenna consisting of a loop or coil of wire depends on the direction from which the waves are arriving

We noted that reception by an open antenna is usually calculated from the strength of the electric field, although the same answer would be given by the magnetic field. The opposite applies to loop antennas. Taking as an example a single square turn of wire as in Figure 17.19, one can see that waves arriving broadside on, whatever their polarisation, induce no net e.m.f. in the antenna. A vertical electric field would cause e.m.f.s in the portions AB and CD, but these would be exactly in opposition around the loop. It is easier, however, to see that the magnetic field, being in the same plane as the loop, would not link with it at all, and that is sufficient to guarantee no signal pick-up.

From any direction in the plane of the paper—say that indicated by the arrow—the magnetic field is at right angles to the loop, so links it to the maximum extent, provided the width is not more than half a wavelength. Although the e.m.f.s from A to B and C to D are equal (and most of the time opposite, if the loop is small compared with a wavelength) they are out of phase, so there is a net e.m.f. around the loop.

At intermediate angles the results are intermediate, so the polar diagram is very like that of a dipole at right angles to the loop.

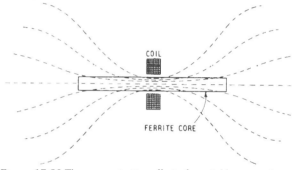

Figure 17.20 The concentrating effect of a suitable magnetic core on the magnetic field of the waves greatly increases the effectiveness of a small coil as an antenna

The size of a loop antenna in a portable broadcast receiver is, at most, very small compared with the wavelengths to be received, so the signal e.m.f. obtained by it is relatively weak. If, in order to embrace as much field flux as possible, antenna wire is wound around the set just inside its outer covering, it also embraces a good deal of assorted material, much of which causes considerable losses of the kind referred to in Section 8.15. If, however, the winding is provided with a high-permeability core, the magnetic flux over a wide area tends to pass through the core, as shown in Figure 17.20, just as electric current favours a low-resistance path. In this way a coil less than 2.5 cm in diameter is equivalent to a very much larger one without the core. It is essential, of course, for the core to be of low-loss material; the non-conducting magnetic materials known as ferrites are used. The tuning inductance so created is too much for very high frequencies, so receivers with facilities for v.h.f. usually include a telescopic vertical antenna.

Chapter 18 Detection

18.1 The Need for Detection

Turning to the reception end in Figure 1.11, we see a number of boxes representing sections of a receiver. Although all of these sections are included in most receivers, they are not all absolutely essential. So we begin with the detector, because it is the one thing indispensable, for reasons now to be explained.

At the receiving antenna, the modulated carrier wave generates e.m.f.s whose waveforms exactly match those of the currents in the sending antenna. The frequency of those currents, as we know, is so high as to be inaudible. It has been chosen with a view to covering the desired range most economically (as discussed in the previous chapter), and bears little relationship to the frequencies of the sounds being transmitted. In order to reproduce these sounds, it is necessary to produce currents having the frequencies and waveforms of the sounds. For example, given the amplitude-modulated carrier wave shown in Figure 15.1c (or Figure 18.1a), the problem is to obtain from it Figure 15.1a. The device for doing so is the *detector*. It is an example of a type of circuit which extracts the modulation frequency from a modulated carrier as the desired product, and is now commonly known as a demodulator. (This term was formerly used for an effect that removed interfering modulation as an undesired audible product; see Section 22.10. This is a logical use of the word; cf 'decontamination').

For receiving frequency-modulated signals (Section 15.4) it is necessary also to include in the demodulator means for converting the frequency variations into the amplitude variations on which the detector works. This device—the *discriminator*—is considered later on in this chapter.

18.2 The Detector

The simplest method of detection is to eliminate half of every cycle, giving the result shown in Figure 18.1b. At first glance, this might not seem to differ fundamentally from a, for the original frequency appears to be still there, though at half strength. It is quite true that the original frequency is there, and means have to be provided for getting rid of it. The fundamental difference is that whereas the average value of each cycle in a is zero (because the negative half cancels the positive half), in b it is proportional to the amplitude (at that moment) of the modulated carrier wave, which in turn is proportional to the instantaneous value of the a.f. modulating signal. The dashed line in Figure 18.1b indicates the average level of the current, and it can be seen to vary at the same frequency and with the same waveform as Figure 15.1a. There is admittedly some loss in amplitude, but we shall see that the types of detector in common use manage to avoid most of this loss and give an output almost equal to the peak values.

The process of suppressing half of each cycle, converting alternations of current into a series of pulses of unidirectional current, is known by the general term *rectification* already met in Section 5.5. The detector usually consists of a rectifier adapted to the particular purpose now in view, and may be taken to include means for extracting the desired a.f. from a mixture like Figure 18.1b. It should be noted that this rectified signal contains not only r.f. and a.f. components, but also a unidirectional component (d.c.). That is, the desired a.f. current does not alternate about zero, as in Figure 15.1a, but about some steady current—positive or negative, depending on which set of half-cycles was eliminated by the rectifier. Although this d.c. is unnecessary and in fact undesirable for reproducing the original sounds, it has its uses as a by-product (Section 22.11).

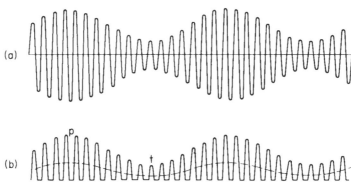

Figure 18.1 (a) Sample of an amplitude-modulated carrier wave, similar to Figure 15.1c. The average value of each cycle is zero: but if one half of each is eliminated (b) the average value of the other halves, shown dashed, fluctuates in the same manner as the modulating waveform (e.g., Figure 15.10)

18.3 Rectifiers

A perfect rectifier would be one that had no resistance to current flowing in one direction, and an infinite resistance to current in the opposite direction. It would, in fact, be equivalent to a switch, completing the circuit during all the positive half-cycles in Figure 18.1a, thus enabling them to be passed completely, as in b; and interrupting it during all the negative half-cycles, suppressing them completely, as also shown in b. For very low-frequency alternating currents such as 50 Hz it is possible to construct such a switch, known as a vibrator rectifier, but any device involving mechanical switching is impracticable at radio frequencies.

The two main types of electronic diode rectifier have already been considered in Chapter 9. In the early days of radio, point-contact semiconductor rectifiers using various minerals were much used under the name of crystal detectors. Between the world wars these were almost entirely superseded by thermionic valves, but were revived and improved during the Second World War as the point-contact silicon diodes used for radar. Now, radio detectors are nearly always semiconductor diodes.

In considering resistance we began with the current/voltage graphs of ordinary conductors (Figure 2.3), and noted that because such graphs were straight lines the conductors described by them were called linear, and that since linear resistances were covered by the very simple relationship known as Ohm's law there was really no need to spend any more time drawing their graphs. Then we came to valves, and again drew current/voltage graphs, such as Figure 9.3, and found that they were curved, indicating non-linear characteristics, not conforming to Ohm's law (except in the modified and limited sense of Figure 11.4, and then only approximately). All non-linear conductors are capable of some degree of rectification, and all rectifiers are necessarily non-linear in the above sense; that is, their resistance depends on the amount of voltage or current, instead of being constant as in Ohm's law.

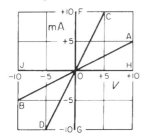

Figure 18.2 Characteristic curves of various linear resistances. The line COD is seen to represent 500 Ω. If the resistance to negative voltages is different from that to positive, as shown, for example, by the line COB, the result is a partial rectifier

Here now in Figure 18.2 are some more current/voltage characteristics, the lines COD and AOB representing (as can be calculated from the current and voltage scales) 500 Ω and 2000 Ω respectively. As we noticed before, the steeper the slope the lower the resistance. To go to extremes, the line FOG represents zero resistance, and HOJ infinite resistance; so FOJ is the graph of a perfect rectifier. COB is an example of a partial rectifier—a resistance of 500 Ω to positive voltages and 2000 Ω to negative voltages. Now apply an alternating voltage to a circuit consisting of any of the foregoing imaginary resistances in series with a fixed resistance of 1000 Ω (Figure 18.3a). The voltage across the 1000 Ω—call it the output voltage—is equal to the applied voltage only if R is zero. Both voltages are then indicated by the full line in Figure 18.3b. If R were 500 Ω (COD), the output would be represented by the dashed line CD, and if 2000 Ω by the dashed line AB.

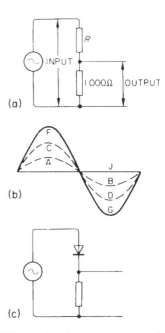

Figure 18.3 The results of applying a sinusoidal voltage to a circuit (a) consisting of any of the resistances represented by Figure 18.2 (denoted by R) in series with 1000 Ω are indicated by (b), which is lettered to correspond with the various characteristics in Figure 18.2. When R is a rectifier it is represented as in (c)

If now a rectifier is substituted for R, as shown in Figure 18.3c, the positive and negative output voltages are unequal. For example, the perfect rectifier (FOJ) gives the output FJ in b; and the partial rectifier gives CB. The average voltage, taken over each whole cycle, is, of course, nil when R is not a rectifier. Using the perfect rectifier, the average of the positive half is 63.6% of the maximum (Section 5.4), while the negative is nil; so the average rectified current taken over a whole cycle is 31.8% of the peak value. Using the rectifier with the graph COB, the negative half-cycle (B in Figure 16.3b) cancels out half of the positive half (C), which is two-thirds of the input; so the rectified result is less than 11% of the peak input.

18.4 Linearity of Rectification

Whether complete or partial, however, the degree of rectification from rectifiers having characteristics like those in Figure 18.2 would be independent of the amount of voltage, provided, of course, that the rectifier was arranged to work exactly at the sharp corner in its characteristic, whether that happened to correspond with zero voltage, as in Figure 18.2, or not. So if we were to draw a graph showing the net or average or rectified voltage output against peak alternating voltage input to a rectifier of this kind we would get a straight line, such as the two examples in Figure 18.4, corresponding to characteristics FOJ and COB in Figure 18.2. These are therefore called linear rectifiers, it being understood that the term is in reference to a graph of this kind and not of the Figure 18.2 kind.

It should be clear from Figure 18.1b that it is necessary for the rectifier to be linear if the waveform of the modulation is not to be distorted. It is therefore unfortunate that in reality there is no such thing as a perfectly linear rectifier. The best one can do is to approach very close to perfection by intelligent use of the rectifiers available. The weak feature of all of them is the more or less gradual transition from high to low resistance, exemplified by the typical 'bottom bend' in Figure 9.3, which causes weak signals to be rectified less completely than strong ones. Such a characteristic is represented in Figure 18.4 by the

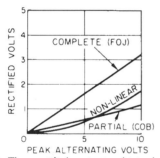

Figure 18.4 The rectified outputs obtained by applying alternating voltages to various rectifiers are shown in these characteristic curves, lettered to correspond with Figures 18.2 and 18.3

curved line, and is explained in greater detail later (Section 18.11). Semiconductor diode characteristics are considerably better, if we allow for the offset or displacement from zero, especially with silicon (Section 9.11).

18.5 Rectifier Resistance

There were two reasons for the poor results of rectifier COB in Figure 18.2 compared with FOJ. One was its *forward resistance* (500 Ω), which absorbed one-third of the voltage, leaving only two-thirds for the 1000 Ω load. The other was its finite *backward resistance* (2000 Ω), which allowed current to pass during the negative half-cycles, so neutralising another one-third of the voltage.

With such a rectifier it would clearly be important to choose the load resistance carefully in order to lose as little as possible in these ways. As it happens, 1000 Ω does give the largest output voltage in this case, the general formula being $\sqrt{(R_F R_B)}$, where R_F and R_B are respectively the forward and backward resistances of the rectifier.

As compared with crystal diodes, the vacuum diode does have one point in its favour. Although it is not perfect in the forward direction, for it always has some resistance, it is practically perfect in the reverse direction; that is to say, it passes negligible backward current, even when quite large voltages are applied. So even if, owing to a high forward resistance, the forward current were very small, the output voltage could, theoretically at least, be made to approach that given by a perfect rectifier, merely by choosing a sufficiently high load resistance. Silicon diodes are now available with very high backward resistance, and lower forward resistance than thermionic types. They have a sharper rectification bend, and of course need no cathode heater. But they need a little forward bias.

Although there are limits to the load resistance that can be used in practice, the loss of rectification can usually be made insignificantly small. In fact, by a very simple elaboration of Figure 18.3c—a capacitor in parallel with the load resistance—it is possible to approach a rectification efficiency of 100% instead of the 32% that is the maximum without it, and so do nearly three times better than our 'perfect' rectifier. This point deserves closer consideration.

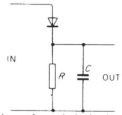

Figure 18.5 Diode rectifier with the load resistor, R, shunted by a reservoir capacitor, C

18.6 Action of Reservoir Capacitor

The modified rectifier circuit is now basically as Figure 18.5. Suppose, in order to get the utmost rectified voltage, we made the load resistance R infinitely great, and that the backward resistance of the diode was also infinite. The capacitor C would then be in series with the diode, with no resistor across it. To simplify consideration we shall apply a square wave instead of a sine wave; amplitude 10 V (Figure 18.6a). We shall also assume that its frequency is 500 kHz, so that each half-cycle occupies exactly one-millionth of a second (1 μs). At the start C is uncharged, and therefore has no voltage across it. It can acquire a charge only by current flowing through the rectifier, which offers a certain amount of resistance, and so when the first positive half-cycle arrives its 10 V at first appears wholly across that resistance, as shown by the full line V_D in Figure 18.6b. The current driven by this voltage starts charging the capacitor, whose voltage thereby starts to rise towards

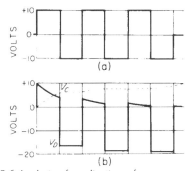

Figure 18.6 Analysis of application of a square alternating voltage (a) to the circuit Figure 16.5. The dashed line in (b) represents the voltage across C; the full line, that across the diode

10 V, as shown by the dashed line. It is clear that the voltages across rectifier and capacitor, being in series, must add up to give 10 V so long as that is the applied voltage; and the full-line curve in Figure 18.6b has been drawn accordingly. We have in fact exactly the same situation as the one we discussed in connection with Figure 3.6, so no wonder the curves have the same shapes as there. The greater the resistance of the rectifier and the capacitance of the capacitor, the slower the rate of charge, just as with a very large balloon inflated through a very narrow tube. As we noted with Figure 3.6, the capacitance in microfarads multiplied by the resistance in megohms is the time constant—the number of seconds required for the capacitor voltage V_C to reach 63% of the applied voltage. Suppose that the rectifier resistance (assumed constant) is 0.01 MΩ and the capacitance 0.0001 µF. Then the time constant is 0.000 001 second or 1 µs. In this case that very conveniently happens to be the time occupied by one half-cycle of the 500 kHz applied voltage. So at the end of the first positive half-cycle V_C is 6.3, while V_D has dropped to $10 - 6.3 = 3.7$. Then comes the negative half-cycle. The diode ceases to conduct, and while the capacitor therefore cannot charge up any more it likewise has no conducting path through which to discharge, and so remains at 6.3 V until the second positive half-cycle. Meanwhile, V_C (6.3 V) plus V_D must now be equal to the new applied voltage, −10. The voltage across the rectifier must therefore be −6.3. Another way of looking at this is to say that the input changes from +10 V to −10 V, so V_D becomes 20 V more negative than +3.7 V, i.e., −16.3 V.

The net voltage applied to the capacitor when the second positive half-cycle arrives is $10 - 6.3 = 3.7$ V, so at the end of this half-cycle the voltage across the capacitor will increase by 63% of 3.7, or 2.3 V, which, added to the 6.3 it already possessed, makes 8.6. The charge thus gradually approaches the peak signal voltage in successive cycles, while the average voltage across the rectifier falls by the same amount, as shown in Figure 18.6b. The rectified output voltage therefore approaches 100% of the peak applied r.f. voltage. A similar result is obtained (but more slowly) with a sine-wave signal.

Because C maintains the voltage during the half-cycles when the supply is cut off, it is called a *reservoir capacitor*.

18.7 Choice of Component Values

This, of course, is excellent so far as an unmodulated carrier is concerned. When it is modulated, however, the amplitude alternately increases and decreases. The increases build up the capacitor voltage still further; but the decreases are powerless to reduce it, for the capacitor has nothing to discharge through. To enable V_C to follow the modulation it is necessary to provide such a path, which may be in parallel with either the capacitor or the diode, so long as in the latter case there is a conducting path through the input to complete the discharge route. The resistance of the path should be low enough to discharge the capacitor C at least as fast as the modulation is causing the carrier amplitude to decline. The speed of decline increases with both depth and frequency of modulation, and calculating the required resistance properly is rather involved, but a rough idea can be obtained as follows.

Suppose, for example, the highest modulation frequency is 10 000 Hz. Then the time elapsing between the peak of modulation and trough (p to t in Figure 18.1b) is half a cycle, or 1/20 000th of a second (50 µs). If C had to discharge at least 63% of the difference in carrier amplitude in that time, the time constant would have to be not more than 50 µs, and to avoid appreciable loss and distortion of top frequencies had better be less. Of the two factors making up the time constant CR, R is decided on the grounds just about to be considered, leaving us with the choice of C to make the time constant suitable.

We have no doubt realised by now that the discharge resistance R needed to enable the rectified output to follow the modulation is the same thing as what previously we regarded as the load resistance. If the load is the final outlet of the receiver, such as a pair of headphones, then its impedance is limited in practice to a few kilohms. More often, however, the detector is followed by an amplifier to increase the power to a level suitable for working a loudspeaker or television tube. With an f.e.t. amplifier the input impedance is many megohms—too high and indefinite to serve as the discharge path. So something of the order of 0.5 MΩ is chosen for R. This suits the thermionic diode but is on the high side for semiconductor diodes. The value of C that makes CR equal to 50×10^{-6} s is 100 pF. For good reproduction of the highest audio frequencies either C or R or both should be reduced. The limit begins to be reached when C discharges appreciably between one r.f. peak and the next, causing a loss of detection efficiency at all modulation frequencies. The choice of CR is most difficult when the carrier-wave frequency is not many times greater than the top modulation frequency, and for this reason (among others) the carrier frequencies chosen for 'hi-fi' broadcasts are always very high (v.h.f.).

If a bipolar transistor is used as the first modulation amplifier, its input resistance is normally of the order of one or two kilohms, or rather more if it is increased by means of an unbypassed emitter biasing resistance (Section 12.9). To maintain a sufficiently high reservoir action at the lowest r.f. (which we shall see in Chapter 21 is usually 470 kHz or thereabouts) C must be much greater, usually

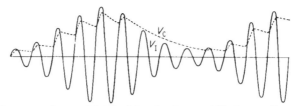

Figure 18.7 V_I represents a modulated r.f. input voltage and V_C the resulting output voltage

about 10 nF, which with 4 kΩ gives a time constant of 40 μs.

The diode is chosen for a forward resistance that is small and a backward resistance large compared with R; for R equal to a few kilohms this requirement is readily met by germanium junction types.

18.8 The Diode Detector in Action

As the action of the deceptively simple-looking circuit in Figure 18.5 is not very easy to grasp, and is also most important, it will be worth spending even more time on it. Figure 18.7 is a diagram which, compared with Figure 18.6, is speeded up so as to show some modulation, and uses proper sine waves for the carrier. Also, the voltage V_C across the capacitor—which, of course, is the output voltage—is drawn together with the r.f. input voltage V_I. Notice that whenever V_I is more positive than V_C the diode conducts and C charges rapidly, because the time constant is CR_F, R_F (the forward resistance of the diode) being much less than R. Directly V_I drops below C_V, C discharges through the only path available (R) with the much longer time constant CR. It is too long in this case, because V_C loses contact with V_I altogether during the troughs, and its waveform is obviously a distorted version of the modulation envelope. But that is because for the sake of clearness the carrier frequency shown is excessively low for the modulation frequency.

Note too that in addition to the desired frequency, V_C has also a saw-tooth ripple at the carrier frequency. Not only has C improved the rectification efficiency but it has also got rid of most of the unwanted r.f.

The third component making up V_C is a steady positive voltage (zero frequency), represented by the average height of the dotted line above the zero line.

18.9 The Detector as a Load

In Chapter 20 we shall see that the Q of receiver tuning circuits is very important. The detector receives its input from such a circuit, across which it is connected. Doing so is bound to reduce the Q of that circuit, and how much it does so can be calculated (Section 8.16) if we know the resistance of the detector. But do we?

With a perfect diode in series with a resistance R, as was assumed in Section 18.2 and shown in Figure 18.8a, the resistance during one set of half-cycles is R and during the other set is infinite, or, with a practical diode, so much larger than R as to have negligible effect. We know (Section 14.5) that the energy-storing capability of a circuit having a high enough Q to be useful for tuning enables its waveform and amplitude to continue almost unchanged even though the circuit receives its maintaining energy in pulse form. The same applies to the removal of energy by resistance. So a resistance load that exists only during every alternate half-cycle is virtually the same as half the load all the time. The power, or rate of energy, delivered to a resistance R is V^2/R, V being the voltage developed across R (Section 2.17). So half the load due to R all the time is the same as the load due to $2R$ all the time.

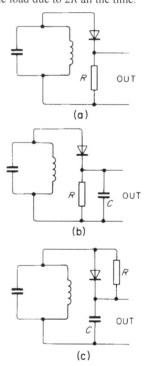

Figure 18.8 The principal varieties of series diode detector circuit

What about the reservoir or peak detector, Figure 18.8b? This looks difficult to assess because, as Figure 18.7 shows, the tuned circuit is unloaded during most of each cycle, but for an unspecified period at each positive peak it is connected by the diode to R and C in parallel. But if we assume that the nearly steady voltage across C is equal to the peak input voltage (in practice, with suitable component values, it nearly is), we can get at it this way: Denote the resistance which, if connected across the tuned circuit in place of the detector, would have the same loading effect, by R'. If V is the r.m.s. voltage supplied by the tuned circuit, the output voltage according to the above assumption is $\sqrt{2}V$. As R is the only component dissipating power, that power will be $(\sqrt{2}V)^2/R = 2V^2/R$. But from the way in which we have defined R' this power is also equal to V^2/R':

$$\frac{V^2}{R'} = \frac{2V^2}{R} \qquad \text{So } R' = \frac{R}{2}$$

Provided that the input circuit is a low-resistance d.c. path, one end of R can be transferred to the other end of it as shown in Figure 18.8c, but now (assuming C has negligible impedance to the carrier) R is across the r.f. input all the time, in addition to the detector loading as already calculated. So R' is now R and $R/2$ in parallel, which is equal to $R/3$, the loading effect being 50% more than in circuit b. This arrangement is therefore seldom used.

It should hardly be necessary to say that when an output of the opposite polarity is required the diode is simply reversed.

The above calculations of loading effects were all based on the assumption of a tuned input circuit. For some purposes (in measuring r.f. voltages, for example) a peak type of diode detector may be applied to a non-resonant circuit. This has no stored energy to supply the extremely non-uniform load, which requires strong pulses of charging current at each alternate peak of the input voltage. If the input source has much resistance these pulses cause a more or less considerable drop in voltage and therefore a loss of output voltage.

There is a drop in voltage even with a tuned circuit if its dynamic resistance is not small compared with the equivalent resistance of the detector. The few-kilohm load where the detector is followed by a transistor loads a tuned circuit excessively, so a suitable step-down transformer ratio from the tuned circuit is chosen.

18.10 Filters

The outputs from detector circuits b and c in Figure 18.8 are something like the dashed line in Figure 18.7. Both types of output contain the desired modulation signal, and both contain also a unidirectional voltage—the average level of the output waveform in the diagram. We shall see in Chapter 22 that good use of this by-product can often be made, but it is usually undesirable in the amplifier that follows the detector. It can easily be removed by a blocking capacitor (Section 14.6). In the outputs of the b and c circuits, there is also a trace of the original r.f., which appears as the notches in the dashed lines in Figure 18.7. If r.f. were allowed to go forward into the amplifier it might upset its working conditions, or even find its way back at amplified strength and cause self-oscillation. So it is desirable with b and c to remove or at least reduce it.

Devices for separating signals of different frequencies are known as *filters*, and are themselves a vast subject. Those used in detectors are usually of the very simplest types, however. All filters depend on the fact that inductive and capacitive impedances vary with frequency. The simple circuits of Figures 6.10, 7.4 and 8.1 can all be used as filters.

Take Figure 6.10: C and R in series. At very low frequencies C offers a high impedance compared with R, and nearly all the voltage applied across them both appears across C. As the frequency rises the impedance of C falls, until, at a very high frequency, most of the voltage appears across R. The dividing frequency, at which the voltages across C and R are equal to one another (and to the input voltage divided by $\sqrt{2}$), is given by $X_C = 1/2\pi f C = R$, so $f = 1/2\pi CR$. (Notice that again we have CR cropping up.)

Suppose the lowest carrier frequency to be handled by a detector is 150 kHz, and the highest modulation frequency is 10 kHz. One might make the dividing frequency 15 kHz. Then $CR = 1/(2\pi \times 15\,000) = 10.6 \times 10^{-6}$. The separate values of C and R depend on the general impedance of the detector circuit. Irrespective of this, and assuming that the filter output is not loaded, Figure 18.9 shows how much of the filter input from the detector would get through at frequencies from 1 to 1000 kHz. There is very little loss at even the highest modulation frequencies, but the residual r.f. is substantially reduced, especially when, as is usual in broadcast receivers, the lowest r.f. is about 470 kHz.

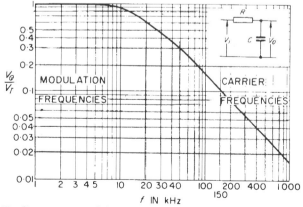

Figure 18.9 The filtering action of CR is shown here as a graph of V_0/V_I against frequency

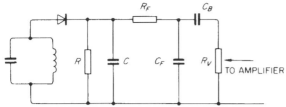

Figure 18.10 Complete detector circuit, comprising Figure 18.8b, Figure 18.9 (inset) and a.f. volume control with blocking capacitor C_B

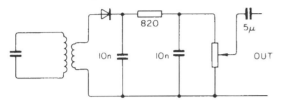

Figure 18.11 Another detector circuit for transistor receiver, with typical component values. Note that for brevity it is usual to omit symbols for the basic units (Ω, F, etc.) and show only multipliers (μ, n, etc.) where needed

Logically the detector circuit would now be as in Figure 18.10, in which R_F and C_F are the filter components, C_B the blocking capacitor, and R_V the volume control which is usually placed here, between the detector and the a.f. amplifier, which is often a transistor with an input resistance of only a few kilohms. A usual value for R_V is 5 kΩ, so the load resistance at maximum volume may be only about 2 kΩ. R_F should therefore be only a fraction of this if it is not to reduce the signal appreciably. R still further adds to the load, but quite often it is removed and $(R_F + R_V)$ made to do its duty in which case C_B is transferred to the output side of R_V. Figure 18.11 shows a typical detector for a transistor broadcast receiver. Note the step down from the tuned circuit to match the comparatively low impedance of the whole thing. With an f.e.t. or valve amplifier the loading problem is much relieved by its virtually infinite input resistance. Resistances are usually about 100 times larger and capacitances 100 times smaller.

18.11 Detector Distortion

In discussing rectifiers we noted that they had to be non-linear conductors in order to rectify, but that if the non-linearity consisted of a sharp corner between two straight portions (e.g., COB in Figure 18.2), the rectifier itself could be described as linear (e.g., COB in Figure 18.4). And that no real rectifiers have perfectly sharp bends, so all are more or less non-linear. Now that we have seen the details of complete detector circuits we are returning to this topic, to consider how to ensure that the non-linearity is less rather than more.

The simple Figure 18.8a rectifier circuit would be the easiest to analyse, but because the reservoir type is the one used almost invariably let us assume it, even though we shall have to do some 'guesstimating'. Figure 18.12 shows the characteristic curve of a germanium diode of a type used for this purpose, and the load resistance (R) is, say, 10 kΩ. Let us suppose that an unmodulated r.f. signal of 1 V peak is applied. And let us guess that with this input the detector has an efficiency of 70%. In other words d.v. output across C and R which back-biases the diode is 0.7 V. So we draw a cycle of the input voltage from −0.7 V as its base line, between A and B. With this voltage across R, the continuous current through it must be 0.07 mA, as

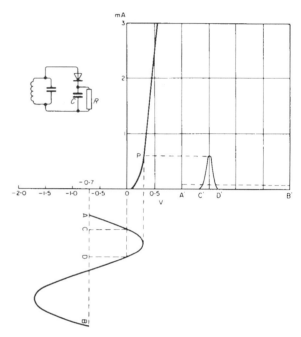

Figure 18.12 Analysis of germanium diode detector performance by means of characteristic curve of the diode

shown dashed to the right of the characteristic curve, from A′ to B′. Current through the diode is limited to the part of the input cycle (C to D) which makes the voltage positive, and by referring the input cycle via the characteristic curve to the current scale we get the diode current waveform. The area under it, being current multiplied by time, is a measure of the charge per cycle going into C. It must be equal to the discharge per cycle, represented by the area under the dashed line A′ B′. It looks roughly right, or perhaps hardly enough; if insufficient, a very slight decrease in negative bias, and therefore in detector efficiency, would correct it.

Now imagine the input cycles slowly changing in amplitude above and below 1 V peak due to modulation. Because the slope of the curve at P is steep, the increases in bias will be very nearly proportional to the increases in peak value. And so will moderate decreases. But if the modulation depth approaches 100%, making the minimum input amplitude approach zero, clearly the decreases in bias will not be in anything like exact proportion. Since the modulation signal we are extracting consists of the ups and downs in bias, when the modulation is deep its waveform will be distorted.

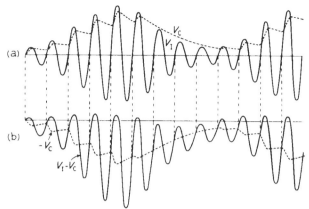

Figure 18.13 In diagram (a), V_I represents a modulated r.f. input voltage and V_C the resulting output voltage. In diagram (b), the full line shows $V_I - V_C$, i.e., the voltage across the diode

Repeating this kind of diagram for different unmodulated amplitudes leads to the conclusion that the greater the amplitude the higher the efficiency and the less the distortion. Further thought should make clear that the loss of efficiency and the distortion are due mainly to the low load resistance necessitating a strong peak of charging current each cycle to keep the bias going for the rest of the cycle, and that better performance all round (including less damping of the input circuit and less loss of signal voltage due to the need for a step down) would be obtained if the load resistance were raised to the hundred-kilohm order as it can be when an f.e.t. or valve amplifier follows.

Both low and high resistance detectors are open to another cause of distortion of deeply modulated signals if the type of circuit in Figure 18.10 is used. Here the resistance to d.c. is R alone, for the other circuit paths are all blocked by capacitors. But a.c. at the modulation frequency has another path in parallel via the low-impedance C_B to R_V, so the resistance to it is lower than the resistance to d.c. This makes the current through the diode swing more widely above and below the mid position during modulation. With 100% modulation this cannot happen at the negative peaks, because the diode current cannot be less than zero. So the negative peaks are clipped.

Looking at Figure 18.12 again we see that if a very weak station is tuned in, its amplitude being 0.1 V or less, detection will be very inefficient because the current passed by the diode at up to 0.1 V is almost negligible. For this reason it is usual to arrange for the diode to have a positive bias of up to about 0.2 V (for germanium) to bring it to the most sharply bent part of the characteristic curve.

In all the foregoing we have assumed that there is no capacitance across the diode. If there were it would form with C a potential divider across the signal source, reducing the efficiency of detection. For this reason junction diodes, which have enough self-capacitance to be appreciable, especially at v.h.f. and u.h.f., may be unsuitable; if so, point-contact types with their very small contact area are used. External stray capacitance must be avoided by attention to the circuit layout.

18.12 Shunt-diode Detector

Let us return to Figures 18.5 and 18.7 for a moment. The voltage across the diode must, of course, be equal to V_I minus V_C, and this is plotted in Figure 18.13 together with Figure 18.7 for reference. The signal across the diode, like that across the reservoir capacitor, has a.f. and z.f. components, and therefore could be used as the output voltage of the detector. The circuit diagram of such a detector is shown in Figure 18.14. Because the diode is

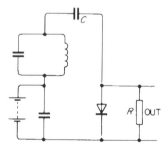

Figure 18.14 Shunt-diode detector circuit

connected directly across the output, it is known as a shunt-diode detector. (The circuits described earlier, Figure 18.8, are, of course, series-diode detectors.) The reactance of C at radio frequencies is small compared with the value of R, and nearly the full r.f. voltage from the LC circuit appears at the output. Thus the shunt-diode detector has a much greater r.f. ripple output than the series-diode circuits, making the use of an r.f. ripple filter essential. Moreover capacitor C effectively connects R across the resonant circuit as in Figure 18.8c so that the loading effect on the tuned circuit is equivalent to a resistance of $R/3$. It is not surprising, therefore, that the series-diode circuit is normally preferred. However, the shunt-diode circuit has one advantage. The reservoir capacitor C acts as a z.f. block between the tuned circuit and the diode, making the shunt-diode circuit useful where the LC circuit is at a d.v.

potential which would upset the action of the diode if it reached it. This occurs, for example, when the tuned circuit is directly connected in the collector or anode circuit of an r.f. amplifier. The battery in Figure 18.14 is intended to simulate these conditions.

18.13 Television Diode Detector

Some liberties can be taken in the value of the time constant for an a.m. sound detector because selectivity requirements (Section 15.9) in the receiver cause the r.f. circuits to restrict the pass-band and limit the upper audio frequencies to 4 kHz or so. Even if the receiver can accept the whole of the transmitted bandwidth, the upper modulation frequencies, being harmonics, are of low amplitude and give correspondingly shallow modulation.

In television, modulation depths up to 100% are possible at the highest video frequencies; in fact the test card includes a grating which does just this. So in the UK system the detector must be capable of demodulating a 5 MHz video signal at full modulation depth. A repeat of the calculation made earlier, but substituting 5 MHz for 10 kHz, suggests that the time constant should be no greater than 20 ns. In fact higher values can be used because the resulting distortion of the waveform is imperceptible on the screen. A commonly used value is 33 ns and, as the load resistance is made 3.3 kΩ, this gives a reservoir capacitance of 10 pF. Care must be taken to reduce to a minimum stray wiring and component capacitances which can easily amount to a few pF.

18.14 F.M. Detectors

The performance of all the detectors considered so far depends hardly at all on the frequency of the carrier wave between very wide limits. Consequently they treat a frequency-modulated carrier wave (Section 15.4) as if it were unmodulated. In order to receive f.m. signals it is necessary either to convert the variations of frequency into corresponding variations of amplitude so that an ordinary detector can be used, or else use a different kind of detector altogether.

The first method can be employed in a very simple manner with an ordinary a.m. receiver by adjusting the tuning so that instead of the incoming carrier-wave frequency coinciding with the peak of the resonance curve (Figure 8.5) it comes in on one of the slopes. As its frequency rises and falls by modulation, the amplitude of response rises and falls. But because the slope of the resonance curve is not quite linear, the amplitude modulation is not a perfect copy of the frequency modulation, so there is some distortion. More serious still, the potential benefits of f.m. are not achieved, because the receiver is open to amplitude modulation caused by interference. And both selectivity and efficiency are poor.

Receivers designed for f.m. reception therefore include means for rejecting amplitude modulation and responding only to frequency modulation. The first of these functions is performed by a limiter and the second by a *discriminator*. In one of the most popular types of f.m. detector—the ratio detector—these functions are combined. However, for a first approach it will be easier to consider a system in which they are separate.

In connection with oscillators we saw in Section 14.9 that if the amplitude of oscillation exceeds a certain amount the transistor or valve cuts off the positive or negative peaks or both. A limiter is usually one of these devices so arranged that it does this at anything above quite a small signal amplitude, so that however much greater the incoming signal may be it is cut down or limited to the same size. Variations in amplitude are therefore almost completely removed, and with them much of the effects of interference.

18.15 Foster-Seeley Discriminator

Fig 18.15a shows the essentials of the Foster-Seeley or phase discriminator. It consists of two Figure 18.8b type a.m. detectors back to back, both receiving the incoming r.f. carrier-wave voltage da from a tuned circuit $L_p C_p$, and each receiving in addition half the voltage from the tuned circuit $L_s C_s$. L_p and L_s are the primary and secondary windings of an r.f. transformer, and because both windings are exactly tuned to the carrier wave and are coupled only very loosely the voltages across them are 90° out of phase,

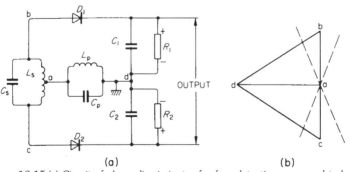

Figure 18.15 (a) Circuit of phase discriminator for f.m. detection, arranged to bring out the principle: L_p and L_s are the primary and secondary windings of the input r.f. transformer; (b) phasors of voltages marked, for carrier-wave frequency. Broken lines indicate the effect of f.m. deviation

as indicated by the phasors da, ab and ac in Figure 18.15b. The total r.f. input to diode D_1 is therefore db, and to diode D_2 it is dc (C_1 and C_2 being r.f. short-circuits). These inputs are equal, so the rectified voltages set up across the load resistors R_1 and R_2 will also be equal. But as they are connected in opposite polarity, the net output is nil.

When the carrier wave is frequency-modulated it swings off the resonance peak and, as explained in Section 8.11, the impedance of a tuned circuit changes from resistance to inductance on one side and capacitance on the other. The phase swings correspondingly, as indicted by the broken lines in Figure 18.15b, changing the one-to-one ratio of db to dc to something alternately greater and less. The value of db–dc therefore becomes alternately positive and negative at the frequency of modulation, and the modulating signal is reproduced at the output. Practical phase-discriminator circuits differ in detail from Figure 18.15 but the main principle is the same.

18.16 Ratio Detector

The ratio detector circuit combines the function of limiter with that of discriminator, so is usually preferred, in spite of slightly higher distortion. It looks very much like the phase discriminator, but one of the diodes is reversed, so when the carrier is unmodulated the total output from the diodes is not zero but is equal to the sum of the two. When modulation takes place this sum tends to remain constant but the ratio between the two rectifier outputs varies at audio frequency. The output of the detector can therefore be taken from either of the two reservoir capacitors as shown in Figure 18.16. $R_1 R_2 C_1 C_2 C$ constitute the load circuit of a diode detector; thus the voltage across it (the sum of the voltages across C_1 and C_2) equals the peak r.f. input to the diodes, and the tuned circuit feeding the detector is effectively damped by a resistance equal to $(R_1 + R_2)/2$.

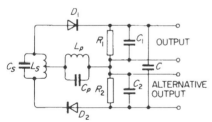

Figure 18.16 Circuit of ratio detector, much used in f.m. receivers

Ideally the r.f. input to the detector should be of constant amplitude but any noise signals present are superimposed on the f.m. carrier causing unwanted amplitude modulation. To reduce the effects of such noise signals, the ratio detector is designed to ignore short-duration increases and decreases in the amplitude of the input signal. To do this the diode load resistors R_1 and R_2 are given low values so that the tuned circuit feeding the detector is heavily damped. This means that the gain of the r.f. stage driving the detector is low. The time constant of the diode load circuit is made about one second by the addition of the capacitor C. As a result it takes appreciable time for the voltage across C to rise to equal the peak value of the r.f. input to the detector.

Now suppose interference causes a momentary increase in the peak amplitude of the r.f. input to the detector. Because of the long time constant the voltage across C cannot instantaneously rise to equal the new peak value and, as a result, the diodes are driven heavily into conduction and their low forward resistance increases the already heavy damping of the tuned circuit, momentarily reducing the gain of the previous stage and minimising the effect of the interfering signal.

If there is a momentary reduction in the peak value of the r.f. input, again the long-time-constant load circuit cannot react instantaneously to the change. Thus the diodes are cut off, removing the damping imposed by the diode load on the tuned circuit. This causes the gain of the previous stage to increase momentarily, offsetting the effect of the reduction in input amplitude. In fact the removal of the damping can result in overcompensation and a common technique is to include low-value resistors in series with the diodes to limit the increase in gain, the resistance values being adjusted empirically to give optimum a.m. rejection. Thus the inclusion of the long-time-constant circuit enables short-term changes in input-signal amplitude to be minimised: in fact the circuit behaves as a dynamic limiter—dynamic because it can respond to slow changes in input amplitude but not to very rapid ones.

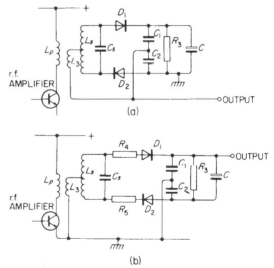

Figure 18.17 Practical ratio detector circuits, (a) unbalanced and (b) balanced

Figures 18.15 and 18.16 are simplified to emphasise the principles of the circuits. In practical versions L_p cannot normally be directly connected to L_s because it is in the collector circuit of an r.f. amplifier and is at a different d.v. from the remainder of the detector circuit. This difficulty can be overcome as illustrated in Figure 18.17, which gives circuit diagrams of practical ratio detectors. The r.f. input for the centre point of L_s is obtained from a coil L_3 tightly coupled to L_p. Circuit *a* is unbalanced, one side of the load circuit being earthed, and the output is taken from C_2. Circuit *b* is balanced, the midpoint of the diode load circuit being earthed. The output is taken from C_1 but it could equally well be taken from C_2. This circuit also includes resistors R_4 and R_5 to optimise the a.m. rejection. In both circuits the two load resistors R_1 and R_2 are replaced by a single equivalent resistor R_3. Notice, too, that C_p has been omitted; it is not essential to tune the primary circuit to the centre frequency.

18.17 Quadrature Detector

We now turn to a type of f.m. detector using entirely different principles. It is variously described as a balanced, symmetrical, product or quadrature detector, and its highly complex circuit is encountered only in integrated circuit form; see Section 24.7. The principle is, however, straightforward and will be described with the aid of simplified diagrams. Two inputs are required with a particular phase relationship between them, and one method of obtaining suitable inputs is shown on the left in Figure 18.18. One input is from the preceding r.f. amplifier and the other is derived from it by the network C_1LCC_2. C_1 and C_2 are very small capacitors and their reactance at the carrier frequency is large compared with the impedance of LC even at resonance. Thus the network is effectively capacitive and the current in it leads the applied voltage (input 1) by nearly 90°. At the resonance frequency of LC its impedance is resistive and thus the voltage generated across it (input 2) also leads input 1 by 90°. At frequencies above resonance the impedance of LC is capacitive and input 2 therefore lags on its phase at resonance. At frequencies below resonance the impedance of LC is inductive and input 2 leads on its phase at resonance. LC is adjusted to resonate at the centre frequency of the f.m. signal to be demodulated.

Now suppose the two inputs are applied to separate limiters (Figure 18.18). Their outputs are two square waves and the displacement between them is 90° at the centre frequency, greater at higher frequencies and less at lower frequencies. These outputs are applied to a complex symmetrical circuit represented in Figure 18.18 by two transistors in series and a load resistor. The transistors are normally cut off but are driven into conduction by the square waves. But because the transistors are in series, current can flow through them and the load resistor only

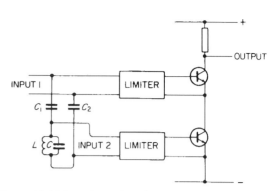

Figure 18.18 Basic form of quadrature detector

when both bases are simultaneously driven positive by the square waves. Thus current flows only during the period of overlap of the pulses as shown in Figure 18.19. The output from the transistors thus consists of constant-amplitude carrier-frequency pulses, the duration of which increases with increasing frequency. The output is, in fact, a pulse-duration-modulated signal. The energy content of the pulses depends on their duration, and the variation in energy can be turned into a variation in d.c. by passing the pulses through a low-pass filter to remove carrier-frequency components. The resulting signal is the required modulation-frequency output.

18.18 Phase-locked-loop Detector

This f.m. detector also uses different principles and again is a complex circuit encountered only in integrated-circuit form. As long ago as 1944 it was realised that if an oscillator could be synchronised by an f.m. signal so that its frequency faithfully followed the variations of the applied carrier, such a circuit could be expected to have two useful properties. Firstly, the amplitude of the oscillator output could be many times that of the applied signal, implying a useful degree of voltage gain. Secondly, the oscillator output amplitude could be independent of that of the applied signal provided this is sufficient to give effective synchronisation: in other words the circuit should give effective amplitude-limiting. Thus the oscillator could be used as a source of amplified and limited f.m. signal which could then be applied to any type of f.m. detector. Used in this way, of course, the oscillator is not in itself a detector but a source of signal for a detector.

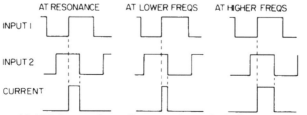

Figure 18.19 Pulse waveforms illustrating the method of operation of the quadrature detector

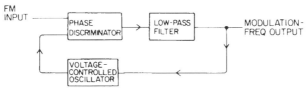

Figure 18.20 Block diagram of a phase-locked-loop f.m. detector

The phase-locked-loop circuit is a more recent application of the principle. It uses a phase discriminator, and a circuit of the type illustrated in Figure 18.18 is commonly employed. But the second input is derived, not from the first input as in the quadrature detector, but from an oscillator forming part of the circuit. See Figure 18.20. The oscillator is so designed that its frequency can be controlled by the value of a voltage applied to it. For example, the oscillator could be a Hartley or Colpitts type in which part of the capacitance of the LC circuit is provided by a capacitance diode, the d.v. input to which therefore controls the oscillation frequency. In the phase-locked-loop circuit the control voltage is the output of the phase discriminator and so we have a kind of negative feedback. Any variation in the frequency of the input signal causes an output from the phase discriminator which alters the oscillation frequency so as to minimise the difference between its frequency and that of the input. In this way the phase of the oscillator is locked to that of the input and follows any variations in it. The output of the phase discriminator contains the required modulation-frequency component, but it is usually passed through a low-pass filter to minimise any r.f. components. This type of detector has the advantage that its performance does not depend on the amplitude/frequency characteristics of an LC circuit and can therefore be expected to be superior to that of detectors such as the Foster-Seeley, ratio and quadrature types.

18.19 Counter Discriminator

We have by no means exhausted the subject of f.m. detectors. For example, there is the counter type, in which the cycles of the incoming f.m. signal are reduced to unidirectional pulses whose shape and duration are unaffected by changes in signal frequency. The number of them occurring per second, and therefore their average voltage, is, of course, directly proportional to the signal frequency. This linear relationship can be very accurately maintained, ensuring freedom from distortion, but the upper practical limit of carrier-wave frequency is only a few hundred kilohertz. As Chapter 21 will show, this is less of a handicap than might be supposed.

Chapter 19

Audio Frequency Amplification

19.1 Recapitulation

Amplification has already been considered in a general way and we have studied the aids used for designing amplifiers: characteristic curves and equivalent circuits. At the receiving end the r.f. signal picked up by the antenna usually needs amplification before being presented to the detector, which the last chapter has shown needs at least a volt or two if it is to do its job well. Besides amplifying, the receiver r.f. amplifier has to select the desired signals and exclude all others. And although it can be allowed to maltreat the waveform of the r.f. cycles, the waveform of the modulation must be carefully preserved. With f.m. this is no problem, but a.m. demands some care.

Our detailed study of transistor and valve parameters and equivalent circuits was for the most part based on neglecting reactances, so that phase relationships were just plus or minus. To what is called a first order of approximation, audio-frequency amplification can be treated on this basis, so we shall now proceed to do so, leaving receiver r.f. amplification (which certainly cannot be) until Chapter 22. Besides radio receivers, a.f. amplifiers are used in hearing aids, tape and disk recording and reproduction, and sound reinforcement.

The main problem is not in providing any desired amount of amplification; it is in doing so without perceptible distortion or extraneous noise. Perceptible, note. It is the human ear that judges the sound resulting from a.f. distortion, so it is important that the amount of distortion as well as the amount of amplification is reckoned as nearly as possible as the listener judges it. There are two main kinds of distortion: one we shall call *gain/frequency distortion** and occurs when the amplification or gain is not the same at all frequencies, so that sounds of different pitch are not reproduced at the correct relative strengths; the other is *non-linearity distortion* (or waveform distortion). In general it is more unpleasant than gain/frequency distortion. But first we should consider how amplitude and frequency are judged by hearers.

---

* BSI calls this *attenuation distortion*, a most misleading term because it fails to indicate that the effect depends on frequency, and in this book we shall therefore use the self-explanatory term gain/frequency distortion.

19.2 Decibels

Suppose a 0.5 V signal is put into an amplifier which raises it to 12 V. This 12 V is then put into another amplifier which raises it to 180 V. Which amplifier amplifies more; the one that adds 11.5 V to the signal, or the one that adds 168 V? Put this way, the figures tend to suggest the second. But if we work out A, the voltage amplification ratio (Section 11.5) we find it is 24 for the first and only 15 for the second. This method of reckoning is the right one because we do not judge by the *amount* by which a sound is increased in strength, but by the *ratio* in which it is increased. Doubling the power into a loudspeaker produces much the same impression of amplification whether it be from 50 to 100 mW or from 500 to 1000 mW.

This is like the way the ear responds to the pitch of sounds. While it is pleased by combinations of sounds whose frequencies have certain fixed ratios to one another, it can find no significance at all in any particular additive frequency intervals, say an increase of 250 Hz. Musical intervals are all ratios, and the significant ratios have special names, such as the octave, which is a doubling of frequency wherever on the frequency scale it may come. The octave is subdivided into tones and semitones.

The same principle is appropriate for the strength of sound or electrical signals, and amplification—or indeed any change in power—is reckoned in ratios. Instead of the 2 to 1 ratio, or octave, a power ratio of 10 to 1 was chosen as the unit. The advantage of this choice is that the number in such units is simply equal to the common logarithm of the power ratio. It was named the bel (abbreviation: B) after A. Graham Bell, the telephone inventor. So an amplifier that multiplies the power by 10 can be said to have a gain of 1 bel. A further power multiplication by 10, giving a total of 100, adds another bel, making 2 bels.

The bel, being rather a large ratio for practical purposes, is divided into 10 decibels (dB). One-tenth of a 10 to 1 ratio is not at all the same as one-tenth of 10; it is 1.259 to 1. If you have any doubts, try multiplying 1 by 1.259 ten times.

So a gain of 1 dB is a 25.9% increase in power. The voltage or current increase giving a 25.9% increase in power is 12.2% *provided that the resistance is the same for both voltages (or currents) in the ratio*, because power is proportional to the square of voltage or current (Section 2.17). The fact that a voltage or current decibel is a smaller ratio than a power decibel does not mean that there are two different sizes of decibels, like British and American gallons; it arises because the ratio of volt or amps is smaller

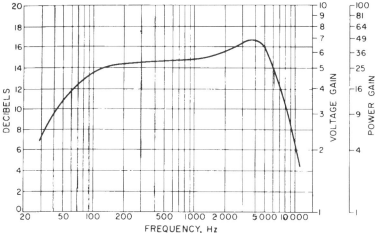

Figure 19.1 A typical a.f. gain/frequency curve, with logarithmic scales in both directions

than the ratio of power in the same situation. Decibels are fundamentally *power* ratios.

If 1 is multiplied by 1.259 three times, the result is almost 2, so a doubling of power is a gain of 3 dB. Doubling voltage or current quadruples the power, so is a gain of 6 dB. A table connecting decibels with power and voltage (or current) ratios is given in Appendix E. Since the number of bels is $\log_{10} P_2/P_1$, or $2 \log_{10} V_2/V_1$ (where P_2/P_1 is the power ratio and V_2/V_1 the voltage ratio) the number of decibels is $10 \log_{10} P_2/P_1$ or $20 \log_{10} V_2/V_1$. If Figure B.5a is a scale of dB, then b is the same scale in power ratios. The dB scale, being linear, is obviously much easier to interpolate (i.e., to subdivide between the marked divisions). Figure 11.6 would be improved if the logarithmic voltage-ratio scale were replaced or supplemented by a uniformly divided scale with its 0 at 1 and its 20 at 10, which would indicate dB.

An advantage of working in dB is that instead of having to multiply successive changes together to give the net result, as with the A figures, they are simply added. A voltage amplification (A_V) of 24 is +27.6 dB, and $A_V = 15$ is +23.5 dB, so the total ($A_V = 360$) is +51.1 dB. The + indicates amplification or gain, attenuation or loss shows as a –. For instance, if the output voltage of some circuit were half the input, so that the power was one-quarter—i.e., a power 'gain' of 0.25—the 1 to 4 power ratio would be denoted by the same number of dB as 4 to 1, but with the opposite sign: –6 dB. Zero dB, equivalent to $A = 1$, means 'no change'. The power losses in lines (Chapter 16) are reckoned in dB; if the line is uniform, the loss in dB is proportional to its length.

An important thing to realise is that decibels are ratios, not units of power, still less of voltage. So it is nonsense to say that the output from an amplifier is 30 dB, unless one has also specified a reference level such as 0 dB = 1 mW. Then 30 dB, being 1000 times the power, would mean 1 W.

To the general public, decibels are units of loudness, but they can be used as such only if a standard reference level is understood. If they only knew, dB can just as well be used for betting odds!

19.3 Gain/Frequency Distortion

The amount of gain/frequency distortion in any piece of equipment is shown by means of a graph in which gain (or loss) is plotted against frequency. For the reasons we have just been considering, both scales are usually logarithmic, the gain scale being linear in dB and the frequency scale usually running from 20 Hz to 20 kHz in three decades (i.e., 10 : 1 ranges).

Figure 19.1 is a typical example. A curve having the same shape at a higher or lower level would mean the same amount of distortion. If a linear A scale were used the shape would vary with the level and so be misleading. An ideal a.f. gain/frequency curve would of course be flat or level over the whole range of frequencies to be reproduced.

As a guide to interpreting frequency characteristics, it is worth noting that 1 dB is about the smallest change in general level of sound that is perceptible. If the loss or gain is confined to a small part of the audible frequency range, several dB may go unnoticed. So far as audible distortion is concerned, therefore, there is no sense in trying to 'iron out' irregularities of a fraction of a decibel. A peak of one or two decibels, though unnoticeable as gain/frequency distortion, may, however, cause non-linearity distortion by overloading the amplifier. An average fall of 10 dB over all frequencies above, say, 1 kHz would be easily noticeable as muffled and indistinct reproduction. A rise of the same amount, or a falling off below 1 kHz, would be heard as thin shrill sound.

A cheering thought is that gain/frequency distortion in one part of the system can usually be compensated in another by deliberately introducing distortion of the opposite kind, emphasising the weakened frequencies. But such methods, called *equalisation*, should not be relied on more than necessary, because they may cause difficulty in using one piece of equipment with others.

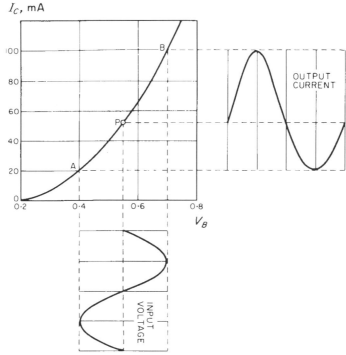

Figure 19.2 Showing the waveform distortion introduced by one type of non-linear characteristic curve

Means for the listener to adjust the frequency curve to suit his taste or to counteract distortion are called *tone controls* and in much audio equipment it allows up to 15 dB gain or loss at the lowest and highest audio frequencies. The controls are usually labelled *bass* and *treble*.

In hi-fi equipment the shape of the audio response may be controlled by a so-called *graphic equaliser*, in which the audio spectrum is divided into a number (e.g., 10) of contiguous frequency bands which are individually amplified, individual gain controls for each band permitting any desired shape of frequency response to be obtained. The gain controls are usually slider types so arranged that a level response is obtained when the sliders are in line. Indeed it is because the positions of the sliders indicate the shape of the frequency response currently selected that the facility is termed *graphic*.

19.4 Non-linearity Distortion

Non-linearity is indicated by a graph of output against input. A diagonal straight line passing through the origin, as in Figure 2.3, represents perfect linearity, and means that any given percentage increase or decrease of input changes the output in exactly the same proportion. Transistors are examples of non-linear devices, as shown by the curvature in their characteristics (Figure 10.12, etc.).

Judging from an output-input graph, the most obvious effect of non-linearity might seem to be that the volume of sound produced would not be in exact proportion to the original. It is true that reducing the signal strength at the detector to one-tenth is likely to reduce the a.f. output to far less than one-tenth, because of the bottom bend (Section 18.11). But that is about the least objectionable of the effects of non-linearity. What matters much more is the distortion of waveform. Because the waveforms of a.f. signals correspond to those of the original sound waves, we must preserve their form with the utmost care. Otherwise the sound will be harsh and we may not be able to distinguish one musical instrument from another (Section 1.3).

19.5 Generation of Harmonics

As far back as Figure 1.2c we saw that when sine waves having frequencies in the ratios 1:2:3, etc., are added together the result is a non-sinusoidal waveform which repeats at its lowest frequency—the one represented by 1. In general the waveforms generated by musical instruments, buzzers, sirens, etc.—and, because of distortion, electronic oscillators—are more or less non-sinusoidal. It was shown by Fourier that repetitive waveforms can be analysed into sine waves having exactly multiple frequencies as above. These component waves are called *harmonics*, but the one of lowest frequency (which is also the frequency of the composite waveform) is usually called the *fundamental* rather than the first harmonic. The composite waveform depends on the number, amplitudes and relative phases of the harmonics. Let us now apply the idea of harmonic analysis to the output waveform of an amplifier.

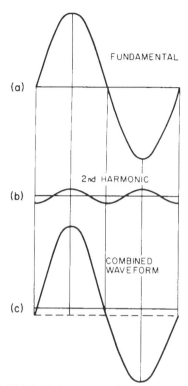

Figure 19.3 Method of estimating the harmonic distortion in the Figure 19.2 output

Figure 19.2 is a typical collector-current/base-voltage curve. If the load resistance were small compared with the output resistance of the transistor, the output voltage would be nearly proportional to I_C. Suppose P is the working point, from which a sinusoidal input signal swings the base between 0.4 V and 0.7 V. Plotting the output current waveform, we see that whereas the positive peak is 48 mA the negative peak is only 32 mA. This cannot, therefore, be a true sine wave, for which the peaks are equal. The average is 40 mA. In Figure 19.3 a sine wave of that size has been plotted at *a*, a second harmonic having an amplitude equal to half the difference between average and actual peak output (4 mA) is shown at *b*, and *c* is the result of adding *a* and *b* together. It is almost identical with the output in Figure 19.2. We conclude that characteristic curves of the Figure 19.2 type, in which the slope increases all the way, generate second harmonics. Unless they are exactly the shape called parabolic they also generate other even-number harmonics, which account for any slight differences between the Figure 19.2 output and Figure 19.3*c*. In this example the amplitude of the harmonic (Figure 19.3*b*) is 0.1 of that of the fundamental (*a*) so the second-harmonic distortion is said to be 10%. Without drawing any waveforms we can find the percentage from the dynamic characteristic curve by joining the extreme points of the swing, A and B, by a straight line and expressing the height of its centre above P as a percentage of the vertical height of B above A.

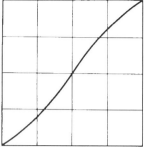

Figure 19.4 In contrast to the type of characteristic curve in Figure 19.2, which generates even-number harmonics, this double bend generates odd-number harmonics

In the same way one can show that a characteristic curve having a double or S bend, seen in Figure 19.4, generates third and other odd harmonics. (Note that because the centre of the curve lies on the diagonal between its ends the percentage of even harmonic, found by the method just described, is nil.) This type of curve often results from attempting to get as much power as possible out of a transistor or valve. Try drawing a load line across Figure 10.15, for example, and deriving from it a curve of output voltage against input voltage, V_G. Most output/input characteristics show a combination of both types of curvature, but usually one or other predominates. Unless there are sharp bends (which should always be avoided) the percentages of harmonics decrease as their order (i.e., number of times their frequency is greater than that of the fundamental) increases. Nevertheless, it is possible to select a particular harmonic by use of a tuned circuit resonant at that frequency; this is the basis of the *frequency multiplier*.

The ear tolerates a fairly large percentage of second-harmonic distortion, less of third harmonic, and so on. A fraction of 1% of the higher harmonics such as the eleventh introduces a noticeable harshness; the reason being that a second harmonic differs by an exact octave from the original tone, whereas the high harmonics form discords.

19.6 Intermodulation

Although for simplicity the distortion of a pure sine wave has been shown, the more complicated waves corresponding to musical instruments and voices are similarly distorted. So when one is dealing with waveforms other than sinusoidal it is usual to perform a harmonic analysis, and then tackle them separately on the basis of simple sine-wave theory; like the man in the fable who found it easier to break a bundle of faggots by undoing the string and attacking them one by one. Each harmonic current and voltage can be calculated as if it were the only one in the circuit—*provided that the circuit is linear*. But just now we are studying non-linear conditions, in which it is not true to say that two currents flowing at the same time have no effect on one another. We shall see in Section 21.3 that when two voltages of different frequency are applied together to a non-linear device the resulting current contains not only the original frequencies but also two other frequencies, equal to the sum and difference of the original two. For some purposes that is a very useful result, but extremely undesirable in an a.f. amplifier. It means that all the frequencies present in the original sound—and there may be a great many of them when a full orchestra is playing—interact or intermodulate to produce new frequencies; and, unlike the lower harmonics, these frequencies are generally discordant. Even such a harmonious sound as the common chord (C, E, G) is marred by the introduction of discordant tones approximating to D and F. When non-linearity distortion is severe, these intermodulation or combination tones make the reproduction blurred, tinny, harsh, rattling, and generally unpleasant. And once they have been introduced it is practically impossible to remove them.

19.7 Allowable Limits of Non-linearity

Whenever one attempts to get the greatest possible power from a transistor or a valve, one comes up against non-linearity. So the problem tends to be most acute in the output stage. If the signal amplitude is reduced, so that it operates over the most nearly linear parts of the characteristics, the distortion can be made very small, but then the output will be uneconomically small in relation to the power being put in. It is not a question of adjusting matters until distortion is altogether banished, because that could be done only by reducing the output to nil. It is necessary to decide how much distortion is tolerable, and then adjust for maximum output without exceeding that limit.

Fixing such a limit is no simple matter, because it should (in reproduction of music, etc.) be based on the resulting unpleasantness judged by the listener. Different listeners have different ideas about this, and their ideas vary according to the nature of the programme. And, as has just been pointed out, the amount of distortion that can be heard depends on the order of the harmonic (second, third, etc.) or of the intermodulation tones, as well as on the percentage present.

Various schemes for specifying the degree of distortion in terms that can be measured have been proposed, but the nearer they approach to a fair judgement the more complicated they are; so the admittedly unsatisfactory basis of percentage harmonic content is still commonly used. Preferably the percentage of each harmonic is specified, but more often they are all lumped together as 'total harmonic distortion'.

19.8 Phase Distortion

The main result of passing a signal through any sort of filter, even if it is only a resistance and a reactance in series, is to alter the amplitudes of different frequencies to different extents (Figure 18.9). If it discriminates appreciably between frequencies within the a.f. band it introduces gain/frequency distortion, which, of course, may be intentional or otherwise. It also shifts the phases to a differing extent; see Figure 6.10, for example. If the signal is a pure sine wave, the phase shift has no effect on the

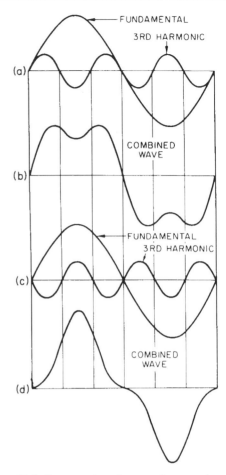

Figure 19.5 Showing that the waveform resulting from adding together two component waves is greatly altered by shifting the relative phase of the components

waveform; but few people spend their time listening to continuous pure sine waves.

To take a very slightly more complex example, a pure wave is combined with a third harmonic, shown separately in Figure 19.5a. The waveform of this combination, obtained by adding these two together, tends towards squareness, as shown at b. Suppose now that after passing through various circuits the relative phases of the component waves have been altered, as at c. Adding these together, we see the resulting waveform, d, which is quite different from b. Yet oddly enough the ear, which is so sensitive to some changes of waveform (such as those caused by the introduction of new frequencies), seems quite incapable of detecting the difference between d and b, or indeed any differences due to phase shifts in continuous waves.

Generally, it can be said that if care is taken to avoid phase distortion, gain/frequency distortion will take care of itself.

19.9 Loudspeakers

Having considered distortion we begin now to apply this knowledge to a.f. amplifiers. Their end product is usually sound from a loudspeaker, which forms the load for the final or output stage of the amplifier. Until we know the requirements here we cannot tell how many and what stages will be needed between the a.f. signal source—detector, tape head, microphone, etc.—and the output. So let us begin at the end with the loudspeaker. Its object, of course, is to convert the electrical power delivered to it from the output stage into acoustical power having as nearly as possible the same waveforms.

A great many types have been devised, but the commonest one in use today is the moving coil. As the name suggests, it works on essentially the same principle as the moving-coil meter (Section 4.2). The mechanical force needed to stir up air waves is obtained by the interaction between two magnetic fields. One of the fields is unvarying and very intense, and the other is developed by passing the signal currents through a small movable coil of wire.

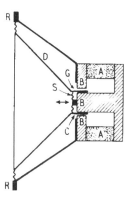

Figure 19.6 Cross-section of a moving-coil loudspeaker

Figure 19.6 shows a cross-section of a typical speaker, in which the steady magnetisation is provided by a ring-shaped permanent magnet A. Its magnetic circuit is completed by iron pole pieces BB, except for a narrow circular gap G. The high permeability of the iron enables the magnet to set up a very intense magnetic flux across the gap. In this gap is suspended the moving coil C, through which the alternating signal currents from the amplifier are passed. The mechanical force is proportional to both the flux density in the gap (which is large and constant) and the current in the moving coil. As the latter is alternating, the force is alternating, acting along the directions of the double-headed arrow in Figure 19.6 and so tending to move the coil to and fro in the gap. The coil is attached to a conical diaphragm D, which is flexibly supported at its centre by a 'spider' S and at its circumference by a ring R, so the vibrations of the coil are communicated to the diaphragm and hence to the air.

It is more difficult to avoid distortion due to the loudspeaker than any other component in the whole system. Just as inductance and capacitance in a circuit without much resistance have a frequency of resonance, at which current to and fro builds up, so inertia and springiness in a mechanical system without much friction tend to make it resonate at a particular frequency. With the moving system of a loudspeaker this frequency is often in the region of 80 Hz; its bad effects can be alleviated by a low output-stage resistance. To reproduce the lowest frequencies fully, it is necessary to move a large volume of air. If this is done by permitting a large amplitude of movement, either a very large magnet system is needed, or there is non-linearity due to the coil moving beyond the uniform part of the magnetic field in the gap. If it is obtained by using a very large cone, this makes it impossible for the cone to vibrate as a whole at the high audio frequencies and they are badly reproduced. The design is therefore a compromise. For the highest quality of reproduction it is usual to have two or more loudspeaker units of different sizes, each reproducing a part of the whole a.f. band and fed from the output stage via filters, known as *crossover filters*, which ensure that each unit receives only that part of the a.f. range it is designed to reproduce.

When the diaphragm in Figure 19.6 is moving, the air in front is compressed and the air behind is rarefied. If the period of one cycle of movement is long enough to give the resulting air wave time to travel round its edge from front to back, these pressures tend to equalise and not much sound will be sent out. To prevent this loss, evidently worst at the lowest frequencies, the loudspeaker is mounted so that it 'speaks' through a hole in a *baffle*. This consists of a piece of wood or similar material, flat or in the form of a cabinet, designed to lengthen the air-path from front to back of the diaphragm and so to ensure that the bass is adequately radiated. Some cabinets are completely sealed so that sound from the rear of the diaphragm is not radiated but absorbed in packing material. In other designs the cabinet is provided with a forwards-facing vent which enables it to act as a kind of squat organ pipe resonating at a very low frequency. Such a vented enclosure enables the low-frequency response of the loudspeaker to be usefully extended. Alternatively a horn may be used.

In a large building use is often made of the same principles as with antennas for directing the radiation in useful directions, instead of towards the roof, from which confusing echoes would be reflected. If a number of loudspeakers are mounted above one another in a column-shaped enclosure their combined output gives reinforced sound horizontally and some degree of mutual cancellation at other angles, owing to phase differences. Such an arrangement is known as a line-source loudspeaker.

As a load, a moving-coil loudspeaker can be regarded very roughly as a resistance, usually in the range 4–16 Ω. For a better approximation a certain amount of inductance in series should be assumed. Of the resistance, part is the resistance of the wire in the moving coil, and the remainder (the useful part) is analogous to the radiation resistance of an antenna (Section 17.8).

The maximum amount of a.f. power needed to drive a loudspeaker depends on two things: the maximum volume of sound required, and the efficiency of the speaker. Well-designed horn types have efficiencies ranging up to as much as 50% but usually a lot less, and with domestic

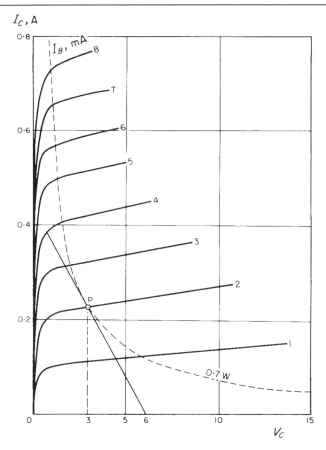

Figure 19.7 I_C/V_C curves of a small a.f. output transistor, showing power-limit curve (dashed) and Class A load line

cabinet types figures more like 5% are usual. The ear is far more sensitive to middle frequencies, say 300–3000 Hz, than the lower ones, so small portable speakers give an apparent loudness far greater than their limited power input might suggest. The maximum 'reasonably distorted' power varies from about $1/3$ to 1 W for portable sets, up to about 5 W for ordinary domestic receivers, and from 10 W upwards for 'hi-fi'.

The only significant rival of the moving-coil speaker is the electrostatic type, which is basically a capacitor. It consists of a pair of large flat metal plates close together, one fixed and rigid, the other flexible. The output of the audio amplifier is connected between them, together with a constant p.d. of the order of a thousand volts which creates a steady attraction. The signals generate variations in this, causing the sound reproduction, which can be the best of any due to absence of resonances, and the wide range of audio frequencies, especially at the high end, reproducing transients especially faithfully. The thin, flat shape is pleasing too. So it is favoured by many hi-fi enthusiasts. But its undoubted merits are at the expense of high price, the provision of the high polarising d.v., low efficiency, and awkward load characteristics.

19.10 The Output Stage

In considering how much power can be got out of a transistor or valve, regard has to be paid to the maker's maximum ratings. These can be quite complicated, but the most important is the amount of power that can safely be dissipated continuously by the device. Figure 19.7 shows a set of I_C/V_C curves for a small output transistor rated at 0.7 W. This power could be 0.7 A at 1 V, 0.35 A at 2 V, and so on for any number of combinations of output current and voltage. If we plot a sufficient selection of them we can draw the dashed curve through them to represent all such points. (A curve of this shape, by the way, is known as a hyperbola.) Every point above it represents more than 0.7 W, so is out of bounds except transiently during a signal swing. But not all the area below it is allowable; there are maximum voltage and current ratings too, which depend on other conditions but which for simplicity we shall assume are 25 V and 1 A.

Consider Class A amplification, in which the current is made to swing equally above and below the working point. The average current therefore remains the same whether there is any signal or not. The power dissipated in the transistor is actually less when it is working, but as it could

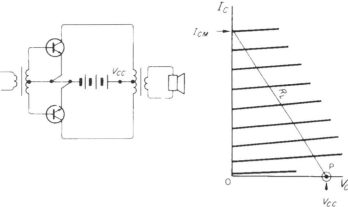

Figure 19.8 Basic circuit diagram and ideal transistor characteristics for Class B output stage

be 'silent' for indefinite periods we cannot take advantage of this to put the working point above the dashed curve. The limit we can go to is putting it on the curve. The problem is to fit a load line on this point in such a way as to represent the maximum power output, while at the same time avoiding distortion by seeing that the I_B curves are cut at equal intervals all the way. To get any reasonable output we shall have to amend these instructions to allow a reasonable amount of distortion. If the load line is nearly vertical we shall get a big swing of current but very little voltage, and vice versa with a nearly horizontal line.

It looks as if we cannot do very much better than the load line shown, in which the current and voltage swings, peak to peak, are 0.38 A and 5 V, giving a power output (Section 5.4) of 0.38 × 5/8 = 0.24 W. The feed is 0.225 × 3 = 0.675 W, so the efficiency is 0.24/0.675, just over 35%, which is well below the 50% for ideal Class A, even though it looks as if we are accepting a fair amount of distortion. The resistance represented by the load line is 13 Ω, so loudspeakers of 10–16 Ω could be worked directly from this transistor. This sort of investigation cannot be much more than a rough approximation, particularly if we remember that at top and bottom frequencies a loudspeaker load line is an ellipse instead of a straight line, which is valid only for a pure resistance.

The positive half-cycles of i_B in Figure 19.7 obviously yield smaller half-cycles of output than the negative ones. If, however, two transistors are connected in push-pull (Section 11.10) the one smaller and one larger half-cycle are added together both for positive and negative, so this particular form of distortion (second-harmonic) is largely cancelled out and the nearly undistorted output is more than twice the equivalent output from one transistor only. The end-to-end load resistance should be made double that suitable for one.

While the heavy power consumption of Class A amplifiers may perhaps be tolerable in amplifiers worked from the mains, and at least has the merit of being constant, it is quite intolerable when batteries are used, their cost per unit being enormously greater. The particularly undesirable feature is the continuance of full power consumption during quiet periods, so that the average efficiency is very low indeed.

19.11 Class B Amplification

The inadequate power output indicated when the transistor curves in Figure 19.7 are examined on a Class A basis prompts us to look into other methods of operation. Class C, however is out, because the half-cycles are so distorted. In Class B the amplifying transistors are biased close to cut-off, so that the quiescent power consumption is nearly zero, yet yield complete half-cycles for putting together to make whole ones. At all volumes—not only at maximum—most of the power expended is returned usefully. Translating this ideal into practice has its problems of course, as we shall see.

The load-line procedure we used with Class A does not work with Class B, because the working point is near cut-off, where the power input is very small. In fact, it is usually referred to as the quiescent point. And even when a load line has been drawn it does not directly indicate the power which the transistor will be called upon to dissipate. So we must try a different approach.

Figure 19.8 shows the basic circuit diagram and ideal characteristics with zero saturation voltage and no bottom bend. So the quiescent point (P) can be placed on the zero-I_C level. The circuit diagram shows that when one transistor is fully conducting, so that the signal voltage (= V_{CC}) exists across its half of the output transformer, an equal and opposite voltage exists across the other half, raising the collector voltage of the cut-off transistor to $2V_{CC}$. Therefore (subject to minor adjustments for factors we have neglected) a supply voltage V_{CC} up to half the maximum V_C rating for the transistor can be allowed. Available battery voltages must of course be considered. On these lines we decide the position of P.

The position of the other end of the load line (I_{CM}) must not be higher than the maximum peak current rating for the transistor. If we go to this limit, the load line can be drawn and its resistance ascertained (= V_{CC}/I_{CM}). The peak signal voltage is V_{CC} and peak current I_{CM}, so (if we remember that r.m.s. values are peak values divided by √2) the maximum signal power is $V_{CC}I_{CM}/2$.

One transistor delivers this for only half of each cycle, so this is the power output *per pair*.

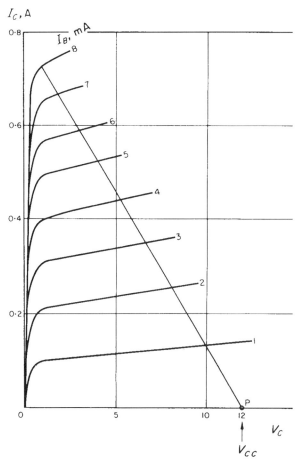

Figure 19.9 Same characteristic curves as in Figure 19.7, but Class B load line

It is often helpful to have the output power, P_0, in terms of load resistance, R_L, rather than I_{CM}. Substituting V_{CC}/R_L for I_{CM} in $V_{CC}I_{CM}/4$ we get

$$P_0 = \frac{V_{CC}^2}{4R_L}$$

Notice that in the foregoing analysis the transistor characteristics do not come into it at all, only the maximum ratings. Notice too that if you short-circuit the output, so that $R_L = 0$, the output power appears to go up to infinity, provided that there is sufficient input drive. Of course it doesn't in practice; for one thing, the drive would not be sufficient. But it is near enough to the truth to make it unwise to risk an output short-circuit, unless a fast-acting fuse is fitted in that circuit (or there is an adequate supply of spare transistors). If V_{CC} is given, then the smaller the transistor (in terms of maximum power rating) the higher the load resistance must be.

The load-line resistance R_L refers to the resistance the transistor sees across its half of the output transformer. If the turns ratio between that half and the secondary winding is 1 : 1, this resistance (with an ideal transformer) is equal to that of the loudspeaker. But usually the transformer ratio is stated in terms of the whole primary; and because there is a 2 : 1 turns ratio between that and each half, the equivalent loading across the whole primary is $4R_L$ (Section 7. 11). In Class A push-pull, both transistors are working during both half-cycles, instead of one being cut off, so the whole-primary resistance is effectively $2R_L$.

In Figure 19.9 a more practical load line is shown, though still with zero quiescent current, on the same curve sheet as Figure 19.7. The resistance it represents is 15 Ω, and the signal peak is 11 V 0.72 A, giving an output power per transistor of practically 2 W. The input is $12 \times 0.72/\pi = 2.75$ W, so with this type of signal the dissipation is just over the rated 0.7 W which compares with the calculated value of 0.5875 W for ideal conditions.

The next thing is to find the power input to each transistor under these ideal conditions. It is equal to the supply voltage (V_{CC}) multiplied by the average current. The average value of half a sine wave is $2/\pi$ times the peak value, I_{CM} in this case. So the total input is $2V_{CC}I_{CM}/\pi$ over the active half-cycle and $V_{CC}I_{CM}/\pi$ over a whole cycle. The efficiency is the signal power per transistor divided by the feed input: $V_{CC}I_{CM}/4 \div V_{CC}I_{CM}/\pi$, $= \pi/4$, which is 0.785, or 78.5%. This compares with 50% for ideal Class A. The power left to be dissipated by the transistor is the input power less the output power, which works out at 0.068 $V_{CC}I_{CM}$.

This power could possibly be found to exceed the rating for the transistor. But it assumes a continuous sinusoidal signal at maximum volume. The programmes that people listen to are not usually of this nature, and even loud sustained music has an average power appreciably less. The specifications of amplifiers of this kind should therefore state the type of programme to which the claimed output refers.

19.12 Distortion with Class B

In small-signal transistors the I_C/I_B relationship (h_{fe}) is remarkably linear, as we saw in Figure 10.11, whereas I_C/V_B is very non-linear because the working range is near the bottom bend. Power transistors on the other hand are worked high up the input curve, where I_C/V_B tends to be linear and I_C/I_B shows a marked falling off as I_C increases. This is noticeable in Figure 19.9, where the spaces between the I_B lines get progressively smaller as one moves up the load line. Incidentally, these characteristics depend hardly at all on V_C, so the published no-load curves are reasonably valid for loaded conditions. If Figure 19.9 had been plotted with V_B curves instead of I_B, the tendency would be for them to become farther apart rather than closer together. Unlike Class A push-pull, Class B does not compensate for distortion of these kinds, which blunt or sharpen both half-cycles equally if the two transistors are alike. The driver (the stage of amplification that provides the input signal for the push-pull stage) actually supplies both voltage and current, and with a bit of luck the two tendencies might at

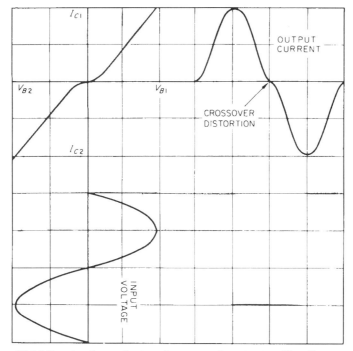

Figure 19.10 Showing the production of crossover distortion in a Class B output stage

least partly cancel out. If luck fails, a more scientific remedy awaits in Section 19.15.

But in any case the bottom bend remains. If therefore the two transistors were biased exactly to cut off, this bend would be reproduced in both half-cycles, as shown in Figure 19.10, where the two I_C/V_B curves are drawn in opposite directions to represent the push-pull connection. The resulting *crossover distortion* is particularly objectionable because it generates high-order harmonics (Section 19.5).

The remedy is to bias the transistors to about the middle of the bottom bends, as in Figure 19.11. If the biasing has been well judged the two output waveforms will add up to a very acceptable enlarged copy of the input. The crossover region is shown inset on a larger scale. Obviously the success of this method depends on how accurately the biasing has been set in the first place and kept there in spite of changes in transistor characteristics, due mainly to temperature changes. It is not much good having a beautifully undistorted output immediately after switching on if severe distortion sets in when the transistors have warmed up. This is where we must look at actual circuits.

19.13 Class B Circuits

What might be called the conventional push-pull circuit is shown in Figure 19.8; whether it is Class A, B or C (or one of their hybrids) depends on the amount of bias used. Unlike the tuned r.f. input and output transformers used in r.f. amplifiers, a.f. transformers are designed to give uniform results over the whole a.f. band (or as much of it as the designer can manage for the money). The input transformer is needed to apply positive signal to one transistor while the other is receiving negative signal. In Class A both are working all the time, each delivering full cycles; in Class B each works half time, alternately delivering half-cycles. An output transformer is needed to combine the outputs in the correct phase to give to the load. With some transistors it is also needed to match their ideal load resistance to the low loudspeaker impedance. As we have seen, the most suitable load resistances for transistors are often in the same range as those of loudspeakers. A.f. transformers—and especially output transformers—are heavy and expensive if they are to be good, and if they are not good they cause both distortion and loss, so it seems a pity to use one merely for combining the outputs from the two transistors. And it is in fact unnecessary to do so.

If Class B transistors are fed in series, as in Figure 19.12, instead of in parallel as in Figure 19.8, the two outputs flow through the load alternately in opposite directions, so reconstituting the whole waveform. Here two p-n-p transistors are shown; for simplicity the bias arrangements are again omitted. An intriguing feature of this circuit is that the input signal for the upper transistor is applied between its base and collector whereas that for the lower transistor is between base and emitter. This might suggest a degree of asymmetry in an amplifier where both transistors are required to contribute equally to the common output. We must not forget, however, that bipolar transistors are current-operated devices and the current gain of a common-collector stage is almost the same as that of a common-emitter stage.

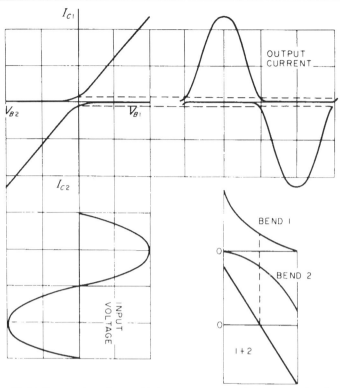

Figure 19.11 Showing how, by suitable biasing, crossover distortion can be eliminated

To avoid the need for a centre-tap on the d.c. power supply it is usual to put a high-C blocking capacitor in series with the loudspeaker, which can then be connected to either end of the supply (Figure 19.14).

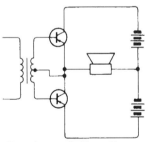

Figure 19.12 One form of Class B output circuit. Its disadvantage is the centre-tap on the power supply

As regards the input transformer, it too can be dispensed with by making use of what is called a phase splitter as the driver. Figure 19.13 shows one type, sometimes called the concertina circuit (when the zigzag symbol is used for resistors), in which outputs of opposite phase are obtained from resistors across each side of the transistor. When transistors are to be driven, the outputs have to supply current as well as voltage, so R is needed to equalise the output resistances of the driver, since the output resistance at the emitter is relatively low (Section 12.9). Phase splitters are essential for transformerless push-pull using valves, but when transistors are used for the output stage advantage can be taken of the fact that they are available in both p-n-p and n-p-n forms. If one of each is used their inputs can be driven with the same polarity, the reversal of phase being brought about by the opposite polarities of the output transistors themselves. They are termed a complementary pair, and Figure 19.14 shows the basic circuit.

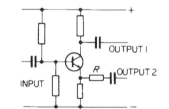

Figure 19.13 Phase-splitter circuit for driving a push-pull stage

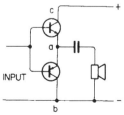

Figure 19.14 Complementary pair method of doing without a two-phase input

In this particular formation, with the input returned to either *b* or *c*, both transistors are working in common-collector mode, so the input signal voltage would have to be actually greater than the output. To avoid this disadvantage by using the common-emitter mode it would seem that the input should be returned to the emitters, point *a*. But as this point is alternating at the full output signal voltage the driver would have to be coupled to the output stage by a transformer, which is what we are trying to avoid. A way out of this difficulty is shown in Figure 19.15, where the input voltage is developed across R_1, the

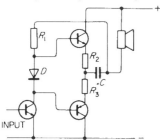

Figure 19.15 Transformerless combined driver and output stage using complementary transistors

top end of which is returned to output emitters signalwise via *C*, while the required d.c. feed is obtained via the loudspeaker. The two output transistors receive the usual current-adjusting emitter bias by means of R_2 and R_3, which because of the large currents flowing through them are typically only an ohm or two. Base bias is supplied by the voltage drop across *D* (Section 13.7). This voltage changes comparatively little over quite a wide range of current. One silicon diode usually provides enough voltage (about 0.6 V) for a pair of germanium transistors; if they are silicon, two diodes in series will be needed.

19.14 Compound Transistors

A convenient method of obtaining a higher current gain than is provided by a single transistor is by combining two as shown in Figure 19.16, forming what is known as a compound pair, or Darlington pair. The essential feature is that the emitter current of the first transistor is the base current of the second, so the current gain of the pair is equal to the product of the current amplification factors of the two transistors; i.e., $h_{fe1} \times h_{fe2}$. This result is achieved with a minimum of components, so the idea is used for many purposes besides power amplifiers.

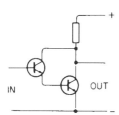

Figure 19.16 Compound pair of transistors for multiplied current gain

Biasing arrangements are similar to those for one transistor, but of course the actual bias voltage required—the difference between V_{B1} and V_{E2}—is equal to the sum of the two separate bias voltages.

In Figure 19.16 the pair is shown in common-emitter mode. When used in common-collector mode the feedback effect (Section 12.9) also is multiplied, so that the input resistance is exceptionally high.

19.15 Negative Feedback

The effect on the apparent resistance of a circuit of injecting into it a voltage or current proportional to the voltage or current applied was explored in Chapter 12, where in most cases the injected signal was in opposition to the applied signal. In amplifiers this is called negative feedback. It is often used intentionally to increase or reduce the input or output resistance of an amplifier. It is also much used for reducing distortion. Transistor characteristic curves are never quite linear, and with Class B there is the risk of crossover distortion. The variation of loudspeaker impedance with frequency introduces gain/frequency distortion too, and sometimes the gain of the transistors falls off at the highest audio frequencies. Loudspeaker resonances cause sounds of certain frequencies to be unduly prolonged.

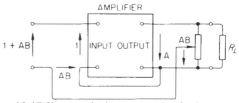

Figure 19.17 Showing the basic principle of voltage negative feedback

Figure 19.17 shows an amplifier in which a fraction (*B*) of the output voltage is fed back in opposition to the input. (Note that with this method of application the amplifier must be inverting.) The voltage gain of the amplifier itself is denoted by *A*. This means that for every signal millivolt (say) applied to the input terminals, *A* millivolts appear between the output terminals, and *AB* millivolts are fed back to the input. The signal source is therefore required to supply not 1 mV but $(1 + AB)$ mV.* If we denote the overall gain with feedback by A', then

$$A' = \frac{\text{output}}{\text{input}} = \frac{A}{1 + AB}$$

If *AB* is so large that 1 is negligible in comparison with it, then

---

* On the ground that *AB* is the measure of negative feedback it would seem logical to regard *B* as negative; the overall input is then $(1-AB)$ which comes to the same thing as the above if you remember that *B* is negative. Treating *B* as positive when feedback is negative may be illogical, but when we are talking only about negative feedback it is probably less confusing than having to remember to use a double negative every time.

$$A' \simeq \frac{1}{B}$$

The great significance of this is that B depends on a potential divider or something of the kind that can be perfectly linear, so the non-linearities involved in A are more or less removed. Also the effects of temperature on the characteristics of transistors, and the very large variations among samples of the same type, are largely overcome.

These major advantages have to be paid for by the increased input to be provided. As a general rule the distortion and overall gain are both reduced by the same divisor, $(1 + AB)$. The overall gain can usually be restored by, say, an extra stage of amplification. But the rule is misleading if it makes us suppose that we need not bother much about avoiding distortion in the basic design because it can always be linearised by negative feedback. The fallacy of this can be seen by imagining a Class B amplifier in which, owing to incorrect bias, both transistors were momentarily cut off at the crossover. The value of A could be normal over the rest of each cycle, but at the crossover A' would be $0/(1 + 0)$, so the gap would be just as bad as without feedback; in fact, even worse because of the sharper transitions between linear and non-amplifying regimes (Section 19.5). Negative feedback can make a good amplifier better, not a bad amplifier good.

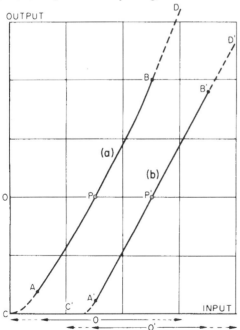

Figure 19.18 (a) Typical output/input characteristic with slight curvature. (b) The same after applying negative feedback

To illustrate this, Figure 19.18 shows at a an output/input curve that is perceptibly but not grossly non-linear between A and B—it would in fact cause about 5½% second-harmonic distortion. P, the centre of the input swing, is the working point. Curve b is the result of applying sufficient negative feedback to reduce the gain (reckoned as the average between A and B) to one-tenth, but for ease of comparison the input scale has been reduced to one-tenth, as if a distortionless 20 dB amplifier had been added before the input. Also for clarity the curve has been shifted to the right, and the corresponding input swing around P′ as working point is from A′ to B′. The positive and negative output half-cycles are now much more nearly equal than in a. The second-harmonic distortion is too small to read off the curve, but if calculated would be found to be one-tenth as much as in a.

Suppose now that the input is increased by 50% so as to include the dotted portions. In contrast to the practically linear D′A′, the bit from A′ to C′ would cause a more drastic increase in distortion than AC because it contains a more acute change of slope where it meets the axis. So one has to be careful, when using an amplifier with negative feedback, not to push the volume control too far.

What about gain/frequency distortion? Suppose that in the example just considered the gain falls off at the highest a.f. to the extent of 6 dB, or 50%. At the lower frequencies, where the full gain operates, the reducing factor of the feedback is $1/(1 + 9) = 0.1$ as we assumed. When A is halved this is $0.5/(1 + 4.5) = 0.091$. So the relative gain at the top frequencies, with feedback, is $0.091/0.1, = 0.91$, or -0.8 dB, which is hardly noticeable. Negative feedback therefore tends to flatten out irregularities in the gain/frequency characteristic.

19.16 Phase Shift with Feedback

The main cause of gain/frequency distortion in amplifiers is reactance. In Section 18.10 we considered the use of a resistance-capacitance potential divider as an elementary sort of filter for removing r.f. signals from the output of a detector. Either kind of reactance causes an increasing loss at either high or lower frequencies, depending on where it comes in relation to resistance; both kinds of reactance together can cause either a loss or gain around one particular frequency, since they form a tuned circuit. In Section 6.7 and elsewhere this frequency discriminating effect was shown to be accompanied by a phase shift. So gain/frequency distortion is accompanied by phase distortion. As negative feedback reduces gain/frequency distortion it seems reasonable to infer that it reduces phase distortion too. And so it does, but with important reservations as we shall see.

For manufacturing reasons, inductance tends to be avoided in modern a.f. circuits. Where a.f. transformers are still used they are analysed into their equivalent circuits (Section 7.10) for calculating frequency effects, which can be done by drawing phasor diagrams for a number of typical frequencies. There is inevitably some inductance in loudspeakers, some of it purely electrical and some simulated by the motion of the coil in the field of the magnet. But most phase shifts in circuits are due to capacitances, intentional or otherwise.

Audio Frequency Amplification 175

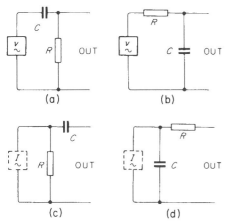

Figure 19.19 The four basic combinations of C and R so often found in signal circuits

Figure 19.19 shows at *a* and *b* the two basic combinations of capacitance and resistance with voltage input and output, and at *c* and *d* with current input and output. Diagram *a* might be a resistive loudspeaker and its blocking capacitor as in Figure 19.14; *b* we have come across in Section 18.10; C in *d* might be stray capacitance across a load resistance, but is not likely to be significant at a.f. other than in high-resistance valve circuits.

All of these oft-recurring combinations can be treated simultaneously by relating all frequencies to what we could call the turning frequency (f_t) because it marks a transition in the frequency characteristic from resistance to reactance regimes. As in Section 18.10, $1/2\pi f_t C = R$, so

$$f_t = \frac{1}{2\pi CR}$$

In Figure 19.19 C can be regarded at high frequencies as a short-circuit, so can be more or less ignored in *a* and *c*; similarly in *b* and *d* at low frequencies, which make it almost an open-circuit.

Figure 19.20 shows the results of phasor diagrams as curves of relative output against frequency relative to f_t, with and without feedback in which $AB = 10$. The 11-fold loss resulting from this feedback is compensated, for easy comparison. Curves are shown for one Figure 19.19*b* or *d* combination and also for two having the same CR value. The curves for Figure 19.19*a* and *c* are a left-to-right mirror image of Figure 19.20. So all combinations of C and R and values of CR are covered. Figure 19.21 shows the corresponding phase shifts. Figure 6.10 shows that in Figure 19.19*a* the phase of the output voltage is ahead of the input, so it is regarded as positive. In *b* the output lags, so the phase shift is negative. A dual phasor diagram would show the shift to be positive in *c* and negative in *d*.

Figures 19.20 and 19.21 show clearly the effect of feedback in extending the flat part of the frequency characteristic curves; with one CR the extension is 11-fold—the same as the reduction in gain, $(1 + AB)$. Note that at f_t, where the loss without feedback is 3 dB, the phase shift is 45°. The same applies to $(1 + AB)f_t$ with feedback. But with two CR units the picture changes somewhat. The curves for no feedback are obtained simply by doubling the dB and phase-angle figures, but this time negative feedback does not increase the frequency coverage 11-fold.

Gain/frequency and phase distortions are certainly much reduced, but beyond about $2f_t$, the phase shift takes a steep plunge; meanwhile the output actually goes higher than at the 'flat' frequencies. We see that at the peak the corresponding phase shift is 90°. Feedback here can hardly be described as either negative or positive. Most of the suppressed gain of the amplifier is therefore released. At still higher frequencies the phase shift approaches 180°, so feedback is largely positive, but meanwhile the loss due to the CR units is so great that the overall output curve continues to drop away steeply.

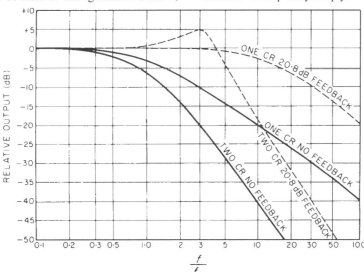

Figure 19.20 Relative output plotted against frequency (relative to the turning frequency f_t) for one and two CR circuits with and without 10 : 1 (= 20.8 dB) negative feedback

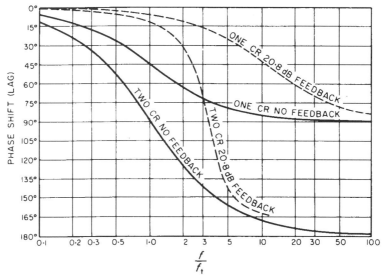

Figure 19.21 Phase shift graphs corresponding to Figure 19.20

At about $1.7 f_t$ the phase shift per CR unit is 60° and the gain is still quite high, so if three units were used the total phase shift would be 180° and the conditions for self-oscillation (Section 14.5) would be fulfilled. The amplifier would in fact have turned into the oscillator circuit of Figure 14.15. Such an amplifier would be described as unstable.

A single LC combination can cause a phase shift of nearly 180°, which is another reason for avoiding L in amplifiers. But even where there is none, the existence of C between stages, and in the transistors themselves, makes it difficult to prevent instability if one attempts to reduce distortion throughout by making the feedback loop embrace the whole amplifier. It is not uncommon for oscillation to take place at some frequency too high to be audible itself but only as distortion or loss of output. With amplifiers of very high gain care has to be taken to avoid instability due to unwanted feedback via stray capacitances or impedance common to high and low level signal circuits. And one is only too familiar with the howl set up when a microphone picks up too much of the output from its loudspeaker; in this case the feedback is acoustical.

So far it has been assumed that B is the same at all frequencies. But sometimes it is designed to vary with frequency, usually by including capacitance as well as resistance in the feedback path. Obviously if negative feedback is reduced at some frequencies the gain at those frequencies is increased. That may be the object, or a modification of the value and phase angle of B may be needed to improve stability.

19.17 Input and Output Resistance

The theory of the effects of feedback on input and output resistance is given in Section 12.5; now we can glance at it again with particular reference to amplifiers. Usually we would be glad to increase the input resistance of a bipolar transistor so as to make it as nearly as possible voltage-operated, more like an f.e.t. or valve. We saw in Section 12.9 how a great increase is obtainable by working it in common-collector (or emitter-follower) mode, and in Section 19.14 a further increase by using a pair of transistors in this way. The effect is produced by what amounts to total negative feedback; *all* the output voltage is fed back to the input. So the input voltage to cause a given input current must be increased $(1 + A)$ times and the apparent input resistance is therefore $(1 + A)$ times greater.

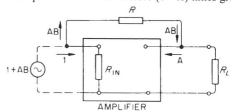

Figure 19.22 Showing the basic principle of shunt-injected negative feedback

In Figure 19.17 some of the output voltage was fed back in series with the input. An alternative method in which output signal is fed back in parallel is shown in Figure 19.22. It is sometimes called shunt-injected feedback. Treating the system as a current amplifier we again call the input 1, this time, say, 1 µA. With a current amplification A, there is an output of A microamps in the direction shown. Of this, a fraction B, controlled by R, is fed back to the input. (The fact that the arrows seem to show it being fed forward is because it is negative feedback.) So for every microamp going into the amplifier the signal current source has to supply $(1 + AB)$ microamps. The current gain (neglecting the fraction of output bypassed via R) is therefore reduced to $A/(1 + AB)$ as in Figure 19.17, and because the current taken from the source is $(1 + AB)$ times as much as is actually going into the amplifier, the input conductance seems to be $(1 + AB)$ times as great, and the resistance is that much less.

Output resistance, as we saw in Section 12.7, is measured by substituting a signal generator for the load and short-circuiting the input voltage generator, the only input then being what is fed back. This is not difficult to do theoretically by calculation, and the result is that voltage feedback reduces the apparent output resistance of an amplifier. But the extent of this effect is greater than the factor $(1 + AB)$, in which A is the gain with load connected. The effect of voltage feedback is to divide the output resistance by $(1 + A_0B)$, where A_0 is the voltage gain with output terminals open-circuited.

So far we have considered circuits in which the negative-feedback signal is derived from the output-load terminals themselves (Figure 19.22) or from a potential divider across the load (Figure 19.17). In such circuits the feedback signal (whether voltage or current) is proportional to the output voltage; such systems are known as *voltage negative feedback*.

There is an alternative method of deriving the negative-feedback signal in which it is taken from a resistor connected in series with the load as shown in Figure 19.23.

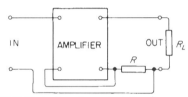

Figure 19.23 Illustrating the principle of current negative feedback. R must be small compared with R_L

Its resistance is made small compared with the load and thus the negative-feedback signal (whether voltage or current) is proportional to the current in the load. Such a system is known as current negative feedback. Current feedback is similar to voltage feedback in its effects on gain and distortion but, unlike voltage feedback, it increases the effective output resistance of the amplifier by an amount $(A + 1)R$. Such feedback is therefore unlikely to be used in the output stage of an amplifier driving a loudspeaker where a low output resistance is desirable as explained later. Current feedback can be and is used in small-signal stages where it can be introduced very simply by omitting the bypass capacitor normally connected across the bias resistor in the emitter or source circuit. The signal now generated across this resistor acts as feedback, reducing gain and distortion, and increasing the output resistance of the stage.

An emitter resistor can simultaneously provide voltage and current feedback. With respect to an output taken from the collector, it gives current feedback as just described, increasing the output resistance. With respect to an output taken from the emitter, however, the emitter resistor gives voltage feedback as described above for the emitter follower, the output resistance being reduced. This feedback thus causes a disparity between the output resistances at collector and emitter, an effect requiring compensation in a concertina phase splitter (Section 19.13).

The effect of voltage and current negative feedback on input resistance depends on the way in which the feedback signal is injected into the input circuit. If it is connected in parallel with the input, as in Figure 19.22, it reduces the effective input resistance but if it is injected in series with the input (as by omitting the emitter bypass capacitor) it increases the input resistance.

When a short-lived signal (called a *transient*) having a frequency at which the loudspeaker it is fed into resonates, the loudspeaker continues to vibrate and emit sound after the signal has ceased. This effect is called *transient distortion*. The oscillation of the moving coil in the strong magnetic field (Section 19.9) generates an e.m.f., and as it is in parallel with the output of the amplifier it will make a current flow therein. How much current depends on the output impedance. If it is very small—say a fraction of the loudspeaker's own impedance—the current will be correspondingly large and will quickly damp out the undesirable oscillation. This is obviously a good thing, which can be achieved by quite a moderate amount of negative voltage feedback.

The 100% negative feedback in emitter followers and cathode followers not only greatly increases the input resistance but reduces the output resistance. So although the voltage gain is less than 1 the current gain can be very large. These devices are therefore largely used as a sort of current transformer to feed low-impedance loads from high-impedance sources. They have two great advantages over inductive transformers: the voltage loss is very small; and the effective frequency band can be very large, including zero frequency. Figure 19.22 can be adapted to illustrate 100% negative feedback by transferring R_L to the R position so that all the output signal is fed back.

19.18 Linearising the Input

The advantages of negative feedback in a.f. amplifiers are so valuable that it is practically always used. The number of ways in which it can be applied are too many to show here, but armed with the foregoing principles we ought to be able to follow most of the circuit diagrams we see.

Many of the low-power transistors used in the initial stages of amplifiers have very linear I_C/I_B characteristics, as in Figure 10.11, and very non-linear I_C (or I_B)/V_B, as in Figures 10.12 and 10.13. Obviously they should be worked from signal current sources rather than voltage. Where the impedance of the actual source was not sufficiently high to qualify it as a current source, a common practice was to bring it up to that standard by adding resistance. Figure 19.24 shows an input-stage circuit (feeds omitted) in which

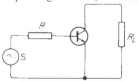

Figure 19.24 One way of linearising a transistor is by means of a relatively high resistance R in series with the input

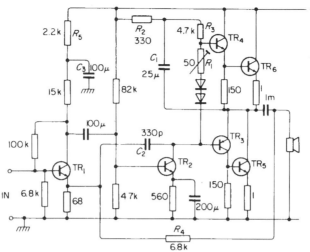

Fig 19.25 A typical complete audio amplifier circuit

the transistor is assumed to have a perfectly linear I_C/I_B curve (i.e., constant h_{fe}). With the voltage signal source S connected direct there would be serious distortion because of the very non-linear input resistance (I_B/V_B curve). Inserting a relatively large linear R makes the total resistance nearly linear and therefore the input current, and therefore also the output current, nearly proportional to the input signal voltage. But of course most of the signal is lost in R.

If instead negative voltage feedback were used the result would be just the same; the voltage drop due to I_B flowing through R would be replaced by a voltage drop due to $h_{fe}I_B$ flowing through part of R_L. But whereas the series-resistance method linearises only the input circuit, and depends on linear I_C/I_B (which is now less often found), feedback can be applied to as much of the amplifier as one wants and corrects all non-linearities, so has virtually superseded series resistance.

19.19 Some Circuit Details

To gather together the fragmentary studies of this chapter and to illustrate some further features, a complete a.f. amplifier circuit is shown in Figure 19.25. Typical component values are given only to indicate the orders of magnitude to be expected. In such diagrams the main unit abbreviations are often omitted, as the component symbol implies it; for example, '1 m' for the loudspeaker blocking capacitor means 1 mF, = 1000 μF.

The output stage follows the lines of Figure 19.15—a complementary pair—each half being made up of two transistors as suggested in Section 19.14. The two biasing diodes are supplemented by a resistor R_1, the special symbol for which indicates that it can be preset to any value up o 50 Ω in order to minimise crossover distortion. A slightly different arrangement is used for feeding the collector of TR_2; instead of coming from the positive supply rail (as it is called) via the loudspeaker, it comes via R_2, and the output signal voltage is added via C_1, so that the signal potential at the top end of R_3 is about the same as TR_2 applies at the bottom end. R_3 is thus made to look to TR_2 like a very high resistance, and a good stage gain is obtained. This particular method of applying the principles of Section 12.5 is called (following American language) bootstrapping.

The emitter resistor of TR_2 is bypassed to prevent negative feedback here, but the emitter resistor for TR_1 is left unbypassed because it forms part of the overall feedback loop right from the loudspeaker via R_4. C_2 provides subsidiary feedback to ensure stability at very high frequencies. Biasing follows normal practice (Section 13.4). R_5 and C_3 together are what is known as a decoupler, the purpose and working of which are explained in Section 28.7.

An interesting development in the design of transistor power amplifiers is illustrated in simplified form in Figure 19.26. A low-power class A amplifier drives the

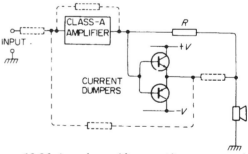

Figure 19.26 An a.f. amplifier providing an example of current 'dumping' and of feed-forward error correction. The dashed lines indicate negative feedback loops

loudspeaker via a low-value resistor R, across which a basic push-pull Class B stage is connected. The Class B stage has no forward bias and, for small signal inputs to the amplifier the voltage across R is insufficient to overcome the transistor offset voltage, so that the Class B stage does not conduct and contributes nothing to the amplifier output. But as the signal input is progressively increased, the Class B stage comes increasingly into operation and, at full output, supplies most of the power to the loudspeaker. Because of the absence of standing forward bias, the Class B stage introduces considerable crossover distortion which can, however, be virtually eliminated by the use of the two negative feedback loops shown in dashed lines, one embracing the Class A stage and the other the entire amplifier. The Class B stage is regarded as a current dumper*; alternatively the whole amplifier can be regarded as an example of a process known as 'feed-forward error compensation,' the Class A stage supplying the output at low signal levels where the dumper stage cannot operate satisfactorily; i.e., the Class A stage corrects the dumper-stage error.

19.20 Noise

While the gain of an amplifier such as the one in Figure 19.25 is sufficient to give full output from a radio detector or some types of record player pick-up, it is usually not enough for the better types of pick-ups and microphones or for tape playback heads. Amplifiers for many other purposes also have to deal with even smaller signals, perhaps only a few microvolts. Designing for the required gain is easy; the real problem is *noise*. This word is used to refer to electrical disturbances which, in sound reproducers, cause audible noise; the same disturbances are also objectionable when the end product is not sound—they mar the picture in television, for example.

Some noise is picked up from external electrical equipment. There is hum from the public a.c. supply. And anything that causes abrupt changes of current generates noise over a wide frequency band; such things as switching, thermostat contacts, appliances using commutator motors, and faulty lighting. Such disturbances can be excluded from a.f. amplifiers by properly arranged earthing, electric and magnetic screening, filters and careful layout of wiring. Printed wiring and integrated circuits (Section 24.7) are less likely to pick up induced noise.

But when all such noise has been suppressed or excluded there is a never entirely avoidable residue generated in the amplifier circuits themselves. An electric current is not a smoothly continuous flow; it is made up of individual electrons, and even when there is no net current in any one direction the electrons in, say, a resistor are in a continuous state of agitation. Their random movements, like the aimless jostling of a crowd which as a whole remains stationary, are equivalent to tiny random currents, which give rise to small voltages across the resistor. Quite a moderate gain, such as can be obtained with three or four stages, is enough to enable them to be heard as a hissing sound, or seen on the television screen as animated graininess. Only the part of an amplifier at or near its input is a noise problem, because all the other parts are followed by less amplification.

This electron agitation increases with temperature and can only be quelled altogether by reducing the temperature to absolute zero ($-273°$ C) which is hardly practical for ordinary purposes, though for special purposes such as reception from space vehicles the important parts are immersed in liquid helium ($-269°$ C). It is known as thermal agitation or *Johnson noise*. Although the voltages occur in a completely random fashion, so that the peak values are continually fluctuating, their r.m.s. value over a period of time is practically constant; somewhat as the occurrence of individual deaths in a country fluctuates widely but the death rate taken over a year changes little. Since the frequency of the jostlings is also completely random, the power represented by it is distributed uniformly over the whole frequency spectrum. By analogy with light, it is often called *white noise*. So if the amplifier is selective, accepting only frequencies within a certain band, B hertz wide, the noise is reduced. The formula is

$$V = 2\sqrt{(kTBR)}$$

where V is the r.m.s. noise voltage, k is what is called Boltzmann's constant and is 1.37×10^{-23}, T is the temperature in kelvins ($°$ C + 273) and R is the resistance in ohms. For example, suppose the resistance of a signal source is $0.1\,M\Omega$ and the full a.f. band (20 kHz) is accepted. Room temperature can be taken as 290 K. The noise voltage is then $2\sqrt{(1.37 \times 10^{-23} \times 290 \times 2 \times 10^4 \times 10^5)}$ = $5.6\,\mu V$. If the total voltage gain were one million (120 dB) this noise (5.6 V) would be very disturbing.

A somewhat similar cause of noise, usually called *shot effect*, is due to the current in diodes, transistors and valves being made up of individual electrons. It is not so easily calculated as Johnson noise because it depends on the structure of the device, space charges and other things, but like Johnson noise it is proportional to $\sqrt{B}$. The r.m.s. noise current tends to be proportional to $\sqrt{I}$, where I is the current flowing through the device, so in the first stage this current should be made small, so long as it is not reduced past the point where the gain of the device begins to be reduced faster. The aim is the highest possible signal/noise ratio. Special silicon bipolar transistors are made for low noise, and in general f.e.ts are low-noise devices. Transistors have a low-noise band from moderately low a.f. up to a certain frequency depending on the design. Above that frequency the signal/noise ratio falls off, and at very low frequencies there is what is called *flicker noise*, or $1/f$ noise because it is inversely proportional to frequency.

The noisiness of a device or stage, or even of a whole amplifier or receiver, is specified as its noise factor or noise figure, which is defined in various ways, of which a simple one is

$$\frac{\text{input signal/noise power ratio}}{\text{output signal/noise power ratio}}$$

It is usually expressed in dB; 0 dB is perfection, and the higher the figure the worse the noise.

---

* An unfortunate term suggesting that the current of this stage is thrown away uselessly, whereas at high signal levels it supplies most of the wanted output.

19.21 Another Box Trick

In Section 12.5 we introduced a box with two terminals whose electrical properties were not what would be expected from its contents. Here is another example of such a box. This time a sinusoidal voltage applied to the terminals gives rise to an input current in quadrature with the voltage. We can therefore say that the box behaves as a reactance—capacitive if the current leads the voltage, and inductive if it lags. But the box contains no inductance or capacitance to provide the reactance, the box behaviour being brought about by a common-emitter amplifier with a phase-shifting negative feedback loop.

A typical circuit of this type is shown in Figure 19.27. Suppose C is given a value such that its reactance at the operating frequency is small compared with R. The current in RC is then in phase with the collector voltage. The voltage across C lags this current by 90° and so also does the base signal and the collector current. Thus the collector current lags the collector voltage by 90° and the collector/emitter path of the transistor behaves as an inductor. The magnitude of the inductance can be calculated in the following way.

If the collector voltage is v_C, the current in RC is given approximately by v_C/R and the voltage across C is $v_C/R\omega C$. The collector current is thus $g_m v_C/R\omega C$ where g_m is the mutual conductance of the transistor. If we represent the effective inductance of the collector/emitter path by L_{eff}, the collector current is also given by $v_C/\omega L_{eff}$. Equating the two expressions for collector current we have

$$\frac{v_C}{\omega L_{eff}} = \frac{g_m v_C}{R\omega C}$$

from which $L_{eff} = RC/g_m$. For given values of R and C, L_{eff} is inversely proportional to g_m. As we saw in Section 10.9 the mutual conductance of a transistor varies with the mean collector current and so, by adjusting the mean current (by alteration of base bias) we can control the value of L_{eff}. Such circuits, known as reactance stages, can be used as modulators in f.m. transmitters. They were also used for controlling the oscillator frequency in automatic frequency control systems before the introduction of varactor diodes.

If R and C are interchanged and if C is chosen to have a reactance large compared with R, the transistor behaves as a capacitor of value $g_m RC$.

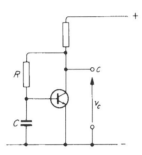

Figure 19.27 A simple reactance stage in which the transistor behaves as an inductor

Chapter 20
Selectivity and Tuning

20.1 Need for R.F. Amplification

In Section 18.11 we saw how the effectiveness of detectors deteriorates quite rapidly as the r.f. input is reduced much below about 1 V. Signals picked up by a receiving antenna are seldom more than a few millivolts and are often only microvolts, so (except from a powerful transmitter at short range) they are too weak to operate a detector satisfactorily. Amplification of the r.f. signals is therefore a practical necessity in order to give a receiver adequate *sensitivity*, which, for an a.m. sound receiver, is usually expressed as the input microvoltage needed to yield a specified a.f. output (often 50 mW) at a specified modulation depth (usually 30%).

Another practical necessity is *selectivity*, which as we saw in Section 8.6 is the ability of a receiver to discriminate in favour of the wanted signal against all others, some of which may be much stronger and only slightly different in frequency. As will have been gathered, the chief means for obtaining selectivity is the tuned circuit. Although tuned circuits are not necessarily associated with amplification they almost invariably are, for reasons that will appear. Meanwhile we shall do well to make a closer study of selectivity.

20.2 Selectivity and Q

We saw in Chapter 8 that both selectivity and sensitivity are improved by increasing the Q of a tuned circuit. Figure 8.5, for example, shows how the response at the frequency to which a circuit is tuned increases in direct proportion to Q, whereas the response at other frequencies is less affected; in fact, at frequencies far off tune it hardly changes at all. The problem of r.f. amplification might therefore seem to be simply one of getting a high enough Q. We found (Section 8.4) that the Q of a circuit is the ratio of its reactance (either inductive or capacitive) to its series resistance, which is the same thing as the ratio of its dynamic resistance to its reactance:

$$Q = \frac{X}{r} = \frac{R}{X}$$

Now the amount of reactance is decided by the requirements of tuning, so attention must be confined to reducing the series resistance, r. This will have the effect of increasing the dynamic resistance, R. We noted (Section 8.14) that r is a sort of hold-all figure including everything that causes energy loss from the tuned circuit; not only the resistance of the coil of wire, but loss in the solid core (if any), in adjacent metal such as screens, and in dielectrics associated with the circuit. These can all be reduced by careful choice of materials and careful design, but size and cost limit what can be done, so the result is a Q seldom very much better than about 200.

In Chapter 14 we saw how the loss that damps out oscillations in tuned circuits can be completely neutralised by energy fed to it at the right moment in each cycle (positive feedback), so that oscillations once started continue indefinitely. This would not do in a receiver, because we want the strength of oscillation in the tuning circuit to depend all the time on what is coming from the distant transmitter. But if the positive feedback is reduced, the amount of energy fed back will partly neutralise the losses. Since we are already using the symbol r to include not only resistance in its strictest sense but all other causes of loss, there is no reason why we should not also make it take account of a source of energy gain. In this sense, then, positive feedback reduces r and therefore increases Q.

We can put this to the test, using an oscillator of the type shown in Figure 14.4 by increasing the distance between the two coils just sufficiently to make the continuous oscillation stop. The tuned output circuit will then behave exactly as if its resistance were abnormally low; any small voltage injected in series with it at its resonance frequency is magnified much more than if the transistor were not working. The same result is obtained if the tuned circuit and the reaction coil are interchanged; in this way an input tuning circuit likewise can be given a sharper resonance.

In Figure 20.1 are plotted some of the resonance curves that might be obtained, using different amounts of positive feedback on a circuit tuned to 1 MHz. The '$Q = 40$' curve represents a rather poor circuit with no feedback, and the others with progressively increased amounts. To raise the Q from 40 to 8000 necessitates very careful adjustment of the coupling, as it is then very near the self-oscillation point, and a slight change in supply voltage or in transistor temperature would either carry it past this point or in the opposite direction where Q would be substantially less. In other words, the adjustment of coupling is extremely critical. In practice it would be difficult to maintain the Q close to such a high value for long, and one would be wise to be content with rather less. Even so, the benefit that can be obtained by feedback is so great that it could not be shown clearly on a graph like those that were used in Chapter 8; Figure 20.1 has therefore been drawn to a logarithmic voltage scale, so compared with Figure 8.5 the increase is much greater than it looks.

The required selectivity and sensitivity seem now to be readily obtainable. If only a simple carrier wave were being received, that might be true. But it is the modulation of the carrier that conveys the desired information (using that word in its widest sense). If the carrier is frequency-modulated (Section 15.4) it is obvious that a definite band

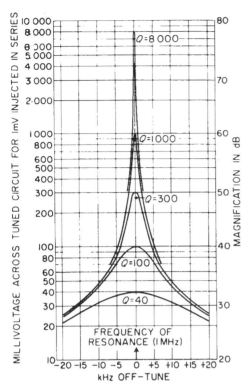

Figure 20.1 Showing how r.f. voltage across a tuned circuit varies with frequency at different Q values, obtained by adjusting positive feedback

of frequencies, with the carrier frequency at its centre, must be received to an equal degree. The need for covering a band of frequencies when the carrier is amplitude-modulated is less obvious, but it none the less exists.

20.3 Oversharp Tuning

If we look back at Figure 20.1 we see that the curve for Q = 8000 fails conspicuously to meet this requirement even for a.f. modulation (Section 15.8) and does not begin to meet the much more demanding needs of video modulation (Section 15.9). The sidebands corresponding to a modulation frequency of 5 kHz are shown to be received only to the extent of 1.3% as much as the carrier wave and the lowest a.f. sidebands. The high notes are quite literally tuned out by overselectivity. Higher pitched sounds still would be even more drastically attenuated. The result would be indistinct speech and boomy music. Even with a Q as low as 100, 30%, or 3 dB, of 5 kHz modulation would be lost, but this would not be too much to restore by an equaliser (Section 19.3).

Because the effective frequency bandwidth for a given Q is proportional to the carrier frequency, whereas the bandwidth required is unaffected by it, the sideband cutting effect is less at higher carrier frequencies and greater at lower carrier frequencies.

With frequency modulation there is the same need to preserve sidebands, but the results of failing to do so are different. Figure 15.7 might certainly suggest that if the bandwidth were limited to 150 kHz all except perhaps the lowest audio frequencies would be lost. They would, however, all be represented at least to some extent by the sidebands up to 150 kHz. But this graph represents 100% modulation, whereas most of the time the percentage is much less. The main effect of sideband cutting is therefore to limit the depth of modulation—and hence the resulting a.f. amplitude—that can be correctly reproduced. In other words, non-linearity distortion, in contrast to gain/frequency distortion of a.m. signals.

This undesirable effect is at least partly counteracted by the limiter in an f.m. receiver, which tends to bring all r.f. signal amplitudes down to one level, but it could not cope with really severe sideband cutting.

Any hopes we may have entertained of achieving adequate sensitivity and selectivity quite simply by pushing up Q are therefore bound to fade.

20.4 A General Resonance Curve

Whether we are considering channel separation (Section 15.9) or loss of sidebands, then, the basis of reckoning is 'frequency off-tune'. To distinguish this particular frequency from others we shall denote it by f'.

The ideal shape for a tuning curve would seem to be one which was level over a frequency band wide enough to include all the wanted sidebands and then cut off completely to exclude everything else. None of the resonance curves we have seen so far looks anything like this—see Figure 20.1, for example. And except for Figure 8.6 they are arranged for comparing sensitivity rather than selectivity. We need to take the response at resonance as the starting point and see how much it falls at off-tune frequencies. This falling-off can be reckoned in dB. An alternative scale for any who may still have difficulty in thinking in dB is in what we shall call S numbers, which are the numbers of times by which response off-tune is less than that at resonance. We now want to know what determines S at any given f'—in general; not like Figure 8.6, which applies to only two values of Q and one carrier frequency.

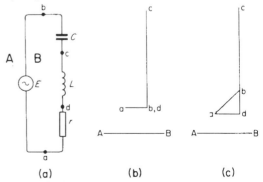

Figure 20.2 (a) A tuned circuit, with its phasor diagrams (b) at resonance and (c) slightly off resonance

To discover this we go back to tuned-circuit fundamentals. Figure 20.2a is a tuned circuit containing a source of e.m.f. E. It could be a signal picked up inductively by L, but it is shown separately to keep things clear. The phasor diagram b is drawn according to the rules in Chapter 5, beginning with the current AB. It shows the condition of resonance, in which X_L and X_C, and therefore the voltages across L and C, are equal and opposite. So the whole of E ($= ab$) is available for driving current through r. $AB = ab/r$, in fact. If f_r is the frequency at resonance, X_L ($= -X_C$) $= 2\pi f_r L$.

Next suppose that the frequency of E increases from f_r to $(f_r + f')$. X_L increases in direct proportion to $2\pi(f_r + f')L$; so the increase is $2\pi f'L$. Provided that f' is small compared with f_r, X_C decreases by practically the same amount. So the total reactance, $X_L + X_C$, has increased from zero to $4\pi f'L$. Figure 20.2c shows the corresponding phasor diagram. Because we have to start with the current before we know what it is going to be, we draw AB the same as at b. As, however, it is E that we are assuming to be constant, the current must actually decrease in the same ratio as the total circuit impedance is increased by altering the frequency. At f_r this impedance is r, and at $(f_r + f')$ it is $\sqrt{[r^2 + (4\pi f'L)^2]}$ (Section 6.8).

The output, or response, of the tuned circuit is the voltage developed across C by the current. (The whole loss resistance r is assumed to be in the inductor.) Now X_C is not quite the same at $(f_r \pm f')$ as at f_r, as the phasor diagrams show; but within our assumption that $f' \ll f_r$ the difference can be neglected. Our selectivity number, S, is the ratio of response at f_r to response at $(f_r \pm f')$, and we now see that this is equal to the current ratio, which is the reciprocal of the impedance ratio. That is to say,

$$S = \sqrt{[r^2 + (4\pi f'L)^2]}/r = \sqrt{\left[1 + \left(\frac{4\pi f'L}{r}\right)^2\right]}$$

The only variable part of this is $f'L/r$, so if we know or assume the L/r ratio for a tuned circuit we can find S for any values of f' we like and plot a selectivity graph. Whether we will care to undertake this work after looking at the formula above and realising that the result holds good for only one value of L/r depends on having a suitable calculator. If, however, S is plotted against $f'L/r$ a curve is obtained which can be adapted for any L/r ratio merely by dividing the horizontal scale numbers by it. Figure 20.3 is that curve. It shows how much the response of any one tuned circuit falls off as the frequency is altered from resonance. Using a logarithmic scale for $f'L/r$ as well as for S has the advantage that the greater part of the curve is a straight line with a 1 : 1 slope, or 6 dB per octave as it is usually specified. This curve is in fact the same as those in Figures 18.9 and 19.20 and like them has a 'turning

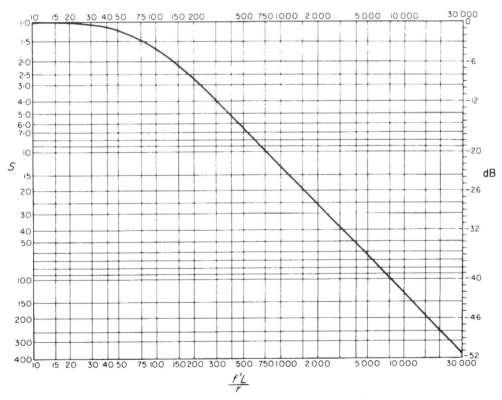

Figure 20.3 Generalised resonance curve for any single tuned circuit. To make it refer to any particular circuit, divide the numbers on the horizontal scale by L/r (or by $Q/2\pi f_r$) to convert them to a scale of f'. The units are: L in µH; r in Ω; f' in kHz; f_r in MHz

frequency', f_t, at which the straight slope continued would meet the zero-dB line. f_t is the f' that makes the reactance ($4\pi f'L$) equal to the resistance (r), so occurs where $f'L/r = 1/4\pi = 0.08$. (It is 80 in Figure 20.3 because of the units used, listed in the caption.) As before, the loss at f_t is 3 dB. What is called the bandwidth (f_B) is the frequency band between the -3 dB points; that is to say, $2f_t$. From the foregoing, f_B is $r/2\pi L$. And because $Q = 2\pi f'L/r$, f_B is also equal to f_r/Q, which is a very useful result. We also find that $Qf'/2\pi f_t$ is the same as $f'L/R$ in Figure 20.3.

There is no need to show the other half of the selectivity curve, because it is practically the same so long as $f' \ll f_r$. If it is not, then a lack of symmetry does appear at frequencies far from resonance, but within our assumption we can regard f' as either + or –.

As an example of using Figure 20.3, suppose L is 150 μH and r is 24 Ω, so that $L/r = 6.25$. 200 on the $f'L/r$ scale then means 32 kHz, at which S is 2.7, or $-8½$ dB. Or suppose we wanted to find the resistance of a coil to give a certain selectivity, say -20 dB at 40 kHz off-tune. The curve shows that for this $f'L/r$ is about 800, so, at $f' = 40$, L/r is 20. If L is 5000 μH then r should be 250 Ω. The loss of 7.5 kHz side frequency due to such a coil is found by looking up S for $f'L/r = 7.5 \times 20 = 150$, namely about 7 dB.

So far from a tuned circuit offering any choice of shape to combine good selectivity with retention of sidebands, what we have found is that there is only one shape, and the most we can do is slide it horizontally along the f' scale until the best compromise is obtained; L/r is then indicated. Once we have decided on the response at any one off-tune frequency, the response at all others is thereby decided too, whether we like it or not.

20.5 More than One Tuned Circuit

Let us now consider what happens when additional tuned circuits are used. If we can assume that these circuits do not react on one another, then it is simple. If one circuit reduces signals at a certain off-tune frequency 3 times (relative to the response at resonance), and another circuit is somewhat more lightly damped and reduces them 4 times, then the total relative reduction is 12 times. If a third circuit gives a further reduction of 3 times, the total is 36. In other words, an r.f. amplifier incorporating these three tuned circuits amplifies this certain frequency 36 times less than it amplifies the frequency to which all three circuits are tuned. In decibels, the three circuits cause reductions of 9.5, 12 and 9.5 dB; the total is found by simple addition: 31 dB.

If the problem is simplified by assuming all the circuits have the same L/r, then the curve in Figure 20.3 is made to apply to a pair of circuits by squaring S; to three circuits by cubing it; and so on. Or by multiplying the dB scale by 2, 3, etc.

Comparison between one and several tuned circuits can be made in different ways. First of all we shall compare them on the basis of equal gain, in order to dispose finally of the question of positive feedback versus r.f. amplification. Taking some typical figures, suppose that the r.f. amplifier is a single stage giving a gain of 50 times, with input and output tuned circuits for which $L/r = 10$. Squaring the S values in Figure 20.3, we get curve a in Figure 20.4. To obtain equal gain by means of feedback, it would be necessary to apply enough of it to multiply Q by 50. This would multiply L/r by the same factor. So curve b is drawn for a single tuned circuit with $L/r = 500$. We see that the modulation frequency corresponding to a 5 kHz note is reduced 3 dB by the r.f. amplifier; that is to say, it retains about 70% of its relative strength. But the receiver with feedback reduces it by 30 dB, leaving only about 3%. True, it gives excellent selectivity; but it would take all the life out of music and make speech difficult to follow. At 9 kHz off-tune the r.f. amplifier is very unselective. But, significantly, at the more remote frequencies its selectivity is beginning to overtake its rival.

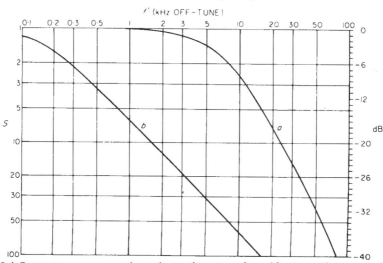

Figure 20.4 Comparison on an equal-gain basis of a stage of amplification with two tuned circuits (curve (a)) with a single tuned circuit given a high magnification by positive feedback (b)

The excessive sharpness of b in Figure 20.4 was a result of having to make the tuned circuit provide as much gain as a stage of amplification. But now let us disregard gain and compare one circuit with several on a basis of equal selectivity. Suppose the intention is to reduce signal voltages at two channels off-tune (18 kHz) to one-hundredth (40 dB) relative to the on-tune voltage. Figure 20.3 shows that the required L/r for a single circuit would be $8000/18 = 445$; nearly as sharp as b in Figure 20.4. With two circuits, S for each would be -20 dB, so $L/r = 800/18 = 44.5$. With four circuits, S would be -10 dB, so $L/r = 13$. With six, S would be $-62/3$ dB, and $L/r = 8.4$. The overall response at other frequencies can be derived from Figure 20.3 by taking S^2 or twice the number of dB for two circuits, and so on. Plotting the results for one and six circuits on one graph, we get Figure 20.5, which shows very clearly that a single tuned circuit, whether its extreme sharpness is necessitated by the requirements of gain or of selectivity, can fulfil these requirements only at the cost of drastic cutting of the wanted sidebands, and that a large number of relatively flat circuits is much to be preferred. Extensions of the curves show, too, that the selectivity at frequencies beyond the chosen reference point of 18 kHz goes on increasing much more rapidly with many circuits than with few.

Lastly, we can compare one circuit with six on a basis of equal retention of sidebands. Figure 20.5 shows that with six circuits the loss of 5 kHz is just over 6 dB. The effect of flattening the single tuned circuit until its loss at this point is the same is represented by sliding curve a along until it coincides with b there (position c). This shows how poor the single-circuit selectivity would then be, in spite of the appreciable sideband cut.

The way things are going, it looks as if six or even more tuned circuits would hardly give a fair approximation to the ideal with which we started Section 20.2. And here it is necessary to remember that the way in which we have been deriving the overall results of several circuits depends on their not reacting on one another, for that would modify their response curves. The usual way of using such circuits is as couplings between one stage and the next in an r.f. amplifier. Here it is difficult to avoid enough coupling via the amplifying devices to cause appreciable interaction. Anyway, the use of anything like six r.f. stages, even if technically feasible, would provide far too much gain, so most of it would have to be thrown away. But is inter-circuit coupling necessarily bad?

Introducing coupling between tuned circuits makes it more difficult to predict their combined resonance curve. But by a rather more mathematical investigation than is appropriate in a book of this kind it can be shown that beneficial results are obtainable in this way, as we shall now see.

20.6 Coupled Tuned Circuits

There are many ways of coupling two tuned circuits; Figure 20.6 shows some of them. The feature common to all is a certain amount of shared (or mutual) reactance. In type a it is capacitance, C_m; in b, self-inductance, L_m; and in c, mutual inductance, M. Fortunately for calculation, the result does not depend (at any one frequency) on the kind of mutual reactance but only on its amount; so we can denote it simply by x. As a matter of fact, method c is nearly always used, because no extra component is needed and the coupling can be adjusted simply by varying the spacing of the two coils. Just as L/r is the factor determining the selectivity of a single tuned circuit, x/r determines the modifying effect of the coupling on the combined resonance curves of two similar circuits.

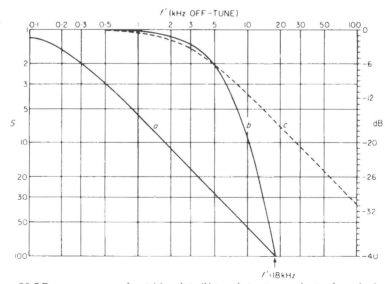

Figure 20.5 Response curves of one (a) and six (b) tuned circuits on a basis of equal selectivity at 18 kHz, showing how the six preserve sideband frequencies much better. Curve (c) refers to a single tuned circuit with its resonance flattened to give equal response at 5 kHz off-tune

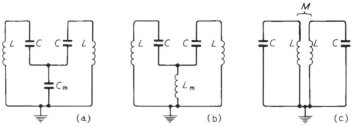

Figure 20.6 Three methods of coupling two tuned circuits: (a) by capacitance common to both, C_M; (b) by common inductance, L_M; and (c) by mutual inductance, M

This result can be shown most clearly in the form of the overall resonance curve for two circuits, (a) with no coupling, except through a one-way electronic device, and (b) with reactance coupling. The question at once arises: with how much? The shape of the curve is different for every value of x. Since the primary purpose of the coupling is to transfer the signal from one circuit to the other, a natural coupling to try for a start is the amount that transfers it most effectively. This is not, as might at first be supposed, the greatest possible coupling. It is actually (with low-loss circuits) quite small, being simply $x = r$.

For this value, called the *critical coupling*, the curve is Figure 20.7b. The curve applies to a pair in which the two circuits have the same L/r; if there is any difference, an approximation can be made by taking the mean L/r. Curve a, for uncoupled circuits, is the same as Figure 20.3 but with all the S figures squared. Although the difference in shape is not enormous, it is quite clear that the curve for the coupled circuits is considerably flatter-topped. But it is also less selective. The advantage gained by the coupling is more obvious if the resistance of the coupled circuits is reduced to half that of the uncoupled circuits, as in Figure 20.7c. If, for example, the L/r of each of the uncoupled circuits represented by a were 10, the L/r for each of the coupled circuits represented by c would be 20, and the off-tune or sideband frequency treated equally by both combinations (represented by the intersection of a and c) would be 6 kHz. Although the coupled circuits would thus retain sideband frequencies up to 6 kHz better than the uncoupled circuits they would also give substantially more selectivity.

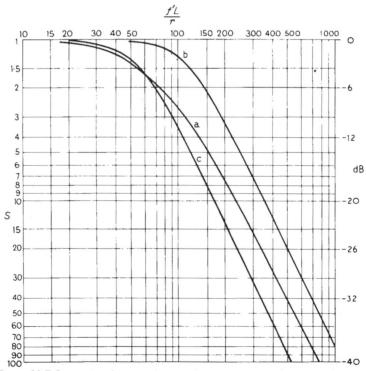

Figure 20.7 Comparison between the overall resonance curve for a pair of similar tuned circuits, (a) uncoupled, and (b) 'critically' coupled. Curve (c) is the same as (b) but shifted to the left to represent a halving of resistance

20.7 Effects of Varying Coupling

Even so, an adjacent-channel selectivity of less than 8 is hardly an adequate projection against interference equal in strength to the wanted signal, still less against relatively strong interference. We may reasonably ask whether any other degree of coupling than critical would give a still better shape; for, if so, any loss in signal transference caused thereby would be well worth while, as it could easily be made up by a little extra amplification. Figure 20.8 shows what happens when the coupling between two tuned circuits is varied. The curves have been plotted to a linear frequency scale both sides of resonance, to show the kind of shapes seen on an oscilloscope screen (Section 23.4).

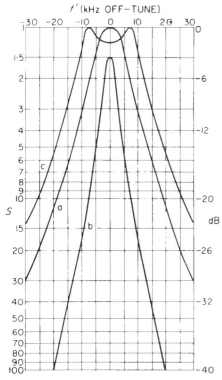

Figure 20.8 Changes in resonance curve of a pair of coupled circuits caused by varying the coupling: curve (a) is for critical coupling; (b) is for less than critical; and (c) for more

Reducing the coupling below critical, by making x less than r, not only reduces the signal voltage across the second circuit but also sharpens the peak, so is wholly undesirable. Increasing the coupling splits the peak into two, each giving the same voltage as with the critical coupling at resonance; and the greater the coupling the farther they are apart in frequency. Here at last we have a means by which the selectivity can be increased by reducing the damping of the circuits, while maintaining sideband response at a maximum. Coupling pairs of circuits are one type of what are known, for obvious reasons, as *bandpass filters*. There is a limit, however, because if one goes too far with the selectivity one gets a result something like Figure 20.9, in which a particular sideband frequency is accentuated excessively. Symmetrical twin peaks are in any case difficult to achieve in practice and still more difficult to retain over a long period, because the coupling adjustments are so extremely critical. Moreover, they make the set difficult to tune by ear because one tends to tune it to the frequency that gives the loudest sound, which in this case is not the carrier frequency.

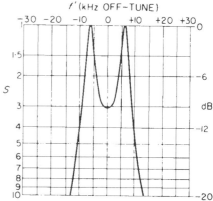

Figure 20.9 The result of trying to maintain sidebands when using excessively selective coils in a band-pass filter—the 'rabbit's ears' produced by over-coupling

Perhaps the best compromise is to have one pair of circuits critically coupled, and sufficiently selective to cause appreciable sideband cutting; and the other pair slightly over-coupled, to give sufficient peak-splitting to compensate for the loss due to the first pair. The ultimate cut-off will then be very sharp and give excellent selectivity.

Mention should also be made of yet another technique for combining a flat top with sharp cut-off. It is called *stagger tuning* and has certain practical advantages, particularly for v.h.f. and u.h.f. A number of single or double circuits are peak-tuned to slightly different frequencies in the pass band, as shown in Figure 20.10. This technique is extensively used in television receivers to achieve the wide bandwidth and particular shape of resonance curve required. See Section 23.7.

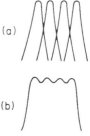

Figure 20.10 'Stagger-tuning', in which the separate tuned circuits are set to 'peak' at slightly different frequencies (as shown at (a)), so that the combined result (b) has a fairly flat top and sharp cut-off

All the resonance curves so far discussed in this chapter have been produced by *adding* the responses of individual or coupled LC circuits. Sometimes a particular type of curve is best produced by *subtracting* the response of an LC circuit. For example a very sharp cut-off at one end of the resonance curve can be achieved by using an LC circuit tuned to a frequency just beyond the cut-off point and arranged to give maximum rejection at its resonance frequency. This idea, too, is widely used in television receivers; e.g., to minimise sound pick-up in the vision tuned amplifier.

20.8 Practical Tuning Difficulties

It is easy enough to talk about choosing L/r ratios and adjusting couplings to give the best possible response curves. It is not usually very hard to achieve something like the desired results at one frequency. But the demand for receivers permanently tuned to one channel is not large. It is when one attempts to vary the tuning over wide bands of frequency that the troubles really begin. Means must be provided for varying the tuning capacitance or inductance of every circuit continuously and exactly together by one control. Because of the limited frequency band that can be covered in one sweep (Section 8.7) switching of inductors is nearly always needed to cover several bands.

It would be extremely difficult to maintain all the couplings at exactly the right adjustments and practically impossible to keep L/r always right. When, as is usual, a coil is tuned by a variable capacitor, L of course remains the same over the whole band, but r is much less at the lowest frequency (maximum capacitance) than at the highest. So the selectivity varies accordingly.

Even if some way out of this varying selectivity trouble were found, it would only be to run into another. For if L/r is kept constant over a band of frequency, the dynamic resistance R—which as we know is equal to L/Cr (Section 8.12)—must be inversely proportional to the tuning capacitance, and therefore directly proportional to the square of the frequency. This means very much less amplification at the low frequency end of the band than at the high frequency end.

Tuning by varying L instead of C could theoretically avoid both these difficulties at once, if r were made to vary in proportion, for then L/r and L/Cr would both be constant. But this is easier said than done. And we have not begun to think about keeping all this lot stable at all frequencies. So it is not surprising that attempts to produce a really practical high-performance receiver on the lines of variable-frequency r.f. amplification have failed. The generally accepted solution is the subject of the next chapter.

Chapter 21 The Superheterodyne Principle

21.1 A Difficult Problem Solved

Fortunately it is possible to combine the advantages of fixed-tuned selectivity, using a sufficient number of tuning circuits carefully adjusted once and for all in the factory, with the ability to receive over wide ranges of frequency. It is done in the great majority of present-day receivers, under the name of supersonic heterodyne—usually abbreviated to superheterodyne or just 'superhet'. A device called a frequency-changer shifts the carrier wave received from any desired station, together with its sidebands, to whatever fixed frequency has been chosen for the selective circuits. This frequency is called the *intermediate frequency* (i.f.), and the following choices are typical:

For receiving—	i.f.
Sound broadcasts (a.m.) on low, medium and high frequencies	470 kHz
Sound broadcasts (f.m.) on very high frequencies	10.7 MHz
Television broadcasts {vision / sound} on v.h.f. and u.h.f.	{33.5 MHz / 39.5 MHz}

The frequencies of all incoming signals in any one of these categories are shifted by the frequency-changer to the same extent. So if, for example, the original carrier-wave frequency of the wanted programme was 865 kHz and it had been shifted to 470 kHz, adjacent-channel carrier waves at 856 and 874 kHz would be passed to the i.f. amplifier as 461 and 479 kHz.

After the desired signal has been amplified at its new frequency and at the same time separated from those on neighbouring frequencies, it is passed into the detector in the usual way, and the remainder of the receiver is in no way different from a 'straight' set.

21.2 The Frequency-changer

The most important thing to grasp in this chapter is the principle on which the frequency-changer works. All else is mere detail.

Suppose a sample of the carrier wave coming from the desired station during one-hundred-thousandth of a second to be represented by Figure 21.1a. As it contains 10 cycles, the frequency must be 1000 kHz. Similarly *b* represents a sample of a continuous wavetrain having a frequency of

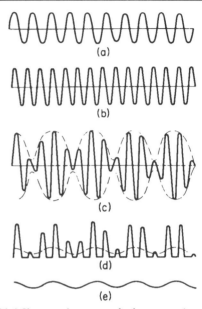

Figure 21.1 Showing the action of a frequency-changer

1450 kHz, which is being generated by a small oscillator in the receiver. Now add the two together. Since the local oscillation alternately falls into and out of step with the incoming signal, it alternately reinforces and weakens it, with the result shown at *c*: waves varying in amplitude at a frequency of 450 kHz. This part of the process is often referred to as *beating* or *heterodyning b* with *a*, and 450 kHz in this case would be called the *beat frequency*. Being above audibility, it is described as supersonic.* The same kind of thing takes place audibly when two musical notes of nearly the same pitch are sounded together.

Notice particularly that what is happening at 450 Hz is only amplitude variation (or modulation) of waves of higher frequency. *No signal of 450 kHz is present*, for each rise above the centre line is neutralised by an equal and opposite fall below the line. (Figure 15.5d is another example of the same thing, except that it is made up of waves of three frequencies, all higher than the frequency at which its amplitude is varying.) The average result, when applied to a circuit tuned to 450 kHz, is practically nil. We have been up against this kind of thing already (see Figure 18.1) and the solution is the same now as then—rectify it.

---

* This term is retained in this context although the word for such frequencies has meanwhile been changed to 'ultrasonic'.

After eliminating the negative half-cycles (*d*) the smoothed-out or averaged result of all the positive half-cycles (shown dashed) is a signal at 450 kHz; that is to say, the difference between the frequencies of *a* and *b*. The mixture of 450 kHz and the original frequencies is then passed to the i.f. circuits, which reject all except the 450 kHz, shown now alone at *e*.

So far, the result is only a 450 kHz carrier wave of constant amplitude, corresponding to the 1000 kHz carrier wave received from the antenna. But if this 1000 kHz wave has sidebands, due to modulation at the transmitter, these sidebands when added to the 1450 kHz oscillation and rectified give rise to frequencies which are sidebands to 450 kHz. So far as the i.f. amplifier and the rest of the receiver are concerned, therefore, the position is the same as if they were a 'straight' set receiving a programme on 450 kHz.

Any readers who, in spite of Figure 15.6, have difficulty in seeing amplitude modulation as sidebands, may find Figure 21.2 helpful. This shows a rather longer sample of the carrier wave (*a*), the individual cycles being so numerous that they are hardly distinguishable. And this time it is fully modulated; the sample is sufficient to show two whole cycles of modulation frequency. The local oscillation, which is certain to be much stronger, and of course is unmodulated, is shown at *b*. When *a* is added to it (*c*), nothing happens to it at the troughs of modulation, where *a* is zero, but elsewhere the cycles of *a* alternately strengthen and weaken those of *b*, causing its amplitude to vary at the beat frequency, in the manner shown. The amplitude of beating is proportional to the amplitude of *a*, and the frequency of beating is equal to the difference between the frequencies of *a* and *b*. The next picture, *d*, shows the result of rectifying *c*. We now have, in addition to other frequencies, an actual signal at the beat frequency (or i.f.), shown at *e* after it has been separated from the rest by the i.f. tuned circuits. The result of rectifying *e* in the detector is *f*; and after the remains of the i.f. have been removed by the detector filter the final result (*g*) has the same frequency and waveform as that used to modulate the original carrier wave.

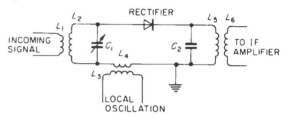

Figure 21.3 Simple type of frequency-changer, in which the incoming signal and the local oscillation are added together by connecting their sources in series, and then rectified

Figure 21.3 shows an elementary way in which the process just described can be carried out. The incoming

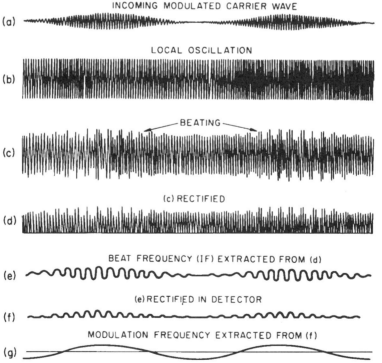

Figure 21.2 In this diagram the sequence of Figure 21.1 is elaborated to show what happens when the incoming carrier wave is modulated. The additional lines, (*f*) and (*g*), show the extraction of the modulation frequency

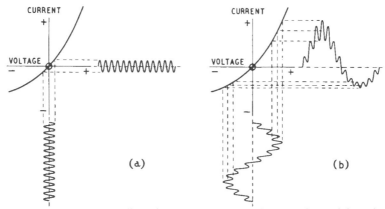

Figure 21.4 Showing the effect of adding a strong signal to a weak signal through a non-linear device such as a transistor. The amplitude of the weak signal is varied (or modulated) by the strong

signal and the local oscillation are added together by connecting their sources in series with one another, and the two together are fed into a rectifier—some kind of diode. C_1 is used for tuning L_2 to the signal, so as to develop the greatest possible voltage at that frequency. There is, of course, no difficulty in getting an ample voltage from the local oscillator. C_2 serves the double purpose of tuning the transformer primary (L_5) to the intermediate frequency, and of providing a short-circuit for the two input frequencies, which are not wanted there.

21.3 Frequency-changers as Modulators

If the rectifier in Figure 21.3 had been omitted, or replaced by an ordinary linear resistor, no new signal frequency would have been created. In linear circuits, currents of different frequencies can flow simultaneously without affecting or being affected by the others. In non-linear circuits that is not so. A rectifier is an extreme kind of non-linear resistance. Figure 21.4 shows the current/voltage characteristic curve of a milder kind. In diagram *a* a single small alternating voltage—frequency, say f_1—is applied, and the resulting current waveform is derived and shown to the right of the graph. Such a small portion of the characteristic curve is not very different from a straight line, so the distortion is only slight. In any case each of the output cycles is the same as all the others. Diagram *b* shows the result of applying the same voltage plus another, which has been chosen to have a much larger amplitude and lower frequency (f_2) to make it easily distinguishable. Its positive half-cycles carry the f_1 signal up the curve to where the slope of the curve is greater (i.e., the resistance is less), so the f_1 current cycles are larger than in *a*. The f_2 negative half-cycles carry the f_1 signal to where the slope is less (i.e., the resistance is greater), so the f_1 current cycles are smaller than in *a*. The f_2 signal has left its mark on the f_1 signal, by amplitude modulating it. The depth of modulation obviously depends on the degree of non-linearity of the circuit.

We already know that the process of amplitude modulation creates signals having frequencies different from both the modulating and modulated frequencies; to be precise, these frequencies are equal to the sum and difference of f_1 and f_2—see Figure 15.5 again. These are not necessarily the only new frequencies created—that depends on the kind of non-linearity—but we shall confine our attention to them.

In Figure 21.1 the difference frequency is clearly visible, especially at *e*, after it has been extracted from the mixture. The fact that in *d* there is a frequency equal to that of *b plus* that of *a* is not at all obvious, but its existence can be shown mathematically, or by actual experiment—picking it out from the rest by tuning a sharply resonant circuit to it.

The purpose of the local oscillator can now be seen as a means of varying the conductance of a part of a circuit carrying the incoming signal, amplitude-modulating it and thereby creating signals of at least two new frequencies.

Either of these could be chosen as the i.f. in a superhet, but the lower or difference frequency is the one usually favoured, because it is easier to get the desired amplification and selectivity at a relatively low frequency. See also Section 21.6 for a more important reason.

21.4 Types of Frequency-changer

A frequency-changer therefore consists of two parts: an oscillator, and a device whereby the oscillation modulates or varies the amplitude of the incoming signal. This device is commonly called the mixer. If the word 'mix' is understood in the sense a chemist (or even an audio engineer) would give it—just adding the two things together—then it is a very misleading word to use. Simply adding voltages or currents of different frequencies together creates no new frequencies, just as merely mixing oxygen and hydrogen together creates no new chemical substance. So from this point of view 'combiner' would be a better name for the second part of a frequency-changer. It can also be fairly termed the modulator.

The oscillator presents no great difficulty; almost any of the circuits shown in Chapter 14 could be used. It is, of course, necessary to provide for varying the frequency over a suitable range, in order to make the difference between it and the frequency of the wanted signal always equal to the fixed i.f.

The simplest type of modulator or 'mixer' is the plain non-linear element—a diode—already seen in Figure 21.3. Frequency-changers essentially of this type are used in radar and other receivers working on frequencies higher than about 800 MHz, for reasons which will appear shortly. But they have certain disadvantages which have resulted in their having been superseded for most other purposes. For one thing, adjustment of C_1 is liable to affect the frequency of the oscillator. The local oscillations on their part are liable to find their way via L_2 and L_1 to the antenna, causing interference with other receivers. In short, the signal and oscillator circuits interact, elsewhere than in the proper direction—through the rectifier. And the rectifier does not amplify; on the contrary, it and its output circuit tend to damp the signal presented to it by the tuned circuit L_2C_1.

The principle of a commonly used type of frequency changer is shown in Figure 21.5. A transistor triode is used

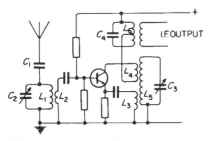

Figure 21.5 Typical transistor frequency-changer circuit for medium and low r.f.

both as oscillator and modulator sometimes described as a self-oscillating mixer. L_1 and C_2 tune the antenna and input circuit. If a wire antenna is used, C_1 limits the amount of extra capacitance the antenna can add to C_2. More often L_1 is itself the antenna, wound on a ferrite rod (Section 17.19). To suit the relatively low input impedance of the transistor, L_1 and L_2 act as a step-down transformer. L_5C_3 is the oscillator tuning circuit, coupled by L_3 and L_4 to emitter and collector in such phase as to maintain oscillation. This has sufficient amplitude to swing the transistor alternately between cut-off and the steep part of its output/input curve, fully modulating the relatively weak signal across L_2 and thereby producing sum and difference frequencies in the collector circuit. A fraction of a volt is enough for a transistor, but the more gradual control of the gate in an f.e.t. calls for several volts. When C_3 is adjusted to make the difference frequency equal to the chosen i.f., the i.f. part of the mixture of signals is selected by L_6C_4 and passed on from there to the i.f. amplifier.

This is a simple and economical type of frequency-changer, but combining the two functions in one device makes it difficult to ensure the best operation for both at all times, and especially to avoid the circuit interactions already mentioned. So where the best performance is desired separate devices are preferred.

So far the modulation has been effected by adding the signal and oscillation together and passing both through the non-linear device. This is therefore called the *additive* principle. If, however, the device that amplifies the signal has another electrode which directly controls its gain and to which the oscillator output is applied, the signal is thereby modulated without being swung between extremes of the transfer characteristic. The amplitude of the signal is, in effect, being multiplied by the alternating voltage of the oscillation, so this is called a *multiplicative* type of frequency-changer. Figure 21.6 shows one of several types of multiplicative frequency-changer—a dual-insulated-gate

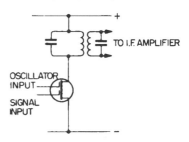

Figure 21.6 One of several types of device which can be used for frequency-changing—a dual-insulated-gate f.e.t.

f.e.t. The signal input is applied to the gate nearer the source, and the oscillator input to the other gate. Both input signals affect the conductivity of the channel and thus modulate the current in it. In this way the two inputs are effectively multiplied, producing, at the output of the transistor, components at the sum and difference of the two input frequencies. The difference component is selected by the tuned transformer in the drain circuit and is passed on to the i.f. amplifier for further amplification.

21.5 Conversion Conductance

The effectiveness of an ordinary amplifying or oscillating type of transistor is, or can be, expressed as the mutual conductance, in milliamps of signal current in the output circuit per volt of signal applied at the input. In a frequency-changer, however, the current at signal frequency is a waste product, to be rejected as soon as possible; what we are interested in is the milliamps of i.f. current per volt of signal-frequency. This is known as the *conversion conductance*, denoted by g_c. Obviously it does not depend on the device alone but also on the oscillation voltage. If this is too small, it does not vary the mutual conductance over its full range; consequently the signal current is modulated to only a part of its full amplitude, and the i.f. current is less than it need be. The conversion conductance reaches its maximum when the local oscillation is enough to bring the mutual conductance to zero during its negative half-cycles and to maximum during as much as possible of its positive half-cycles. The positive part of Figure 21.2c is therefore reproduced with the full g_m, while the negative part is completely suppressed; the net result is the same as with a perfect rectifier, Figure 21.2d, The incoming signal, however weak, is swept right

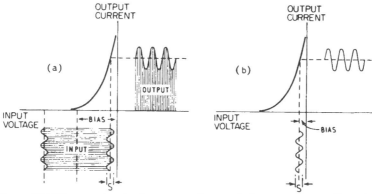

Figure 21.7 Diagram showing how conversion conductance compares with mutual conductance

round the bend by the oscillation. That explains why a superhet is sensitive to the feeblest signals, which, if applied straight to the detector, would make no perceptible impression.

In Figure 21.7a, a small sample of oscillation plus signal (such as Figures 21.2c or 21.1c) is shown in relationship to the input voltage scale of an active device curve. (The fact that the oscillation and signal may actually be applied to different terminals does not affect the essential principle.) Since the i.f. fluctuations are due to the signal alternately strengthening and weakening the local oscillation, their amplitude is, as we have seen, equal to that of the signal. Neglecting the effect of the load in the output circuit, as we generally can with these types of device, corresponding fluctuations in output current half-cycles are practically equal to those that would be caused by an i.f. voltage of the same amplitude as the signal voltage amplified straightforwardly, as indicated at *b*. If the i.f. current amplitude in *a* were the same thing as the amplitude of the i.f. envelope, this would mean that the conversion conductance was practically equal to the mutual conductance. But, as is shown more clearly in Figure 21.1*d*, the fluctuation, in *mean* value of the half-cycles is $1/\pi$ or only about 32% of the peaks (Section 18.3). So g_c is only about one-third of g_m under ideal conditions, and in practice is between one-third and one-quarter.

The output resistance of the type of transistor likely to be used as a frequency changer is so much higher than any normal load resistance (R) that the voltage gain is approximately equal to $g_m R$. The corresponding quantity in a frequency-changer, the *conversion gain*, is equal to $g_c R$ and is the ratio of output i.f. voltage to input r.f. voltage. Even though a frequency changer gives only about a quarter of the gain of a comparable r.f. amplifier, it makes a very welcome contribution to the gain of a receiver, in addition to its main function.

Gain is not everything, however. At the input to any amplifier it is noise that decides how much gain can usefully be sought (Section 19.20). A frequency-changer is inherently more noisy than a comparable straight amplifier. In transistors, noise depends mainly on the construction, and they range from very bad to very good. If a high sensitivity is required, combined with a good signal/noise ratio, there should be at least one low-noise stage of amplification at the original signal frequencies. The part of the receiver that deals with those frequencies is known as the *preselector*.

21.6 Ganging the Oscillator

We have seen that the intermediate frequency is equal to the difference between the signal frequency and the oscillator frequency. With an i.f. of 470 kHz, for example, the oscillator must be tuned to a frequency either 470 kHz greater or 470 kHz less than the signal.

Suppose the set is to tune from 525 to 1605 kHz. Then, if higher in frequency, the oscillator must run from (525 + 470) to (1605 + 470), i.e., from 995 to 2075 kHz. If, on the other hand, the oscillator is lower in frequency than the signal, it must run from (525 − 470) to (1605 − 470), or 55 to 1135 kHz. The former range gives 2.09 as the ratio between highest and lowest frequency; the latter 20.6. Since even the signal-circuit range of 3.06 is often quite difficult to achieve, owing to the irreducible capacitance likely to be present in a finished set, the oscillator range from 995 to 2075 kHz is the only practical choice.

The next problem is to control both signal-frequency circuit tuning and oscillator frequency with one knob. Because there must be a constant frequency difference between the two, it hardly seems that what is called a ganged variable capacitor—consisting, in effect, of two identical capacitors on the same spindle—would do. As we have just seen, the ratio of maximum to minimum capacitance is quite different. This difference, however, can easily be adjusted by putting a fixed capacitance either in parallel with the oscillator capacitor to increase the minimum capacitance, or in series to reduce the maximum. The ratio of maximum to minimum having been corrected in either of these ways, correct choice of inductance for the oscillator coil will ensure that it tunes to the right frequency at the two ends of the tuning scale.

In the middle of the tuning range, however, there will be an error. The oscillator tuning capacitance will be too large or too small, depending on whether a parallel or series capacitor is used to restrict the oscillator range. It is found that a judicious combination of the two methods, using a small parallel capacitor to increase the minimum a little, and a large series capacitor to decrease the maximum a little, will produce almost perfect 'tracking' over the whole waveband.

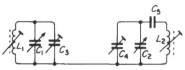

Figure 21.8 Preselector and oscillator tuning circuits arranged to maintain a constant frequency difference

Figure 21.8 shows these tuning arrangements applied to one frequency band. $L_1C_1C_3$ is a preselector tuning circuit; there is often at least one other, either at the input and output of a stage of amplification or loosely coupled to one another by reactance. Sometimes L_1 can be preset to the designed value by the screw adjustment of the magnetic core, which is what the symbol indicates. C_1 is one section of the ganged tuning capacitor, and C_3 is a preset *trimmer* to bring the total effective capacitance to the designed value when C_1 is at its minimum, since in this condition the frequency is critically dependent on stray circuit capacitance. C_2 is another section of the tuning capacitor, identical to C_1, for varying the oscillator frequency; C_4 is the oscillator trimmer, and C_5 the series capacitance or *padder*. Sometimes (especially if L_2 is not adjustable) it can be preset by a small variable capacitor in parallel with a fixed one.

For each further frequency band additional inductors are fitted, the unused ones being short-circuited by the band switch to prevent them from acting as resonators.

Although the frequency band used for v.h.f. sound broadcasting is much wider than the medium and low frequency bands in terms of hertz, when reckoned as the ratio of maximum to minimum—which is what counts in designing the tuning system—it is narrower.

Tuning in simple radio and television receivers is achieved by varying the capacitance or the inductance of the LC circuits. In normal use receivers are left tuned, for most of the time, to one of a small number of local stations, and a desirable feature is a simple means of selecting the desired signal which avoids the need for adjusting the tuning control. Even before the Second World War many sound receivers had push-buttons for selecting desired signals. In most sets pressing a button rotated the tuning capacitor to the appropriate tuning position. Since then much effort has been devoted to electronic methods of channel selection and most make use of variable-capacitance diodes or *varactors* as tuning elements.

As we saw in Section 9.11, the depletion layer between the two conducting regions in a junction diode is like the dielectric in a capacitor, but it has the useful property that its thickness—and hence the capacitance—depends on the applied voltage. If the signal frequency and the oscillator circuits of a receiver are tuned by varactor diodes, these can all be controlled by a single voltage via a network which ensures that the tuned circuits track accurately. Thus the receiver can be tuned by adjusting this voltage and, by switching the tuning voltage between the values appropriate to wanted channels, preset tuning can readily be achieved. Push-buttons can be used to switch the tuning voltage, but in modern television receivers the switches are electronic (Chapter 25), consisting of transistors in integrated circuit form (Section 24.7) which operate when a finger touches an electrode known as a touch sensor. Alternatively the electronic switches can be operated by infra-red or ultrasonic radiation from a remote-control unit, which can also adjust other functions, e.g., picture brightness.

21.7 Whistles

A superheterodyne is liable to certain types of interference from which an ordinary set is free. The most noticeable result of these is a whistle, which changes in pitch as the tuning control is rotated, as when using a 'straight' receiver in an oscillating condition, but generally less severe. There are many possible causes, some of which are quite hard to trace.

The best-known is *second-channel* or *image* interference. In the preceding section we have seen that it is usual for the oscillator to be adjusted to a frequency higher than that of the incoming signal. But if, while reception is being obtained in this way, another signal comes in on a frequency higher than that of the oscillator by an equal amount, an intermediate frequency is produced by it as well. If the frequency difference is not *exactly* the same, but differs by perhaps 1 kHz, then two i.f. signals are produced, differing by 1 kHz, and the detector combines them to give a continuous note of 1 kHz.

An example will make this clear. Suppose the signal desired works on a frequency of 200 kHz, and the i.f. is 470 kHz. When the receiver is tuned to 200 kHz the oscillator is 470 kHz higher, 670 kHz, and yields a difference signal of 470 kHz, to which the i.f. amplifier responds. So far all is well. But suppose now a signal of 1137 kHz is also able to reach the frequency-changer. It will combine with the oscillation to produce a difference signal of 1137 – 670, = 467 kHz. This is too near 470 kHz for the i.f. amplifier to reject, so both are amplified together by it and are presented to the detector, which produces a difference frequency, 3 kHz, heard as a high-pitched whistle. Slightly altering the tuning control alters the pitch of the note, as shown in the table opposite.

Since both signals give rise to carriers falling within the band to which the i.f. amplifier must be tuned to receive one of them, this part of the set can give no protection against interference of this sort. So the superhet principle does not allow one to do away entirely with selective circuits tuned to the original signal frequency.

Another form of interference, much more serious if present, but fortunately easy to guard against, is that due to a signal operating within the i.f. band itself. Clearly, if it is able to penetrate as far as the i.f. tuning circuits it is amplified by them and causes a whistle on *every* station received. Again, a good preselector looks after this; but the 525 kHz end of the medium frequency band may be dangerously close to a 470 kHz i.f. If so, a simple rejector circuit tuned to the i.f. and placed in series with the antenna will do the trick. It is known as an i.f. trap.

Set tuned to:	Oscillator at:	I.F. signal due to 200 kHz signal (wanted)	I.F. signal due to 1137 kHz signal (interfering)	Difference (pitch of whistle)
kHz	kHz	kHz	kHz	kHz
195	665	465	472	7
196	666	466	471	5
197	667	467	470	3
198	668	468	469	1
198½	668½	468½	468½	0
199	669	469	468	1
200	670	470	467	3
201	671	471	466	5
202	672	472	465	7
203	673	473	464	9

An example of how second-channel interference is produced

The foregoing interferences are due to unwanted signals. But it is possible for the wanted signal to interfere with itself! When its carrier arrives at the detector it is, of course, always at intermediate frequency. The detector, being a distorter, inevitably gives rise to harmonics of this frequency (Section 19.5). If, therefore, these harmonics are picked up at the antenna end of the receiver, and the frequency of one of them happens to be nearly the same as that of the signal being received, the two combine to produce a whistle. It is easy to locate such a defect by tuning the set to two or three times the i.f. The cure is to by-pass all supersonic frequencies that appear at the detector output, preventing them from straying into other parts of the wiring and thence to the preselector circuits or antenna.

Oscillator harmonics are bound to be present, too, and are a possible cause of whistles. Suppose again the receiver is tuned to 200 kHz and the i.f. is 470. Then the oscillator frequency is 670, and that of its second harmonic is 1340 kHz. If now a powerful local station were to be operating within a few kHz of 870, its carrier would combine with the harmonic to give a product beating at an audio frequency with the 470 kHz signal derived from the 200 kHz carrier. Such interference is likely to be perceptible only when the receiver combines poor preselector selectivity and excessive oscillator harmonics.

Excluding most of these varieties of interference is fairly easy so long as there are no overwhelmingly strong signals. But if one lives under the shadow of a transmitter it is liable to cause whistles in a number of ways. Besides those already mentioned, an unwanted carrier strong enough to force its way as far as the frequency-changer may usurp the function of the local oscillator and introduce signals of intermediate frequency by combining with other unwanted carriers. Any two signals on frequencies whose sum or difference is nearly equal to the i.f. may cause interference of this kind. Here again, preselector selectivity is the cure, but at sites exceptionally close to a source of interference a special trap circuit may have to be added to reduce the effects.

The foregoing list of possible causes of interference is by no means exhaustive, but it is only in exceptional situations that whistles are conspicuous in a receiver of good design.

Chapter 22
Radio-frequency and Intermediate-frequency Amplification

22.1 Amplification So Far

We have examined methods of audio-frequency amplification at some length in Chapter 19. Broadly, they consist of amplifying devices—such as transistors—connected together by couplings which are sometimes called *aperiodic* because they are designed to have no natural period or resonance but to treat all frequencies alike, at least those within the audio band. Since then we have assumed the practicability of amplifiers that respond only to the very limited band of frequencies included in one sound radio channel or the much wider (but, in proportion to the carrier frequency, still small) band in a television broadcast channel. In these the inter-stage couplings consist of tuned circuits. For an i.f. amplifier these couplings are preset, whereas for amplification at the original signal frequency they have to be varied by the user if he wants to receive on more than one channel. Although 'r.f.' includes i.f., where both are mentioned in the same context it can be taken that it means the original incoming signal frequency.

22.2 Active-device Interelectrode Capacitances

The electrodes of active devices, together with their connecting leads, act as tiny capacitors, with effects on the behaviour of the circuit as a whole. These capacitances are normally only a few picofarads and, to get some idea of what they mean in practice let us calculate the reactance of 4 pF at various frequencies (Section 6.3). The reactance

Frequency	Reactance
10 kHz	4 MΩ
100 kHz	400 kΩ
1 MHz	40 kΩ
10 MHz	4 kΩ
100 MHz	400 Ω
1 GHz	40 Ω

varies from 4 MΩ to 40 Ω and suggests that these small capacitances have negligible effect on the operation of active devices at audio frequencies. But at much higher frequencies the reactance is low enough to play a vital part. This is illustrated in the v.h.f. oscillator circuit (Figure 14.13a) in Section 14.6. Figure 22.1a shows in dashed lines the three internal capacitances c_{be}, c_{bc} and c_{ce} of a bipolar transistor as if they were external components. (Note that the fact that they are actually internal is indicated by a small c in conformity with the usual convention for active devices.)

From Figure 22.1b, which shows the internal capacitances in the equivalent current-generator circuit, we can see that in a common-emitter amplifier c_{be} is across the source of input signal and c_{ce} is across the output of the amplifier. So, if the transistor is used as a tuned amplifier,

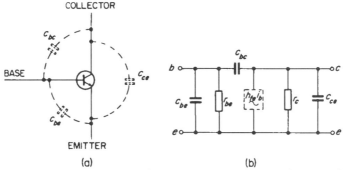

Figure 22.1 (a) Bipolar transistor showing capacitances between the terminals. (b) Corresponding equivalent current-generator circuit

these internal capacitances can form part of the tuning capacitances of the input and output tuned circuits, and are no disadvantage except that they limit the frequency to which one can tune with a given inductance.

c_{bc} is, however, different. This is because neither of its 'plates' is earthed and, more significantly, this capacitance provides a signal feedback path between the output and the input terminals. Although we introduced this explanation by reference to bipolar transistors, all active devices have an internal capacitance between input and output terminals. In a bipolar transistor used in the normal amplifying mode the base-collector junction is reverse-biased, and as we saw in Section 9.11 such a junction behaves as a capacitor. In a junction-gate f.e.t. the input junction is again reverse-biased and so provides a capacitive link with the channel and hence with the drain terminal. In an insulated-gate f.e.t. the gate terminal is a plate capacitively coupled to the channel and thus the drain. So what follows applies to all types of active device.

In Chapter 12, and especially Section 12.5, we saw how the value of a resistance as calculated from the voltage and current applied at one end can be quite different from its own real value if the other end is fed with a current or voltage directly proportional to that applied. We were then considering only resistances (or conductances) and only plus and minus, so that the only effects were increases or decreases in value. In this chapter, however, we have to take into account reactances as well as resistances, and consequently all possible phase relationships as well as just 0° and 180°.

22.3 Miller Effect

To take a simple case first, consider Figure 22.2a, in which C is a capacitor hidden in our now well-worn black box, in series with a special sort of meter that measures the charge passing into the capacitor when the switch is closed. As capacitance is equal to the charge required to raise the p.d. across it by 1 V (Section 3.2) we have here a means of measuring C. But suppose (b) that when we connect our 1 V to one end of C a battery giving A volts is applied at the other end with polarity as shown. The charging voltage is $A + 1$, so the charge is $A + 1$ times as much as that due to 1 V only. Knowing nothing of the hidden battery, we would suppose the box contained a capacitance equal to $(A + 1) C$, as at c. So far as external electrical tests can tell, the boxes b and c are identical.

This is exactly what happens in an amplifier having a voltage amplification A, if the output load is purely resistive at the frequency concerned—it could be the dynamic resistance of a tuned circuit. As A might be 50 or more, an internal input/output capacitance of only 2 pF could have the same effect as one of 100 pF. This would be serious enough, but at a slightly different frequency or output tuning the phase angle could shift nearly 90° one way or the other. The effect can be considered by analysing the feedback into two parts, one exactly negative (180°) and the other 90° different. The 180° component magnifies the internal capacitance as explained. In so far as the output load was capacitive, however, the current fed back to the

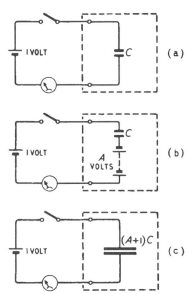

Figure 22.2 Showing how the introduction of an extra voltage, such as that fed back from output to input in an active device, can be equivalent to a multiplied capacitance

input terminal would be in such a phase as to appear to input voltages as conductance, which would have a loading or damping effect. An inductive output load, on the other hand, would feed back current in the opposite phase, giving the effect of negative conductance. If this was enough to offset the loss conductance of the input circuit, oscillation would result.

These feedback effects via the internal input/output capacitance of an active device are named after their discoverer, J. M. Miller. With tuned amplifier couplings, Miller effect causes at least tuning difficulties and can even cause oscillation. Now we can see that if resistance coupling were used at radio frequencies the inevitable stray capacitance across it would lead to a capacitive component of feedback current tending to increase the input conductance—which one tends to assume is practically zero with an f.e.t. So Miller effects are, in general, bad.

22.4 High-frequency Effects in Transistors

There are two other causes of apparent input conductance, which with normal design do not seriously impair the usefulness of active devices as amplifiers at frequencies below about 50 MHz, but need to be considered at very high frequencies. One is inductance of the emitter or source lead (Figure 22.3). This is common to both input and output circuits and produces the effect of input conductance by feedback.

The other is the time taken for charge carriers to travel between emitter and collector in a bipolar transistor—known as *transit time*. Besides the ordinary capacitive input current due to the input voltage across c_{be}, an additional current is induced by the movements of charge carriers across the base region. A small fraction of these is retained in the base region, but the remainder form a moving space charge which must be matched by a change of charge on the base, constituting a capacitive current. But at very high

Figure 22.3 Feedback due to emitter lead inductance and base-emitter capacitance of a transistor

frequencies the transit time is appreciable compared with the time of one cycle, so a phase lag develops in the response, which changes a capacitive current into a conductive one. This effect is, of course, lessened by reducing the thickness of the base region which, in transistors intended for r.f. application, is of the order of 0.001 mm. This would lead to unacceptably high capacitance were it not that the low power involved in receiver r.f. amplifiers enables the area to be correspondingly minute. By feats of technology, types of transistors are manufactured that give useful gain not only at v.h.f. but u.h.f. Types designed for high-power a.f. output stages necessarily have relatively very large junction areas causing high capacitances, and thicker bases causing greater delay effects, so some falling-off in performance may be detectable even at the top end of the a.f. band.

Before leaving the subject of transit time it is worth mentioning that certain types of microwave device make use of it to achieve amplification: these are briefly described later in this chapter.

The combined influence of internal capacitances and delayed response is very difficult to represent in an equivalent circuit that is both simple and accurate enough for practical purposes, even though 'accurate' is not a very demanding word where transistors are concerned. There is also the complication that depletion-layer capacitance depends on the applied voltage (Section 9.11). The h parameter diagram (Chapter 12) can be used, but the values of the parameters are no longer simple numbers; they are 'complex' numbers, having both magnitude and phase angle. The algebra is the same in form, but the in-phase (positive or negative) and at-right-angles (or quadrature or reactive) components have to be kept in separate accounts, as explained in mathematical books on a.c. And the complex h parameters vary with frequency.

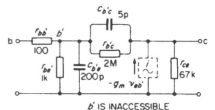

Figure 22.4 Hybrid π transistor equivalent circuit

For this introduction, at least, a clearer picture of the behaviour of a transistor at high frequencies is presented by what is called the hybrid π equivalent circuit, Figure 22.4. A great advantage is that its parameters are roughly constant with frequency. Typical values are shown for an r.f. transistor under working conditions. Another advantage is that these parameters are fairly simply related to the complex h values. Over a small band of frequency such as the 470 kHz i.f. band the hybrid π circuit of Figure 22.4 can be replaced by a slightly simpler equivalent in which $r_{bb'}$ is eliminated by allowing for it in the values of the other components. However, to see how the more comprehensive circuit works in practice let us draw phasor diagrams for it at two widely different frequencies.

22.5 Transistor Phasor Diagrams

Figure 22.5 shows the Figure 22.4 equivalent circuit fitted with a resistive load of 20 kΩ and labelled for phasors. Let us first draw diagrams for a frequency low enough for the capacitances to be neglected. As usual, the best starting place is the output, and to get realistic proportions let us calculate actual values, assuming an output ec of 1 V.

Because the input voltage will be relatively small, we choose a scale that brings only 1/400th of ec into the diagram, Figure 22.6. We calculate and plot AF, FE and DA, which are 50, 15 and 0.5 µA respectively. For this purpose $r_{b'e}$ is negligible compared with $r_{b'c}$. So the current generator has to supply 65.5 µA, and making use of the generator equation given we find eb' to be 65.5/40 = 1.64 mV. DC is zero in this case, so DB is 1.64/1 = 1.64 µA. That completes the current diagram and reveals AB as 0.5 + 1.64 = 2.14 µA, from which bb' follows as 0.214 mV, to complete the voltage diagram. The input voltage, eb, is 1.64 + 0.214 = 1.854 mV, so the voltage gain (ec/eb) is 1000/1.854 = 540, the current gain is 23.4, and the power gain 12 600, or 41 dB. Note that $r_{b'c}$ causes some reduction in gain by negative feedback ($BD < BA$). The input resistance, eb/AB, is 866 Ω.

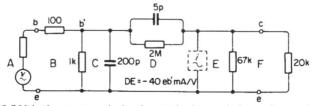

Figure 22.5 Hybrid π circuit marked with typical values and phasor-diagram labels

Figure 22.6 Low-frequency phasor diagram for Figure 22.5

Let us now see what happens at 470 kHz. At this frequency the reactance of $c_{b'e}$ is 1.7 kΩ. And $c_{b'c}$ is 68 kΩ, which in parallel with $r_{b'c}$ is very nearly equivalent to the same capacitance in series with 2.3 kΩ of resistance. The total impedance (Section 6.8) is hardly more than 68 kΩ. We begin as before with 1 V of ec, and assuming the same load we get the same AF and FE (Figure 22.7). But DA is

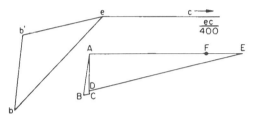

Figure 22.7 470 kHz phasor diagram for Figure 22.5

very different both in magnitude and phase angle. The feedback impedance is still large enough for $r_{b'e}$ to be neglected in comparison with it, but is much less than before, so DA is 1000/68 = 14.7 µA; and by drawing the impedance triangle (or by calculation) we find that it leads ec by 88°. Plotting it fixes D and gives us ED as 67.5 µA, the total generator current. Using the equation again we find eb' to be 1.7 mV, so DC is 1 µA leading it by 90°. CB is 1.7 µA, in phase, making BA 16 µA and bb' 1.6 mV. The completed diagram tells us that eb is 2.7 mV, so the voltage gain is down to 370 and the current gain more drastically down to 3.1. The power gain is not simply A_vA_i this time, because there is a phase difference of 37° between input voltage and current. The input power is therefore not $eb.AB$ but $eb.AB \cos 37°$ (Section 6.9), so the power gain is $A_vI_v/\cos 37°$, = 1450, or 31.6 dB. The ratio eb/AB indicates an input impedance of 170 Ω, which cannot be accounted for by the values of $r_{bb'}$, $r_{b'e}$ and $c_{b'c}$ alone, but clearly points to Miller effect. The apparent value of $c_{b'e}$ is in fact 3.2 nF; 16 times its real value.

Nevertheless this transistor is capable of useful i.f. gain, and in a suitable circuit could go considerably higher. The data we have gathered by drawing the phasor diagrams will help us soon in designing an i.f. amplifier.

22.6 Limiting Frequencies

The very important transistor parameter that we have so often been using in our studies at low frequencies is the short-circuited-output current gain, h_f, or h_{fe} when applied to the common-emitter mode in particular. It should be interesting to see what happens to it when account is taken of high-frequency effects. This is very easy, after what we have just been doing, because the short-circuited-output condition enables Figure 22.4 to be drastically simplified.

Because r_{ce} is short-circuited it can be removed for a start. And as there can be no output voltage across a short-circuit there is no feedback via $c_{b'c}$ and $r_{b'c}$. Instead, they are simply in parallel with $c_{b'e}$ and $r_{b'e}$, and their values are normally such that they too can be removed. And since h_{fe} is a current ratio, $r_{bb'}$ does not come into the problem. So the circuit boils down to Figure 22.8, in which $c_{b'e}$ is the only high-frequency influence.

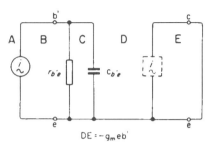

Figure 22.8 This greatly simplified equivalent circuit often gives a useful approximation to the high-frequency performance of a transistor

In Figure 22.5 the output current DE is given in terms of eb', equal to the impedance of $r_{b'e}$ and $c_{b'e}$ in parallel (which we can call $z_{b'e}$) multiplied by the input current AB. This part of the circuit is the same as Figure 6.11, and the calculation of $z_{b'e}$ is as shown in connection therewith (Section 6.10). As with the other combinations of a resistance and a reactance we have encountered, what happens at different frequencies can be seen most clearly if we consider two ranges of frequency and the point at which they meet. But first let us get h_{fe} in terms of $z_{b'e}$

$$h_{fe} = \frac{DE}{AB}$$

$$= \frac{g_m b'e}{AB}$$

$$= \frac{g_m AB z_{b'e}}{AB}$$

$$= g_m z_{b'e}$$

This means simply that, if g_m is constant, h_{fe} is directly proportional to $z_{b'e}$. Over the range of frequency low enough for the effect of $c_{b'e}$ to be negligible, h_{fe} is constant and is represented by a horizontal line. That is the low-frequency h_{fe} we have been using hitherto. At frequencies high enough for the effect of $r_{b'e}$ to be negligible (because practically all the current is going via $c_{b'e}$) $z_{b'e}$, and therefore h_{fe}, is inversely proportional to frequency so is represented by a line sloping at 6 dB per octave (Section 20.6). The two lines join at the turning frequency (f_t) at which $r_{b'e}$ and $x_{b'e}$ (= $1/2\pi c_{b'e}$) are equal. Because the numerical value of $z_{b'e}$ at this frequency is equal to $r_{b'e}/\sqrt{2}$ there is a 3 dB loss there. We should by now be quite familiar with this kind of curve, which is the same as in Figures 18.9, 19.20 and 20.3.

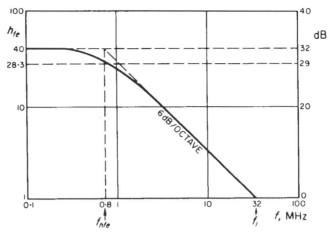

Figure 22.9 Gain/frequency graph of a transistor derived from Figure 22.8, by which its frequency parameters can be defined

Figure 22.9 shows it again, calculated for the values given in Figure 22.5. As g_m is there 40 mA/V and $r_{b'e}$ is 1 kΩ, the low-frequency h_{fe} is 40, or 32 dB. The frequency at which $c_{b'e}$ (200 pF) has a reactance of 1 kΩ is almost 0.8 MHz. That is what we have been calling the turning frequency, but in this particular context it is called the cut-off frequency and (in the common-emitter mode) it has the symbol f_{hfe} or f_b. As we see, it is the frequency at which h_{fe} has dropped 3 dB below its low-frequency value. The phasor diagram for this point is shown in Figure 22.10, for comparison with Figure 22.7. Because $h_{fe} = DE/AB$ it implies a phase angle as well as a numerical ratio.

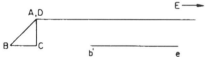

Figure 22.10 Simplified high-frequency phasor diagram for a transistor, derived from Figure 22.9

If the 1-in-1 (6 dB per octave) gradient is drawn from this point until it meets the 0 dB line it does so at 40 × 0.8 = 32 MHz. This frequency also has a special symbol, f_1, because at it the value of h_{fe} is 1 (or 0 dB) so there is no current gain. A more commonly used symbol for it is f_T, but this one is sometimes taken to mean the product of low-frequency h_{fe} and bandwidth. As the bandwidth is regarded as the frequency band up to the −3 dB point, this interpretation of f_T gives it the same value: 40 × 0.8 MHz in this example. But this is only so in this rather over-simplified theory of the relationship between h_{fe} and frequency, and in practice the different definitions of f_1 and f_T may result in slightly different values.

The same zero-dB frequency is also normally nearly the same as that at which h_{fb} is 3 dB down. It therefore is a measure of common-base bandwidth. And as the low-frequency h_{fb} is almost equal to 1, the gain × bandwidth definition of f_T leads to about the same f_T for common-base as for common-emitter. Although therefore a transistor in common-base gives no current gain and so less voltage and power gain than in common-emitter, what it has it maintains up to a much higher frequency: about h_{fe} times higher, in fact.

22.7 An I.F. Amplifier

For this first attempt at designing an i.f. amplifier we take single tuned circuit couplings, such as are used in most portable transistor receivers. We start from the bandwidth, f_B, which we found in Section 20.4 is equal to the resonant frequency (470 kHz in this case) divided by the Q of the tuned circuit. A practical bandwidth is 4 kHz each side of resonance, or 8 kHz altogether. So the required Q is 470/8; say 60. This, it should be understood, is the Q under working conditions, loaded by the output resistance of the preceding transistor and the input resistance of the following one. At the moment we do not have either of these data for the transistor represented by Figure 22.5. If we can assume that the load is 20 kΩ resistance we have the input *impedance* at 470 kHz, 170 Ω, but Figure 22.7 shows that the input current leads the input voltage by 37°. A little calculation on the lines of Chapter 6 shows that this is equivalent to 216 Ω in parallel with 1.23 nF. The output resistance is more difficult. Another phasor diagram is needed to derive it from Figure 22.5, and we would have to know the signal source impedance as well. If we assume it is made equal to the input resistance, the output resistance comes out at about 25 kΩ, in parallel with capacitance.

Although the tuning product LC is fixed for us by the resonance frequency (Section 8.7) we are free to choose the ratio L/C. If C is small, so that most of it is made up of transistor and wiring capacitances, it cannot be relied upon to be constant. About 250 pF total is a typical choice. The corresponding L is 460 μH, and its reactance at 470 kHz is 1.36 kΩ. The Q of the tuned circuit itself is ideally infinite, in practice it might be 140. The corresponding dynamic resistance (Section 8.10) is 190 kΩ. When loaded ($Q = 60$) it will be 82 kΩ, so the combined input and output loading is 144 kΩ (for 144 and 190 in parallel are 82). If we suppose that input and output loadings are equal, then each must be 288 kΩ. For 25 kΩ to load the tuned circuit to that extent it must be connected across only part of the tuning coil, the fraction of turns being $\sqrt{(25/288)}$, $\simeq 0.3$ (Section 7.11). Similarly the following transistor input should be connected across $\sqrt{(0.216/288)}$, $\simeq 0.027$. The usual way in which this is done is shown in Figure 22.11. The symbol for the inductor shows that the value of L is preset by what is called *permeability tuning*, which usually takes the form of a ferrite core that can be screwed in and out of the coil to vary the proportion of it in the magnetic field.

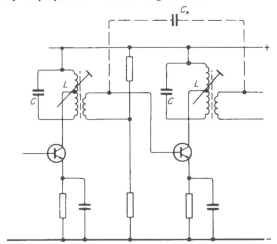

Figure 22.11 Circuit diagram of typical i.f. amplifier for 470 kHz

Although this design would work after a fashion, the transistors (if correctly represented by Figure 22.5) really are not quite good enough for it at this frequency. There is effectively a nearly 11:1 voltage step-down from one transistor to the next, necessitated by the low input impedance due to Miller effect. Because even a smaller amount of Miller effect has objectionable results (when setting up the tuning, for example) it is sometimes worth while to defeat this effect by applying external feedback equal and opposite to that caused by the internal transistor capacitance—a process known as *neutralising*. This can be done by connecting a neutralising capacitor C_n between the transistor base and a point where the output signal voltage is inverted with respect to the collector voltage. A suitable point is the secondary winding of the collector transformer (Figure 22.11).

Because the internal feedback is not purely capacitive, to neutralise it completely a certain amount of resistance needs to be put in series with C_n. If exactly the right values are chosen, the transistor should work entirely as a one-way device. This superior form of neutralisation has been given the even grander name of *unilateralisation*.

An alternative to neutralising is to limit the gain from the transistor to a value at which the signal fed back to the input via the internal capacitance is so small compared with the signal input that it has negligible effect on it. The gain can be reduced by 'tapping down' the base and collector connections to the tuned circuit more than suggested above. Even with gain thus limited it is still possible to obtain all the amplification needed in a vision i.f. amplifier from three bipolar transistors.

For a better combination of selectivity and sideband retention, and especially for getting the flat top needed for vision i.f. amplifiers, double tuned circuits with suitably adjusted couplings (Section 20.6) are used, as in Figure 22.12. In receivers working at one i.f. for a.m. and another for f.m., the i.f transformers are usually connected with

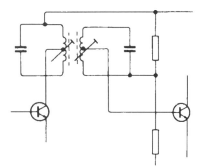

Figure 22.12 Alternative i.f. inter-stage coupling, double-tuned

primaries in series and secondaries in series, no switching being needed. So far from tapping down being needed at the high i.f. and wide bandwidth necessary for vision, not only is the whole winding usually in series with the collector but sometimes there is even a damping resistance in parallel.

In both a.m. and f.m. i.f. stages the LC tuning circuits are sometimes supplemented or even substituted by ceramic filters. These are similar in principle to the quartz crystals described in Section 15.8 and consist of a small ceramic disk with electrodes deposited on each side. As the frequency and bandwidth are determined by the dimensions, these are fixed, normally at about 470 kHz or 10.7 MHz. Ceramic filters combine good i.f. tuning with small size.

Another type of filter now used in i.f. amplifiers is the surface-acoustic-wave (s.a.w.) device. By means of a transducer* the signal is transmitted along the surface of a piezo-electric material in the form of minute deformations of the surface. By shaping the surface suitably the effect of a bandpass (or other type of) filter can be achieved and the signal can be recovered by use of another transducer. Extremely small robust filters with stable characteristics can be produced by this means.

22.8 R.F. Amplification

Amplification at the original signal frequency, if that is lower than about 30 MHz, is now seldom used except in high-performance 'communications' receivers; but when it is used the techniques are similar to those for i.f., with the

---

* A transducer is, in general, a device able to convert an input signal in one form of energy into a corresponding signal in another form. Thus a microphone is an acoustic/electrical transducer. The transducers in the s.a.w. filter are electrical/mechanical.

addition of variable tuning. This is done by one or more extra sections on the ganged tuning capacitor. Because the i.f. amplifier is relied upon for most of the selectivity, the preselector tuning circuits do not have to be very sharp, so slight errors in ganging are not serious. So long as the tuning capacitor sections are reasonably alike, all that is needed is to make the inductances and minimum capacitances equal. The only difficulty arises in the first tuned circuit, to which the antenna is coupled; this problem is considered in Section 22.12.

Preselector (r.f.) amplifiers are most used at frequencies higher than about 30 MHz, because noise generated in the frequency-changer tends to be greater the higher the frequency, and signals tend to be weaker. So the all-important signal/noise ratio is usually not good enough without some amplification at the original frequency. This is rather awkward, because the difficulties in achieving such amplification increase with frequency. And of course only low-noise devices can be considered.

Feedback via the internal base/collector capacitance of transistors makes the common-emitter tuned amplifier useless at v.h.f. or u.h.f. But the common-base circuit is suitable, and to understand this we must know a little about the principles of screening.

Feedback occurring via a capacitance can obviously be avoided by getting rid of the capacitance. The capacitance between any two objects can be eliminated by placing between them as a screen an earthed metal sheet of sufficient size. The operation of such a screen can be understood by considering Figure 22.13a, in which a stray capacitance is represented by two plates A and B

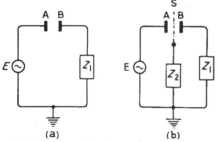

Figure 22.13 Illustrating the theory of screening

separated from one another by an air-space. The alternating e.m.f. E will therefore drive a current round the circuit Earth—E—A—B—Z_1—Earth. Across Z_1, which is an impedance of some kind between B and earth, the current will develop a potential difference, and this p.d. will be the voltage appearing on B as a result of current passing through the capacitance AB.

At b a third plate S, larger than either of the two original plates, is inserted between them in such a way that no electric field passes directly from A to B. We now have no direct capacitance between A and B, but we have instead two capacitances, AS and SB, in series. If an impedance Z_2 is connected between S and earth the current round the circuit Earth—E—A—S—Z_2—Earth will develop a p.d. across Z_2. Since Z_2 is also included in the right-hand circuit the voltage across it will drive a current round the circuit Earth—Z_2—S—B—Z_1—Earth, and this will give rise to a potential on B. So far, S has not screened A from B; there remains an effective capacitance between them which, if Z_2 is infinitely large, amounts to the capacitance equivalent to that of AS and SB in series.

Now imagine Z_2 to be short-circuited. Current will flow round the first circuit, but since there is now no impedance common to both there will be no driving voltage to produce a current in the latter. No matter what alternating voltages are applied to A, none will appear on B, even though large currents may flow via S to earth. The effective capacitance between A and B has therefore been reduced to zero, and B is completely screened from A.

It is very important to note that S is only effective as a screen if it is connected to earth (or to a constant-potential point in the circuit) by a negligible impedance.

Now in a common-base amplifier it is feedback via the internal emitter/collector capacitance which could cause difficulties. But the base region, sandwiched as it is between the emitter and collector, and earthed, acts as a screen permitting stable r.f. amplification. Use of the common-base mode results in a relatively low input impedance and absence of current gain. But because the dynamic resistances of tuning circuits at these frequencies are also low they are less affected by the low input impedance than in higher r.f. practice, and even a moderate gain is worth having, in terms of signal/noise ratio.

Figure 22.14 shows a preselector stage suitable for the 87 to 95 MHz f.m. broadcasting band. Because this band is all within 4½% of its centre frequency a fixed-tuned input circuit, damped as it is by the low input impedance of a common-base transistor, covers it adequately. The output circuit is inductance tuned, and tapped down at the output so as to have a high enough Q to give a useful contribution to selectivity.

For low noise, f.e.t.s are better than bipolar transistors. If used in common source mode they need neutralising, but can be used unneutralised in common-gate; for example, an f.e.t. can take the place of the bipolar transistor in Figure 22.14.

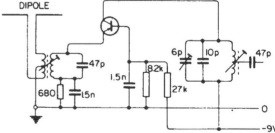

Figure 22.14 Common-base preselector stage suitable for v.h.f.

An f.e.t. common-gate stage is often combined with an f.e.t. common-source stage to form what is known as a cascode amplifier, which gives low noise and high overall gain (Figure 22.15).

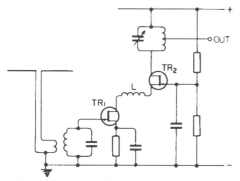

Figure 22.15 Cascode v.h.f. preselector stage using f.e.t.s

Here the output f.e.t. (TR$_2$) is working as a voltage amplifier in common-gate mode and therefore presenting the common-source TR$_1$ with such a low load impedance that it gives little or no voltage gain. Miller effect in TR$_1$ is thus too small to be troublesome. But its own input impedance is high, so the input tuned circuit gives good voltage magnification, and TR$_1$ good current gain. Some inductance (L) is sometimes included to separate the stray capacitances of the two f.e.t.s and prevent them from reducing the gain at very high frequencies; the way it does this will be explained in connection with Figure 24.10.

Bipolar transistors can be used in cascode, but their noise contribution is likely to be rather higher.

The need for amplification at v.h.f. and u.h.f. usually goes along with wide bandwidth; for example, television and radar. Because of the difficulties of unilateralisation at such frequencies, a common policy is to use transistors with the lowest possible feedback capacitances in conjunction with stagger tuning (Section 20.7) or overcoupling. Both of these types of tuning help to give the desired wide flat top to the frequency response, and at the same time reduce feedback, either because the gain is less or because the input and output circuits are not tuned to exactly the same frequency.

Another feature of u.h.f. amplifiers is that the inductances required for tuning are too small to be provided in the form of coils, so tuned circuits in the form of quarter-wave 'lines' as described in Section 16.10 are used.

Figure 22.16 shows a typical arrangement, in which L_1 is the input tuned circuit consisting of a quarter-wave rod which is the inner conductor and the surrounding metal screen (shown dashed) is the outer conductor. It is inductively coupled to the u.h.f. antenna feeder on one side and to the emitter of the common-base transistor on the other. The collector is connected to a tapping on the output tuned circuit L_2. The d.c. is fed through a choke coil that prevents the signal voltage from being short-circuited. Note that all the feeds pass through the screen via 1 nF 'feed-through' capacitors which effectively earth all signals to the screen. L_2 is loosely coupled to L_3 through a hole in the screen, so that the pair are a band-pass filter represented in conventional symbols by Figure 20.6c. Its output is inductively coupled to the common-base frequency changer. L_4 is the oscillator tuning circuit, back-coupled to the emitter via C_2. The u.h.f. and i.f. signals in the collector circuit are separated by a choke coil, and the i.f. signals are developed across L_5. All four line circuits are tuned by a four-section ganged capacitor C_{1a-d}.

In modern tuners the capacitors C_{1a} to C_{1d} are likely to be replaced by varactors controlled from a single voltage as described in Section 21.6.

22.9 Screening

The basic principle of screening was explained in the previous section with particular reference to active device

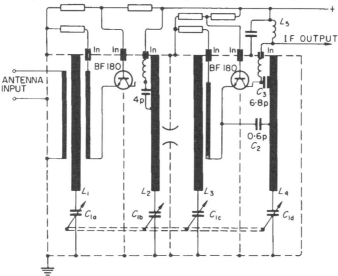

Figure 22.16 Complete circuit diagram of u.h.f. 'front end' (preselector and frequency-changer stages) using 'line' tuning

electrodes; the same principle applies to the external screening of circuits. The object is to prevent capacitive coupling between two circuit elements by interposing a conducting sheet which is either directly earthed or has such a large capacitance to earth that it is effectively at zero signal potential. Metal sheet or foil is the usual screening material, but its effectiveness is not greatly reduced by perforations, and for some large-scale purposes wire netting is good enough. Because even very small stray capacitances have sufficiently low impedance at very high frequencies to form undesired circuit paths, the need for screening generally increases with frequency. The complete screening of the u.h.f. circuits in Figure 22.16 is an example of this.

One cannot always assume that because two points are connected to such a screen they must both be at the same signal potential. If a substantial signal current flows between points some distance apart there may well be enough impedance between them to give rise to an appreciable signal p.d. and therefore a risk of undesired coupling. So care has to be taken in laying out u.h.f. circuits.

Signal leakage via d.c. or relatively low-frequency leads is another possibility to watch. Figure 22.16 includes two examples of u.h.f. chokes for this purpose. Sometimes the lead itself can be given sufficient inductance by threading it through one or more ferrite beads designed for this application.

So far we have considered screening as a means of confining electric fields. The fact that they are often effective in confining magnetic fields too should not prevent us from distinguishing clearly between the two kinds of screening. The essential feature of electric screening is negligible impedance to a point of zero potential. A thin sheet of copper or other low-resistance metal connected to earth or its equivalent is usually effective. The effect of stray magnetic fields is to induce e.m.f.s in any conductors encountered. Magnetic screens work on two different principles. One kind surrounds the source of the field with such a low magnetic impedance— *reluctance* is the term—that practically all the magnetic flux passes through it in preference to the high-reluctance space outside, as in Figure 22.17. To ensure the low reluctance the screening material must have a high permeability. Special alloys such as mumetal are used for this purpose.

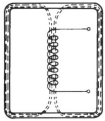

Figure 22.17 Showing the principle of one method of magnetic screening

The other type of screening works by forming short-circuited paths magnetically coupled to the source of the field, which induces currents in them, and these currents set up magnetic fields in opposition to the inducing fields. It is in fact an example of transformer action. The important property of the screening material is not permeability (which would impede its action) but, as with electric screening, conductivity. There is no need for a magnetic screen to be earthed, but there must be continuous low resistance around the paths of induced currents (which is not essential for electric screening). If the 'can' in Figure 22.17 was made of copper instead of magnetic alloy it would have currents induced in it in planes parallel to the turns of wire in the coil, and the net external field would be greatly reduced. If earthed, it would also be an effective electric screen. If it had a saw cut made across it, dividing it into two paths and they were both earthed, it would still be an electric screen, but the magnetic screening would be almost completely destroyed if the saw cut interrupted the induced currents. Thus a vertical slot parallel to the coil axis (Figure 22.17) would prevent magnetic screening whereas a horizontal slot would hardly affect it.

So it is possible to screen both kinds of field, or either of them alone.

In general the permeability screen is the more effective magnetic type at low frequencies, including z.f., and the conductive screen at high frequencies. The permeability type increases the inductance of the enclosed inductor, especially if it is shaped so as not to be a good conductive screen. Provided that this is not objectionable, the screen need not be widely spaced from the inductor. A closed magnetic core acts as a magnetic screen by short-circuiting the magnetic flux. A conductive screen reduces the inductance and tends also to increase its resistance, the closer it is to the inductor. So the inductor itself should be designed to have as little external field to screen as possible.

22.10 Cross-modulation

The emphasis on distortion in the chapter on a.f. amplification has been notably lacking in the present chapter. The reason is that so long as the shape of the modulation waveform is preserved, the shapes of the individual r.f. cycles that carry the modulation are unimportant. They could all be changed from sine waves to square waves or any other shape, and indeed this is actually done in f.m. receivers by the limiter, which chops all the cycles down to a constant size, regardless of distortion. That would not be allowable in an a.m. receiver, of course (the purpose of the limiter being in fact to remove any a.m.) but so long as the relative sizes of the cycles were preserved they would still define the original modulation envelope. A perfect detector cuts off half of every cycle, yet this drastic r.f. distortion introduces no distortion of the modulation. It is the imperfect or non-linear detector (Section 18.4) that distorts the modulation by altering the relative proportions of the r.f. cycles. But unless the modulation is fairly deep it is surprising how little it is distorted by even a considerable curvature in the r.f. amplifier's output/input characteristic. It must be remembered that except for the output of the stage feeding the detector the amplitude of the r.f. signals is so small that it is unlikely to experience appreciable non-linearity.

There is an important exception to this. If signals very much stronger than those to which the receiver is tuned are not very widely different in frequency they may get some way into the receiver before being tuned out by its selectivity. In a superhet most of the selectivity is concentrated in the i.f. amplifier, for the reasons given in Section 20.8. So such signals may well be present in the first stage at considerable amplitude. What happens there is very similar to the frequency-changer action illustrated in Figure 21.4*b*, the interfering strong signal now playing the part of the local oscillation. If a transistor handling these signals simultaneously is appreciably non-linear within the amplitude of the strong signal, other signals at sum and difference frequencies will be created and may cause whistles as already mentioned at the end of Section 21.7. But unlike the local oscillations generated in the frequency-changer, the unwanted signals will usually be *modulated*.

Figure 22.18 shows a characteristic curve, with O as the working point. The amount of non-linearity between say A and B is so slight that even if as large a section of the curve as that was being covered by the wanted signal in the first

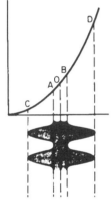

Figure 22.18 Showing how the average slope of an amplifier curve varies during each modulation cycle of a strong signal

stage the distortion of its modulation would be negligible. But now consider what would happen if there was also a strong modulated signal as depicted below, sweeping between C and D at the peaks of its modulation and between A and B at the troughs. The small wanted signal would be swept with it. Owing to the markedly different slopes at C and D, the output amplitude of the wanted signal would be forced to vary according to the modulation of the unwanted signal. In other words the wanted signal is to some extent modulated by the unwanted modulation. So even if the unwanted signal is removed entirely by subsequent selectivity it has left its mark indelibly on the wanted carrier which, with its two-programme modulation, passes through the receiver and both are heard together.

If the wanted transmission were switched off the programme of the interfering station would cease with it, proving that the interference was due to *cross-modulation* and not to insufficient overall selectivity.

Clearly the remedies for cross-modulation are two-fold: avoid excessive non-linearity in the early stages, and provide enough selectivity to protect them against overwhelmingly strong interference. Such interference is unlikely on the v.h.f. and u.h.f. bands, so the problem seldom intrudes there, which accounts for the fact that sometimes there is no variable tuning at all before the first stage (Figures 22.14 and 22.15, for example), but broadcast receivers for the medium frequency band (especially if used with large antennas) and communications and other specialist receivers must be designed with cross-modulation in mind.

It is fortunate that when a carrier wave is tuned in it affects the detector in such a way that if some interfering modulated carrier wave is present its modulation is rendered more or less ineffective, i.e., the carrier is truly demodulated. In this way the selectivity of the receiver is apparently increased. With sharp tuning, the percentage modulation of the wanted a.m. signal is also reduced, reducing detector distortion. So it may pay to tune sharply in the interests of selectivity and detector linearity, and make up for the resulting weakening of outer sidebands by upper frequency compensation in the a.f. department.

22.11 Automatic Gain Control

Receivers for sound are usually designed for a certain signal amplitude at the detector, such that at a moderate depth of modulation (say 30%) there is enough a.f. gain to give maximum output, plus a bit in reserve. The volume control, to enable the output to be adjusted between more-than-maximum and zero, is usually placed between the detector and the first a.f. stage. Ideally, all carrier waves to which the receiver is tuned would arrive at the detector at the same amplitude. This implies that the r.f. gain is always adjusted inversely to the strength of the signal as it comes from the antenna. This adjustment is not one that the user of the receiver wants to be bothered to make, so most receivers are designed to do it automatically, by means of what is called *automatic gain control* (a.g.c.). To get close to the ideal some fairly complicated circuitry is needed. Here we shall look at only the simplest methods.

The general principle is to use the rectified carrier wave voltage part of the detector output (the d.v.) to alter the bias of one or more of the preceding stages in such a way as to reduce their gain. The choice of stages calls for careful consideration. Control of preselector stages, if any, is most desirable, but the process usually increases non-linearity so the risk of cross-modulation must be watched. The type of frequency changer that combines oscillator and 'mixer' is not available for a.g.c., because oscillation might be stopped thereby, or at least mistuned. The last i.f. stage is handling signals at their maximum strength, so must not be allowed to become non-linear enough to distort the modulation. In most transistor receivers it is customary to control only the first i.f. stage.

Usually the rectified carrier-wave voltage is applied to the amplifier in the polarity that reduces output current, to which the mutual conductance and consequently gain are roughly proportional. To prevent the control from being

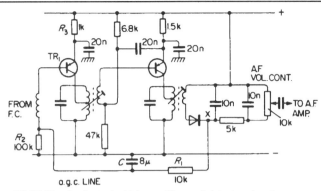

Figure 22.19 Circuit of a typical i.f. amplifier and detector showing a.g.c. system

upset by modulation of the carrier, the detector output is first put through a filter with a time constant long enough to remove even the lowest modulation frequencies but not so long as to cause a noticeable time lag on tuning from one carrier wave to another.

Figure 22.19 shows the parts involved, with typical values. Two i.f. stages are shown, the second followed by a diode detector, with filter for removing the residual i.f. The rectification of the carrier wave makes the point X more positive than the positive side of the power supply. The modulation is removed by the filter made up of R_1 and C, and the positive potential reaches the base of TR_1 via the secondary coil of the tuned i.f. transformer at the frequency-changer output. Note that as these are p-n-p transistors they normally work with a small negative bias; for TR_1 it comes from the negative side of the power supply via R_2. R_1, R_2 and R_3 have been chosen so that with no output from the detector the bias for TR_1 is about the best for good i.f. gain. The a.g.c. voltage, by making the bias less negative, reduces the collector current and gain. Because only a fraction of a volt is needed to reduce the collector current from its initial value to nearly zero, control is very 'steep', not requiring a very large signal voltage at the detector, and one-stage control is sufficiently effective. This type of circuit in which received signals cause the a.g.c. transistor to be biased back is known as *reverse control*.

A.g.c. systems are often designed so that only when an adequate working voltage at the detector is exceeded does it reduce the gain. This feature is called—not very aptly, for time is not involved—*delayed* a.g.c.

Note that the a.g.c. potential divider system in Figure 22.19 applies a small forward bias to the detector diode, sufficient to bring the working point to the sharpest point on its characteristic curve (see Figure 9.14) and thereby increase its response to weak signals.

Oddly enough, a.g.c. is possible with transistors by applying control bias of the opposite polarity to that just indicated. For this *forward* a.g.c. (as it is called) to work, there must be a suitable amount of resistance in the collector circuit. So when a.g.c. bias is applied and increases the collector current, the voltage reaching the collector falls and as the transistor begins to run into saturation its gain falls fairly steeply. At the same time its output resistance falls and damps the tuned circuit into which it works, increasing the bandwidth and sideband acceptance. The associated loss of selectivity can be accepted because it takes place only when a comparatively strong carrier wave is tuned in and the gain is low, so interference is unlikely. Special r.f. transistors are required to obtain satisfactory results from forward control. The characteristics of reverse-control r.f. transistors are not suitable.

A.g.c. is hardly necessary for f.m., because the limiter irons out the differences in strength between incoming signals. In the exceptional cases where signals of such widely different strengths are encountered that the limiter would have difficulty in coping with the full range, the nearly constant d.c. output of the f.m. detector would obviously be ineffective for a.g.c. purposes but the long-time-constant circuit of a ratio detector (Section 18.16) provides a convenient source of a.g.c. voltage.

In the vision section of a television receiver the provision of a.g.c. is complicated by the fact that the vision signal has no constant mean carrier amplitude; it varies with the average light and shade in the picture. So if a.g.c. is applied in the same way as for sound, it tends to brighten scenes that are supposed to be dark and darken scenes that are supposed to be bright. In designs where trouble is taken to exclude this defect, certain parts of the composite vision signal that are not reproduced on the screen, and which represent the strength of the incoming signal, are selected by what is called a gating circuit for use as a.g.c. bias.

22.12 The Antenna Coupling

The receiving antenna can be regarded as a signal generator and the receiver as the load, and the signal voltage is fixed by the incoming field strength and the nature of the antenna, so the maximum power law (Section 11.8) does apply. But the transfer of power has to be compatible with selectivity and easy tuning.

Three main classes of antenna have been considered in Chapter 17: (1) the dipole, in single or multiple form, which works at or near its resonance frequency, decided by its dimensions; (2) the loop or coil, which usually takes the place of the first tuning inductor in the receiver and is tuned by a variable capacitor; (3) the wire-and-earth

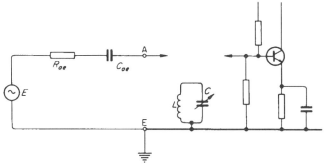

Figure 22.20 The three sections involved in coupling the antenna to a receiver: the antenna, the tuning circuit and the amplifier input

antenna, which can be almost any size and is usually worked below its resonance frequency, so has capacitive reactance. Antennas in class 1, which normally are connected to the receiver via a cable, can be regarded as resistances equal to the characteristic resistance of the cable, such as 70 Ω. Class 2 antennas are of course essentially inductances, but when tuned can be regarded as high (dynamic) resistances. Antennas of class 3 may be anything from a few feet of wire indoors to a large outdoor structure, with an impedance of about 20–1000 pF in series with perhaps 50–5000 Ω resistance. And in conjunction with multi-band receivers they may be required to work at almost any radio frequency. So the most suitable coupling arrangements are rather difficult to determine! We shall now consider this most difficult of the three cases.

The nature of the problem is illustrated by Figure 22.20, in which R_{ae} and C_{ae} represent the series resistance and capacitance of the antenna (which vary somewhat with frequency) and E the e.m.f. it picks up from space. On the right there is the first active device (a common-emitter transistor in this example), and L and C are available for tuning purposes. The problem is how to interconnect these three systems.

Ideally, in order to put into effect the maximum-power law any reactance in the antenna's impedance should be tuned out by reactance of the opposite kind, leaving only resistance. This resistance should then be connected to the transistor in such a way that it matches (i.e., is equal to) its input resistance. As it is very unlikely that the two resistances would happen to be equal, some impedance-transforming (or matching) device such as a transformer is indicated.

Basically this is the problem we have already met in connecting one amplifier stage to another. But in this case it is complicated by a number of things: whoever designs a receiver to work from an open antenna usually has no control over or knowledge of the electrical characteristics of those antennas that may be used; the arrangement often has to work well enough over a very wide range of frequency; and there are the selectivity and tuning considerations already mentioned but not yet considered in detail.

First let us assume ideal conditions: a purely resistive antenna and transistor input, and a perfect transformer to couple them (Figure 22.21a). If the transformer ratio is denoted by $1:n$, from the load's point of view Figure 22.21b is the same as a. The maximum power is delivered to R_L (and therefore the greatest voltage, V, set up across it) when $n^2 R_{ae} = R_L$. So $n = \sqrt{(R_L/R_{ae})}$, and

$$V_{max} = \frac{E}{2}\sqrt{\left(\frac{R_L}{R_{ae}}\right)}$$

If you assume some values for R_L and R_{ae} and calculate n required to give V_{max}, and then calculate V for values of n each side of the optimum, you will find that V falls very little below V_{max} even when n is as much as 30% above or below its best value. In other words, n is not critical. Which is just as well, for practical cases are far from ideal.

Where does the tuned circuit come into it? If it was omitted, so that there was no selectivity in front of the first stage of amplification, serious cross-modulation would be likely, especially if the antenna was a large one. The tuned circuit should interfere as little as possible with the ideal conditions shown in Figure 22.21 so far as the wanted

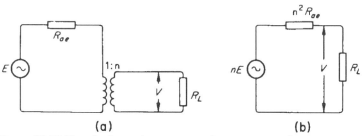

Figure 22.21 Use of r.f. transformer to match antenna impedance to receiver input impedance

signals are concerned, and short-circuit all others as much as possible. It would do this if it was connected in parallel with either primary or secondary side of the transformer in Figure 22.21. But if R_{ae} and R_L were not unusually high they would probably reduce the Q of the tuned circuit so much that its selectivity would be poor, unless the L/C ratio was so low that C would be right out of line with the capacitances used to tune the other circuits in the receiver. If the receiver is to be used by unskilled persons we can assume that only one tuning control is allowable. So C has to be provided by a section on the ganged tuning capacitor (Section 21.6).

The solution is the same as we used for our i.f. amplifier: tap the input or output circuits or both across part of the tuning circuit, or use separate windings. In other words, combine tuning circuit and matching device. Figure 22.22 shows what we could now have, where both varieties of

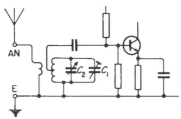

Figure 22.22 Coupling antenna to receiver by means of a combined tuning circuit and impedance-matching transformer

method are included. Except that the tuning circuit is fed from an antenna instead of a transistor, the circuit is identical in principle to an i.f. amplifier (Figure 22.19). But in practice it is very different, because of the wide range of frequencies (instead of one fixed one), the wide range needed for C_1, dictated by the requirements of the ganged tuning capacitor, and the great range of possible antenna impedances. So far we have conveniently been assuming that they are all at least resistive. While that is roughly true of antennas in class 1, open-wire antennas usually have substantial capacitance, and because the amount is unknown (it could be hundreds of pF) and varies with frequency and with the size of the antenna connected, and even a few pF can mistune the circuit seriously, a major problem exists.

Both over-coupling and under-coupling between antenna and tuning circuit reduce signal transfer, but over-coupling increases the disturbance to tuning and reduces the selectivity (both for that reason and by reducing Q), whereas under-coupling improves both tuning and selectivity. So the usual policy is to under-couple considerably, say by tapping well below the optimum even for the largest antenna. Signal strength is then likely to be poor with a small antenna. One solution, not much used now, is to provide alternative tappings. In another (Figure 22.23a) the primary (L_1) of a transformer is designed so that with the antenna capacitance it resonates at a frequency below the lowest in the range of the receiver. The larger the antenna, the more the resonant frequency of this coil departs from that of the secondary and so the less the capacitance that is transferred into it. The greater signal loss is compensated by the stronger signal the larger antenna brings in. Signal transfer tends to be poor at the highest frequencies, at which L_2C_2 is most out of tune with the antenna, so a small capacitor C_3 is sometimes included to help things along. Another method (Figure 22.23b) has the merit of simplicity and of ensuring that whatever antenna is used the capacitance it adds to the tuned circuit must be less than C_4. And the range of signal strength brought in by antennas of different sizes is to some extent levelled out, for the larger it is the more its coupling is reduced by the capacitance C_4 (which is usually only a few pF) being so much less than its own. At the same time the effects of diverse antenna resistances are reduced.

Providing for v.h.f. dipoles is comparatively easy because their resistance is specified, their reactance is fairly small or even zero, and the frequency ratio is usually quite limited.

Portable receivers with internal antennas that are in effect large tuning coils are easier still, for the designer has everything under his control.

22.13 Klystrons and Travelling-wave Tubes

We saw in Section 22.4 that the time taken by charge carriers to cross the base region of a transistor limits the upper frequency at which it can be successfully used as an amplifier. But, as mentioned in that section, there is a class of r.f. amplifying devices which relies on transit time for its

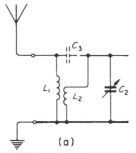

(a)

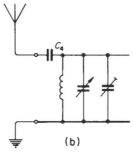

(b)

Figure 22.23 Two other methods of antenna coupling: (a) a relatively high-inductance primary coil (L_1) augmented if necessary by top capacitance (C_3); (b) small series capacitance

operation. We have already encountered one of these devices; it is the klystron described in Section 14.14. These devices need a significant transit time because, during it, the electron beam is gathered into bunches which generate the output of the tube.

Another microwave device depending on transit time is the travelling-wave-tube (t.w.t.). This uses the same principle, i.e., the electron beam is bunched by the electric field set up by the input signal but, whereas in the klystron the electric field is confined to a small area between the buncher grids, in the t.w.t. the field occupies the entire length of the tube. In one type the input signal passes along a helix (Figure 22.24) the pitch of which is so chosen that the effective velocity of the input radio wave along the tube axis is equal to the velocity of the electron beam. Wave and beam thus move together along the axis and the direct interaction between them causes the beam to gather into bunches which, interacting with the helix again, generate the output of the tube.

This is a very simplified account of a basic t.w.t. There are highly complex types using means other than a helix to slow down the radio wave and these require a book of this

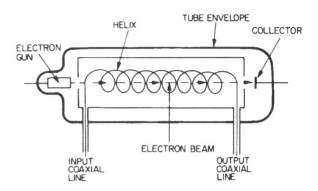

Figure 22.24 Simplified construction of a travelling-wave tube

size to do them justice. All we need to know is that t.w.t.s can be used as r.f. amplifiers between 200 and 60 000 MHz, that the output can be between a milliwatt and a kilowatt, gain between 30 and 50 dB and that they have a wide bandwidth. They are extensively used in microwave links.

Chapter 23 Electronic Imaging: Video

23.1 Description of Cathode-Ray Tube

The preceding chapters have covered the general principles underlying all radio systems (and incidentally quite a number of other things such as deaf aids and r.f. cookers), together with sufficient examples of their application to provide at least the outlines of a complete picture of the reception of sound programmes. There remains one device which is so important for such purposes as television, radar and computing that this book would be incomplete without a section on it: the cathode-ray tube which, when used to display television images, is known as a picture tube.

In many respects it resembles a valve; the differences arise from the different use made of the stream of electrons attracted by the anode. Both devices consist of a vacuum tube containing a cathode for emitting electrons, an anode for attracting them, and (leaving diodes out of account) a grid for controlling their flow. But whereas valves are designed for using the electron stream (or anode current) externally, cathode-ray tubes use it inside by directing it against one end of the tube, where its effects can be seen. This controlled stream of electrons was originally known as a cathode ray; hence the name of the tube.

Figure 23.1 shows a section of a c.r. tube. The collection of electrodes at the narrow end is termed the *gun*, because

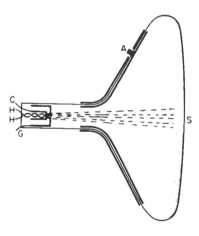

Figure 23.1 Section of cathode-ray tube for television. H = heater; C = cathode; G = grid; A = anode; S = fluorescent screen

its purpose is to produce the stream of high-speed electrons. There is the cathode with its heater, and close to it the grid. Although the names of these electrodes follow valve practice, their shapes are modified; the emitting part of the cathode is confined to a spot at its tip, and the grid is actually more like a small cup with a hole in the bottom. Provided that the grid is not so negative with respect to the cathode as to turn back all the emitted electrons, a narrow stream emerges from this hole, attracted by the positive anode voltage, which for television is usually between 10 and 25 kV. The anode current is comparatively small; seldom much more than 1 mA and often much less. Many different shapes of anode have been favoured, almost the only feature common to all being a hole in the middle for allowing the electrons, greatly accelerated by such a high voltage, to pass on to the enlarged end of the tube. Usually the anode is extended in the form of a coating of carbon inside the tube, as shown in Figure 23.1. One might perhaps expect the electrons to be attracted straight to the anode itself instead of going on to the screen. One reason why they do not is that during most of the journey they are surrounded by the anode so there is no difference of potential or electric field.

As with valves, the grid is normally kept negative and takes negligible current.

Although the electron target is generally called the screen, the name has nothing to do with screening as we considered it in Section 22.9. The glass is coated with a thin layer of a substance called a *phosphor*, which glows when it is bombarded by electrons. This effect is known as *fluorescence*, and is essential in a c.r. tube; but there is also an effect, called *phosphorescence*, which is a continuance of the glow after bombardment. Phosphorescence depends on the choice of phosphor, and with some dies away in a few thousandths of a second, and with others (for 'long after-glow' tubes) lasts as much as a minute. The colour of glow also is determined by choice of phosphor. The brightness of the glow is controlled by the grid bias, which varies the anode current and hence the rate at which the electrons bombard the screen. In c.r.t.s for oscilloscopes and plain television the phosphor layer is continuous and of uniform thickness over the screen area. The different form of construction needed in picture tubes for colour television is described later.

When the electrons have done this bombarding they have to be withdrawn from the screen, or they would soon charge it so negatively that the following electrons would be repelled from it. The bombardment dislodges electrons

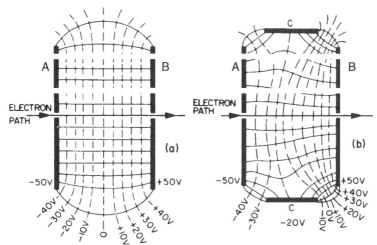

Figure 23.2 Diagrams to illustrate electric focusing

(known as secondary electrons) from the screen and enables them to be emitted. Enough secondary electrons are drawn away to the anode to balance those arriving from the gun. In modern television tubes the phosphor is covered with an extremely thin film of aluminium, which provides a conducting path to the anode. The film is too thin to stop the electrons from reaching the phosphor through it, and it is sufficiently mirror-like to reflect the glow and thereby increase its brightness as seen from outside the tube.

The only thing that can be done with the tube as described so far is to vary the brightness of the patch of light on the screen by varying the grid bias. To make it useful, two more things are needed: a means of *focusing* the beam of electrons so that the patch of light can be concentrated into a small spot; and a means of *deflecting* the beam so that the spot can be made to trace out any desired path or pattern. Both of these results are obtained through the influence of electric or magnetic fields. The fact that electrons are so infinitesimally light that they respond practically instantaneously, so that the trace of the spot on the screen faithfully portrays field variations corresponding to frequencies up to millions per second, is the reason for the great value of the cathode ray tube, not only in television but in almost every branch of scientific and technical work.

In picture tubes both focusing and deflection can be done by magnetic fields, but as the electric methods are rather easier to understand we shall consider them first.

23.2 Electric Focusing and Deflection

When considering electric fields we visualised them by imagining lines of force mapping out the paths along which small charges such as electrons would begin to move under the influence of the field forces (Section 3.3). For example, if a difference of potential were maintained between two parallel circular plates, A and B in Figure 23.2a, the lines would be somewhat as marked—parallel and uniformly distributed except near the edges. As an electron moves from, say, A to B, its potential at first is that of A and at the end is that of B; on the journey it passes through every intermediate potential, and the voltages could be marked, like milestones, along the way. If this were done for every line of force we could join up all the points marked with the same voltage, and the result would be what is called an equipotential line. If potential is analogous to height above sea-level (Section 2.14) equipotential lines are analogous to contour lines.

In Figure 23.2a the equipotential lines (dashed) are labelled with their voltages, on the assumption that the total p.d. is 100 V. An important relationship between the lines of force and the equipotential lines is that they always cross at right angles to one another.

Now introduce a third electrode, C in Figure 23.2b, consisting of a cylindrical ring maintained at say −20 V. As C is a conductor, the whole of it must be at this voltage; so the equipotential lines between C and B must be very crowded. Along the axis between A and B, where C's influence is more remote, they open out almost as they were in a. When the lines of force are drawn it is seen that in order to be at right angles to the equipotential lines they must bend inwards towards the centre of B.

This is the principle employed in electric focusing. The electron beam is assumed to be travelling from left to right through apertures in electrodes A and B. At least one extra anode (corresponding to C) is adjusted to such a voltage as to make the convergence bring all the electrons to the same spot on the screen. The process is analogous to the focusing of rays of light by a lens, and in fact the subject in general is called electron optics. Just as a lens has to be made up of more than one piece of glass if it is to give a really fine focus, so a good electron lens generally contains at least three anodes at different voltages. The focus is usually adjusted by varying one of the voltages. There are too many different varieties of electron lenses in c.r. tubes to illustrate even a selection, but Figure 23.3 shows one of the simpler arrangements, with a potential divider for maintaining the electrodes at suitable voltages.

Electric deflection is quite simple. If two parallel plates are placed, one above and the other below the beam as it

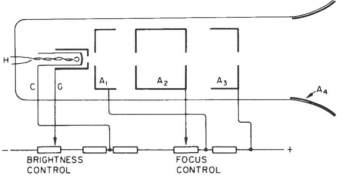

Figure 23.3 C.r. tube gun with electric focusing, including potential divider for feeding and control

emerges from the gun ($Y_1 Y_2$ in Figure 23.4), and there is a difference of potential between the plates, the electric field between them will tend to make the electrons move from the negative to the positive plate. This tendency, combined

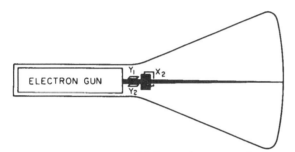

Figure 23.4 Arrangement of deflector plates in a cathode-ray tube

with the original axial motion, will deflect the beam upwards or downwards, depending on which plate is positive. A second pair of plates, $X_1 X_2$, is placed on each side of the beam for deflecting it sideways. By applying suitable voltages between Y_1 and Y_2 and between X_1 and X_2 the spot of light can be deflected to any point on the screen.

23.3 Magnetic Deflection and Focusing

Magnetic deflection depends on the force which acts on an electric current flowing through a magnetic field. In this case the current consists of the electron beam, and the field is produced by external coils close to the neck of the tube. The force acts at right angles to the directions of both the current and the field (Section 4.2), so if the coils are placed at the sides of the beam as in Figure 23.5, so as to set up a magnetic field in the same direction as the electric field between the X plates in Figure 23.4, the beam is deflected, not sideways, but up or down. A second pair of coils, above and below, provides sideways deflection.

With electric deflection, the angle through which the beam is deflected is inversely proportional to the final anode voltage. So if, say, 100 V between the plates is sufficient to deflect the spot to the edge of the screen when the voltage on the final anode is 1000, raising the anode voltage to 2000 makes it necessary to raise the deflecting voltage to 200. With magnetic deflection, the angle is inversely proportional to the square root of the anode voltage; so if, say, 20 mA in the deflecting coils was sufficient in the first case, it would have to be raised to only 28.3 mA in the second.

Raising the anode voltage thus necessitates more deflecting current or voltage, but it brightens the glow on the screen and generally improves the focus.

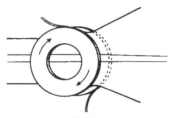

Figure 23.5 Arrangement of deflector coils around a cathode-ray tube

For focusing magnetically, it is necessary for the direction of the field to be along the tube's axis. Electrons travelling exactly along the axis are therefore parallel to the field and experience no force. Those that stray from this narrow path find themselves cutting across the field and being deflected by it. That much is fairly obvious, but it is not at all easy to predict from first principles the final result of such deflection. Actually, by suitably adjusting the field strength, the electrons can be made to travel in spiral paths which converge to a focus in the desired plane. In picture tubes the magnetic field can be produced by a coil wound around the neck of the tube, in which case the focus is adjusted by varying the direct current through the coil. Alternatively one can use a ring-shaped permanent magnet, with a movable yoke for focus control. As with electric focusing, the beam is subjected to the focusing field before being deflected.

A slightly different principle is used in the long magnetic lenses employed with television camera tubes. The tube, including its deflecting coils, is surrounded for the whole of its length by the focusing coil. The electrons again pursue spiral paths which bring the beam to a focus at regular spacings along the length of the tube. By adjustment of the direct current in the focusing coil (or one of the accelerating voltages of the tube) one of the focus points can be made to coincide with the target. In long magnetic lenses focusing and deflection occur together.

23.4 Oscilloscopes

To the electronics engineer a cathode-ray oscilloscope is at least as essential as X-ray equipment to the surgeon. In fact there is scarcely any branch of science or technology in which oscilloscopes are not immensely helpful; and not least, of course, digital computers (Chapter 26). Some are quite simple; others extremely elaborate.

In order to use a c.r. tube at all there must of course be a power unit to provide the high anode voltage, a variable grid bias (usually labelled 'Brightness', Figure 23.3), a variable focusing anode voltage, and of course cathode heating.

It is obvious that an electrically deflected c.r. tube with these auxiliaries can be used as a voltmeter, by applying the voltage to be measured between a pair of deflection plates and noting how far the spot of light moves across the screen. It has the advantage of being a true voltmeter, since current is not drawn from the source of voltage being measured; but of course the anode voltage must be maintained at a constant value or the reading will be inaccurate.

If the deflector voltage alternates at any frequency above a few cycles per second, the movement of the spot is too rapid to be followed by eye; what one sees is a straight line of light, the length of which is proportional to the peak-to-peak voltage. But the possibilities of the c.r. tube are more fully realised when voltages are simultaneously applied to both pairs of plates. For example, if a source of test input voltage to an amplifier is applied to one pair and the output is applied to the other, the appearance of the line or trace on the screen is very informative. If the amplifier is linear and free from phase shift, it is a diagonal straight line. Non-linearity shows up as curvature of the line, and phase shift as an opening out into an ellipse.

A particularly useful range of tests can be performed if to the horizontally deflecting (or X) pair of plates is applied a voltage that increases at a steady rate. The c.r. tube then draws a time graph of any voltage applied to the other pair (Y plates), and in this way its waveform can be seen. The usual procedure is to arrange that when the 'time' voltage has swept the spot right across the screen it returns it very rapidly to the starting point and begins all over again. If the cycles of the waveform to be examined are all identical and the start of every sweep is made to occur at the same phase of the waveform, the separate graphs coincide and appear to the eye as a steady 'picture'. The equipment for producing this repetitive time voltage is called a time-base or sweep generator.

Another practically essential facility in an oscilloscope is a system of adjustable deflection-plate bias voltages for enabling the starting point of the spot of light to be placed anywhere on the screen. Usually there is an X-shift control and a Y-shift control.

Magnetic deflection also can be applied to oscilloscopes by means of external coils as in Figure 23.5, but seldom is, because of the current required and the variableness of coil impedance with frequency. In television, however, deflection is practically always magnetic.

The usefulness of an oscilloscope would be very severely limited without amplifiers to enable adequate deflection to be obtained from weak signals. Ideally, such amplifiers would give a known constant linear gain at all frequencies from zero to the highest in any signals likely to be applied, and would introduce no phase shift. Such amplifiers have much in common with video amplifiers and their design, particularly where high gain is required, is not easy. In fact the principles of these amplifiers justify more space than can be given here and are therefore described in the next chapter. The cost of an oscilloscope depends in no small degree on the specifications of its amplifiers, and especially on their maximum frequency. On some of the cheaper models there is no X amplifier, the assumption being that either the time base will be used, or some other source of comparable voltage.

Because of the difficulties of providing high gain without appreciable distortion of any kind, high deflection sensitivity of the c.r. tube is desirable; but this conflicts with the essential requirements of good focus and bright spot trace, which call for high anode voltage and consequently low deflection sensitivity. Tubes for oscilloscopes therefore often include a feature called *post-deflection acceleration* (p.d.a.), a name that really explains itself. In such a tube the voltage on the anode that accelerates the electron beam ready to pass between the deflection plates may be as low as 1 kV, enabling good deflection sensitivity to be obtained; after deflection the beam is further accelerated by an anode held at 10 kV or even more. Sometimes the acceleration is imparted progressively by a very high-resistance spiral of carbon deposited on the inside of the flared portion of the tube and acting as a continuous potential divider between the final anode and the pre-deflection anode.

23.5 Time Bases

The ideal waveform for the time-base (X-plate) deflecting voltage (or current, if magnetic deflection is used) as described in the preceding section is shown in Figure 23.6, which explains why it is called a sawtooth waveform. The most important feature is the constant slope of the sweep, necessary if the time scale of the graph is to be linear.

Time-base generation is so important in nearly all applications of c.r. tubes that it deserves some attention

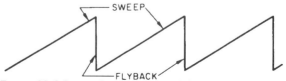

Figure 23.6 Sawtooth waveform used for c.r. tube time-base deflection

now, at least in general principle, although the details come within the scope of Chapter 25. The subject is a very large one, and hundreds of time-base circuits have been devised. The basis of many of the methods is to charge a capacitor at a controllable rate through a resistor, and then discharge it quickly by short-circuiting it. With an ordinary resistor the charging is not linear; it follows the exponential curve shown in Figure 3.6b, the reason being that as the voltage across the capacitor rises the charging voltage falls off and so does the current.

There are two main ways of overcoming this defect: one is to convert the non-linear sawtooth into a linear one, and the other is to use some sort of resistor that passes a constant current regardless of the decline in voltage across it. Often a combination of these methods is used.

In some designs the flyback is brought into action by the time-base voltage itself which, when it reaches a predetermined value, causes a low resistance to occur suddenly across the charged capacitor. In others the flyback is initiated independently. A truly instantaneous flyback would necessitate discharge at an infinite rate, implying an infinitely large current, which is impossible. In practice an appreciable time is required. So in oscilloscopes no attempt is made to use every cycle of a signal waveform to build up the stationary trace on the screen. Figure 23.7a shows, for example, four successive cycles of a waveform to be examined. Below at b we see a linearly rising deflection voltage which continues slightly beyond the duration of a, to make sure that the whole cycle is included. The flyback then occurs. During it a negative bias is automatically applied to the grid of the c.r. tube to suppress the beam. At the start of the third cycle of a the time base comes into action again and the waveform is reproduced on the screen; and so on.

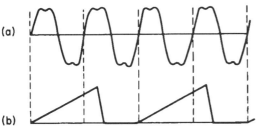

Figure 23.7 (a) Four cycles of a waveform that is to be exhibited on an oscilloscope screen. To make sure of including the whole cycle, only alternate cycles are projected, the others being suppressed to allow time for flyback (b)

To make the time base begin always at exactly the same phase in the signal cycle, so that a steady trace appears on the screen, some kind of synchronising device is needed. Here again, many techniques have been worked out. The simplest method is to arrange for a controllable amount of the signal voltage to be fed to the time-base generator at such a point that it prods it into action.

If magnetic deflection is used, as in television, a sawtooth current is required. The voltage waveform needed to cause such a current would itself have a sawtooth form only if the impedance of the deflection coils were pure resistance, but they are bound to have some inductance as well, and at high frequencies the inductance may predominate. As we saw in Figure 4.10, the rate at which current grows in an inductance is proportional to the voltage across it. So if we want the current to grow at a constant rate we have to apply a constant voltage.

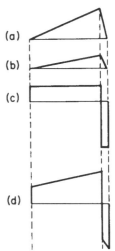

Figure 23.8 (a) Required waveform of magnetic deflection coil current. The voltage waveform to overcome the coil resistance is shown at (b), the voltage for the coil inductance at (c), and the total voltage at (d)

In general the deflection coils have both resistance and inductance, so if a in Figure 23.8 is the required deflection current the part of the applied voltage needed for the coil resistance would have the form shown at b, the part needed for the inductance at c, and the total applied voltage at d. Quite a large amount of ingenuity has been expended on means for economically generating acceptable approximations to the ideal waveforms (Section 23.10).

23.6 Application to Television

For television transmission, as we saw in Section 1.8, images are regarded as made up of small areas called picture elements. For high definition a very large number of elements is needed and we can obtain an estimate of the number by considering the structure of the television picture. Suppose the elements are square and are arranged in 600 horizontal rows. The number of elements in each row depends on the ratio of the width to the height of the picture—the *aspect ratio* as it is called. This is 4 : 3 to agree with cinema film practice. Thus there are 600 × 4/3 = 800 elements per row and the total number of elements in the whole picture is 600 × 800 = 480 000. Pictures are transmitted at the rate of 25 per second and the total picture element frequency is 25 × 480 000 = 12 000 000 per second. If light and dark picture elements alternate, each pair necessitates one cycle of signal; so the frequency of whole cycles is 6 MHz.

In practice this calculation has to be slightly modified to take account of other things, but it does give some idea of the problem.

These *video-frequency* (v.f.) signals cannot be directly radiated, because they are liable to have any frequency from at least 6.0 MHz down to almost zero; so they must be used to modulate a carrier wave just like a.f. signals. But for reasons discussed in Chapter 15 it is necessary for the carrier-wave frequency to be a good many times higher than the highest modulation frequency. That is why television cannot be satisfactorily broadcast on frequencies less than about 40 MHz, and the bandwidth is not a mere 20 kHz or so as in sound broadcasting but extends to several megahertz.

The way in which the elements are scanned in television was described in Section 1.8 and is illustrated in simplified form in Figure 1.8. It should be clear that scanning of this kind can be achieved in a cathode-ray tube by means of a high-frequency time base connected to the horizontally deflecting plates or coils, and a low-frequency vertical time base. The spot then traces out a succession of lines, and if its brightness is meanwhile being controlled by the v.f. signals in proportion to the brightness of the corresponding point in the scene being broadcast, a picture will be produced on the screen of the c.r. tube.

To cover all the picture elements, the scene has to be scanned in some hundreds of lines; the greater the number of lines the better the possible definition of the picture. Clearly the scanning at the receiving end must be in step with that in the camera, so a standard number of lines must be agreed upon. Several different line standards are or have been used; for the sake of example the 625-line standard used in Britain will be assumed. The corresponding number of times the complete picture is scanned is 25 per second, so the line frequency is $625 \times 25 = 15\,625$.

Twenty-five pictures per second is a low enough frequency to cause noticeable flicker. To avoid this the scene is actually scanned 50 times per second, first the odd lines only, then the even lines. So although the complete picture frequency is still 25, the frequency of the vertical time base is 50 and flyback occurs after every 312½ lines. This subdivision of the scanning is called *interlacing*, and

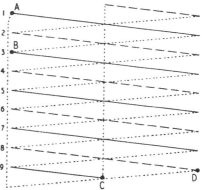

Figure 23.9 Nine-line interlaced 'raster'. The dotted lines are flybacks, which normally are invisible

each scan making up half the complete picture is called a *field* (rather confusingly, because it has nothing to do with electric or magnetic fields).

A scanning diagram showing all 625 lines would be difficult to follow, so Figure 23.9 shows a *raster* (as it is called) of only 9 lines. The principle is the same, however. Beginning at A, the line time base scans line No. 1, at the end of which the flyback (shown dotted) brings the spot of light back to the left-hand edge. But not to A, for the field time base has meanwhile moved it downward two spaces to B. In the same way all the odd lines are scanned, until No. 9, halfway along which, at C, the field time base flies back to the top of the picture. Although for clearness it is shown as doing so vertically, this is not practicable as it would mean infinitely high speed. In a 625-line system a time of up to 31 lines is allowed, so fewer than 600 lines are actually available for the picture. During field and line flyback the cathode ray is cut off, so nothing appears. The line time base can now fill in all the even lines, shown dashed.

An essential addition to the system is some means for ensuring that all the receiving scanners work in exact synchronism with the scanner in the camera; otherwise the picture would be all muddled up. Synchronising signals are transmitted during the flyback periods of the camera

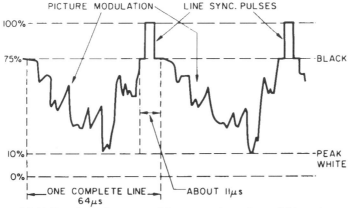

Figure 23.10 Complete television modulation signal waveform of 2 lines duration

scanner, and the receivers separate these from the picture signals and apply them to the time base generators to control their exact moments of flyback.

23.7 Characteristics of Television Signals

Two sample lines of combined signals, as used to modulate the carrier wave, are shown in Figure 23.10. This is the so-called negative modulation used with the 625-line standard, in which the minimum carrier-wave strength (10% of maximum) represents the brightest parts of the picture and 75% represents zero brightness, or black. The 75% to 100% strength is reserved for synchronising signals. (405- and 819-line systems used positive modulation in which these levels were reversed.) Each line in a 625-line signal lasts for 1/14 625 s, or 64 µs. Of this, about one-sixth of the time is used for flyback and for emitting a single-pulse line synchronising signal, which itself lasts for only about 6 µs.

In Figure 23.11 eight lines are shown but to concentrate attention on the line sync. signals the picture modulation between them is represented merely by the letter P. The rising front of each sync. pulse, marked by a dot, is used to make the receiver line time base transfer the spot of light to the left-hand edge of the picture ready to start the next line.

Looking again at Figure 23.9 we see that the last line of the first (and all odd) fields must be interrupted exactly half way in order to ensure accurate interlacing. So at this point, marked X in Figure 23.11b, line pulses begin to occur twice as often. The line time base ignores these extra pulses and continues to respond only to those marked in the diagram with a dot. The six half-line pulses after the last half-line of picture are used to mark time (for less than 0.0002 s!) in preparation for the field sync. pulses, which consist of six much wider pulses as shown. The rising fronts of those marked with a dot act also as line sync. signals, for the line base must be kept running steadily even when nothing is appearing on the screen. The field pulses make the spot return to nearly the top of the screen ready to begin an even-numbered field, and after a few more half-line pulses normal service is resumed (Figure 23.11a).

Following this field the flyback takes place at the end of a full line, so the field sync. signals are as shown in Figure 23.11c, which needs no further explanation.

An important point is that modulation of a television carrier wave does not take place equally above and below an unmodulated level as in the transmission of sound. During intervals between scenes the sync. signals must be continued steadily, and their peaks indicate maximum (100%) carrier level. If the receiver is to reproduce the black level correctly it must be able to distinguish a carrier wave at 75% of maximum strength from lower strengths, even when it is unmodulated by a picture (as when a blank screen is shown). In other words, it must be able to respond to the zero-frequency (d.c.) part of the received signal.

We can now make a more accurate estimate of the upper video frequency. The duration of the visible part of a line is about 52 µs, so the number of elements per second is $800 \times 10^6/52 = 15.4 \times 10^6$. On the assumption that the maximum video frequency is half the frequency of the picture elements, it amounts to $15.4/2 = 7.7$ MHz. In this calculation we assumed the elements to be square—tantamount to assuming that the definition is equal in the horizontal and vertical directions. The maximum frequency allowed in the 625-line specification is 5.5 MHz, so the horizontal definition is necessarily less than the vertical.

If amplitude modulation were used in the same way as for sound broadcasts, even this reduced v.f. would occupy a radio bandwidth of $5.5 \times 2 = 11$ MHz. And room must be made for the accompanying sound, and also some separation from signals of other stations. So the total width of a television channel would be so great that there would be insufficient channels even in the considerable v.h.f. and u.h.f. bands available. Moreover the gain of each amplifier stage in the receiver is inversely proportional to the bandwidth (Section 22.6), so many stages would be needed. We have seen how amplitude modulation at any one frequency results when a carrier wave and two other waves, one at a lower frequency and the other at a higher, are added together, as in the phasor diagram Figure 15.6. If we omit one of the two sideband phasors there is still amplitude modulation, and when rectified in the receiver

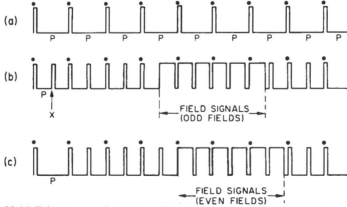

Figure 23.11 Television synchronising pulses. The rises used for line synchronising are marked with a dot; P denotes a line of picture signals. Line pulses only are shown at (a); the modifications required at the end of each field for field synchronisation are as at (b) and (c)

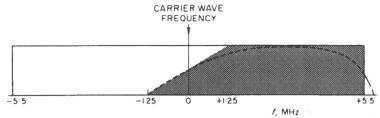

Figure 23.12 The ideal television receiver tuning acceptance is shown shaded; the broken line indicates a more practical performance

the modulation frequency is still extracted. We can easily see that there would be some distortion of the modulation waveform, but at low modulation depths it would be very slight.

It has been found that little is lost if the receiver is made sufficiently selective to tune out most of one sideband. The tuning is adjusted by the makers so that the carrier-wave frequency is halfway down the lower-frequency slope of the overall selectivity curve, as shown by the broken line in Figure 23.12, where the theoretical receiver acceptance is shaded. So the receiver bandwidth problem is almost halved. Note that at very low video side frequencies (those closest to the carrier frequency) reception is practically ordinary double sideband. Up to 1.25 MHz one side becomes progressively weaker but the other progressively stronger, so that the total continues to be the same, as it does also above 1.25 MHz where reception is entirely single-sideband.

Radiating the whole of the lower sideband would merely broaden the channel unnecessarily, so all of it beyond 1.25 MHz is more or less filtered off at the sending end to give a *vestigial* sideband. The frequencies broadcast are therefore as shown in Figure 23.13. (The received part is again shaded.) The width of an entire channel is thus reduced to 8 MHz. Note that these diagrams do *not* show the actual amplitudes at the various frequencies in the broadcast signal; they naturally depend on the scene being televised at any moment, and usually fall off with increasing video frequency.

Either a.m. or f.m. can be used for television sound; in the 625-line standard it is f.m. The v.f. modulation is always a.m.

The specification shown faces a receiver with quite a difficult job, for the vision section ought to respond as in Figure 23.12 over a 6.75 MHz band and yet reject the associated sound only 0.5 MHz higher—and the adjacent-channel sound 0.75 MHz lower. Not infrequently receivers fail to do this, and the reproduced picture is visibly disturbed by the sound signal.

Ignoring the quite considerable problems at the sending end, let us now take a broad look at what a television receiver must include.

23.8 Television Receivers

Figure 23.14 is a block diagram of a typical plain receiver for 625 lines. Usually there is a single preselector stage to ensure that the signal is strong enough when it comes to the relatively noisy frequency-changer. We have already seen an example of this part of a television receiver; Figure 22.16. It should be designed to amplify the full channel frequency band of 8 MHz equally; for example, from 470 to 478 MHz for the lowest-frequency u.h.f. channel, No. 21. If this is done, the adjacent-channel selectivity is likely to be hardly noticeable, but fortunately the range of reception on such frequencies is limited to the fairly local area and any other transmitters in that area would be allotted widely different frequencies. Change of channel at v.h.f. was usually obtained by switching different pre-tuned inductors into circuit, but with the transmission-line tuners used for u.h.f. this is not practicable and tuning is by ganged capacitor or varactor.

The frequency changer shifts the whole frequency pattern of the channel to the i.f., usually 33.5 MHz for vision carrier and 39.5 MHz for sound carrier. Because the

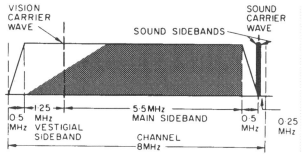

Figure 23.13 Diagram of the frequency allocations in one complete 8 MHz television channel. The shaded portion is again the ideal receiver acceptance

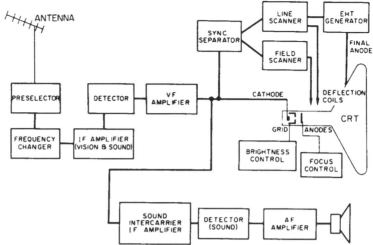

Figure 23.14 Block diagram of typical television receiver (plain)

amplifier has to cover such a wide band in relation to this i.f., the gain per stage is limited and several stages are needed. Surface-acoustic-wave (s.a.w) filters (see end of Section 22.7) are now normally used for television i.f. amplifiers, as they are simple, compact, robust, and facilitate provision of suitable pass-band shape. In our example, vision and sound continue together through the detector and even the v.f. amplifier. By that stage the frequencies present are as shown in Figure 23.13, if the vision carrier is taken as zero frequency. The sound carrier wave, beating with the vision carrier, yields a 6 MHz signal, with the sound sidebands around it. The vision detector, in fact, acts as a second frequency changer for the sound signal, having an output i.f. of 6 MHz, which is amplified in the v.f. stage along with the video frequencies. By means of various tuned rejector circuits (often called traps) the vision and sound signals are separated from one another. For the principles of the v.f. amplifier see Chapter 24.

The sound signals may be further amplified at what is called the intercarrier i.f. of 6 MHz before being passed through the f.m. detector (limiter and discriminator) to the a.f. amplifier and loudspeaker. All this follows the principles described in earlier chapters.

Alternatively the sound and vision may be separated earlier—perhaps somewhere along the main i.f. amplifier— or there may be a separate intercarrier detector for sound.

If the v.f. signals are applied to the cathode of the c.r. tube, less signal voltage is needed than if the grid were used, because making the cathode more positive increases the negative bias and reduces the first anode voltage (relative to cathode), both of which reduce brightness. Moreover the polarity of the picture signals that is correct for the cathode normally happens also to be right for dealing with the synchronising signals. So cathode modulation of the beam is usual, the grid being reserved for control of general picture brightness by the viewer; there is no scarcity of voltage for that.

Because of the polarity of cathode modulation just referred to, the vision detector and v.f. amplifier must be arranged so that an increase in radio signal amplitude makes the cathode more positive, the 75% (black-level) amplitude being just enough to extinguish the spot altogether. The sync. pulses then make the cathode more positive still, so are unable to produce any visible effect, being 'blacker than black'. The same applies to a positive modulation system if the detector is connected the opposite way.

23.9 Synchronising Circuits

A third outlet from the v.f. amplifier goes to the sync. separator, which removes the picture part of the signal, leaving only the sync. signals. It consists of what looks like a perfectly ordinary stage of resistance coupled amplification, Figure 23.16, for example. As shown, the input is the video signal with the sync. pulses positive-going. (It would have to be reversed for a p-n-p transistor.) The d.c. component is removed by C, with the result that in a waveform diagram there will be equal areas above and below the zero-signal line, marked O in Figure 23.15a, which show two sample lines of video signal, the first corresponding to a bright scene and the second to a dark one. Thus the sync. pulses move up and down with scene brightness as shown in Figure 23.15a, and a bias setting

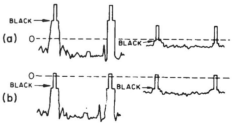

Figure 23.15 Showing television signal voltage levels with (a) and without (b) d.c. restoration

suitable for removing all but the sync. signals during a bright scene would remove the sync. signals too during a dark one. But it is possible to put this right by what is known as d.c. restorer action (Figure 23.15*b*). R_1 in Figure 23.16 is given an unusually high value, so that the

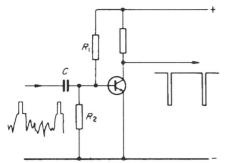

Figure 23.16 Example of a sync. separator circuit

base-to-emitter diode is on its bottom bend. It therefore rectifies the input signal as in a shunt-diode detector (Section 18.12), pushing it down until the whole of the picture part the signal is cut off. Only the sync. pulses appear at the output, where, owing to inversion by the amplifier, they are negative-going. The d.c. restorer action also automatically ensures that any excess pulse amplitude is cut off by the excess bias it develops.

If an f.e.t. is used, no bias at all is needed, as the desired range of operation as a pulse amplifier is obtained between zero and cut-off bias.

Next, the field sync. signals have to be sorted out from the line signals which function at regular 64 µs intervals all the time. The principle of this sorting out can be understood by referring back to our capacitor-charging experiment, Figure 3.5. Line signals (Figure 23.11*a*), which consist of brief pulses separated by relatively long intervals, can be simulated by flicking the switch alternately between A and B, the B periods being (say) 10 times as long as the A periods. The charging voltage thus varies as in Figure 23.17*a*. The voltage across the capacitor has hardly begun to rise by the time the charging voltage drops back to zero. So it never gets a chance to build up (*b*). But if the A and B periods are reversed to simulate a train of field pulses (*c*) the capacitor charge builds up as shown at (*d*). The corresponding voltage in a television receiver is fed into the field time-base generator to bring the flyback into action.

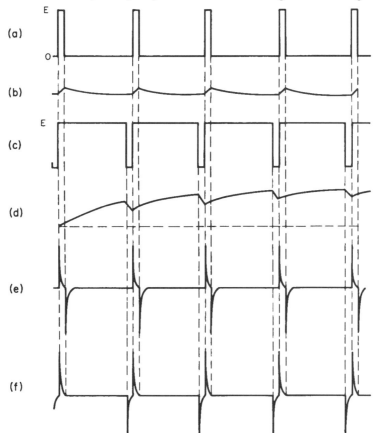

Figure 23.17 The normal narrow line sync. signals (*a*) do not build up a field sync. signal (*b*), but the wide signals (*c*) do (*d*). Both *a* and *c* yield line sync. pulses in the same phase ((*e*) and (*f*))

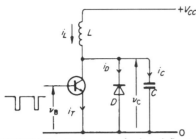

Figure 23.18 Ideally simple circuit for line deflection current shaping

The difference between the odd and even fields, necessary for interlacing, has been explained in connection with Figure 23.11.

As the dots in Figure 23.11 indicate, line synchronising is effected by the rising fronts of the pulses. Apart from the half-line pulses, which the line time-base generator ignores (see Section 24.6 with reference to Figure 24.16) these fronts are equally spaced whether the pulses themselves are narrow or broad. To prevent the different widths of the field signals from disturbing the steady rhythm of the line time base, the return strokes of all the pulses are eliminated before the line signals are fed to the line time-base generator. This is done by applying the combined sync. signals to a series capacitance followed by a shunt resistance, but this capacitance is smaller and it is the voltage across the resistance that is used to initiate the flyback. Figure 23.17e shows this voltage, which has the same shape (speeded up) as in Figure 3.6a showing the current in the circuit and therefore also the voltage across the resistance. Substituting field sync. signals alters the pattern to Figure 23.17f, but there is no difference in the rising pulses, and the falling ones are eliminated by biasing off.

If the flyback for every line were initiated directly by its own individual sync. pulse, and spasmodic interference was coming in along with the sync. signals, the lines affected would be displaced and the picture torn apart horizontally. To avoid this, a system called flywheel synchronisation is commonly used. Without going into details one can say that the line time base is allowed to run continuously, producing pulses of its own at every flyback. These are compared with the incoming sync. signals in what is called a phase detector. If both occur at the same frequency, nothing happens. But directly the line generator gets out of step, the phase generator produces a voltage, the polarity of which depends on whether the generator is running fast or slow. This voltage is applied to the generator in such a way as to correct its speed. The time constant of the system is chosen so that correction is gradual; occasional mutilation of the line sync. pulses by interference then has little or no effect, as there is an averaging out of the control.

23.10 Scanning Circuits

The line time base is the greater problem because of the greater amplitude required and its much higher frequency,

and the correspondingly short time—only a few microseconds—for the current through the deflection coils to be changed from maximum one way to maximum the other way during flyback. This is not remarkably fast by oscilloscope standards, but television is faced with the demand for maximum screen size combined with minimum distance from back to front. That means a very much larger angle through which the beam has to be deflected, and therefore a correspondingly larger effort from the time-base generator (or scanner). Electric deflection is not practicable, so substantial *current* has to be reversed very rapidly. And as this is only one of many functions in a complete receiver, only a limited amount of power and space can be allowed for it.

The subject is extremely complex, and circuit diagrams of the many varieties of line scanner that have been used usually give little clue to the exact working conditions. In spite of the wide variations in detail, the basic principle is the same for all. Instead of providing a new lot of energy for each stroke—15 625 every second—energy is made to circulate to and fro, only the losses needing to be made up.

The basic method is so like the generation of continuous oscillation as described in Section 14.2 that it might be wise at this point to re-read that section. The example now to be described is the simplest of all, but it should be understood that its working has been idealised and that in practice there are many complications.

Figure 23.18 shows the circuit. The type of transistor is one that can pass a heavy current, say 10 A, and withstand a relatively high collector voltage, say 150 V—but not both at the same time, as we shall see. Because it can pass current only in one direction, whereas the current in the deflection coils is required to alternate equally in both directions, a diode D is connected in parallel to provide a path for the opposite current. Although the line deflection coils are normally connected to a transformer winding and include some resistance, they are represented ideally by an inductance L. C is an energy-storing capacitor, which is effectually in parallel with L because the V_{CC} supply is shunted by a capacitor of negligible impedance.

The base input consists of negative-going pulses as shown in Figure 23.18 and also in Figure 23.19a, corresponding approximately to line sync. pulses. All the time between them the base is positive. Let us begin halfway through this inter-pulse phase. The transistor is bottomed, passing current with hardly any emitter-to-collector voltage (v_C). Practically all of V_{CC} is therefore across L, so the current through it grows at a steady rate. Figure 23.19b shows v_C almost zero, and at c the current i_T growing steadily. Because v_C, such as it is, is positive, no current i_D can flow through D, and only a very small charging current i_C flows into C, so the voltage across it (v_C) rises almost imperceptibly—see Figure 23.19d, e and b. The current through the deflection coils is therefore practically i_T, so the spot on the screen moves steadily from the centre towards the right-hand edge.

By the time i_L has become enough to deflect the spot to the edge, the input pulse arrives and cuts off i_T. This moment corresponds to the point a in Figure 14.2 in a practically identical circuit. What happens is that i_L (which cannot change suddenly because of the inductance L) is diverted to C and starts charging it—note the sudden jump in i_C, equal to the drop in i_T. The voltage v_C builds up to a positive peak, where momentarily it is neither rising nor falling, so i_L and i_C must be zero. This corresponds to b in Figure 14.2. C then starts discharging, so i_C (and therefore i_L) reverses. And so v_C and i_C continue to follow the same half-sine-wave shape as in Figure 14.2a to c. This makes the spot on the screen (extinguished by a negative pulse on the grid of the c.r. tube) fly back to the left-hand edge). From there things are different, at first because of D. Directly the negative half-cycle of v_C begins, D conducts and being practically a short-circuit it holds v_C nearly constant at a very small negative value. The voltage across L is therefore again virtually equal to V_{CC}. So i_L again rises steadily, carrying the spot from left to right. Because v_C hardly varies, C hardly charges at all, and i_L and i_D are almost the same current.

This state of affairs continues until about half-way between the input pulses, by which time the transistor is once more ready to conduct. Directly v_C becomes positive it does so, taking over from D. This is where we came in, so the whole programme repeats, continuously.

The current supplied by V_{CC} is i_L, and according to Figure 23.19f it is entirely a.c., so can be supplied by a large capacitance across the V_{CC} source. No net d.c. power is needed. This ideal cannot of course be achieved in practice; we have already noted the resistance of the coils as one cause of loss. But the power to be supplied is only a fraction of what would be needed if it were not for the energy-storing ability of C.

That is not the only economy effected by this type of circuit. We have noted the rapid change in i_L during flyback, giving rise to a peak of v_C. The inductance L must be quite small if this voltage peak is to be within the limit of what the transistor can stand. That is to say, it must have only a few turns of wire and the current i_L must be large in order to deflect the beam sufficiently. As was mentioned earlier, the actual deflection coils are connected across a low-voltage transformer winding, through which the rapid change in current occurs at flyback. Another winding having a very large number of turns of fine wire steps up the voltage peak to something like 16 kV, and a high-voltage low-current rectifier suppresses the reverse voltage so that the d.c. output is available for the final anode of the c.r. tube. Because of the small current load, sufficient smoothing is obtained by the capacitance (of the order of 1.5 nF) between the inner conducting coating of the tube and an external one laid on for that purpose. Economy again!

There is also a need for other voltages greater than V_{CC}; these too can be obtained in similar manner from this line output transformer. With these omitted, the line output stage is likely to look something like Figure 23.20. D_2 is the e.h.t. (extra high tension) rectifier. Alternatively the e.h.t. and the lower high voltages for focusing, c.r.t. final anode, etc., can all be obtained from a lower-voltage flyback transformer combined with a voltage multiplier, as in Figure 27.10.

Owing to the comparatively slow rate at which current in the field deflection coils changes, their inductance is less significant than their resistance. The voltage needed has

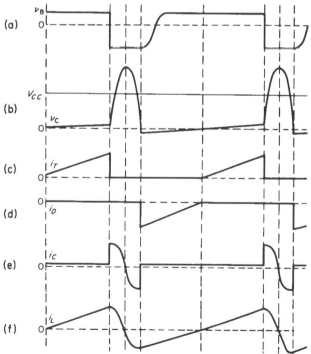

Figure 23.19 Waveform diagram for Figure 23.18

therefore a mainly sawtooth form and the stage supplying the coils is basically an amplifier of such a waveform provided by the field scanning generator. Neither field nor line deflecting currents would be sufficiently linear without means for linearising, details of which can be found in books on television receiver design.

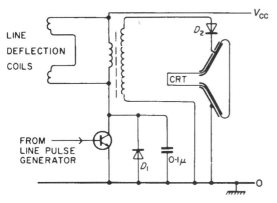

Figure 23.20 Showing how the type of line scanning circuit shown in Figure 23.18 is adapted to produce also e.h.t. supply for the c.r. tube

23.11 Colour Television

Reproduction of scenes in colour, by photography, printing and television, relies on the fact that any colour can be acceptably made up from three primary colours in the right proportions. The most apparently straightforward way of applying this fact to television is to transmit three complete pictures—one in each primary colour—in place of each one in the system we have just been studying. Assuming for the moment that pictures in the three separate primary colours (to each of which the term 'monochrome' *correctly* applies) can be reproduced on a single screen within 1/25th sec, 'persistence of vision' makes them appear collectively as a picture in full colour.

Although this scheme has been successfully demonstrated, the practical objection is that because everything has to be done three times as fast the bandwidth required is three times the already awkwardly wide requirement for plain television. Alternatively, the three primary colours could be broadcast concurrently on three channels, but the total frequency band would of course still be three times that for one ordinary channel. This factor of 3 could no doubt be cut down to something slightly less by taking advantage of the fact that with colour a rather lower standard of definition is acceptable. Even so, the economic disadvantages of this approach are so considerable that very much more complicated systems have been devised in order to avoid it and keep colour television within the standard channels. These are the systems in actual use. Although there are several, the differences between them are only in details which are too involved to include in a book of this kind. An essential requirement of any practical colour system is that it be *compatible*; that is to say, colour programmes must be receivable on non-colour receivers just as if they were uncoloured, and uncoloured programmes receivable as such on colour receivers.

The basic problem—to restate it in a different way—is to convey extra picture information (colour) without widening the frequency band. A solution might appear impossible, for a fundamental law of communication (Hartley's law) is that the product of bandwidth and time needed to convey a given amount of information is constant. So if you try to convey more information in a given time you have to use more bandwidth. This law assumes that the information is coded in the most economical manner. The compatibility requirement is the apparently insuperable difficulty, for it rules out any more economical coding of the basic picture (without colour), even if that were possible. So no room seems to be left for colour information.

The solution is to superimpose the colour information on the plain picture information (*luminance*) within the same frequency band, coding the colour information so unobtrusively that it does not ruin the picture. Provided that the normal standard of luminance definition is preserved, quite a low definition of colour is acceptable, as infants demonstrate with their painting books. The colour v.f. is therefore limited to about 1 MHz. And it is put high up in the luminance v.f. band, centred on a subcarrier-wave frequency of approximately 4.43 MHz, because the interference with the luminance information is least noticeable in this region of frequency. The subcarrier itself certainly would be very noticeable if it were broadcast, but it is suppressed at the transmitter and restored in the receiver. Figure 23.21 shows, for comparison with Figure 23.13, the resulting channel structure. By the way, the technical term for the colour information is *chrominance*, and it comprises two parts: *hue* and *saturation*. Hue is the colour itself—red, blue, etc.—and saturation is the intensity of the colour. In an uncoloured picture all colours have zero saturation, and 'pastel shades' are low in saturation compared with 'geranium red'. Saturation is conveyed by the amplitude of the chrominance signal, and hue by its phase displacement. The whole subject, theory and practice, is a vast one but we can give some basic information on the ingenious method adopted to code and transmit the chrominance information in the NTSC and PAL colour television systems. The colour information is received from the colour cameras, slide scanners, telecines or video tape recorders in the form of red (R), green (G) and blue (B) signals (Section 15.7). These must be impressed on the 4.43 MHz subcarrier in such a way that they can readily be separated at the receiver. The first point to realise is that it is not necessary to transmit all three colour signals. These make up the plain, i.e., luminance, signal (Y) which is transmitted in the normal way by amplitude modulation of the main carrier. So we need transmit only two colour signals on the subcarrier; the third can be obtained by arithmetical operations on these and the Y signal. In fact the signals used to modulate the subcarrier are not chosen from the R, G and B signals themselves but are (R-Y) and (B-Y), known as *colour-difference* signals. The R and B signals can be obtained simply by adding the Y signal to the colour-difference signals, and the addition is often achieved in the picture tube itself by applying the Y

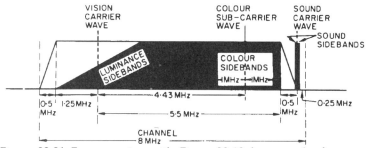

Figure 23.21 For comparison with Figure 23.13 here are the frequency allocations for a colour television signal

signal to the cathodes, which is the normal technique in plain receivers, and the colour-difference signals to the control grids.

The problem has now been resolved into that of transmitting two signals independently on one subcarrier. To do this, *quadrature* modulation is used. The subcarrier is first split into two components with a phase difference of 90° between them. One component is then amplitude-modulated by the (R-Y) signal and the other by the (B-Y) signal in circuits which suppress the carrier. The resulting modulated signals are now recombined to form a single suppressed-carrier signal carrying both modulation waveforms. This is the signal mentioned above, which represents colour saturation by its amplitude, and hue by its phase angle. It is superimposed on the luminance signal, but because of the limited bandwidth of the colour information its sidebands are confined to the upper end of the video band as shown in Figure 23.21.

It is not possible to detect a suppressed-carrier signal by the types of a.m. detector described in Chapter 18. The

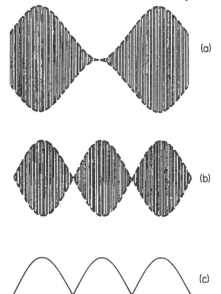

Figure 23.22 Sinusoidal modulation of a carrier (*a*) with carrier retained and (*b*) with carrier suppressed. (*c*) shows the result of detecting a suppressed-carrier signal with a conventional a.m. detector

reason for this can be seen from Figure 23.22, which shows the results of sinusoidal modulation of an r.f. signal with carrier retained *a* and carrier suppressed *b*. Signal *b* applied to a normal a.m. detector would produce an output consisting of two peaks in the same direction for each cycle of modulating waveform, representing gross distortion (Figure 23.22*c*). This failure occurs because, if the diode is arranged to be driven into conduction by positive r.f. peaks, it cannot reproduce the negative half-cycles of the modulation waveform, these being carried solely by negative r.f. peaks.

This difficulty can be overcome and distortionless detection achieved by arranging for the diode to be driven into conduction, not by the modulated-r.f. signal itself, but by a separate carrier-frequency input signal exactly synchronised with it. Because of the need for the second input, such detectors are known as synchronous, and to permit synchronous detection of the colour-difference signals in a television receiver a sample of the unmodulated subcarrier is included in the transmitted waveform. It consists of eight cycles of subcarrier located in the blanking period following each line sync. signal. At the receiver this 'colour burst' is used to synchronise an oscillator the output of which is applied to the (R-Y) detector and, after 90° phase shift, to the (B-Y) detector.

We can now see the reason for the 90° phase shift. In a diode detector with an *RC* load circuit the diode conducts for only a small fraction of each r.f. cycle when the input signal is passing through its peak value. In fact the diode is often said to 'sample' the peak values of the input signal. Figure 23.23 illustrates this for synchronous detection of, say, the (R-Y) signal and the dashed lines indicate the sampling instants. At these instants the carrier component of the (R-Y) signal is at its peak value but, because of the quadrature relationship, that of the (B-Y) signal is passing through zero and so has no effect on detection of the (R-Y) signal. Similarly in the (B-Y) detector, the (R-Y) signal has no effect. So both colour-difference signals can be recovered without mutual interference. From these signals and Y it is possible to obtain (G-Y) which is needed for application to the picture tube. The colour-difference signals need amplification, but the video amplifiers require only a limited frequency range (up to 1 MHz) compared with the full range (up to 5.5 MHz) for the Y signal.

To display the colour signals as a picture we need a device which accurately superimposes the red, green and blue areas on the plain image. It would be possible to use a separate picture tube for each of these signals, and to combine all the images by mirrors and lenses but this would be a cumbersome and costly arrangement, and normally a single tube is used. To form its screen, three types of phosphor are needed, one which glows red under

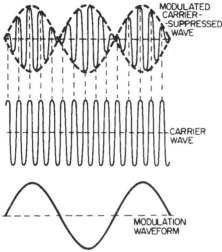

Figure 23.23 Action of a synchronous sampling detector in detecting carrier-suppressed signal. The dashed lines indicate the sampling periods

electron bombardment, the second green and the third blue; and the phosphors must be in the form of areas with minute width (less than 0.1 mm) to permit good picture definition. Three electron guns are required (or one gun firing three beams), and behind the screen a mask is needed to ensure that the beam from one particular gun, no matter to what extent it is deflected, always strikes red phosphor areas, the beam from the second gun strikes only green phosphor areas and the beam from the third gun strikes only blue phosphor areas. The use of a mask to screen each beam from the areas it must not strike is known as the shadowmask principle, and is used in all colour picture tubes although there are many different forms of construction. Figure 23.24 shows one arrangement known as the precision-in-line (p.i.l.) tube, in which the phosphors are in the form of narrow vertical stripes on the screen (arranged in the order R, G, B), the shadowmask consists of vertical bars, and the electron guns are in a horizontal row, the beams passing through the apertures in the mask to strike the appropriate phosphor stripes.

23.12 Television Camera Tubes

So far in our account of television we have not mentioned one very important application of cathode-ray tubes—in television cameras where they are required to produce a picture signal from the optical image of a scene.

The earliest type of camera tube, the *iconoscope*, was a cathode-ray tube in which the electron beam was deflected so as to scan a light-sensitive target. This consisted of a sheet of insulating material backed by a conductive layer (the signal plate) and carrying on its scanned face a mosaic of very small individually insulated areas of photo-emissive material. The optical image was focused on the light-sensitive face and electrons were released from the individual areas in direct proportion to the brightness of the image there. In this way a positive charge image of the scene was built up on the scanned face of the target and it grew during the field period, charging up the capacitance between the mosaic and the signal plate. The scanning beam neutralised the charge image, generating a signal in the backing plate. The succession of discharges constituted the required picture-signal output.

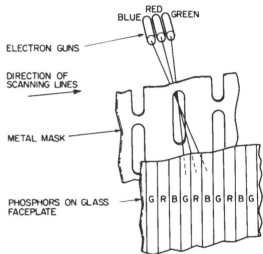

Figure 23.24 The principle of the precision-in-line (p.i.l.) picture tube

Iconoscopes had a number of disadvantages. They were insensitive and they generated spurious output signals as a result of secondary emission from the target. Moreover, the need to project both the optical image and the scanning beam on to the same face of the target resulted in an inconvenient camera shape.

The introduction of the *orthicon* tube solved many problems. In these the target was a very thin layer of photo-conductive material, backed by a transparent signal plate. Thus the optical image could be focused on one face whilst the electron beam scanned the other. The scanning beam landed on the target at just sufficient strength to neutralise the positive charge image, generating the picture signal at the output at the signal plate. A low-velocity scanning beam was essential to avoid secondary emission from the target. An important feature of the tube was that the scanning beam returning from the target having lost electrons was amplitude-modulated by the required picture signal and could be directed into the input of an electron multiplier conveniently of circular design and surrounding the electron gun. Thus was produced the *image-orthicon* tube, a very sensitive tube, used successfully for many years in plain television broadcasting.

Because the image-orthicon was a large tube, it was not convenient for use in a colour television camera where three or four tubes were required. In these the optical image is split into its red, green and blue components, each of which is focused on the target of a separate camera tube. Small photo-conductive tubes known as *vidicons* are generally used. The luminance signal could, of course, be produced by combining the outputs of the three tubes, but often a separate fourth vidicon is incorporated in the camera to provide this.

It would be very difficult to use a three- or four-tube assembly in a compact camera, such as that in a camcorder. Fortunately it is possible to obtain a colour picture signal from a single camera tube. One way of doing this is to deposit on the outer face plate of a vidicon an array of very thin vertical stripes of translucent colour-filtering material in the order green, cyan, clear, green, etc. The green filter allows only the green component (G) of the incident image to pass through to the photo-conductive layer; the cyan filter permits the green and blue components (G + B) to pass through, whilst the clear stripes permit all three components (R + G + B) to pass through. As horizontal scanning proceeds, therefore, the output signal of the tube samples the G, G + B and R + G + B components in turn. By arithmetical matrixing of these, the R, G and B signals can be obtained.

Modern television and still video cameras have dispensed with cathode-ray tubes and use instead solid-state image sensors. These are described in Section 26.5.

23.13 Application to Radar

If we make a sound and hear an echo t seconds later, we know that the object reflecting the wave is $167t$ metres distant, for sound travels through air at about 335 metres per second and in this case it does a double journey. Radar is based on the fact that radio waves also are reflected by such objects as aircraft, ships, buildings and coastlines. If the time taken for the echo to return to the point of origin is measured, the distance of the reflecting object is known to be $150\,000t$ km away, the speed of radio waves through space being almost 300 000 km per second. Since distances which is useful to measure may be less than a kilometre, it is obvious that one has to have means for measuring very small fractions of a second. In radar, as in television, time is measured in microseconds (μs).

This is where the cathode-ray tube again comes in useful. With a time base of even such a moderate frequency as 10 kHz one has a scale that can be read to less than 1 μs, and a much faster time base can easily be used if necessary. The time between sending out a wave and receiving its echo can be measured by making both events produce a deflection at right angles to the time base line. This can best be understood by considering a typical sequence of operations.

We begin with the spot at A in Figure 23.25, just starting off on a stroke of the time base. Simultaneously a powerful transmitter is caused to radiate a burst or pulse of waves. This is picked up by a receiver on the same site, connected to the Y plates so that the spot is deflected, say to B. The pulse must be very short, so that echoes from the nearest objects to be detected do not arrive while it is still going

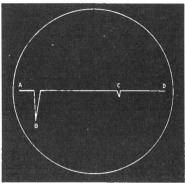

Figure 23.25 Appearance of trace on cathode-ray tube in one type of radar receiver

on. In practice it may have to be a fraction of a microsecond, and as it has to consist of a reasonable number of r.f. cycles, their frequency obviously has to be many MHz. A more pressing reason for extremely high frequency is that it enables one to radiate the pulses precisely in any desired direction (Section 17.18). Typical frequencies are 3 GHz and 10 GHz (10 cm and 3 cm wavelengths).

The time-base voltage continues to move the spot across the screen, and if the receiver picks up one or more echoes they are made visible by deflections such as C. When the spot reaches the end of its track, at D, it flies back to the start; then, either at once or after a short interval, begins another stroke. The transmitter, being synchronised, registers another deflection at B, and the echo is again received at C. The repetition frequency is normally high enough for the trace of the spot to be seen as a stationary pattern.

The time for an echo to return from a distance of one kilometre being $1/149\,896$ s, is equal to the time occupied by one cycle of a 149 896 Hz oscillation. By switching the Y plates over to a 149 896 Hz oscillator, so that the cycles are seen as a wavy line, the time base can be marked off in distances corresponding to kilometres.

This is only a mere outline of one of many ways in which the c.r. tube is used in radar. Figure 23.25 shows what is known as an A-type display. The only direct information it yields is *distance*. For most applications one wants to know the direction of the reflecting object in the horizontal plane, the term for which is *azimuth*. A ship's navigator, for example, in fog wants to be able to see the positions of coasts and other ships all around within a certain range. It is possible to use the A-type display for this by having an extremely directive antenna system and turning it around until the echo is at a maximum; the angle of the antenna then reveals the direction of the echoing object, or target, as it is called.

This early method has been superseded by the plan position indicator (p.p.i.) in which the time base starts from the centre of the screen and moves radially outward like a spoke of a wheel. The antenna rotates steadily and the time base deflection coils rotate in synchronism with it so that the 'spoke' moves around and covers the whole screen in each revolution. Normally the glow is almost completely suppressed by negative bias at the grid of the c.r. tube, so the rotating line can hardly be seen; echo signals are made to reduce the bias, bringing up the glow at the appropriate radial distance and at an angle which indicates the direction of the target. The phosphor of the screen has a sufficiently long after-glow to keep echoes visible until they are renewed the next time round. If there is relative movement between the target and the radar, the echo moves gradually across the screen and the relative speed and direction of motion of the target can be found.

Figure 23.26 shows a typical p.p.i. trace. The range scale is marked by a series of fixed circles. This is the normal form of display for shipborne radar and for airport surface movement control.

Figure 23.26 P.P.I. type of radar display

For military and police purposes the p.p.i. method of detecting and measuring movement is not nearly fast enough, so an entirely different technique is used. Everyone is familiar with the change in pitch of a police siren or other source of pitched sound as it passes along the street. While it is getting nearer the pitch is higher than if the source were kept at a constant distance, because each successive cycle starts from a nearer position and so has a shorter journey and arrives after a shorter interval. When the source is receding, conditions are reversed and the pitch is lowered. This is called the Doppler effect, and applies also to radio waves. A Doppler radar is designed to measure small differences in radio frequency between the signals radiated and the echoes received. A meter indicating this difference is scaled in relative speed towards or away from the observer.

For purposes such as air traffic control, the *height* of the aircraft observed is important. One way of obtaining this information is to use an antenna system that is highly directional in the vertical plane and to move it up and down ('nodding') instead of or in addition to rotating it continuously or from side to side. Figure 23.27 shows that the height of the target aircraft depends not only on the measured angle of elevation of the radar beam (φ) but also on the distance along the beam (given by the echo delay) and at long ranges is further complicated by the curvature of the earth. So determining the height is not altogether simple. The display of information presents problems too. One method is to show the echoes as spots on a graph on the c.r. tube screen, angle of elevation being plotted against range. Direction has to be shown separately by the azimuthal angle of the antenna. Another scheme is to have several beams at different fixed angles of elevation, together wide enough to cover the whole range, with some overlapping of adjacent beams. The whole system rotates continuously in azimuth, and there is a p.p.i. for each beam as well as one showing the combined echoes of them all. Each of the range zones on any of the separate p.p.i.s, between a consecutive pair of the range circles, is marked with the upper and lower limits of height of any echo in that zone, depending on the beam angle for that p.p.i.

Nearly all types of radar use radio pulses, and we have already noted that for short ranges the pulse has to be brief enough to be over by the time the nearest echo could arrive. For an object 50 m away the return journey is 100 m and is done in $1/3\,\mu s$, so the pulse would have to be shorter than that. Short pulses also enable the shape of the target to be traced in greater detail than long ones.

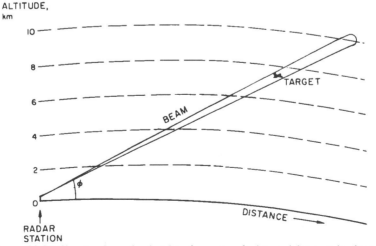

Figure 23.27 Showing how the height of an aircraft detected by a radar beam depends on angle of elevation (φ) and distance

Another basic specification of a radar system is its *pulse recurrence frequency* (p.r.f.). The p.r.f. must not be so great that more than one pulse is radiated in the time represented by the longest range. If that range were, for example, 150 km, the time between pulses would have to be at least 1000 µs, or 1 ms, so that the p.r.f. could not be more than 1 kHz.

So a moderately long-range radar might have a pulse length of 1 µs and a p.r.f. of 1 kHz. The transmitter would thus be radiating during only one-thousandth of the time it was on—a proportion called the *duty cycle*. The pulse power in this example would be 1000 times the average or mean power, which is the figure that counts in reckoning heating of the transmitter equipment. An installation of quite moderate size can radiate a pulse of the order of 1 MW—enormously more than it could do continuously. Pulse radar can thus achieve ranges of hundreds of miles even though the amount of power reflected from a small distant target is only a minute fraction of that radiated, and though only a minute fraction of the total reflected power is received back at the radar.

As in television, the shorter the pulse the higher the maximum modulation frequency and the greater the bandwidth. Even if the pulse has rounded corners instead of being square-cut the bandwidth must be reckoned as not less than $2/T$, where T is the pulse duration; for example, 4 MHz for a 0.5 µs pulse. The design of radar receivers is thus something like that of television. But the incoming signal is far weaker, and usually at a much higher radio frequency. Amplification at that frequency is usually impracticable by ordinary methods, so the signal may be fed straight into a crystal diode, which in conjunction with an oscillator does the job of frequency-changer. There will be two or three times as many i.f. stages as in a domestic television receiver, with an i.f. of the order of 50 MHz, followed by detector and pulse amplifier to bring the signal level up to that needed to deflect or modulate the c.r. tube beam. Pulse amplifiers have much in common with the video amplifiers used in television, and the design principles used in both are described in the next chapter.

The same sharply directive antenna is used for both sending and receiving, with special devices to direct the power along the appropriate feeders and to protect the sensitive receiver from the high transmitter power.

23.14 Application to Computing

At the time of writing (October 1996) most of the world's production of cathode-ray tubes is destined for use in visual display units (v.d.u.s) for digital computers, which are described briefly in Chapter 26. This application is similar to that to colour television, except that the standards are generally more exacting.

Ergonomics dictates these higher standards. While television viewers generally sit several feet from the television screen, for most computing purposes the user sits only inches away because he or she may need to read very small type or attend to minute graphical details that appear on it. For this reason the resolution is normally higher, interlacing with its attendant jitter is not used and the number of frames per second (called the refresh rate) is generally higher. If television standards were used, computer operators would suffer fatigue or eye strain after a few hours of work.

A typical display screen for a personal computer is 800 × 600 pixels with a refresh rate of 75 Hz. A *pixel* is the smallest area in a screen that can be addressed (that is, controlled) and it is usually square with a nominal size of 1/96 inch. The content of the v.d.u. display is mapped into a dedicated area of memory, called screen or video memory, in the computer. Changes to the screen are made by changing the contents of video memory. This memory is scanned by the video circuitry which processes its contents to generate a video signal in a format suitable for the v.d.u. The maximum number of colours in the display depends on the size of the memory allocated. Most modern computers allow a maximum of at least $2^8 = 256$ colours (each pixel being represented by one eight-bit byte) and some computers are fitted with sufficient video memory to allow each pixel to be represented by three eight-bit bytes, one for each of the primary colours red, green and blue. Such a screen may therefore use any of $2^{24} = 16\ 777\ 216$ colours.

23.15 Alternatives to the C.R. Tube

While the cathode-ray tube still reigns supreme in television, radar and computer v.d.u.s, it is not the only means of displaying pictures or text generated or stored electronically.

Light-emitting diodes (l.e.d.s) were described briefly in Section 10.12. Although their principal use is singly as on/off indicators, they are also used to display alphanumeric data. As *seven-segment displays*, they can present numeric data, as shown in Figure 23.28. Integrated circuits using logic elements as described in Section 26.3 are available which take a binary coded input and drive the

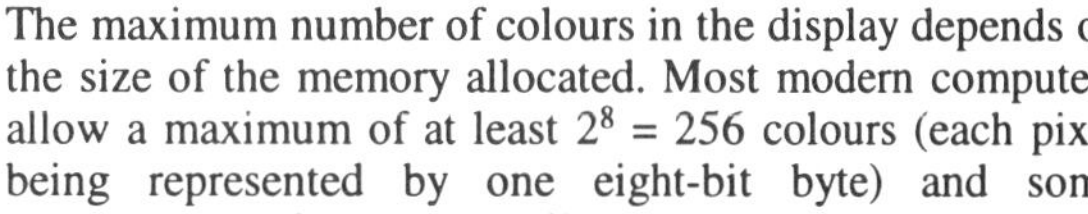

Figure 23.28 Numerals as produced by a seven-segment display.

l.e.d. segments directly to display the required numeral. L.e.d. *matrices* such as that shown in Figure 23.29a display a wide range of alphabetic, numeric and other characters. For ease of packaging and connection, in such units the cathodes of the l.e.d. elements in each row share a common connection and the anodes of the l.e.d. elements in each column share a common connection, as shown in Figure 23.29b. To display any character requires a scanning process rather like a miniature c.r.t. display. Each row of cathodes is energised in turn, the cycle being repeated typically 20 to 50 times per second. While a particular row of cathodes is energised, power is applied only to those columns whose dots in the current row require illumination. Although only one row is lit at a time, at frequencies of 20 Hz and above, to the human eye it will appear that the whole character is illuminated.

Clearly, complex logic circuitry is required to allow a matrix of this kind to display any characters at all—and this

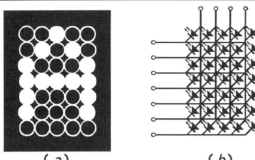

Figure 23.29 (a) External appearance of a 5 × 7 l.e.d. dot matrix displaying the letter 'A'. (b) Its internal structure showing the cathodes of the individual diodes commoned in rows and the anodes commoned in columns. This reduces the number of pins on the device to 12.

circuitry must be multiplied if a row of such matrices is being used, for instance, as a message display system. Such circuitry could be created using the pulse generators and logic gates described in Chapters 25 and 26, but dedicated driver circuitry is available in integrated circuits.

It was thought at one time that large matrices of tiny l.e.d.s might form the basis of display systems for portable television receivers and computers. One difficulty is the range of colours available. While l.e.d.s in colours from red through orange and yellow to green present no problem, blue l.e.d.s pose a challenge. Semiconductor materials that emit blue light are available, but are more expensive and require a higher operating voltage than do the other colours; moreover combining blue with red and green elements on one substrate has proved impractical.

Furthermore all l.e.d. displays suffer the disadvantage of consuming appreciable power, albeit less than that of a similarly sized c.r.t. For obvious reasons this is undesirable in battery-powered equipment. Fortunately for electronics an alternative display system was developed.

Liquid crystal displays (l.c.d.s) employ a different technology. They are used in seven-segment numeric displays in digital watches and calculators and in the larger dot-matrix displays in portable computers. Although more expensive than corresponding l.e.d. displays, their current consumption is minimal because they do not generate light; they simply block it or admit it, as required. They are used in front of either a mirror to reflect ambient light or some other light source within the equipment.

The 'twisted nematic' l.c.d.s most commonly used consist of two thin layers of glass separated by a liquid having a crystalline structure; the liquid crystal is typically 10 µm thick. On each inner surface of the glass the pattern of the required display is printed in a transparent conductive material; this, of course, is supplied with the necessary connections.

In front of the l.c.d. display is one polarised filter and behind it is another, with their planes of polarisation at right angles (Figure 23.30). Because they are at right angles, one might expect that no light from the rear reflector (or backlighting source) would reach the viewer's eye. The liquid crystal, however, has some remarkable optical properties. One is that in the absence of any electrical energisation it rotates the plane of polarisation of any light passing through it by 90°. Thus an unenergised l.c.d. display appears to be quite transparent.

When an electrical charge is applied to the liquid crystal between two of the printed areas, however, that part of the display loses its ability to change the polarisation of passing light. The action of the two polarising filters now prevents light from passing through that area, which to the viewer that area now appears to be black.

The liquid crystal is an insulator and draws no current apart from a minimal leakage current, making it ideally suitable for use in battery-powered equipment such as watches, calculators, laptop computers and portable television receivers. Colour l.c.d. displays use colour filters to provide the colour content of the picture.

A complication is that it is not practical to drive the display using d.c. voltages. A steady voltage across the display damages the liquid crystal. For this reason the drives to the printed areas are a.c., typically pulses at about 20 Hz derived from an oscillator within the equipment. Another disadvantage is the comparatively slow response time. Those who use laptop computers will be familiar with the 'trail' of pointers that lingers on the screen for a few moments after the pointer has been moved rapidly across it.

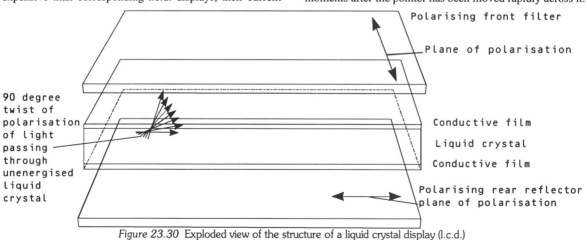

Figure 23.30 Exploded view of the structure of a liquid crystal display (l.c.d.)

Chapter 24

Non-sinusoidal Signal Amplification

24.1 Waveforms

Up to the last chapter, 'signals' were normally sinusoidal. That chapter was an introduction to applications in which sine waves hardly appear, except as radio carrier waves used to carry extremely non-sinusoidal forms. Most of these are of two main kinds. There is the pulse, which is often made to recur at regular intervals, as in Figure 24.1a. If the pulses and intervals are equal, or nearly so, as at b,

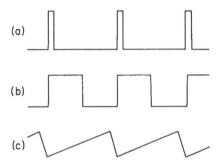

Figure 24.1 Examples of the most important non-sinusoidal waveforms: (a) narrow pulse, (b) square wave and (c) ramp or sawtooth

they are commonly called square waves, but in principle are the same. The next most important waveform is the ramp or triangle, which in recurrent form is commonly called sawtooth (c). We have already seen in Figure 23.8d an example of how other waveforms can be compounded of these two. We have also had some inkling of the ease with which the sharp corners of such ideal forms as in Figure 24.1 can be rubbed off in real circuits with their stray capacitances and inductances—even in circuits where sine waves are well preserved. On the other hand circuits suitable for pulses are quite likely to distort sine waves very seriously. So, having given so much attention to sine waves, we now need to look a little into the very different philosophy of sharp-cornered waveforms, the relative importance of which has for some time been increasing.

In spite of the very different principles and practice ruling sinusoidal and non-sinusoidal signals, there are connecting links in theory, and it is possible in practice to convert one to the other.

As far back as Figure 1.2 we saw how, by adding together several pure sine waves, we could get an obviously non-sinusoidal form. If this form is to repeat exactly, at a frequency f, it must be made up exclusively of sine waves having frequencies f, $2f$, $3f$, $4f$, etc. (but not necessarily all of them); that is to say, a fundamental (frequency f) and its harmonics. Even though Figure 1.2c does show some signs of squarishness, however, we might perhaps doubt that it is even theoretically possible to make a perfect square wave—still less a narrow pulse—out of pure sine waves. We might have even stronger doubts about the possibility of square waves produced simply by switching, as in Figure 3.5, being composed of sine waves.

Admittedly a perfect square wave, with vertically rising fronts, requires an infinite number of harmonics, but they are of steadily decreasing amplitude, so a very good approximation to a square wave can be made within practical limits of frequency. The ideal recipe is to add together all the odd harmonics only, in phase at the start, the amplitude of each harmonic being inversely proportional to its number. So if 1 is the relative amplitude of the fundamental, the amplitude of the third harmonic is $1/3$, the fifth $1/5$, and so on. Figure 24.2 shows the result of adding together a fundamental and its first seven odd harmonics (i.e., up to the 15th) in this way. The flat parts are still rather wavy; higher harmonics are needed to flatten them out.

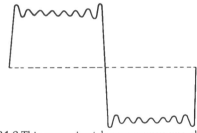

Figure 24.2 This approximately square wave was obtained by adding together a fundamental sine wave and its first seven odd harmonics in the right proportions, in phase at the start

It should now be clearer why a wide frequency band is needed to reproduce square waves without distortion. The result of passing a perfect square wave through a filter that cuts off sharply at a frequency just over 15 times its fundamental frequency would be a distorted square wave like Figure 24.2. In practice the loss of signal with increasing frequency is usually more gradual and the distorted forms not exactly like this. The main defects are: appreciable rise and fall times; 'overshoot', in which the fronts rise too far and then fall back, as in Figure 24.11b; and 'ringing'—damped oscillation, due to circuit

229

resonance. If the low frequencies (mainly the fundamental) are reduced the flat parts drop. Figure 24.3 shows all these defects.

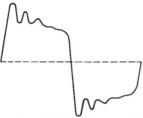

Figure 24.3 Typical result of passing a nearly perfect square wave through circuit having restricted frequency coverage

Narrow pulses are more difficult still to amplify without distortion because the proportions of the higher harmonics are greater. Triangular waves like Figure 24.1c but with equal up and down slopes, on the contrary, are easier because the harmonic amplitudes decline inversely as the square of the harmonic number.

The foregoing are just a few examples of the Fourier principle (Section 19.6) which states that any periodic (i.e., exactly repeating) waveform can be analysed into harmonic sine waves. Put in the reverse way, any periodic waveform can be synthesised by adding together sine waves, as we have been doing in Figure 24.2 for the particular case of square waves. Reference books on signals often include formulae for the most useful waveforms. As with those we have been considering, they are all unending series, but one can stop at the point where a close enough approximation has been obtained.

24.2 Frequency and Time

By now it should have become clear that there are two alternative ways with signals. They can be treated either on a frequency or a time basis. If they are sinusoidal, frequency is much the more convenient. This particular waveform is unique in having only one frequency and in not being altered by any linear circuit composed of *R*, *C* and *L*. All the work we have done (notably in Chapters 6–8) on reactance, impedance, phasor diagrams, etc., may seem to have been quite hard—and it is only an introduction to much more advanced study of this kind—but it is very simple compared with the mathematics of non-sinusoidal signals in circuits. That is one reason why the sine waveform is so much used. Reactance is based on frequency, and by enabling a.c. circuits to be calculated by an extended form of Ohm's law it provides a marvellous short-cut. So much so that when dealing with non-sinusoidal signals it may actually be easiest (as we noted in Section 19.6) to analyse them into their component sine waves, find out what happens to these separately in the circuit, and then put the results together at the end. In general the output waveform will not be the same as at the input (Figure 19.5). This procedure is like running round a tennis ball to avoid having to play it backhand. And in any case it relies on the circuit law (Superposition theorem) that the parts of a total current in a circuit can be calculated

separately, not being affected by the other parts; and this is not true of non-linear circuits. (In frequency-changers, for example, currents obviously behave differently according to whether they are together or separate.)

At the opposite extreme are simple on-off square or pulse waveforms in certain simple circuits. We have considered several of these: the charging capacitor (Section 3.5); the energising inductor (Section 4.8); the oscillator (Section 14.2); and quite recently the more complicated linear time base (Figures 23.18, 23.19).

Significantly, a warning was necessary that the actual behaviour of the last of them is a good deal more complicated than as shown. These were all treated on a time basis. Any reference to frequency was at once converted into time; pulse recurrence frequency, for example. This is very easy because frequency and time are reciprocals of one another: $f = 1/t$.

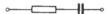

Figure 24.4 Simple combination of circuit elements used as a starting point for study of two alternative approaches to signals

Take the extremely simple circuit shown in Figure 24.4. If a sinusoidal voltage were applied to this, calculation of the resulting current and the separate voltages across *C* and *R* would obviously start with calculating the reactance of *C*, $1/2\pi fC$. That is the frequency approach. If, however, the input were a square wave we would deal with it as in Section 3.5, which would involve time. (Remember, the time constant is *CR*.) It is quite possible, but infinitely tedious, to arrive at the same result by breaking down the square wave into its sinusoidal harmonics, calculating the separate currents according to the impedance of the circuit at each of their frequencies, and adding together all the resulting waves on a base of time. On the other hand, it would be quite a puzzle to deal with the single sine wave by the method that was so easy with the square wave, in terms of the time taken to charge the capacitor, and then convert to frequency.

Because frequency and time are mutually reciprocal, high frequency means short time (per cycle). And we have been learning that the shorter the time in which a pulse occurs the wider the frequency band it occupies; if the pulse is to be perfect, an infinite band. At the opposite extreme is a perfect sine wave, which if it is to have zero frequency bandwidth (because it has only one frequency) must continue in time for ever, because any stoppage or variation in amplitude would be modulation and this introduces sidebands. These extreme comparisons are shown in Figure 24.5*a* and *b*. There is one particular waveform, occupying a central position between these extremes, which has the same shape on both bases (*c*).

These frequency–time relationships are fundamental to communications (Hartley's law, Section 23.11). Now let us look at some of the basic problems encountered in amplifying non-sinusoidal signals. The circuits to be described have applications other than the amplification of the waveforms illustrated in Figure 24.1. For example, they are used in oscilloscopes to amplify the input signal before application to the Y plates, and to amplify the time-base output to the level required by the X plates. Such amplifiers must have an excellent performance because the oscilloscope is used for precise measurements of the distortion in displayed waveforms. Another application for the circuits which follow is for amplifying video signals. The amplifiers used in radars, television studios and in video transmitting and recording equipment also need a very high standard of performance, but a lower standard can be accepted from the video amplifiers in television receivers because a small amount of certain types of distortion does not seriously degrade a television picture.

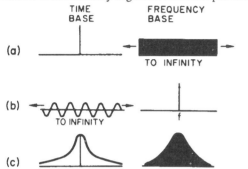

Figure 24.5 Three different signal waveforms as they appear when graphed against time (left) and frequency (right)

24.3 Frequency Range

As already mentioned, if the edges of a pulse are truly vertical the number of harmonics is unlimited and the frequency range infinite. Clearly we cannot use (or even generate) such ideal pulses in practical equipment, which necessarily has a limited bandwidth. We are therefore interested in the effect on the shape of a pulse of eliminating all the harmonics above a particular frequency.

It is to tilt the edges so that they make a small angle to the vertical and to introduce curves at top and bottom as shown in Figure 24.6. There is, in fact, a simple

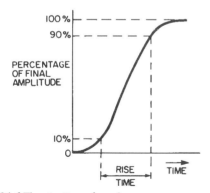

Figure 24.6 The rise time of a pulse

relationship between the steepness of the edges and the upper frequency limit. The steepness of the edges is measured by the time taken for the signal to rise from 10% to 90% of its final value. This is known as the rise time, and is related to the uppermost frequency f_{max} by the formula

$$\text{rise time} = \frac{1}{2f_{max}}$$

If $f_{max} = 10\,\text{MHz}$ the rise time is $0.05\,\mu\text{s}$. There is a corresponding fall time, which is the time taken for the trailing edge to fall from 90% to 10% of the original value.

As a good high-frequency response is needed to reproduce nearly vertical edges, perhaps one could guess that an extended low-frequency response is required to reproduce horizontal parts of waveforms. In fact, if a pulse switches between two truly horizontal levels, then the frequency response must extend down to zero frequency. In other words the amplifier must be capable of amplifying steady voltages. Such direct-coupled amplifiers are difficult to design as explained later (Section 24.6).

24.4 Importance of Phase Response

At a particular instant all the harmonics of a repetitive pulse signal are in phase so that the starts of their positive half-cycles all add to give the steep leading edge of the pulse. As these are being amplified they all suffer some phase shift as a result of the inevitable inductance and capacitance in parts and wiring. But at the output terminals the amplified harmonics must again be momentarily in phase so as to reconstitute the amplified steep edge. It is impossible to design an amplifier to have zero phase shift over a wide frequency band, and so instead we make the phase shift vary with frequency in such a way that all the harmonics have the phase relationship required to produce the near-vertical edge at the amplifier output.

To understand how this is achieved, consider the amplification of a single sinusoidal signal of frequency f and suppose that the amplifier has a phase lag at this frequency of θ. The output signal lags the input signal by $\theta°$, and we can imagine that this is the result of a process of continuous phase retardation which occurs as the signal passes through the amplifier—akin to what happens when a signal passes along a transmission line. Thus we assume that the signal takes a finite time to pass through the amplifier, and it is very easy to calculate this time. The phasor representing a sinusoidally varying signal rotates through 360° in each cycle and thus $360f°$ in each second. The time taken to rotate through $\theta°$ is thus $\theta/360f$ seconds. This is known as the *phase delay* of the amplifier at the frequency f. As a numerical example suppose the lag at 1 MHz is 60°. Then the phase delay is given by $60/(360 \times 10^6)\,\text{s} = 0.17\,\mu\text{s}$. In this way we could calculate the phase delay for all the harmonics of a pulse signal. To reproduce the pulse without distortion all the amplified harmonics must arrive at the output in the same phase relationship as they had at the input. This requires the phase delay to be the same for all harmonics. So phase delay must be constant and independent of frequency over the frequency range occupied by the harmonics. This desirable result necessitates making phase lag directly proportional to frequency. The curve of θ plotted against f should therefore be a straight line passing through the origin, as in Figure 24.7. A simple transistor amplifier with a resistive collector load is likely to have a θ/f relationship similar to that shown in dashed lines. Methods of linearising the curve are considered in the following section.

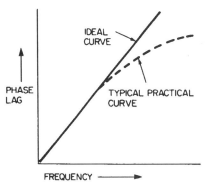

Figure 24.7 Ideal and practical phase/frequency curves for an amplifier

Thus we can conclude that to reproduce pulses without distortion the frequency response of an amplifier must be level and the phase characteristic linear over the frequency band occupied by the pulse.

For an amplifier with simple inter-transistor coupling circuits, there is a mathematical relationship between the frequency and phase response so that, given one, the other can be calculated. Good frequency response and good phase response go together, and if the frequency response is made level over the band occupied by the pulse the phase response is likely to be satisfactory too. But the frequency response should fall off gradually above the upper frequency limit. A sharp cut-off above this limit will give an unsatisfactory phase response.

We have now examined the frequency and phase response of a pulse amplifier. A third important characteristic is the input/output relationship. For an a.f. amplifier this has to be as linear as we can make it, and negative feedback is extensively employed to achieve the required degree of linearity. This is also true of pulse and video amplifiers where very low distortion is essential. As already implied, such strict standards are not necessary in the video amplifiers of television receivers. For example, if the frequency response of such an amplifier falls towards the upper cut-off frequency, the effect is to degrade fine detail in the reproduced picture, giving a soft effect described as poor definition. This can be compared with the effect of 'top cut' in an a.f. amplifier which robs the sound of its distinctness. Moreover strict proportionality between input and output signal may be undesirable. To appreciate this consider a device with a square-law characteristic, i.e., one in which the output is proportional to the square of the input. An a.f. signal which operated over the whole length of such a characteristic would produce very severe distortion at the output. But if a video signal is applied, all that happens is that signals near white level are amplified more than those near black level, i.e., contrast is increased.

Provided no limiting occurs at either end of the characteristic, the detail in the scene is unaffected and many viewers might regard the output as an improvement on the input! Often the amplitude characteristic of a video amplifier is made deliberately non-linear in order to compensate for non-linearity in the characteristic of the picture tube, the light-output/signal-input curve of which is even steeper than a square law curve.

24.5 Obtaining the High-frequency Response

There is no difficulty in choosing transistors capable of uniform service up to 5 MHz; the main problem is the coupling impedance needed to develop the output voltage. We shall see in a moment that the greatest practical impedance is generally small compared with the output impedance of the transistor to which it is connected, so the gain can be regarded as proportional to the coupling impedance. It should therefore be constant over the whole working frequency band, and large enough for satisfactory gain.

Although transformers can be designed to cover a wide band of high frequencies or from, say, 30 Hz to 30 kHz, it is not practicable to cover both at once, and even if it were its size, weight and cost would rule it out. Simple resistance coupling is the ideal solution—or would be but for stray capacitance. Counting the picture tube for which the output of the v.f. amplifier is required, the circuit wiring, and the v.f. transistor itself, this capacitance is usually in the range 15–25 pF. It is shown as C in Figure 24.8. At 4 MHz, 25 pF has a reactance of only 1.6 kΩ, effectively in parallel with

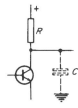

Figure 24.8 The frequency bandwidth of a v.f. amplifier is normally limited by stray capacitance, C

the coupling resistance R. If R also was 1.6 kΩ, there would be a loss of 3 dB at 4 MHz compared with low frequencies (Section 19.16). Increasing R would increase the low-frequency gain, but the comparative fall-off at 4 MHz would be all the greater. Decreasing R increases the frequency at which the reactance of C is equal to it (the 3 dB loss frequency) but decreases the gain of the stage and increases the signal current needed to set up across R the required voltage output, which may be as much as 100 V for a picture tube. To get 100 V output across 1.6 kΩ requires 100/1.6 = 63 mA, which is quite substantial.

The fall-off point can be shifted to a higher frequency, or the allowable R increased, by counterbalancing C by means of inductance. The inductance (L) cannot be connected directly in parallel with C, because at low frequencies its impedance would be so low that there would be very little gain. This objection can be avoided by putting L in series with R, but calculations are more difficult because the equivalent parallel inductance and resistance vary in rather a complicated way with frequency. A useful rule is that by making $L = CR^2/2$ the fall-off at the -3 dB point is reduced to zero. You might perhaps guess that this is the frequency at which C is exactly tuned out by L, but actually there is no such frequency. A substantial part of the benefit is due to the fact that at the frequencies where C appreciably shunts R the presence of L makes the equivalent parallel resistance higher than R. The equivalent parallel inductance does not completely neutralise C at any frequency.

Suppose C is 20 pF and that in the hope that the scheme just described will work we make f_t rather lower than the maximum required frequency—3 MHz, say. $R = 1/2\pi f_t C$, = 2.65 kΩ. So $L = 20 \times 2.65^2/2 = 70\,\mu\text{H}$. Figure 24.9a shows the calculated high-frequency loss without L, and we note that it is 3 dB at 3 MHz, as planned. Curve b shows the result of inserting the calculated value of L, and we see that the -3 dB point has been moved up to about 5.5 MHz. At 3 MHz the expected zero loss is achieved. At slightly lower frequencies there is a flat peak; it is not really enough to worry about, but if you insist on what is called a maximally flat response the choice of L is $0.414CR^2$.

In some wide-band amplifiers inductance is used between the output and input capacitances as shown in Figure 24.10, instead of or in addition to the position just considered, and with or without parallel resistance.

For oscilloscopes and some other purposes a top frequency of 30 MHz or even higher is needed. Use of

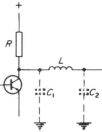

Figure 24.10 An alternative position for the h.f. loss-reducing inductor in a v.f. amplifier. It has the advantage of separating the stray capacitances C_1 and C_2

inductance in both the positions mentioned helps; apart from that the general policy is to reduce the effect of stray capacitances by reducing the coupling resistance. To retain a useful amount of gain, the mutual conductance of the amplifying device must be made high, and in general this means a high output current at the working point. So although the power output required from a very wide-band amplifier may be quite small it is likely to need power types of device and means for getting rid of quite a lot of heat. This is not easy to combine with the necessary very high f_T.

We have dealt with this problem in terms of frequency, but, as pointed out at the beginning of this chapter, in television, radar, pulse telegraphy, etc., we are really more concerned with time—the time taken to jump from black to white on the television screen, for example. As already

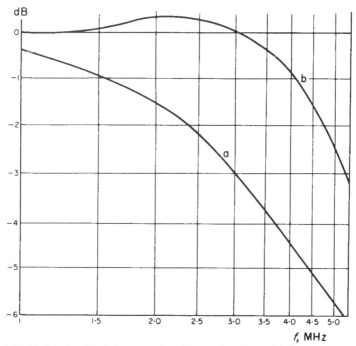

Figure 24.9 (a) Calculated high-frequency loss (due to C in Figure 24.8) in an example of a v.f. amplifier. (b) The result of inserting a suitable amount of inductance in series with R

shown this is measured by the rise time and, from Figure 24.11, we can see that the effect of adding the inductance is

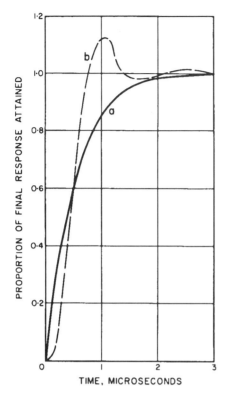

Figure 24.11 As an alternative to the frequency curves in Figure 24.9, the performance of a v.f. amplifier can be expressed on a time basis. The use of inductance is seen here to result in slight overshoot

to reduce the rise time from 1.15 µs to 0.50 µs. This would result in a considerable improvement in picture definition. There is, however, almost 10% overshoot. This would not seriously spoil a television picture, but would disqualify the amplifier for an oscilloscope (Section 23.4) intended to reveal just such shortcomings in transient signals such as pulses.

In place of the inductor in the collector circuit it is possible to apply negative feedback by means of a capacitor in the emitter circuit, and this has the double benefit of improving linearity as well as extending the high-frequency response. All that is necessary is to reduce the value of the bypass capacitor normally connected across the emitter bias resistor (Figure 24.12). At low frequencies the reactance of the capacitor is so high that it does not significantly shunt the resistor, which therefore gives current negative feedback, reducing gain and distortion. But at higher frequencies where the shunt capacitance in the collector circuit begins to take effect, it is arranged that the reactance of the emitter capacitor becomes comparable with the emitter resistor. In this way the feedback is progressively removed, so extending the high-frequency response. So

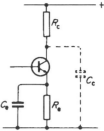

Figure 24.12 Use of frequency-discriminating current negative feedback to compensate for the effects of collector stray capacitance

here we have one RC combination offsetting the effect of another, and it is no surprise therefore to learn that optimum results are achieved by making the two time constants equal. Using the collector component values quoted above, 2.65 kΩ and 20 pF, let us suppose that the emitter resistor is 400 Ω. Then the required emitter capacitance is given by

$$C_e = \frac{R_c C_c}{R_e} = \frac{2650 \times 20}{400} = 132.5 \text{ pF}$$

The voltage gain of the amplifier is given approximately by R_c/R_e, i.e., 6.6, rather less than that of the shunt-inductance amplifier, but of course the feedback amplifier has better linearity and a higher input resistance, which could be useful if the amplifier directly followed the vision detector.

24.6 Operational Amplifiers

The shunt-injected feedback circuit (Figure 19.22) was treated in Section 19.17 as a current amplifier. In practice a signal current source takes the form of a voltage source in series with such a high impedance that the current is determined almost entirely by it rather than by whatever circuit it works into. If this high impedance takes the form of a resistance that is a permanent part of the amplifier, the basis is laid for an interesting and important class of feedback voltage amplifiers.

Figure 24.13 is essentially Figure 19.22 plus this resistance, R_1. If the current gain is sufficiently large the

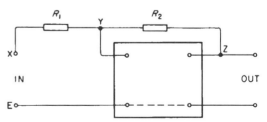

Figure 24.13 The simplest form of operational amplifier circuit

part of the input current that goes into the amplifier at Y is negligible compared with what goes through R_2 to the 'live' output terminal, Z. For the feedback to be negative, the potential of Z has to be opposite to that of Y (relative to E). And because of the high gain of the amplifier it must be very much larger. The only way that these conditions can be fulfilled is for the signal potential to vary as shown in Figure 24.14. If V_x alternates, the line see-saws about point P as the fulcrum, so M. G. Scroggie named Figure 24.13

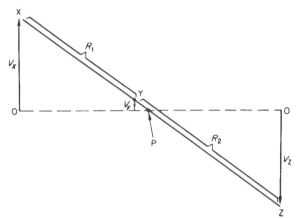

Figure 24.14 Showing the distribution of potential along R_1R_2 in Figure 24.13

the see-saw circuit. If the voltage gain of the amplifier is large enough, the position of Y on this diagram is so near P as to be hardly distinguishable from it and we can say that

$$V_z \simeq -\frac{R_2}{R_1} V_x$$

So the first use of such an arrangement is as a voltage inverter in which the output is almost as nearly equal and opposite to the input as R_2 can be made equal to R_1.

Next, a voltage can not only be inverted but reduced or enlarged by choosing R_1 and R_2 in the desired ratio.

Because these results depend negligibly on the characteristics or changes thereof in the amplifier (always assuming its gain is large), this sort of system is used in computers of the analogue class; that is to say, those in which voltages or currents represent the data in the calculations. They come up for attention in Section 25.2. Because amplifiers of this type can be used for performing mathematical operations they are usually known as *operational amplifiers*.

Another technical term arises from the fact, which Figure 24.14 shows so clearly, that point Y is always very close to zero or earth potential without actually being connected to it. It is therefore called a *virtual earth*, and has found a number of useful applications in circuitry.

Operational amplifiers usually have to work down to zero frequency, so must be direct-coupled throughout. This immediately raises a difficulty. The gain of the amplifier having to be large, every little variation in d.c. conditions in the first stage, due to temperature changes especially, is amplified and will confuse the results. Two matched transistors are likely to vary to nearly the same extent, so the difference between them will be relatively small. Z.f. amplifiers therefore almost always have at least a first stage

that responds to the *difference* between two signals (one of which of course can be zero) and this adds considerably in other ways to its usefulness as an operational amplifier. Figure 24.15 shows the most used differential (or balanced)

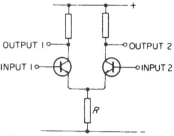

Figure 24.15 'Long-tailed pair' balanced amplifier circuit

amplifier stage, usually called the long-tailed pair, the 'long tail' being the high resistance R. It should have as high a resistance as possible, and to save negative supply voltage an electronic substitute for a high resistance (a 'constant-current source', Figure 28.9) is sometimes used. The idea is to keep the total current to both transistors virtually constant, so that increasing it by a positive signal at one of the inputs communicates via R an equal and opposite signal to the other input. Equal signals of the same polarity at both inputs (called common-mode signals) therefore have very little effect. False signals due to such things as temperature changes are usually of this kind. Ability to distinguish between difference signals and common-mode signals is called *common-mode rejection*. A useful feature of the long-tailed pair is that the second input provides a convenient point at which to inject a negative-feedback signal.

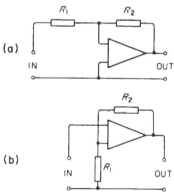

Figure 24.16 Conventional representation of (a) inverting and (b) non-inverting balanced operational amplifiers

Amplifiers are often represented in diagrams by triangles, as in Figure 24.16, in which the direction in which the signals get larger is represented by the direction in which the width of the triangle gets smaller. This seems illogical until we realise that the triangle can be regarded as an arrow head pointing in the direction of signal flow. Balanced amplifiers have two inputs along the base line as shown, and the common output is at the point. (Normally only one of the two available outputs in Figure 24.15 is used.)

Except for its balanced input, Figure 24.16a is the same as Figure 24.13, and Figure 24.16b is a non-inverting version of it. R_1 is still in the input circuit, but it is also in the output circuit, so instead of the overall gain being $-R_2/R_1$, it is $(R_2 + R_1)/R_1$. If R_2 is made zero, we get a 1 : 1 non-inverting condition. Except that the gain is not large enough for operational purposes because it has only one stage, emitter followers are included in this form. Among other things, it can be used to make one point in a circuit maintain virtually the same potential as another to which it is not connected. (The virtual earth is a particular example.) In this way the effect of an unavoidable capacitance can sometimes be circumvented by preventing appreciable differences of potential from occurring across it.

24.7 Integrated Circuits

Because operational amplifiers consist of resistors and silicon transistors and diodes, and involve comparatively low power, the *integrated circuit* form of manufacture is ideal for them. A whole batch of identical amplifiers is laid down on a slice of silicon, the various p and n regions being produced by successive exposures to appropriate vapours which diffuse into the material. The conductive paths for resistors and 'wiring' are formed by metal deposits. The patterns of these actions are controlled by microscopic masks obtained by photographic reduction. Silicon dioxide prevents diffusion, so is laid down on the unmasked areas and can later be removed in preparation for the next process. The fact that pairs of transistors can be laid down very close together in identical circumstances minimises differences in performance that could result in common-mode effects.

A complete amplifier may occupy only one or two square millimetres. In fact miniaturisation has now reached the point where a single integrated circuit or 'silicon chip' (as the non-technical public call them) can contain as many as 2 000 000 transistors, not to mention passive components. The capital cost is large but the unit cost is small, so if large numbers are required they can be made cheaply. Inductors and large capacitors cannot be included, but reactances can be simulated by transistors with feedback phase-shifted by C and R (Section 19.21), and we have seen how negative feedback can be used to obtain precise performance from imprecise active devices. Design philosophy for integrated circuits is therefore quite different from that appropriate to circuits made up of separate components connected together with wire. But the principles on which they work are the same, so are covered by this book and there is no need to spend any more time on integrated circuits (i.c.s) as such.

In practical applications i.c.s usually have to take their places as components in systems where components of other kinds must also be included. If the minute size of the i.c.s is not to be wasted, these other components should be as small as possible. An example of this is the ceramic filter described in Section 22.7. For compactness, stability and ease of production the wiring usually takes the form of a copper sheet laid down on insulating material (*substrate*) and eaten away by acid to the desired pattern (*printed circuit*) rather like the i.c. itself but on a large scale. On this substrate the components, including the i.c.s, are mounted. Inductors of low value can be 'printed' on the sheet as spirals.

24.8 Analogue Computers

Operational amplifiers are the basic components of *analogue computers*, which were popular with engineers until they were all but superseded by digital personal computers (described in Chapter 26). Because operational amplifiers are available as microscopic integrated circuits, equipments using large quantities of them are relatively inexpensive and occupy little space.

In analogue computers voltages are used as analogues of whatever numerical data are being processed. These voltages may be derived directly from transducers on, for example, the industrial plant being controlled. An inverter amplifier (Figure 24.16a) changes the mathematical sign; a potential divider divides; a potential divider in the negative feedback loop of an operational amplifier multiplies. If several input voltages are applied, each through its own resistance equal to R_1 in Figure 24.16a, the total input current at Y is equal to the sum of the separate input currents, so the output voltage is equal to minus the sum of the applied voltages. That gives addition—and subtraction, as negative voltages can be applied. Factors can be introduced by using different values of R_1. Differentiation is accomplished by substituting a capacitor for R_1 and integration by substituting a capacitor for R_2 (Section 25.1). By suitable combinations of operational amplifiers, differential equations are almost instantly solved. Waveform generators or shapers (described in Chapter 25) are sometimes used to introduce mathematical functions corresponding to them.

Although inherently less accurate than digital computers, analogue computers were nevertheless useful for engineering purposes, because few engineering data are known, or need be known, very precisely (think, for example, of the inevitable tolerances in transistors, and the uncertainty of maximum loading on a structure); quantities such as temperature, pressure and velocity can easily be converted into proportional voltages or currents instead of first into numbers; and the effects of varying the data can be observed continuously. Instead of laboriously calculating for several alternative values of each parameter and choosing the best combination, the engineering designer could vary them and see the results at once, even if the mathematical relationships were complex. Or he could hand the whole problem to the computer and get it to indicate the optimum specification.

The basic disadvantage of analogue computers is that any departure from exact proportionality of the voltages to the quantities represented by them introduces error. It is only the use of large amounts of negative feedback which, by nearly eliminating the effects of variations in the active devices in the operational amplifiers, keeps such errors within reasonable limits.

Chapter 25

Electronic Waveform Generators and Switches

25.1 Differentiating and Integrating Circuits

In the previous chapter we introduced the basic non-sinusoidal waveforms and we discussed the problems of amplifying them. In this chapter we continue the story of these waveforms by describing circuit techniques commonly used with them. We also consider methods of generating non-sinusoidal waveforms and introduce switching circuits—a most important type of circuit because they are used in vast numbers in, for example, digital watches, pocket calculators, video games and computers.

We begin by considering differentiating and integrating circuits. These names may seem rather formidable but we are already familiar with the things themselves. The names refer to mathematical operations. If y is some quantity that varies, differentiating is finding the *rate* at which it varies, often (but not necessarily) with time. Integrating is the opposite process: given the rate at which y varies, find y. In Figure 25.1 on the top line we have a y that varies in a whole range of different ways for our instruction. Below it is plotted the rate at which y is varying; the mathematical shorthand for this, when the variation is with respect to time (t) is dy/dt. From a to b, y is increasing at the same steep rate all the time, so dy/dt has a large positive value. From b to c, y is not changing at all, so dy/dt is zero. From c to d it is decreasing steadily, at a slower rate than the increase from a to b, so dy/dt is less and negative. de and ef are curves of the type we found in charging capacitors and switching on inductors, and the interesting thing is that dy/dt has the same shape, though obviously it must be upside down in this case. Lastly y goes through a complete sine wave. We have already dealt with this one in Section 6.2, so we can finish the job by completing the lower line showing that dy/dt is also a sine wave but shifted quarter of a cycle to the left. This is called a cosine wave. Here we get yet another angle on the unique status of this particular waveform; it is the only repetitive one that retains its shape (though not its phase) when differentiated. And, since integrating is the reverse of differentiating, integrating dy/dt in Figure 25.1 (lower line) gives the upper line, in which the sine wave still appears but shifted quarter of a cycle to the right. Recalling Figure 6.4 we can see that capacitance can truly be said to differentiate the applied voltage by the current it accepts. In the same way, looking again at Figure 7.2 we can say that inductance integrates. On the other hand if we start with current we see that, as regards the terminal voltage it sets up, capacitance integrates and inductance differentiates. They do this to any waveform, repetitive or not, but only the sinusoidal repetitive shape and the 'growth' (exponential) shape (e.g., Figures 3.6 and 4.10, and 25.1d to f) are preserved. Other shapes (e.g., Figure 25.1a to d) are drastically changed. We notice especially that our two basic non-sinusoidal shapes, pulse and ramp, are directly related to one another by differentiation and integration.

All this being so, the reason why circuits containing capacitance and inductance are in certain circumstances called differentiating or integrating circuits should now be clearer.

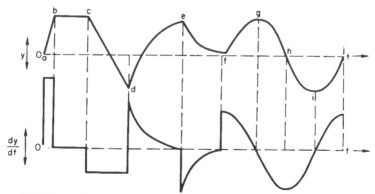

Figure 25.1 The top line shows various ways in which a quantity can vary; below are the corresponding rates of variation

We have already come across some examples here and there. A good one was the line time-base generator (Section 23.10) in which the requirement was a steadily growing current in a pair of coils. The method was to apply a constant voltage (Figures 23.18, 23.19). This is therefore an integrating circuit—getting Figure 25.1a to b (upper line) from the lower line.

Except in this kind of example, where inductance is an essential component, capacitance is usually more convenient. We have just seen that either C or L can both differentiate and integrate. So commonly C is preferred for both purposes. For differentiation, the input is voltage and the output is current; for integration, vice versa. This is all very well if one can accept the change from voltage to current or from current to voltage. But often one wants to keep it all in voltages. A current can be transformed into a proportional voltage by passing it through a resistance. But unfortunately the circuit to which the input voltage is applied is no longer just C but is C and R (Figure 24.4) so it is no longer perfect either as a differentiator or as an integrator.

In Figure 25.2 square voltage waves (V) are applied to this CR circuit. If R is relatively small, then the current will be very much the same as it would through C alone; that is to say, V_D will be an approximate differential of V. If on the contrary R is relatively large the current through it (and therefore through C, which also is large so that its reactance

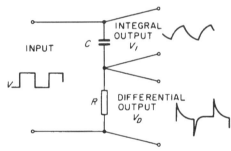

Figure 25.2 Showing the tendency towards integration and differentiation of the CR combination (Figure 24.4)

is small) will be an approximate copy of V, and V_I therefore approximately the integral of V. For differentiating, then, CR (the time constant) should be small, and for integrating it should be large, compared with the time of a cycle of V. In Figure 25.2 CR is medium, being equal to a cycle, so neither function is done well. Ideally, V_D should correspond only to the varying parts of V—the steep rises and falls— and V_I should have steady slopes corresponding to the flats in V. Figure 25.3 shows how the output waveforms depend on the ratio of CR to the square-wave cycle time, T.

An important point to remember is that the sum of V_I and V_D must at every moment be equal to V; obvious, one would suppose, but the necessity is surprisingly often overlooked by students drawing the supposed output waveforms corresponding to given inputs in actual circuits.

Note too that a differentiating circuit tends to discriminate in favour of high frequencies, and an integrating circuit in favour of lower frequencies, so they are often used singly or in combination as filters; for example, Figure 18.10.

The inferior performance due to the presence of an R in the circuit can sometimes be tolerated. We came across an example of this in the CR circuit for sorting out the sync. pulses in a television receiver; the typical waveforms in Figure 23.17 are similar to those in Figure 25.3 and elsewhere.

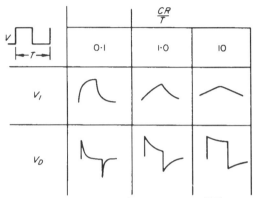

Figure 25.3 How the ratio of time constant (CR) to signal period (T) affects the waveforms in Figure 25.2

A good standard of integration is often wanted, because it is the basis of time-base generators. The curves for V_I in Figure 25.3 show the dilemma: the better the linearity the less the amplitude, and a fairly large amplitude is needed for time bases. Some of the ways of dealing with this problem are included in the next section, but meanwhile let us look at a most important general method for obtaining a high standard of integration and differentiation; so good that it is used in analogue computers (Section 24.8) for performing these two essential mathematical operations. It makes use of two more varieties of the operational amplifiers to which we were introduced in Section 24.6.

There, we saw only the types in which the input and feedback impedances were resistances. In Figure 25.4 a capacitance C takes the place of the feedback resistance. As

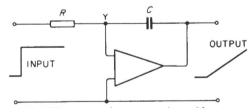

Figure 25.4 This type of operational amplifier is a very precise integrating circuit

before, the gain of the amplifier (represented by the triangle) is sufficient for the input voltage to be relatively very small; so much so that the input point Y is a 'virtual earth', because its potential is hardly distinguishable from earth's. So if a constant input voltage is imposed (indicated by the step input 'waveform') a constant current flows, determined by R. Only a minute part of this current is needed by the amplifier, so the constant current flows into C and the potential across it therefore rises linearly, as indicated at the output. In this way the spoiling effect of resistance in series with C is virtually removed.

Suppose now that the whole thing is reversed; the output becomes the input, to which a linearly rising voltage is fed, as in Figure 25.5. This input, applied to C with its other terminal virtually earthed, ensures a constant input current.

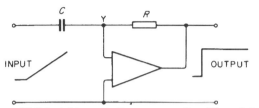

Figure 25.5 And this one is the corresponding differentiating circuit

Flowing through R, this current sets up a constant output voltage, which of course is the differential of the input; and similarly any other input form is differentiated.

25.2 Waveform Shapers

The raw material for most non-sinusoidal waveforms—and sometimes sinusoidal ones—is the pulse or square waveform. It can be produced either by reshaping existing signals or by generating them direct by special oscillators. We consider shapers first.

Available signals are likely to be more or less sinusoidal—the mains supply, for example, or the output of an oscillator—and if the frequency is too low it can be multiplied (Section 19.5). They can easily be converted into nearly square waves by passing them through an amplifier that very readily overloads, or limits. Figure 25.6 shows how simple this is. The positive half-cycles of the output voltage are all removed because the input voltage is then

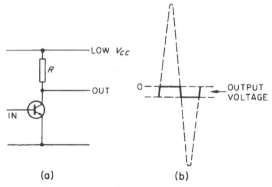

Figure 25.6 Simple type of circuit (a) for obtaining a nearly square wave (b) from a sine wave

negative and allows only leakage current (negligible with silicon transistors) to flow through R. By using a low value of V_{CC} and plenty of input signal the negative half-cycles are nearly all removed by bottoming. However much the amplifier limits, the fronts and backs of the squares obviously must slope to some extent, and the corners are rounded a little by the less than perfectly sharp cut-off in each direction. These defects can be reduced by another stage of amplification. What can be done in this way is limited by circuit and transistor capacitances; put another way, if a really high standard of pulse shape is needed, the circuits that handle it must be of the wide-band v.f. type (Section 24.5).

With the simple arrangement in Figure 25.6a the negative half-cycles of output will be slightly briefer than the positive, because both lie on the same side of the zero line (here, at V_{CC}); this can be corrected by a suitable amount of bias. Different positive/negative (or *mark/space*) ratios than 1:1 can be obtained by biasing nearer to positive or negative peaks, but one cannot go too near without impairing the waveform.

So if a large ratio is wanted—short pulses—the preferred scheme is to follow the limiter by a differentiating circuit and then bias off the unwanted pulses, exactly as described in connection with Figure 23.17 (and, incidentally, in the pulse-counter type of f.m. discriminator, Section 18.19); see Figure 25.7. In the output stage, bias and bottoming are used to cut off not only the pulses of unwanted polarity but

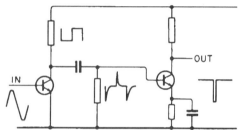

Figure 25.7 Circuit for producing narrow pulses from other waveforms

also for topping and tailing the wanted ones. In situations where clipping off unwanted portions of waveforms is needed without also amplification, suitably biased diodes are often used (Section 25.8).

This pulse shaping is often used with a view to producing the sawtooth form, for an oscilloscope time base for instance. In the preceding section we saw that if a constant voltage or current, corresponding to the flat parts of pulses or square waves, is integrated the result is a linear slope. The television line time base system (Figures 23.18, 23.19) gave us a practical example with a current output. A capacitor is an integrator, but the problem is to keep the charging current constant without using such a large resistance in series that the required input voltage is unreasonably high. What is needed is something that behaves as a very high resistance but does not need a correspondingly high voltage to pass current through it.

F.e.t.s can be used for this purpose because, as we saw in Section 10.10, they have a considerable range of anode voltage in which the current varies little; yet the amount of current, which together with C determines the rate of voltage rise across C, is easily controlled by gate bias. And when studying the characteristics of bipolar transistors in the common-base mode (Section 12.10) we found that collector current is remarkably independent of collector voltage, as indicated by its very high v_c/i_c resistance, at very low total voltages. So if a capacitor C is connected as in Figure 25.8 and is charged at a comparatively rapid rate by

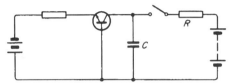

Figure 25.8 Circuit illustrating principles of ramp waveform production by charging and discharging capacitor

closing the switch, opening it allows C to discharge linearly through the transistor. The sequence is then repeated continuously to yield a sawtooth voltage waveform across C. In practice the switch is another transistor controlled by a pulse waveform.

In one respect transistors leave a little to be desired for voltage outputs: the maximum voltages that can be used are low. And step-up transformers for waveforms of this type are not easy to design or make. However, transistors are being developed all the time with higher working voltages.

Operational amplifiers offer a very precise solution, but although their necessarily high internal gain is needed for such applications as computers, a single stage may do as much as is needed for normal time-base purposes. And it gives the inversion required for the essential negative feedback. And so we get a type of circuit that is often called the Miller integrator but which is more appropriately named after its inventor, Blumlein. Figure 25.9 shows a

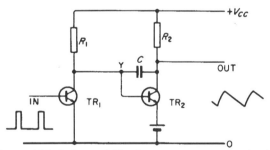

Figure 25.9 A form of Blumlein integrator circuit, closely related to Figure 25.4

transistor version, complete with pulse-operated switch TR_1. During the input pulse TR_1 is 'turned on' and is in the bottomed state, so its collector and the base of TR_2 and the terminal of C connected to it are all held at a potential only slightly above 0, while TR_2 being cut off allows the other terminal of C to be practically at V_{CC} volts. C is therefore charged to nearly that voltage. When the pulse ceases, TR_1 is cut off and TR_2 is turned on, so C discharges through R_1 and TR_2. In the ordinary way the rate of discharge would fall off exponentially as the voltage of C fell, but owing to negative feedback Y is a virtual earth so the current through R_1—and therefore the discharge current, except only for some small variation in the base current of TR_2—is kept constant. To provide for a quick flyback, R_2 is much less than R_1.

25.3 Waveform Generators: Relaxation Oscillators

If a sinusoidal waveform is not required in addition to others developed from it as described in the last section, there is no point in carefully designing an oscillator for producing shapely sine waves only to distort them drastically. Oscillators can be designed to produce square waves or pulses directly; in fact, that is what an oscillator tends to do anyway if it is given its head. Besides dispensing with at least one stage, this method more readily yields a useful range of mark/space ratios.

When considering sine-wave oscillators we found that the most important conditions for least distortion of this waveform and for most precise frequency maintenance were the same: to use only just enough positive feedback to keep oscillation going. It follows that direct pulse generators, which depend on excessive positive feedback, are in general not so good at keeping themselves within very close frequency limits. That does not matter if, as in television, they can be synchronised from without. There is a whole series of types of pulse generator ranging from the completely independent to the completely dependent shaping types already discussed:

1. Independent free-running oscillators establishing their own frequency.
2. Oscillators which without control would operate in group 1, but which are pulled into step by external synchronising signals. (1) and (2) are both classed as *astable* systems.
3. Oscillators which cannot run continuously on their own but which on being triggered from an external source will execute one complete cycle and then wait until the next triggering signal. This class is termed *monostable*.
4. Oscillators similar to (3) except that when triggered they execute only part of a cycle and need another signal to make them complete it. These are called *bistable*. Although so dependent on outside sources, they do at least determine the duration and waveform of their part-cycles, and in this important respect differ from waveform shapers.

All these groups of oscillators are distinguished from the sine-wave variety, in which the transistor works continuously as an amplifier, as *relaxation* oscillators, in which it works intermittently as a quick-acting switch. The frequency is determined not by an LC tuned circuit but by delays due to time constants, either CR or L/C (Sections 3.5 and 4.8) or both. It is the relaxation or gradual discharge of the time-constant elements between the active or switching phases that gave rise to the name for this class of circuits. For an understanding we must once again keep in mind the

Electronic Waveform Generators and Switches

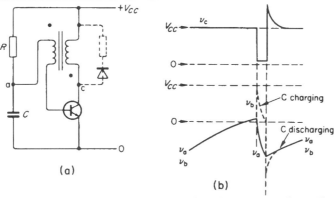

Figure 25.10 Blocking oscillator circuit (a), and relevant waveforms (b)

fundamental principle that the voltage across a capacitor cannot change abruptly but only as a result of current flowing into or out of it for a time, and its dual principle, that the current in an inductor cannot change abruptly but only as a result of voltage across it for a time. Although all relaxation oscillators are basically the same in principle, having a large amount of positive feedback coupling through a time-delay circuit, one particular variety—the *blocking oscillator*—is usually distinguished from the rest.

25.4 The Blocking Oscillator

The distinguishing feature is inductive coupling, which makes blocking oscillators look deceptively like the ordinary oscillators described in Chapter 14; see, for example, Figure 25.10. Attempts, in the interests of brevity, to explain their working along the same lines as ordinary oscillators are likely to mislead, for it is really quite different.

The cycle of oscillation comprises two phases, one of them being a small fraction of the whole period, in which a lot of things are happening very quickly, so let us begin with the other by supposing that C is charged so as to make point a in Figure 25.10a negative. As this point is connected through R to $+V_{CC}$, C gradually discharges as shown by the exponential curve of v_a, which would eventually carry it to the V_{CC} level. The transistor being cut off by this negative bias, the collector potential v_c is the same as V_{CC}, while V_b is the same as v_a. Directly v_a becomes slightly positive, collector current begins to flow through the primary coil of the transformer. The rate of current increase through its inductance sets up a voltage across the coil. Assuming that the transformer has 100% coupling and 1:1 ratio, an equal voltage must appear across the secondary coil. The dots indicate the terminals having the same polarity; at this moment, negative.

So the increasingly positive potential of the base gets a sharp boost from the transformer, which intensely accelerates the action just described. The voltages across the two transformer coils are now equal to V_{CC} less the small collector-to-emitter voltage in the bottomed state. The secondary voltage recharges C through the base-to-emitter resistance, which being much less than R accounts for the relatively steep downward slope of v_a and v_b.

Directly v_b drops back near to 0 it starts to reduce collector current, and this reduction induces transformer voltages of the opposite polarity. Again there is the intensely accelerated action, this time switching off the transistor. The very fast decay of current makes the induced voltage v_c highly positive; so highly, in fact, that a diode and resistor are sometimes connected as shown dashed to save the transistor from being broken down. (Compare the action of a television line oscillator, Section 23.10.) The secondary voltage now makes the base more negative even than v_a, but as it is already negative and non-conducting this is not very important. The high induced voltage has charged stray capacitances, and when it ceases (as of course it very quickly does) these discharge as shown by the downward curve of v_c and the upward curve of v_b.

The exact waveforms depend on many details of transformer and transistor, and in general are less sharp than those shown in Figure 25.10b. Correspondingly the action is far more complicated than the foregoing simplified description. But the main features are present: a pulse of v_c, which can be made as brief as a small fraction of 1 µs by suitable choice of transformer and transistor and C, followed by a relatively long interval determined by CR, during which v_a and v_b are the first half of an exponential curve. The p.r.f. can therefore be conveniently controlled by varying R.

Because R tends to bias the transistor on, this form of blocking oscillator is free-running. But if a synchronising signal is injected at point a, just before the pulse is due, v_b crosses the zero line at once and triggers the transistor into pulse action. To ensure synchronism the free-running p.r.f. is adjusted to be slightly lower than the sync. pulse frequency.

Note that oscillators such as this, whose frequency depends on a time constant, can be synchronised not only to the frequency of the sync. pulses but to submultiple frequencies. Figure 25.11 shows the graph of v_b again, with regular sync. pulses superimposed. It is not until the fifth of these that the voltage is able to rise above zero and start the pulse phase. The oscillator is here acting as a frequency divider.

This circuit or variations of it is commonly used in television receivers to generate the field scanning

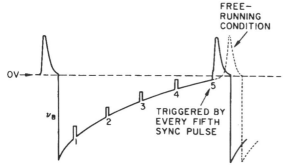

Figure 25.11 Showing combined synchronising and frequency-dividing action

waveform. It must be followed by an output stage, with provision for linearising the sawtooth.

The circuit shown can easily be converted into a monostable one by connecting R across C, or in some other way arranging that the normal base-emitter voltage is zero, so that no action takes place until a triggering pulse is received, of the polarity that turns the transistor on.

25.5 The Multivibrator

The blocking oscillator is a remarkably simple device for obtaining very brief pulses. Transformers do not, however, fit well into modern circuit construction (Section 24.7) so for square waves or longer pulses the choice is usually a circuit in which the signal inversion is obtained by a second transistor. The only other components needed are resistors and capacitors, so a whole pulse generator can be made up as an integrated circuit on a tiny chip. There is a wide choice of circuits, and a wide choice of books about them, so only a very brief introduction is included here.

The most important basic circuit is shown in Figure 25.12 and is still known by its original name of *multivibrator*, given to it in 1918 because it was found to be a convenient source of many harmonics for frequency multiplying purposes. Hundreds of them can be detected if care is taken to obtain a sharp-cornered waveform by minimising stray capacitances. It has since been developed mainly for its waveform as a whole rather than for the separate harmonics. One can easily see that it is just a two-stage resistance-coupled amplifier with the whole of its output fed back to the input. If A is the total gain, then the amount of positive feedback is A times as much as is needed just to maintain oscillation, and as A can easily run into thousands there is a very fast switch-over from each half-cycle to the next, and therefore very steep wavefronts.

When V_{CC} is applied, the inevitable absence of perfect symmetry will ensure that one transistor is slightly ahead in passing collector current. Suppose it is TR_1. Then v_{c1} will be slightly more negative than v_{c2}, and as v_{c1} via C_1 immediately makes v_{b2} more negative, tending to switch TR_2 off and prevent v_{b1} from going negative, TR_1 becomes firmly on and TR_2 firmly off. This is where we come in at the start of Phase 1 in Figure 25.12.

Because TR_1 is bottomed, v_{c1} is only just above zero. And because C_1 has had no time to charge, v_{b2} is pushed down to the same extent. Although in doing so it cuts off TR_2, v_{c2} cannot instantly spring up to $+V_{CC}$, because C_2 has to charge through R_2 and TR_1 (base to emitter). So v_{c2} rises exponentially as shown. v_{b1} is thereby pushed positive, but only slightly and temporarily because of the charging of C_2. When that is complete v_{b1} is held at a small positive bias by current via R_4. This fall-off in v_{b1} reduces the TR_1 collector current slightly, so there is a slight rise in v_{c1}. Meanwhile C_1, held on the left at the near-zero v_{c1}, is charging through R_3, so v_{b2} rises exponentially. Assuming, as is usual, R_3 and R_4 are considerably greater than R_1 and R_2, this growth is comparatively slow.

The end of Phase 1 is brought about by v_{b2} (similarly to v_b in Figure 25.10) becoming positive, starting current in TR_2 and thereby a negative-going v_{c2}, which is communicated by C_2 to the base of TR_1, making v_{c1} rise. The amplification ensures a rapid reversal of conditions, TR_1 being switched off and TR_2 on. By the end of Phase 2

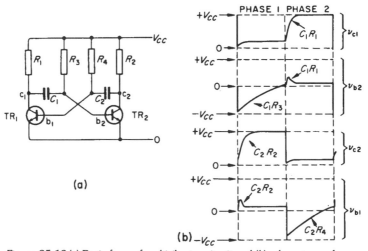

Figure 25.12 (a) Basic form of multivibrator circuit and (b) relevant waveforms

a complete cycle has been accomplished and the whole process repeats continuously.

Phases 1 and 2 being equal indicates that the component values are the same for each half of the circuit, or at least that $C_1R_3 = C_2R_4$, these time constants clearly being mainly responsible for the durations of the phases. Any desired mark/space ratio within reason can be obtained by choosing component values to make these time constants suitably unequal.

Like the blocking oscillator, the multivibrator is naturally free-running but can easily be synchronised by pulses injected almost anywhere in the circuit—usually at one of the bases. Both types of circuit are also self-generators of non-linear ramps. By making R_1 and R_2 as low as possible, a nearly square waveform can be obtained from either collector, and if necessary it can easily be improved by shapers (Section 25.2).

25.6 Bistables

If the two capacitors in the astable multivibrator are replaced by the base-biasing resistors as in Figure 25.13 the circuit ceases to have a self-triggering action and has to be externally triggered both ways. In other words it is stable in both conditions (either transistor conducting) so is termed bistable. It has ceased altogether to be a self-oscillator, and in its basic form has no time constants; these, we recall, served as delay-action self-triggers.

Like the astable, this particular type of bistable was invented as far back as 1918, in valve form of course. It was named after the inventors as the Eccles-Jordan relay, and is now one of the main 'building blocks' in digital computers. Whereas the multivibrator is a two-stage 100% positive-feedback a.c. amplifier (because the stages are coupled to one another by capacitors) the Eccles-Jordan is a d.c. amplifier. Like the multivibrator it cannot maintain a balanced stage with each transistor passing the same amount of current, because the slightest disturbance of the balance, say by circuit noise, starts a movement that is rapidly accelerated by the feedback until one transistor is bottomed and the other cut off. By this time it should be unnecessary to go through the whole sequence in detail.

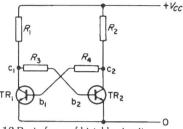

Figure 25.13 Basic form of bistable circuit

The essential feature is that it remains in that state until switched from without.

Suppose TR_1 is the bottomed one. Then c_1 is near enough to zero potential for insufficient bias current to flow through R_3 to make TR_2 conduct. At least, that is the intention, but, to make quite sure, TR_2 is biased off—the usual practice is to connect both bases through a pair of resistors to a negative voltage, as we shall see in a moment. TR_2 being cut off, c_2 is at $+V_{CC}$ potential, which applies a large positive base bias to TR_1, ensuring that it is fully on. And there is no internal reason why this state should ever change.

Now apply a negative triggering pulse to b_1. Provided that it is enough at least to reduce the collector current in TR_1, a positive amplified pulse will be applied to b_2, making TR_2 conduct and reinforce the triggering pulse at b_1, ensuring a speedy and complete switch-over. A negative pulse applied to b_2 restores the status quo. (With p-n-p transistors of course all the polarities are reversed.)

Often it is more useful to be able to reverse state by successive pulses from the same source, like the type of electric light switch in which one push turns the light on and the next turns it off. Figure 25.14 shows this and other practical elaborations. R_7, with the emitter current of the 'on' transistor passing through it, makes both emitters several volts positive from 0; $+V_{CC}$ is increased by the same amount to make V_{CE} the same as before. R_5 extends the R_2R_4 potential divider so that, when TR_2 is on, the base of TR_1 is definitely negative. R_6 similarly continues R_1R_3. C_1 and C_2 are usually fitted to speed up the switch-over by providing a low-impedance path for the high-frequency

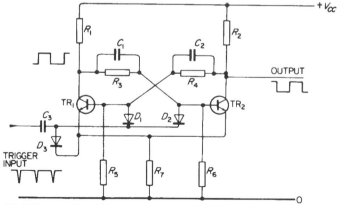

Figure 25.14 Practical bistable circuit, with diode pulse-forming and steering system for alternate on-off

components of the waveform. C_3 and D_3 sharpen the trigger waveform by differentiating action. When TR_1 is on, b_1 is positive to the common emitters whereas b_2 is negative, so D_1 conducts and D_2 does not, and the negative-going pulse affects TR_1 only, switching it off. When the next pulse comes the situation is reversed and TR_2 is switched off.

One other bistable circuit is important enough to be mentioned even in an introduction: the Schmitt trigger, Figure 25.15. Comparing it with the similarly numbered Figure 25.14 we see that it is essentially the same except

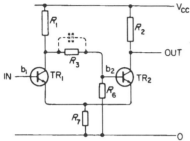

Figure 25.15 Schmitt bistable circuit, with d.c. triggering

that TR_1 is controlled directly instead of by TR_2. The input must therefore be d.c. instead of pulses. If the voltage applied to b_1 exceeds a certain critical positive value V_1, TR_1 is switched on, and TR_2 switches off in the now familiar way. If now the input level is reduced to a certain voltage V_2, less positive than V_1, TR_1 is switched off and TR_2 on. In this circuit R_7 does more than simply create a potential more negative than the emitters, as in Figure 25.14. Because the current through TR_1 is controlled by the input potential, so thereby is the voltage across R_7, and this in turn controls the base-emitter voltage of TR_2. In other words TR_1 and TR_2 are emitter coupled.

25.7 Monostables

The monostable multivibrator comes half-way between the astable multivibrator, which can trigger itself both ways continuously, and the bistable, which can trigger itself neither way so stays put until it is triggered externally. It needs to be triggered from its stable condition, but triggers itself back to the original state after an interval determined by a built-in time constant. So we are not surprised to find that one-half of the basic circuit is the same as in the astable type and the other half is the same as in the bistable type. Figure 25.16 shows quite a practical form of the circuit, in which the negative bias to ensure that TR_1 remains completely cut off when in the stable state is supplied from a negative source instead of by the voltage drop due to emitter current through R_7 in Figures 25.14 and 25.15. A suitable triggering circuit is also shown. For easy comparison with the previous circuits the corresponding components have the same numbers.

Typical waveforms are shown. A sample triggering pulse appears as v_a; its shape is not very important so long as the rising front is reasonably steep. It is differentiated by C_3R_8 which has a relatively short time constant to give the short sharp pulses $v_{a'}$. Only the positive one (v_{b1}) reaches the base of TR_1, the negative one being blocked by the diode D_1. The switch-over started by such a positive pulse sends b_2 sharply negative (see v_{b2}) and correspondingly c_2 sharply positive to form the front of the output waveform. Note that because there is no capacitor C_2 this front rises steeply instead of in a charging curve as in Figure 25.10. The positive voltage thus imparted to the top end of R_4 maintains v_{b1} positive and TR_1 on after the triggering pulse has died away. As soon as C_1 has charged sufficiently to make v_{b2} positive, the switch-over from TR_1 to TR_2 takes place in the same manner as in the astable multivibrator.

There are many uses for this kind of circuit, such as generating pulses of controllable length, reshaping pulses that have deteriorated in transmission, and creating adjustable time delays.

25.8 Clampers and Clippers

Figure 25.17 shows a simple multi-purpose device that we have already met more than once. It goes by various names according to the purpose. One of these, known as clamping, is shifting a wavetrain as a whole, lining up either positive or negative peaks to any desired potential. As it is shown it lines up or clamps the positive peaks to what is arbitrarily

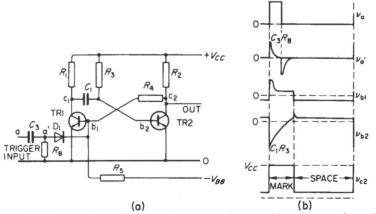

Figure 25.16 Practical form of monostable circuit with trigger (a), and its waveforms (b)

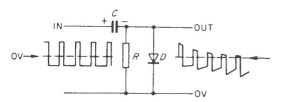

Figure 25.17 Diode clamping circuit

regarded as zero voltage or earth (0 V). To clamp the negative peaks, reverse the diode. To clamp to some other potential, bias the lower end of the diode to that voltage. The action is precisely as already described in Section 18.12. The upward slope of the negative peaks of the output is due to the negative charge on C, accumulated through the diode conducting at the positive peaks, leaking off slightly through R. To minimise distortion of the waveform due to such leakage, the time constant CR should be many times greater than the period of one cycle. While the scheme works more or less with any waveform, one that is very sharply peaked at the polarity of clamping (such as the sync. pulses being treated in this way in Figure 23.15) is likely to have its peaks clipped somewhat because of the very short time available for topping up C.

If the wavetrain is amplitude-modulated, the time constant CR has to be more carefully chosen, according to what is required. If it is made sufficiently long the charge will remain in C not just between one cycle and the next, but between one maximum-amplitude cycle and the next, with perhaps many weaker cycles in between. Figure 25.18 shows this, with modulated sine waves clamped positive of zero. CR in this case must be long compared with T, which is the length of one cycle of the *modulation* frequency.

Note that this output contains the original modulated carrier wave plus a z.f. (d.c.) component. So, as with the unmodulated pulses, the circuit can be used as a d.c.

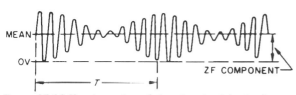

Figure 25.18 Showing action of clamping circuit having long time constant on modulated waveform

restorer. It does not contain the modulation frequencies. To get them, CR must be short enough for the charge on C to leak fast enough to follow the highest-frequency modulation, as explained in Section 18.7. In that role the circuit is a diode detector. Its output contains a z.f. component nearly equal to the *unmodulated* peak value of the carrier. This component is often used for purposes such as a.g.c.

D.c. restoration is often applied, as in television receivers, to the modulation waveform, after the carrier wave has been removed. This was shown in Figure 23.15, where the peaks that are clamped to a fixed potential are the tips of the sync. pulses. In this case the time constant should be long compared with the duration of one line (64 µs in a 625-line system) but could be less than that of one field. If d.c. restoration is applied by rectifying the v.f. picture signal only, after the sync. pulses have been removed (Figure 25.19) the z.f. so obtained should be averaged over the whole picture, so as to be a measure of

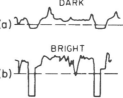

Figure 25.19 D.c. restoration of television signals. When applied positively, (b) yields a much larger bias voltage than (a)

its average brightness, and therefore CR should be longer than 1/25 s but short enough to respond reasonably quickly to changes in brightness of scene; 0.1 s, say.

Note that the diode in Figure 25.17 can be the input of a transistor which is performing some other function such as amplifying.

In all these applications one object is to preserve the wavetrain or signal as far as possible from mutilation. Sometimes, however, one needs to clip off unwanted bits. We have seen in Section 25.2 how to do this by means of transistors. Where their amplification is not needed, diodes can be used instead. The simplest method is to substitute a resistor for C in Figure 25.17; see Figure 25.20, in which two diodes are used in order to clip both half-cycles. The

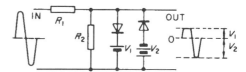

Figure 25.20 Example of clipping circuit for both polarities

batteries represent suitable sources of bias. Until the instantaneous signal voltage exceeds $+V_1$ or V_2 neither diode is conducting and the arrangement is a potential divider reducing the signal in the ratio $R_2/(R_1 + R_2)$. R_2 is not strictly necessary as a component but represents the resistance of whatever is being fed, combined with the leakage of the diodes if that is appreciable. R_1 should therefore not be larger than it need be to make the resistance of the diodes when conducting, plus their bias sources, negligible in comparison with it, so that an appreciable fraction of what is supposed to be removed is not developed across them.

That is the shunt type of clipper. The series type (Figure 25.21) is more suitable for performing the reverse operation: rejecting all of a waveform that does not exceed the bias voltage. Note that this voltage appears at the output all the time. With no bias this clipper is a simple rectifier giving a mean output equal to half the mean value of a half-cycle of the input (Section 28.3).

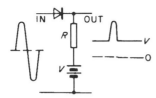

Figure 25.21 Series-diode type of clipping circuit

25.9 Gates and Choppers

In computers and other electronic equipment much use is made of electronic switches or *gates*. They fall into two main classes: *linear* or transmission gates, in which the signal gets through in its original form when the gate is open and not at all when it is closed; and *logic* gates, in which the output is due to the combination of signals present at the input and is not necessarily of the same form. Logic gates are used primarily in digital circuits and are described in the next chapter; we shall consider only linear gates here.

Diodes can be used as linear gates, but as many as six may be needed in specially balanced circuits in order to keep the controlling and controlled signals separate without serious loss of signal. This problem is reduced by using transistors, in which the base is available as a control electrode to render, the collector-emitter path conducting or non-conducting as required. Figure 25.22 shows an n-p-n transistor in this role. So long as the gating voltage V_G is

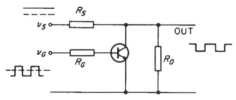

Figure 25.22 This type of gating circuit can be used as a chopper for z.f. (d.c.) amplification

negative the transistor is non-conducting and can be ignored; the signal voltage V_S gets through in the ratio $R_0/(R_S + R_0)$. When V_G is positive the transistor is highly conducting and short-circuits R_0. The transistor has a lower 'off' resistance and a higher 'on' resistance when connected as shown rather than with V_S connected between base and emitter. Only positive signals can be controlled by the form of the circuit shown; for negative signals use a p-n-p transistor, and for both use one of each in parallel, controlled by gate signals of opposite polarity.

Figure 25.22 shows the gate controlling a d.c. 'signal', in which role it is known as a *chopper*. As we saw in Section 24.6, it is very difficult to obtain high gain in a straightforward d.c. amplifier because the slightest change in transistor characteristics or supply voltages in the first stage is amplified too and cannot be distinguished from the signal. So a common technique is to chop the signal into pulses at a frequency suitable for amplifying. A chopper can be regarded as a frequency-changer, changing the signal frequency from zero to an 'i.f.'. Conversely the conventional frequency-changer in a radio receiver can be regarded as a kind of gating circuit.

Linear gates are used also for control purposes where faster action is needed than mechanical switches or relays can provide.

25.10 Design

You may be disappointed not to find component values given for all the circuits in this chapter, or at least detailed instructions on how to calculate them. Component values are seldom given because they depend so much on the precise circumstances and requirements, so without full discussion would be misleading. And full discussion would swell this book excessively and duplicate the adequate literature on the subject. The hope is that anyone who has studied this book will be able to design his own pulse-handling circuits, which are very largely an application of Ohm's law, plus CR time constants.

The electronics side, at least for the less demanding applications, is easy: the power levels are usually low, so problems of heat dissipation, thermal runaway and leakage currents are absent; and the diodes and transistors are used in only two states: fully conducting or non-conducting. In the conducting state the forward bias can be reckoned as 0.2 V for germanium and 0.6 V for silicon. The conducting ('bottomed') collector-emitter voltage, or knee voltage, is usually a small fraction of a volt; if a closer figure is needed the makers' curves or data should be consulted.

If very high-speed operation is needed, then the choice of diodes and transistors is rather important. Here again the makers provide plenty of information. But except as exercises or for specialists the design of pulse circuits is hardly necessary, as most of those likely to be needed are available as integrated circuits. But it does help if one knows how they work—and to provide that knowledge is the purpose of this chapter.

Chapter 26
Digital Techniques

26.1 Analogue and Digital

It would of course be absurd to attempt a comprehensive description of digital electronic techniques—and especially digital computers—in one short chapter. All that can be done is to indicate broadly what they involve and how they make use of the principles we have been studying.

The first thing is to distinguish between *analogue* and *digital*. Analogue computers were described briefly in Section 24.8. In these the magnitudes being processed are represented by continuously variable quantities, such as voltage, current or charge, whereas in digital computers they are represented by numbers, which are not continuously variable. One-third cannot be expressed by any finite number of decimal digits (being 0.333... for ever) but one-third of a distance is theoretically possible. Paradoxically, however, digital computers and calculators are capable in practice of far higher precision than analogue types.

Electrical instruments such as voltmeters illustrate the distinction. The ordinary pointer types are analogue instruments, voltage being indicated in terms of a different quantity: the angle of the pointer. Such voltmeters are simple in principle and have comparatively few parts, but the standard of workmanship goes up so steeply with the degree of precision demanded that they are limited in practice to very moderate precision. Digital voltmeters, in which the reading is exhibited in figures, are very much more complicated and contain very many parts, so for ordinary 1% accuracy cannot compete in price with pointer instruments. But the number of kinds of parts is quite small and they are individually cheap, and the precision of reading can be increased several decimal places by increasing the number of parts of the same few kinds, little or no improvement in quality being necessary.

The old-style rotary-dial telephone is a digital system. Seven twists of its simple mass-produced dial, generating seven groups of up to ten pulses, could be used to select, with absolute precision, any one from ten million different numbers. An analogue scale capable of being read unmistakably to one in ten million would be a formidable task for the instrument maker!

Coming to computers themselves, we may consider that one-time very familiar analogue type, the slide rule, in which numbers are represented by lengths along scales, proportional to the logarithms of the numbers. A digital calculator to perform to the same standard, which is sufficient for most engineering purposes, would be somewhat more expensive. But it could work incomparably faster. And digital calculators can be made incomparably more precise.

26.2 The Binary Scale

We normally use a decimal system of numbers, presumably because we normally have 10 fingers (including thumbs) on which to count. A primitive tribe is reputed to use only three numbers: one; two; plenty. When the need arose for numbers greater than could be represented by a reasonable variety of symbols, the simple and ingenious convention was adopted of placing a symbol one space to the left to mean the number of groups of ten, and a further space to the left to mean the number of ten groups of ten (i.e., hundreds); and so on. Using only ten different symbols, this system extends the range of numbers indefinitely. We may be so used to it that we hardly realise that 47 023 (to take an example of a decimal number) is really shorthand for

$$(4 \times 10^4) + (7 \times 10^3) + (0 \times 10^2) + (2 \times 10^1) + (3 \times 10^0)$$

And of course we shift to the right for subdivisions, 10^{-1}, 10^{-2}, etc., the units position ($\times 10^0$ or 1) being identified when necessary by a dot on its right-hand side.

For the purposes of digital computers this system has the disadvantage of necessitating ten different unmistakable electronic states, such as values of collector current, to represent the ten numerals. Even if extensive negative feedback is used to counteract changes in transistor characteristics and supply voltages, it is very difficult to alter the states at will by the required number of units without risk of error. And even one unit of error is serious when it is in the millions column!

But if the number of different states is reduced from ten to two everything becomes very easy. We need only consider two states: 'off' and 'on'. The last chapter was mainly about circuits that are in an 'off' or an 'on' state, and signals that can be used to change them from one to the other. So long as these signals are not weaker than a certain minimum, their strength is unimportant; and so is the exact amount of current or voltage in the 'on' state.

The symbol corresponding to 'off' is 0, and 'on' is 1. That is not very hard to remember. Except for the smaller

Decimal notation	Binary notation	Decimal notation	Binary notation
0	0	6	110
1	1	7	111
2	10	8	1000
3	11	9	1001
4	100	10	1010
5	101	etc.	etc.

variety of numerals, the procedure is the same as in decimal notation. That is to say, when the variety of numerals runs out (as it does above 1!) the units' position drops back to 0 and 1 is carried. So the binary numbers go up as in the above table.

The binary number 1010111 therefore is shorthand for

$(1 \times 2^6) + (0 \times 2^5) + (1 \times 2^4) + (0 \times 2^3)$
$+ (1 \times 2^2) + (1 \times 2^1) + (1 \times 2^0)$
$= 64 + 16 + 4 + 2 + 1$
$= 87$ in decimal notation

An obvious disadvantage of binary notation is the greater number of digits required; about $3^1/_3$ times as many on average. But this is usually more than offset by the simpler circuitry, even allowing for the decimal-to-binary and binary-to-decimal converters at input and output.

The smallest piece of information (using that word in its technical sense) is a 'yes' or 'no' answer to a question, which is given by 1 or 0 and is therefore equal to one binary digit, commonly abbreviated to *bit*. A specified number of bits, making up a binary number, is a *word*.

The basic operation in digital computers is addition. Multiplication can be regarded as repeated addition; 4 × 3, for example, means 4 + 4 + 4. This is a tedious way for humans to multiply, but not for computers that can do millions of operations per second. Where high-speed multiplication is necessary, provision can, however, be made for the computer to do it directly. 'Long multiplication' is extremely simple in the binary scale. For example, multiply 11010 by 101:

```
   11010
   00000
   11010
10000010
```

In practice one would not write down the second row, or any row corresponding to 0 in the multiplier. As we see, the essential processes are shifting the whole multiplicand successively to the left and adding all the results. Means for doing both these things will appear in Section 26.5.

There are several possible electronic methods of subtracting, corresponding to methods people use. In one, we replace each digit of the number to be subtracted by its nines complement—the number needed to make it up to nine—and add, finally transferring the 1 at the left-hand end to the right-hand end and adding it on; for example, take 108 from 762:

```
      762
      891 (nines complement of 108)
     1653 (sum)
     └──↑      Answer: 654
```

A binary computer similarly adds the ones complement, which it obtains simply by means of a NOT gate (Section 26.3); for example, take 1101 from 10110:

```
    10110
    0010 (ones complement of 1101)
    11000 (sum)
    └──↑   Answer: 1001
```

Division is repeated subtraction. To take a decimal example, 34 ÷ 9; 9 can be taken from 34 three times, leaving 7 over, so the answer is $3^7/_9$. The binary computer has to count the number of times the subtraction can be done before the remainder becomes negative.

26.3 Logic Operations and Circuits

Besides the familiar arithmetical operations, there are several other *logical* operations that are frequently performed on binary numbers. These operations are performed by electronic circuits called *logic gates* which often form parts of i.c.s.

Logic gates produce output signals (usually pulses) when certain combinations of input signals occur. An OR gate, for example, gives an output whenever any one or more of a number of possible input pulses occur. Suppose there are two inputs, A and B; then an output is given by a signal at A *or* B or both. This kind of gate can be devised using one diode per input, as Figure 26.1*a*. *R* must be large compared with the resistance of a forward-biased diode and small compared with that of a reverse-biased diode.

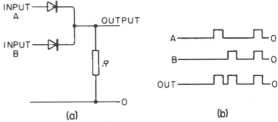

Figure 26.1 A logic OR gate; its output corresponds to a pulse at A or B

There are four possible combinations of input to A and B, and the output is related to them as indicated in the following table, known as a *truth table*, which confirms that this is an OR gate. In this we assumed the inputs to the gate to be positive pulses. It is, however, possible to regard Figure 26.1*b* as trains of broad negative pulses instead of narrow positive ones and if we work out the relationship between inputs and outputs (both regarded as negative pulses), we are in for a surprise. Remember that what was the datum line for the positive pulses is now the operating

Truth Table for an OR Gate

A	B	output
positive	positive	positive
positive	no signal	positive
no signal	positive	positive
no signal	no signal	no signal

level of the negative pulses and vice versa. We can see that we now obtain an output pulse only when there is an input pulse at A *and* B not A *or* B as before. The truth table for negative pulses (given below) confirms that pulses are needed at both inputs (in a multi-input gate, at all inputs) to give an output pulse. A gate with this property is known as an AND gate and the interesting thing is that the circuit of

Truth Table for an AND Gate

A	B	output
negative	negative	negative
negative	no signal	no signal
no signal	negative	no signal
no signal	no signal	no signal

Figure 26.1 can behave as an OR gate or an AND gate depending on the polarity of the pulses used. To understand gate circuits it is therefore essential that the polarity convention should be stated or implicit in any diagrams. The positive convention (usually called 'positive logic') is more widely used than the other.

If the diodes are reversed in Figure 26.1, they conduct on negative pulse inputs and a repeat of the above procedure will show that the circuit behaves as an AND gate for positive logic and an OR gate for negative logic—the opposite of the behaviour described above.

Transistors can also be used in OR and AND gates with the advantage of gain as well if that is needed. Another facility provided by transistors and much used in logic circuits is inversion, which changes 'signal' to 'no signal' and vice versa. Although any common-emitter amplifier inverts, so hardly qualifies as a gate, in the context of electronic logic it is called a NOT gate. Most often, however, it combines the NOT function with an OR or AND gate, making what is termed a NOR or NAND gate.

A simple n-p-n-transistor gate is shown in Figure 26.2 with the relevant waveforms. With both A and B at zero volts the base is sufficiently negative to cut off the transistor and give +12 V output. If either A or B input is made not less than about +3V, the base goes slightly positive, making the transistor conduct and bring the output voltage near zero. To reduce interaction between the input sources and also to reduce loading on them, the input

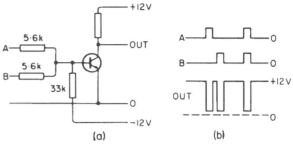

Figure 26.2 A transistor NOR gate; a +12 V output is obtained whenever there is a pulse at neither A nor B

resistors may be replaced by diodes as in Figure 26.1a. For the positive logic convention this gate gives an inverted or NOT signal whenever there is an input at A or B; this is the property of a NOR gate. For negative logic, however, the gate gives a NOT signal only when there is an input at A and B—the property of a NAND gate. So, as expected, the behaviour of this gate depends on the logic convention.

If a p-n-p transistor is substituted in Figure 26.2 and all the polarities consequently reversed, the result is a NAND gate for positive logic and a NOR gate for negative logic. If any input is at zero volts the base is slightly negative which makes a p-n-p transistor conduct and keeps the output voltage only slightly less positive than 0 V. Only if all inputs are positive can the base go positive and cut off the transistor, making the output voltage fully negative.

The simple circuits used to illustrate the principles of logic gates in Figures 26.1 and 26.2 are rarely used in practice. They suffer from a number of practical problems such as poor response times and ambiguity concerning the voltage threshold that distinguishes a '1' input from a '0'. Figure 26.3 shows the internal circuit diagram of a NAND gate on a standard TTL (Transistor/Transistor Logic) i.c. Four such gates are present on each i.c. Note the special symbols for Schottky diodes and transistors. The junctions in Schottky diodes use a mixture of silicon and another element, often aluminium, to give a much lower voltage drop when forward-biased. This is typically about 0.2 V,

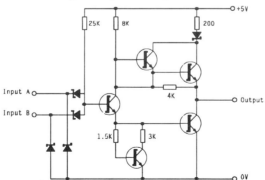

Figure 26.3 The circuit of a NAND gate used on an industry-standard i.c. Note the symbols used for Schottky diodes and transistors. Their significance is explained in the text.

c.f. 0.7 V for a standard silicon diode. A Schottky transistor is a standard silicon transistor in which a Schottky diode has been connected in parallel with the collector-base junction (the diode's cathode to the transistor's collector, an n-p-n transistor being assumed). The purpose of the diode is to prevent the transistor from becoming saturated ('bottomed'). As the collector voltage falls to about 0.2 V below the base voltage, the diode begins to conduct, short-circuiting away the base bias. You may wonder why, in a digital circuit, saturation conditions should be deemed so undesirable as to provoke such rigorous preventative measures. The answer is that it takes appreciably longer to change the state of a transistor under saturation conditions than one which is held short of saturation by a Schottky diode. Appreciably in this case may mean only a few picoseconds, but picoseconds are significant in a logic circuit that is switching at a frequency of 100 MHz. The gate has push-pull output for a similar reason. Since it can both charge and discharge any stray capacitance in the circuitry, it improves the gate's performance in high-speed computing operations.

Obviously we need a set of symbols for these logic gates, not only for brevity but also for clarity because the circuit diagrams of gates do not instantly reveal what kind they are. Various symbols are in use; those in Figure 26.4a are British standard. The number inside the rectangle indicates how many of the inputs have to be 'on' in order to give an 'on' output. Those in Figure 26.4b are American, but are widely used because their contrasting shapes facilitate the rapid understanding of a logic diagrams; the same symbols are used outside electronics, e.g., in critical path analysis.

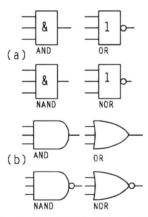

Figure 26.4 Standard symbols for binary logic components: (a) British and (b) American. Inputs are on the left; output on the right

26.4 Adders

Logic gates are the components from which circuits for mathematical operations are constructed. Since the basic binary operation is to add two digits, let us consider next the construction of an adder, itself a building block in many kinds of digital equipment.

Let us call our two digits A and B. There are four possibilities, listed in the following table. On the right we see the results the computer has to obtain. A whole system

A		B		Sum	Carry
0	+	0	=	0	0
0	+	1	=	1	0
1	+	0	=	1	0
1	+	1	=	0	1

of algebra, described by Boole in 1854, has been resurrected in order to arrive at the necessary computer circuitry for all requirements in the quickest and most certain way, but there is not enough space for it here. Instead, a logic circuit for solving our elementary problem

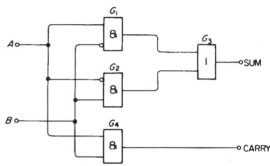

Figure 26.5 A half-adder circuit—a 'building block' in digital equipment

appears ready-made in Figure 26.5. Note that a circle across an input or output means NOT, the operation being performed by an *inverter*, a circuit which outputs the complement of its input. Let us now check that it does do its job, by making a table of inputs and outputs of the four gates for each of the four possible values of A and B. In the circuit, 1 is represented by a pulse and 0 by no pulse (or a space). It is assumed that A and B are presented simultaneously to the system at the terminals so marked.

The B input to G_1, being via a NOT, is shown reversed in the table; likewise the A input to G_2. G_1 and G_2, being AND gates, give a 1 output only when both A and B are 1. Their outputs are the inputs to G_3, which gives 1 when either of them is 1. Its output is the required sum for that particular digital place. But in the case of $1 + 1$ only we need to carry 1, and this result is indicated by the output of the single AND gate, G_4.

Inputs A B	G_1 inputs	G_1 output	G_2 inputs	G_2 output	G_3 output	G_4 output	Answer
0 0	0 1	0	1 0	0	0	0	0
0 1	0 0	0	1 1	1	1	0	1
1 0	1 1	1	0 0	0	1	0	1
1 1	1 0	0	0 1	0	0	1	10

We see then that this logic circuit does all that is required of it for the first (units) digit, but in all higher places there may be a carry from the one below. So a third input must be provided for it and the table extended to include altogether eight possibilities.

In designing an adder for more than one place, two courses are open to us. We can use a single adder to deal with the digit places in succession. This is called the series system and is economical in circuits but wasteful of time. And it must include some means for storing the carried 'bit' produced by the adding operation for one digital place until the next higher one is being dealt with. A bistable is commonly used for this purpose. Suppose TR_1 in Figure 25.13 receives a periodical negative pulse at b_1. If TR_1 was originally 'on', the first pulse would turn it off and subsequent pulses would do nothing. But if TR_2 receives a pulse from some other source (the CARRY output of Figure 26.5, say), turning it off and TR_1 on, the next pulse to TR_1 would turn it off and thereby generate a positive pulse at c_1 and a negative pulse at c_2. Whichever of these was of the appropriate polarity would be fed back into the adding system as a carried digit. The system shown in Figure 26.5 is known as a *half-adder* and is readily available in integrated circuit form. A complete or full adder is one that copes with a carried digit.

The alternative to the series system is the parallel one, in which addition of all the digits takes place at the same time. A full adder is needed for every place except the lowest. Figure 26.6 shows the circuitry for the first two places

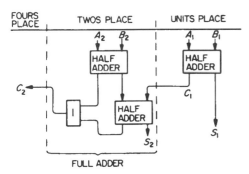

Figure 26.6 Showing how half-adders are combined in a computer

(ones and twos). For higher places the twos circuit is repeated to the left as many times as may be necessary. Each half-adder block comprises the elements in Figure 26.5. We can see at once that any carried 1 resulting from addition of the A_1 and B_1 is added to the sum of A_2 and B_2 and comes out as S_2. If there is a carry from either of the half-adders in the full adder, it comes out as C_2. And so on. To check the performance of the full adder, make a table listing the eight possible combinations of A_2, B_2, and C_1, and with the aid of our table for the performance of a half adder find the corresponding values of S_2 and C_2, comparing these with what you get by doing the sum mentally. They should of course be the same.

26.5 Shift Registers

Another essential component of much digital equipment is the *shift register*. This is an array of linked bistables, as many of them as the application requires, which has the useful property of being able to transfer sequences of bits along the array, one place to the right or left at a time, in response to control pulses.

Figure 26.7 shows this function for a simple example of only five digit places. The number, 10011, is first shown

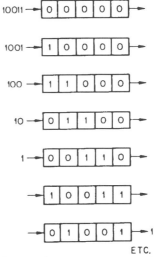

Figure 26.7 Showing the successive steps in the action of a shift register

ready to enter, at the left. One pulse from the 'clock' (Section 26.7) transfers the unit's digit to the first bistable, where it exists as one of its two possible states. Which one it is determines what happens as a result of the next shift pulse. If it is an 'on' or 1 state, as here, it reverses the next bistable from 0 to 1. And so in successive stages the whole word is taken in. The last diagram shows the first digit emerging as a 1 pulse. If the last bistable's output is fed back to the first, as in some shift registers, the word is kept circulating and is thus stored until required and made to emerge by some control signal. Or it may be directly applied to the A input of an adder and added one digit at a time to another numerical word now coming in directly, or shifted one place at a time for multiplication.

For its various functions the basic bistable circuit is subject to slight modifications. The one shown in Figure 25.14 is arranged so that successive input pulses reverse its state. In a shift register type the TR_2 output of the first triggers the TR_1 of the next.

Shift registers serve a wide variety of purposes throughout digital electronics. In computers and electronic calculators they are used in binary multiplication, as described in Section 26.2. For example, multiplying the binary number 11001 (= 25) by 1101 (= 13) involves the following processes. First binary 11001 is multiplied by binary 1000 (the most significant bit of 1101) by using a shift register to shift 11001 three binary places to the left, giving an intermediate result of 11001000, which is stored. The second significant bit of 1101 is also 1, so the same action is taken, but with the shift register shifting 11001 two places to the left, giving an intermediate result of 1100100, which is also stored. The third digit of 1101 is 0; since multiplying anything by 0 yields 0, no action need be taken regarding this digit. The fourth and last digit of 1101 is 1. Since multiplying any value by 1 leaves it unchanged, an unchanged copy of 11001 is also stored as an intermediate result. Finally the three intermediate results are added together as explained in Section 26.4 to give the binary number 101000101 (= 325).

Another useful function of shift registers is to convert serial signals to parallel signals and vice versa. A *serial signal*, as its name suggests, is a sequence of pulses on a single conductor. It is the format in which digital data are received over a telecommunications system or read from a disc-based storage system. For many purposes and especially for data processing, however, digital data are required in parallel format on an array (called a *bus*) of 8, 16 or 32 conductors, each representing a binary value. The incoming serial signal is fed into a shift register having the appropriate number of bits, as described earlier in this section, and the statuses of the array of bistables is then read directly on to the parallel data bus. To convert from parallel to serial format, the process is reversed. The bistables in the shift register are individually set so that they match the word currently on the data bus and then the serial version is extracted from the shift register.

A third important function of shift registers is in the seemingly unlikely world of video imaging; indeed they are the basis of modern television cameras and video still cameras. These units use *charge coupled devices* (c.c.d.s), specialised igfets (Section 10.11) which were originally developed as memories and shift registers. These devices have long channels with the drain and gate connections at the ends and a large number, possibly thousands, of gates spaced at regular intervals along the channel (Figure 26.8). Each gate electrode acts as one plate of a capacitor whose other plate is the substrate forming the channel. These capacitors ('cells') are capable of holding electrical charges; moreover, by the application of appropriate combinations of voltages to the neighbouring gates, it is possible to transfer the held charges along the channel from gate to gate in a similar way to the transmission of bits along a shift register.

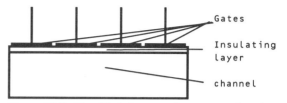

Figure 26.8 Simplified diagram of a section of a charge-coupled device (c.c.d.)

C.c.d.s are only reliable for normal shift register applications at high frequencies, because thermally generated carriers in the channel quickly accumulate within uncharged cells changing them to charged ones; this effect is called the *dark current*. That the effect is accelerated by the presence of light has led to the application of c.c.d.s in video cameras. The image is focused on to an array of c.c.d.s, effectively a mosaic of light sensors. The cells in each c.c.d. are initially uncharged. Within a brief fixed period the charges are shifted out of each c.c.d. in turn and measured; the level of charge in each cell is an indication of the level of light in the area of the picture that was focused on it. Thus a picture signal is built up. For colour television, colour filters are fitted over the c.c.d.s.

C.c.d.s offer many advantages over the camera tubes described in Section 23.12. For example, they require no evacuated glass envelope, no electron gun and no high-voltage supply. The resulting cameras can be made very small, permitting the gathering of pictures in confined spaces where a studio camera could not be installed.

26.6 Counters

Electronic counters are used, for example, in counting articles passing on a conveyor belt, each article generating a pulse by interrupting a beam of light falling on a light-sensitive type of transistor. In high-precision frequency meters a counter is switched on for (say) exactly one second, during which the number of cycles is counted. Or by using a known frequency, an unknown time interval can be measured, also with great precision.

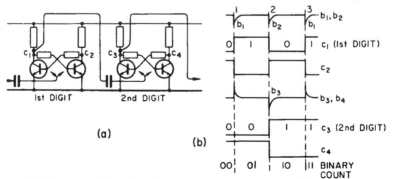

Figure 26.9 (a) A two-digital-place binary counter, using one bistable per place, with (b) waveforms during a count of 3 pulses

We are familiar with mechanical decimal counters, on car dashboards, for example. A mechanical binary counter would have only two symbols, 0 and 1, on each wheel. The first incoming 'pulse' would move it from 0 to 1; the second, back to 0 and at the same time the wheel would pulse the second wheel from 0 to 1. And so on. Obviously we can do the same thing electronically, using a chain of bistables of the Figure 25.14 type in place of wheels. Figure 26.9 shows in some detail how.

At *a* the circuits of the first two bistables in the series are shown in simplified form, omitting for clarity such practical details as the diodes in Figure 25.14 needed to delete positive pulses and steer negative ones to the 'on' transistor. The waveforms at the points marked are shown at *b*. On the top line are the pulses to be counted.

They need not be at regular intervals. 'Voltage positive' is taken to mean 1, so is read from the first collector in each bistable, but equally well it could be 'transistor on' or 'current flowing' and read from the second. The second bistable is triggered at the even-numbered input pulses and indicates the second digit. If each bistable has, say, a lamp to show whether its second half is on or off, and the lamps are placed in a row, right to left, the total of a series of pulses from one or more sources could be read off in binary notation, as shown on the last line in Figure 26.9*b* for the first two. For convenience a binary-to-decimal converter could be interposed.

26.7 Digital Computers

In order to understand what a digital computer has inside it and why, let us consider what we do when we compute mentally or with the aid of pencil and paper. First we have to receive the data and instructions on what to do with them; for example, 'Add 27 to 189 and multiply the result by 17'. The first part of a computer, correspondingly, is an *input unit*, to receive data and instructions in a form the computer can use, such as keyboard, magnetic disc or tape, punched cards, etc.

Next, we perform the arithmetical operations of adding and multiplying. The regular procedure with the above problem would be to add 7 to 9, get 16, put down 6 in the units column, carry 1 to the tens column and add it to 2 and 8, getting 11, put down 1 and carry 1 to the hundreds column, adding it to 1 to get 2. Partial answer: 216. Then we would multiply this 216 by 7 and add it to 2160. Or we might have a ×17 multiplication table or ready reckoner giving the answer directly. The part of a computer that manipulates the given digits to provide a partial or final answer is called the *arithmetic and logic unit*.

This contains circuitry similar to that described earlier in this chapter for performing arithmetical and logical operations. The length of binary word (that is, the number of bits it contains) that can be processed is a function of the processor's architecture; in most current microcomputers at the time of writing it is 32 bits.

The versatility and power of computers stems from their ability to perform long sequences of these comparatively simple arithmetical and logical operations—and also other operations concerned with storing and copying binary numbers—in response to a list of instructions, the *program*. Moreover, it performs them very quickly, moving from one instruction to the next in response to pulses from a 'clock', in practice a crystal oscillator, running at speeds of up to 200 MHz.

The part that uses the given instructions to bring this unit into action in the appropriate ways at the right moments is the *control unit*, and the part that stores partial results until they are needed (taking the place of the paper on which we write them down, or our memory if it is good enough) and such useful information as the multiplication table to shorten frequently needed arithmetical operations, is called the *store* or *memory*.

Finally, the part that converts the answers from computer form to a form that is convenient for the user of the computer, is the *output unit*. Its 'readout' may be displayed on a screen, printed on paper, or sent as electronic signals to control a machine.

These divisions are shown in Figure 26.10.

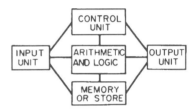

Figure 26.10 A computer can be divided into these five departments

26.8 Stores

Digital computers require two types of store. One is external to the computer proper to hold what might be called archive information, often consisting of vast quantities of data. Typical of the information held in such storage systems are payroll details of the staff of a large organisation.

The capacity of a store could be expressed as the number of bits, i.e., binary digits it can hold, but the bit is a very small unit for mass storage systems and the capacity is usually expressed in bytes, a byte being a succession of bits (usually eight in number) which is treated as a unit in computers but is smaller than a binary word. An eight-bit byte offers $2^8 = 256$ possible values, often expressed as numbers between 0 and 255. For example, a byte might be an alphanumeric character forming part of textual data. A four-byte word could represent the address code for the location of a particular item of information in a store, which will be explained later. The mass storage of a large computer might have a capacity of 20 000 000 000 bytes, usually written 20 Gbytes (or 20 000 Mbytes), typically on one or more 'hard discs', which are described in the next chapter.

One of the disadvantages of tape and disc stores is that it takes some time to find a wanted item of information; in other words, as the computer engineer would put it, 'access time is appreciable.' Such a store is said to have *serial access*.

The other type of store is that required within the computer itself to hold the instructions required by the computer to carry out its allotted task (the program) and the results of intermediate stages of calculation. Stores for this purpose are electronic, such as a matrix of transistor bistables. By arranging these in a rectangular array with leads from every column and every row it is possible to switch a given bistable by signals applied to the appropriate column and row leads. Consequently a wanted item of information can be found very quickly; the information is 'location addressable' and the 'access time' is very short. Such a store is said to have *random* or *direct access* and is known as a RAM (Random Access Memory).

Electronic data storage will be considered in more detail in the next chapter.

26.9 Computer Languages

The central processing unit of a digital computer responds only to instructions in binary form. These instructions are known as 'machine code' or 'machine language'. Programming in machine code is tedious, since every individual processor operation must be specified; more than a dozen of these instructions may be needed to enable a computer to add two numbers.

To simplify programming, so-called 'high-level' programming languages are available. The advantage of these is that they allow frequently used routines (such as mathematical functions, entering data at the keyboard, displaying results on the screen, printing them on paper or archiving them to disc) to be entered as single instructions. One of the easiest of such high-level languages to learn is BASIC (Beginner's All-Purpose Symbolic Instruction Code) in which the instructions resemble everyday English.

Because the central processing unit understands only machine code, however, programs written in high-level languages have to be translated. There are two kinds of high-level language distinguished by the stage at which translation takes place. Many high-level languages (such as 'C', 'C++' and Pascal) are called *compiled* languages. The programmer writes the source code, that is, the program in its high-level language form, using a text-editing program. The resulting text is then analysed by a program called a *compiler* which generates the corresponding machine code. This may require a further operation, *linking*, if the program uses libraries of previously compiled routines. The compiled machine code is then archived and ready for testing. So two separate stages, writing and compiling, are involved in creating the executable version of the program.

The alternative to compilation is called *interpretation*. In interpreted languages, such as most versions of BASIC, there is only one production stage, namely writing the source code in the high-level language. The translation takes place while the BASIC program is running, rather like an interpreter providing a simultaneous translation to a speech in a foreign language. Although the computer appears to execute the source code directly, in fact the central processing unit is running a machine-code program called the BASIC interpreter. This scans through the BASIC instructions, analyses them and executes appropriate machine code routines (mostly held within the interpreter itself). Because this scanning and analysis inevitably take place *in real time*, that is, during the running of the program, a program written in an interpreted language generally runs more slowly than a similar program in a compiled language.

Many of today's computer programs run in an environment commonly called a *graphical user interface*. This uses *icons*, graphical representations of objects and operations, which the user manipulates using a pointing device such as a mouse or trackball. This makes the computer's processing power simple and intuitive ('user friendly'). Sometimes this environment is called the *desktop* because it provides many of the features found on an executive's desktop, such as a clock, calendar, notepad and word processor (in place of a typewriter). Instructions to the computer consist of graphical analogues of everyday operations. For example, to copy a file (an item of stored data) from a diskette to the hard disc the icon representing the file is selected in a graphical display of the diskette's contents and 'dragged' into a graphical display of the hard disc's contents. Most users find this more congenial than earlier practice which might have involved typing an abstruse command such as `COPY A:FRED.DOC C:*.*`.

The creation of programs to use these graphical user interfaces is very complex and a new generation of *application development* software has been designed to simplify it. These make use of the same techniques to build up programs in a modular (*object oriented*) manner. Some of the underlying code is generated automatically by the development software itself, but traditional programming skills are still needed to handle some of the processing tasks. Examples of such development systems include Visual Basic (based on BASIC), Visual C++ (based on C++) and Delphi (based on Pascal).

26.10 Input and Output Devices

The input to a computer can be from any of a wide range of input devices. Most familiar are keyboards resembling those of typewriters, which allow commands and textual and numeric data to be typed in. Another common input device is the *mouse* which moves a pointer around a display screen and allows icons such as images of buttons to be pressed or graphics such as lines and shapes to be drawn in any position. Even printed documents can be input; a device called a *scanner* moves an array of light-sensitive cells (Section 10.12) over the document to produce an electronic image of it called a *bitmap*, a two-dimensional array of data representing the colours of the square dots of which the image is composed. The scanned image may be archived as a bitmap, but programs are available that analyse the bitmap and recognise text in it, generating a textual document that can be edited in a text editor or word processor; this process is called *optical character recognition* (OCR).

Material that already exists as computer data can be input directly from storage media such as tape or disc (considered in the next chapter). Data may also be received directly from other computers. In organisations neighbouring computers are often linked together in *local area networks* (LANs) which allow them to exchange data and share expensive resources such as colour printers. Any computer (or LAN) may also form part of a *wide area network* (WAN); a device called a *modem* (short for modulator/demodulator) enables computers to exchange data over long distances using ordinary telecommunications lines. Such WANs may be nationwide or even worldwide, such as the Internet. In this way any computer can avail itself of the resources of millions of other computers.

The output from a computer usually first appears on the screen of a visual display unit (VDU), i.e., a cathode-ray tube which is also be used for displaying the input data, in the form of text or pictures (*graphics*). It can also be printed on paper using a printer which may employ any of a wide range of technologies. Many computers are also equipped with sound synthesis circuitry which can often reproduce sounds in full stereo with 16-bit accuracy, that is, quality theoretically as good as that from compact disc. This sound may be used to provide background music or sound effects for games programs, but also has serious uses, especially in educational software, such as the learning of foreign languages or music theory.

The ability of computers to receive data from external devices, process it and send the results on to other equipment automatically has revolutionised industry. In the process industries, for example, a computer could monitor inputs from sensors on a process plant and operate controls to, say, initiate heating or cooling of vessels to maintain the process under optimal conditions.

In the broad field of *computer-aided manufacturing* (CAM) computers control machine tools, mechanical handling and inspection equipment to automate whole production lines (or *cells*). One advantage of CAM is its flexibility. Unlike a traditional production line which assembles only one product, under computer control one production line may produce any of a wide repertoire of products in any sequence. A further advantage is the ease with which the manufacturing process can be changed. For example, the machine tool instructions for a new component can be generated directly from engineering drawings produced using a *computer-aided design* (CAD) program. Still further advantages result from automated quality control. If the computer controlling a production cell receives input from an inspection machine measuring, for instance, the sizes of critical holes drilled in the product, it can automatically initiate the replacement of a worn drill bit before the dimensions of the hole are out of tolerance. This form of *statistical process control* eliminates the costly wastage caused when out-of-specification products are scrapped.

26.11 Microprocessors and Microcomputers

Originally the central processing unit of a computer consisted of a complex assemblage of components on one or more circuit boards. Developments in integrated circuit technology made it possible for a processor consisting of an arithmetic and logic unit, control unit and sometimes memory as well to be contained within a single i.c. a few millimetres square. Such i.c.s are known as *microprocessors* and, since they are manufactured in vast quantities, the unit cost is very low.

A general-purpose computer using a microprocessor is known as a *microcomputer*. Although this name may suggest that a microcomputer is capable of performing only the most menial of computing tasks, this is not so; today's microcomputers far exceed in computing power the mainframe machines (the ultra-powerful machines used by banks, utilities and other large organisations to process thousands or millions of records) of a few years ago and are gradually replacing them.

Microprocessors, however, are not confined to general-purpose computing applications. They are by definition general-purpose electronic components. As such they are used increasingly in applications where switching operations must be carried out in sequence or at predetermined times and/or under the control of input signals or stored information. For example, they can control switching in telephone exchanges and machine tools in factories. At home they control washing machines and video cassette recorders. In telephones they make possible press-button dialling, memorise and call favourite telephone numbers, and repeat any call number as often as desired, if not available the first time, by pressing one button.

26.12 Digital-to-Analogue and Analogue-to-Digital Conversion

Despite the apparent complexity of digital techniques (in comparison with the analogue uses of electronics which have occupied most of this book) digital techniques are being used increasingly to handle audio and video material, traditionally the province of analogue methods. For example, the NICAM stereo sound accompanying television transmissions is transmitted in digital format. Some mobile telephones digitise speech for transmission over the radio links. Digital f.m. radio and television broadcasting is proposed. The next chapter describes the use of digital techniques to store audio material, compact disc being a familiar example.

The principal reason for the use of digital techniques to transmit and store material which for decades has been handled acceptably in analogue format is that digital signals are inherently more robust, that is, less prone to corruption. Once the material has been digitised, it consists of a train of pulses representing '1's and '0's. If as a result of some loss in transmission one pulse is a little more feeble than its neighbours, it will probably still be read correctly at its destination and restored automatically to its proper level. Even if it is below the threshold and is read incorrectly, in digitised sinusoidal matter the distortion arising from an error of one bit, or even a brief succession of bits, is likely to pass unnoticed. A similar loss in transmitted analogue material might have catastrophic effects on its quality.

In the following discussion we shall restrict our deliberations to the digitising of audio waveforms. Similar techniques can, however, be applied to video and other analogue material. One consequence of converting an analogue signal, which can have any value of instantaneous amplitude within certain limits, to digital format is a restriction on the range of its possible amplitude values to a finite number, a process known as *quantising*. For example, in digitised audio systems at least 1000 levels are needed if the sound finally reproduced from the subsequent digital-to-analogue converter is to be reasonably free from distortion. High-quality systems use a far greater number of levels (over 16 000 levels in NICAM and compact disc). The number of levels required in the digital signal is of course of fundamental significance in the design of the analogue-to-digital and digital-to-analogue converters.

Another fundamental consideration is the frequency of sampling. A digital representation of an analogue signal source such as audio consists essentially of a series of numbers. Each represents the instantaneous amplitude, effectively a snapshot, of the source audio waveform taken at regular intervals in much the same way as a television picture is regarded as a series of still pictures of its moving subject. To minimise distortion the samples must be taken at such brief intervals that the signal waveform can be reconstructed with reasonable accuracy at even the highest audio frequencies. So if we take our upper frequency limit as 15 kHz (the upper limit of stereo f.m. radio broadcasts) snapshots of the waveform should ideally be taken several times during the 67 μs period of a 15kHz signal. Fortunately we can get away with only two snapshots per period because at such high frequencies the resulting distortion is undetectable by the human ear, the unwanted harmonics being above the audible range.

This poses the question of how to convert digital signals to analogue and vice versa. We shall begin perhaps rather irrationally by considering digital-to-analogue conversion because it is simpler than its opposite and serves well to introduce the concepts involved in it.

In fact a digital-to-analogue converter (DAC) need not contain any active devices at all, as the example in Figure 26.11 shows. This circuit requires the incoming digital signal to be in parallel format; a shift register (Section 26.5) could be used to convert a serial input. This DAC consists only of a *ladder network* of resistors; simple calculations reveal that it will correctly convert its 4-bit input to any of

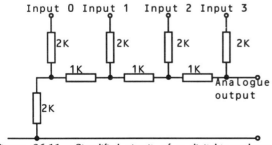

Figure 26.11 Simplified circuit of a digital-to-analogue converter (DAC)

16 possible levels of output voltage. Similar but larger networks could be deduced to convert the 16-bit digital inputs used, for example, in compact disc players or computer sound cards. The only drawback to this circuit is that its accuracy is dependent on the consistency of the levels of the digital input voltages.

An alternative technique is the *ramp generator* circuit. The incoming digital value is fed to a counter clocked from a high-accuracy source such as a quartz-controlled oscillator. This counts down to zero, the time of the count being proportional to the instantaneous audio amplitude. While the counter is counting a capacitor is charged from a constant-current source; its charge voltage, proportional to the count time and therefore to the audio amplitude, is fed to the output of the unit.

Analogue-to-digital conversion is more complex. There are two basic techniques, both having many variations. Continuous ADCs, as shown in Figure 26.12, monitor the incoming analogue signal continuously and produce a continuous digital output which can then be sampled as required. The analogue signal is distributed to one input of each of *n* comparators, *n* being the number of quantising levels required. A *comparator* is a circuit, usually supplied as a collection of several identical specimens on an integrated circuit, which, as its name suggests, compares the voltages on its two inputs. It outputs a logical '1' when the voltage on one input is higher than the other and a logical '0' when the other input is higher. It is in some respects similar to an operational amplifier, described in Section 24.6, except that it is used without external feedback and its output is one bit.

Each comparator in our example compares the incoming analogue signal with a reference voltage derived from a potential divider chain. The outputs of the comparators are fed to an encoder to convert them to binary (parallel) format; a shift register could further convert them to serial format if required. For simplicity the example shown produces only a 2-bit output, which would be of little practical use. Clearly for an ADC of this type to give 1024 quantising levels (10-bit) it would require 1024 comparators; using large-scale integration techniques, such an ADC could be implemented as an integrated circuit.

The other technique used for ADCs is the *serial* system. The incoming analogue waveform is sampled, as described earlier, to provide a series of snapshots at regular intervals. A chopper, described in Section 25.9, might be used for this. The instantaneous voltage sampled is held as a charge in a capacitor until the next sample is taken. The digital output is updated for each sample.

One such ADC is the *ladder network converter* (Figure 26.13). This uses only one comparator; one input is from the capacitor holding the current analogue sample and the other is the reference voltage. This is not static but rises in a series of rapid steps, one for each quantising level, from zero towards the maximum voltage that can be converted.

This rising reference voltage is generated by a high-frequency digital counter whose output is applied to the inputs of a DAC of the type shown in Figure 26.11. The counter begins its count at zero when each sample is taken and continues until the comparator output changes state,

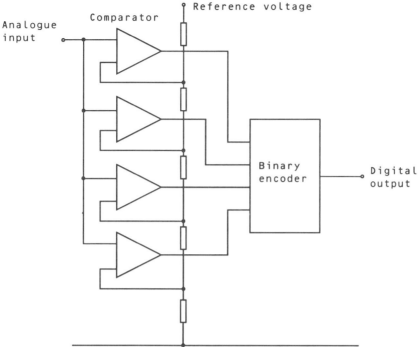

Figure 26.12 One type of analogue-to-digital converter (ADC)

indicating that the voltages on its inputs are equal. The count is then stopped and the current counter reading is sent to the output of the ADC. A shift register can convert it to serial format if required.

These have been necessarily simplified accounts of the ADCs and DACs used by the broadcasting authorities and others for the digital transmission of sound programmes. In practice the circuitry is further complicated by the volume compression and complementary expansion necessary to minimise the noise inevitably introduced by quantising.

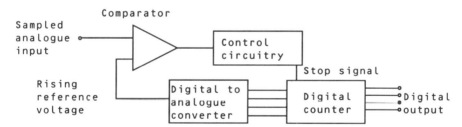

Figure 26.13 Block diagram of a ladder-network analogue-to-digital converter

Chapter 27 Electronic Data Storage

27.1 The Need for Storage

Most of this book has been concerned with the processes and equipment used in the broadcasting and reception of radio and television programmes, these familiar examples serving well to illustrate many of the principles of electronics. The power of radio and television to carry sounds and moving colour images anywhere in the world in a fraction of a second has made these the most potent communications media of the twentieth century.

With the growth of broadcasting, however, there arose a requirement for a means of *storing* the programme material. Indeed there are two such requirements, one at each end of the transmission chain. Firstly, broadcasting stations prefer whenever possible to have their programme material prepared prior to transmission since this eliminates many of the uncertainties (such as the duration of events and the suitability of the subject matter) that inevitably attend a 'live' broadcast. Secondly, members of the listening and viewing public may wish to keep copies of interesting programmes for future reference or may wish to hear or see a programme that is transmitted at an inconvenient time; suitable equipment at the receiving end can receive the programme as it is transmitted and store it for subsequent reproduction.

Another data storage requirement arose as a consequence of the rapid rise in the use of digital computers. This is a requirement to store (and to distribute in a convenient form) large collections of software, which may include both programs and data.

Interestingly the processes involved in electronic data storage are analogous to those in radio transmission and reception. The essential difference is that the material is to be reproduced at a distant *time* rather than a distant *place*, although the medium of storage is often also used as a vehicle of distribution. As in broadcasting, the latter use imposes rigid standards since the medium must give satisfactory reproduction on a diversity of equipment.

The *medium* which physically stores the material corresponds to the hypothetical 'ether' through which electric waves are propagated. Unlike intangible 'ether', however, this is a manufactured article expressly designed for its purpose. Transferring the material to the medium, called *recording* with audio or video material and *writing, saving* or *archiving* in a computer environment, is analogous to transmitting. Reproducing the material from the storage medium, called *replaying* or *playback* with audio or video material and *reading, loading* or *retrieving* in a computer environment, is analogous to receiving.

Today many storage media and methods are available and in this brief summary it is possible to give only broad outlines of the processes involved. For more detailed information the reader should consult specialised literature on the subject.

27.2 Lessons from History

The storage of sounds and moving pictures predates electronics. Edison announced the invention of the phonograph, a machine for recording sound, in 1877. It was a purely mechanical device in which sound waves impinging on a diaphragm set up sympathetic vibrations in it; connected to the diaphragm was a stylus whose excursions caused undulations in a spiral groove being cut in the surface of a rotating wax cylinder; the undulations of this groove, of course, reproduced the waveform of the sound being recorded. For playback the process was reversed, the stylus being drawn along the groove so that its undulations set the diaphragm vibrating, emitting a distorted but nevertheless recognisable version of the original sound. A few years later Edison improved the system by replacing the cylinder with a disc. In so doing he established a recording medium which is still widely used, albeit technologically obsolete, at the end of the twentieth century. Electronics enhanced the quality of reproduction and also made possible a compatible stereophonic system, but the recording itself was still held in the excursions of a physical groove (and variations in its depth in stereo recordings) which followed the waveforms of the recorded sound.

There are many claimants to the invention of the motion picture. Edison demonstrated his 'kinetoscope' in 1888, but the most widely observed centenary of the motion picture industry celebrated the aptly named Lumière brothers who showed movie films in France in 1895. Both systems involved taking a sequence of photographs of the moving subject at brief intervals and, to replay the 'recording', displaying (usually by projection) the sequence at the same speed. As in television, if the interval between frames is suitably brief, the result is the illusion of movement. Today's motion pictures use essentially the same techniques.

Although neither the phonograph nor the motion picture is electronic, both demonstrate the principles which underlie most electronic storage techniques. In both, for instance, the recording medium is required to *move* in order to provide a time base. The gramophone record spins causing the stylus to vibrate in its groove; the cine film threads its way through the projector pausing briefly to allow each frame to be displayed as a temporary still picture. Most electronic storage systems involve either spinning discs or moving tape.

This movement has implications which we have not met before in the pages of this book. Most storage equipment is unlike 'pure' radio and television equipment in that it contains moving parts such as motors and bearings which are properly the province of mechanical engineering. One consequence of this is *wear* both in the drive mechanism (leading to problems of reliability) and, worse still from our point of view, sometimes in the storage medium itself. Wear in the storage medium results in a gradual deterioration in the quality of reproduction of the recorded material, a phenomenon familiar to users of gramophone records. The philosopher's stone for data storage is a system which suffers no such degradation and ideally has no moving parts. Such ideals are rarely attainable in this world, but current technology can approach this ideal, as we shall see later.

Before considering some practical storage systems, let us summarise the characteristics of an ideal storage system. It must interface easily with electronic equipment and offer useful storage capacity at the required bandwidth. The medium must be capable of some permanent (but possibly reversible) change which must be detectable if retrieval is to be possible. The medium should be compact and robust and, ideally, inexpensive.

27.3 Magnetic Storage Systems—Analogue

It is perhaps not surprising that magnetism with its close relationship to inductance (described in Chapter 4) forms the basis of many storage systems. All rely on the ability of an electromagnet both to 'write' magnetic patterns on a disc or tape of suitable material and to 'read' magnetic patterns previously written there.

An electromagnet, as we saw in Section 4.1, consists of an inductor with a suitable core, Figure 27.1. It behaves as a magnet only when electric current flows in the coil. The

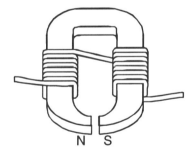

Figure 27.1 Simplified structure of an electromagnet. Use of a horseshoe shaped core, as shown, concentrates the resulting magnetic flux in the gap

magnitude and direction of the current in the coil determine respectively the strength and polarity of the magnetic field produced. If an alternating current, such as the output from an audio amplifier connected to a suitable programme source, is passed through an electromagnet, the magnetic field will have alternating polarity and varying strength; indeed this forms the basis of the operation of most loudspeakers (Section 19.9).

Any magnet can be used to magnetise other objects, provided, of course, that they are made of magnetic material. Magnetic materials behave as though they were composed of elementary regions called 'domains'. Each domain is like a miniature independent magnet; while the material is unmagnetised these domains have random orientation so that their magnetic fields tend to cancel one another giving the material as a whole no overall magnetisation. Under the influence of an external magnetic field, however, the domains align themselves with the newly arrived lines of force. Usefully for our purposes, if they are handled carefully, they retain their new alignment when the external magnetic field is removed.

Now consider what happens when a tape of magnetic material is drawn at constant speed past our electromagnet carrying its audio programme. The tape will become magnetised, but not uniformly so; the strength and polarity of each domain in the tape will depend on the strength and polarity of the current flowing in the electromagnet when that domain passed it and came under its influence. Since that current represented the waveform of the audio material being fed to the amplifier, it follows that the shape of that waveform is now represented in the magnetic strength and polarity of the successive domains in our tape, Figure 27.2. Moreover, this magnetic pattern in the tape is permanent, that is to say, it will stay there until the tape is remagnetised.

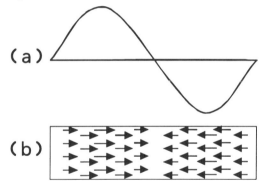

Figure 27.2 (a) One cycle of a sinusoidal, e.g. audio, waveform; (b) how this waveform might be represented in an analogue magnetic recording. The arrows represent individual domains in the tape; their length and direction represent degree of magnetisation and polarity respectively

Now let us consider what happens when our magnetised tape, having been returned to its original position, is drawn once again across the electromagnet at the same speed, but this time with the electromagnet connected in the *input* circuit of the audio amplifier. The tiny magnetic fields of the domains within the tape induce a small alternating voltage in the electromagnet and this, of course, is amplified. And this alternating voltage will have the waveform of the original audio material that was used to make the recording. So the original audio programme will be reproduced by a loudspeaker connected to the output of the audio amplifier.

This account has described only the fundamental principles involved. A practical audio tape recorder that is to offer reasonable quality would do rather more than deposit a raw audio analogue on to the tape. It is true, however, that audio tape recording differs from both radio transmission and video tape recording in that no carrier signal is needed. Mechanical considerations give tape-based storage systems a bandwidth equivalent to 10 octaves so that the whole audio spectrum (taken as 20 to 20 000 Hz) can just be encompassed. It is, however, standard practice to mix an ultrasonic 'bias' signal (typically around 100 kHz) with the audio being delivered to the recording head, that is, the electromagnet that magnetises the tape. The reason for this is that the response of the tape to the magnetising process is not linear. The introduction of the bias signal helps to keep the working point on the most linear part of the transfer curve, so reducing harmonic distortion.

Care is needed in the physical design of the recording head. Because the most recently recorded part of the tape will still be within the magnetic field produced by the record head, the recording held there is liable to be partially overwritten by that currently being recorded, a 'blurring' effect. One way to minimise this is to make the gap in the tape head's core very small so as to concentrate the generated magnetic flux in a confined area (Figure 27.1). Another is to get the tape away from the head quickly by using a high tape speed, which will also extend the high-frequency response. But a high tape speed will use more tape; it also has implications for the design of the moving parts of the machine.

Because the voltage which the recording induces in the winding of the tape reading head during playback is very low, comparable with the thermal noise attending semiconductor operation, considerable amplification is needed to raise it to a level suitable for input to a standard audio power amplifier. Consequently, reproduction from tape suffers from a poor signal-to-noise ratio, the notorious 'tape hiss', unless precautions are taken to minimise it. A reduction in tape hiss can be obtained by attenuating the high-frequency response during playback, but this also attenuates the high-frequency component of the recorded material, giving 'woolly' reproduction. Commercial noise-reduction systems apply a selective expansion to the high-frequency component during recording; on playback a corresponding attenuation of the high-frequency component restores it to its proper level and simultaneously reduces the unwanted hiss, by typically 9 to 10 dB.

Systems which magnetically recorded sound on steel tape or wire had been demonstrated at the end of the nineteenth century, but little serious use was made of them until wire and magnetic tape recorders were employed for military purposes during the Second World War. From these evolved the audio tape recorder in which the recording medium is a thin but robust plastic tape having a magnetic coating, usually of gamma ferric oxide or chromium dioxide. Domestic tape recorders typically use two 'tape heads', that is, electromagnets. The tape first passes an erase head, energised with high-frequency a.c. during the recording process, which removes any previous recording from the tape by randomising the orientation of the domains. The other head transfers the magnetic flux in both recording and playback. Professional machines use separate heads for recording and playback, giving a total of three heads.

Early tape recorders were open-reel machines which were fiddly to operate. In the late 1960s Philips introduced the 'Compact Cassette' which eliminated any need for the user to handle the tape itself by enclosing it in a convenient and robust carrier. The tape in these cassettes is 3.81 mm wide carrying two monaural or two pairs of stereo tracks; running at a speed of $1^7/_8$ inches per second, cassette tapes provide up to two hours of audio. With a combination of high-quality machine and tape, a reasonably flat frequency response from 20 to 20 000 Hz is attainable. For professional purposes open-reel machines are still preferred, a popular tape speed being 7½ inches per second. Recording studios use specialised machines having as many as 24 parallel tracks on wide tape. This allows the output of up to 24 microphones (or other sources) to be recorded simultaneously during a studio session; later the output during playback can be mixed to obtain the most pleasing sound and re-recorded as conventional two-channel stereo.

Video signals, assuming the 625-line system colour television standards used in most of Europe, as we saw in Chapter 23, have a bandwidth of about 5.5 MHz, so present a range of challenges for magnetic recording. Early video recorders were essentially faster versions of audio machines, using higher tape speeds to obtain the necessary frequency response. But high tape speeds, as we have seen, introduce mechanical problems, such as the handling of miles of tape! The problem was solved by mounting multiple heads in the curved surface of a drum which rotates so that the heads scan across the tape in a direction almost opposite to that of the tape movement, Figure 27.3. In this way an adequate head-to-tape speed is achieved using tape speeds no greater than in audio recorders. In the popular VHS (vertical helical scan) standard for domestic video cassette recorders, the tape is 12 mm wide and it runs at $1^7/_8$ inches per second. An advantage of this system is

Figure 27.3 Helical arrangement of tracks on a VHS tape

that during playback the tape can be held stationary leaving the head scanning so that a still picture is produced. Also, recognisable if somewhat distorted pictures can be obtained while the tape is being wound rapidly forwards (*cueing*) or backwards (*reviewing*) which assists searches for an item of interest.

There is a further important difference between video and audio tape recording. The range of frequencies present in a colour video signal is enormous; the highest frequencies are around 5.5 MHz and the lowest around 5 Hz. Because this range far exceeds the 10 octaves that can be handled directly in a tape-based system, a carrier is used. By frequency-modulating the video on to a carrier at 6 MHz, the frequency range to be recorded becomes 0.5 MHz to 11.5 MHz, well within the 10 octaves span.

27.4 Magnetic Storage Systems—Digital

The audio and video tape recorders described above are analogue devices; the strength of magnetisation of the domains on the tape is continuously variable, representing the instantaneous amplitude of the sinusoidal waveform of the material being recorded.

Magnetic storage systems are also widely used to record digital data. In one sense digital recording is simpler than analogue recording. Because each bit (binary digit) has one of only two possible values, usually considered as '0' and '1', the contents of a bit can be stored using only the polarisation of the domains to represent the data, one polarity as '0' and the other as '1', Figure 27.4. The strength of magnetisation is irrelevant as long as it is sufficient to read. This makes digital recordings more robust than analogue ones.

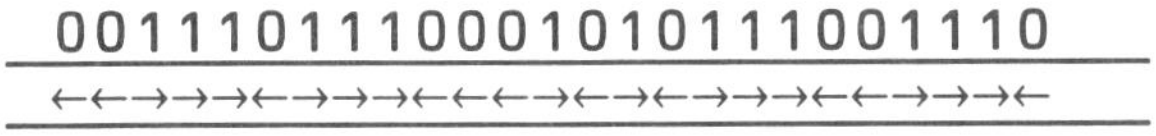

0 0 1 1 1 0 1 1 1 0 0 0 1 0 1 0 1 1 1 0 0 1 1 1 0
←←→→←→→→←←→→←←→←→→→→←←→→→←

Figure 27.4 Representation of binary digital data in a magnetic medium can use polarity of magnetisation only

Digital techniques are sometimes used to store audio material, as in the DAT (Digital Audio Tape) standard which uses a rotating head to scan across the tape as in video recording, and the DCC (Digital Compact Cassette) standard which uses multiple tracks in a tape format physically identical to Compact Cassette. The analogue audio source is converted to digital data before recording and the digital data retrieved during playback is restored to analogue audio using the techniques described in Section 26.12. The advantage gained from these seemingly cumbersome conversions between analogue and digital formats is that digital storage eliminates some of the inherent weaknesses of analogue tape recording, such as the non-linearity of tape response and the poor signal-to-noise ratio, especially in quiet passages.

Tapes are also used for storing computer data. Early mainframe computers used banks of multi-track open-reel tape mechanisms to store vast quantities of data. These have now been superseded by hard discs which are described later. Tapes are still widely used, however, for back-up storage* of computer data. The special cassette tapes, having storage capacities sometimes of several Gbytes, fit in mechanisms called *tape streamers* which are temporarily (and sometimes permanently) connected to the computer. Tape is also used for the distribution of large collections of business software, although in this role it is being supplanted by CD-ROM, which is described in the next section.

Tape format offers high storage capacity, but the physical form of the medium gives it one major disadvantage—a very long access time. In other words, it may take several minutes to wind the tape to the particular item of interest. This is only a mild irritant in a domestic tape recorder when the item is a piece of music or a recorded television programme and it is not important when performing computer back-ups or restorations which always take appreciable time. But for everyday use in computer systems, data need to be retrieved very quickly, ideally in under a second. For this reason, the principal storage systems in modern computers use disc format. The read-write heads are mounted on the end of arms which can be swung to access any part of the disc surface, in much the same way as a record player pick-up.

There are two principal forms of magnetic disc used for computer storage. *Floppy discs* are so named because the original 8-inch and 5¼-inch versions were enclosed in flexible plastic covers. These have now been largely replaced by 3½-inch *diskettes* which are housed in rigid plastic covers, although the term *floppy* is still sometimes, inappropriately, applied to them. The 3½-inch discs carry 80 tracks per side, spaced at intervals of 1/135 inch. Storage capacity is 720 Kbytes to 2.88 Mbytes per disc, depending on disc quality and format. In most computers the disc rotates at a constant 300 r.p.m. during accesses and each track carries the same quantity of data irrespective of its distance from the centre (and therefore its length). Data transfer time is appreciable (as much as 1 minute for 1 Mbyte). At one time floppy discs were the principal means of software storage and distribution for users of microcomputers, but as computers have grown in power and data processing requirements and applications software has grown in size, their storage capacity and low data transfer rate have become inadequate for many purposes and other media offering higher storage capacities, such as CD-ROM, are beginning to replace them.

So-called *hard discs* are now almost universal in computers of all types. A hard disc differs radically from a floppy disc in that it is usually not removable. That is to say, the disc is fixed permanently in its housing inside the computer; indeed it is sometimes called a *fixed disc*. The software it contains has been copied on to it from other sources such as floppy discs, CD-ROMs or over a network; it is not normally supplied with software ready installed. The disc itself is usually a collection of several rigid discs on a common spindle; the discs are double sided, each side having its own read/write head; the arms move in unison and the heads glide above the disc surface on an air cushion a few microns thick, Figure 27.5. The disc and its mechanism are mounted in a hermetically sealed enclosure to prevent the ingress of dust which would make the system unusable. The disc spins, usually at 3600 r.p.m., continuously while the host computer is switched on. Storage capacities available at the time of writing (June

---

* Backing-up, that is, making a copy of valuable computer data on another storage medium, is standard practice in business and government organisations. It is a precaution against the loss of the data if the medium containing the original should fail or become corrupted or if the host computer is stolen or damaged by fire or other hazard. When required, the back-up copy can be restored to the original machine (if available) or to another machine.

Figure 27.5 Interior of a hard disc drive showing the multiple platters and ganged arms bearing the read/write heads

1996) are in the range 500 Mbytes to 20 Gbytes. Access time is typically a few ms and data transfer time is very fast, e.g., 1 Mbyte per second.

One fundamental difference between the disc-based digital storage systems and the other storage systems considered in this chapter is that the rotational speed of the disc is not directly related to the rate at which data is stored or retrieved. The only requirement is that the disc should rotate at least as fast as is necessary for the fastest data transfer; in practice its speed may be many times faster than this. Data is not normally written directly to the disc surface or retrieved directly from it; instead an intermediate *buffer memory* (a block of solid-state RAM, described later) is used. Its function is to eliminate mismatches between the speed of the storage medium and that at which data to be written is made available or at which data being retrieved is required. Material to be written to the disc is written first to the buffer and then as the destination location on the disc surface passes the head, the data transfer begins and proceeds at a rate determined by the disc speed. Similarly, when retrieving data, the buffer is filled with retrieved data; this is read by the destination application at its own pace; when most of the data in the buffer has been read, further data from the medium overwrites that which has been read. In this way the next item of data is always immediately available. The read/write head may linger over the same track for hundreds of rotations of the disc and many transfers of data may take place during this time.

Magnetic storage systems have the advantage of comparatively simple interfacing to electronic circuitry and also that the media are reusable. That is to say, the media can be reused by simply overwriting old, unwanted stored material with new. They have, however, two principal disadvantages. Firstly, all except hard discs involve contact between the delicate surface of the medium and the read/write head, so that the medium itself is subject to wear, causing eventual data corruption or loss of performance. Hard discs are less susceptible to this than are other magnetic media, but they are delicate and, with their constant operation, the bearings or motors fail after a while

(usually several years of service); hence the importance of backing up valuable stored material on another medium. Secondly, as magnetic media are stored, they are inevitably subjected to stray magnetic fields from electricity cables, nearby electrical equipment and even natural sources. These have a demagnetising effect, the magnetic patterns on the media gradually fading away. Careful storage in a cool environment well away from cables and electrical equipment will minimise the losses.

27.5 Optical Systems—Compact Disc (CD)

In Section 10.12 we saw that semiconductor devices are available which can emit and detect light. Consequently it is not surprising that optical techniques now form the basis of many data storage systems. They represent a further step towards the ideal in that the medium is not subjected to any physical wear, reading and writing using light; consequently the stored data does not suffer any degradation. The most widely used optical system is *Compact Disc* (CD).

CD was introduced by Philips in the early 1980s as an alternative to the long-playing record. Within 15 years it had effectively supplanted it. It offered the advantages over that medium of compactness (usually only 5 inches in diameter), vastly superior audio quality, improved signal-to-noise ratio and, since it is a non-contact medium, of not being subject to wear. So, unless it is damaged in an accident, the recording on an audio CD should sound the same after years of use and many thousands of playings as when it left the factory. It is cheap to manufacture and can hold up to 80 minutes of stereo audio.

The medium is digital, the original audio being sampled at approximately 44 kHz and converted to 16-bit numbers. The bits are written on the master disc using a laser beam; a '1' is represented by a pit of diameter approximately 1 µm burned in the surface by the laser and a '0' by a plain surface where the laser beam was turned off. Some 20 000 tracks of pits and spaces are accommodated on the disc surface. Besides the audio data, the tracks contain error checking and timing information. In production discs reproduced from the master the pits are stamped out of thin aluminium foil which is then embedded in rigid transparent plastic.

To replay the disc, a low-power laser beam is trained on its surface from the read head. Light sensors detect whether the laser light is being reflected straight back, indicating that the current location is a 'space', or scattered, indicating that the current location is a pit. Servo systems keep the laser beam in focus on the surface and the head over the required track, Figure 27.6. Playing begins at the centre of the disc and proceeds outwards; nominal rotational speed is initially 500 r.p.m. but decreases to about 200 r.p.m. at the perimeter of the disc to maintain a constant rate of data recovery; in practice higher speeds are often used, the recovered data being stored in a buffer until required for digital-to-analogue conversion.

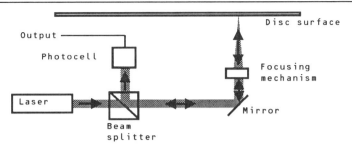

Figure 27.6 The essential processes involved in a CD player. The laser beam is directed via the mobile mirror on to the disc surface. If the beam is reflected (indicating a binary '0') it is detected by the photocell via the beam splitter.

CD is also a popular medium for the storage and distribution of many kinds of digital material. *Photo-CD* is a standard introduced by Kodak for the storage of digitised photographs. Up to 100 photographs can be stored on a CD, each at five different resolutions. To view the photographs the disc must be inserted in either a dedicated Photo-CD player connected to a television receiver or in the CD-ROM drive of a computer having Photo-CD-compatible software.

CD-ROM is a standard which uses the CD format to store and distribute computer programs and data. The capacity of a CD-ROM is about 700 Mbytes, comparable to a hard disc.

So-called *laser discs* were introduced at about the same time as CD. These are 12-inch discs using similar techniques, but they usually store video as well as audio material. Using various methods of data compression, some of which suppress detail that would normally pass unnoticed, it is now possible to accommodate one hour of video (plus high-quality audio) on a 5-inch CD. Considerable data processing is required to reconstitute a motion picture in real time, that is, as the disc is playing. An enhanced CD format under development by a consortium of manufacturers is due for commercial release in 1997; called DVD (Digital Versatile Disc) its capacity of up to 18 Gbytes is sufficient for five hours of high-quality video with CD-quality audio soundtrack.

CD in all its guises has one drawback; it is normally a read-only system, like the gramophone record that it was originally intended to replace. Clearly there are CD recording units. In fact they are readily available and are used not only for creating the masters of CDs for mass production, but also, increasingly, for backing up hard discs and for distributing collections of computer programs and data. Although a CD recorder allows the creation of CDs, it cannot erase a previous recording on a CD and re-use the space that it occupied in the medium as a magnetic storage medium can. Some CD systems are said to be *multi-sessional* meaning that material can be recorded on the CD in one session and more material added to it in subsequent sessions until the disc is full. But material that has been written to the disc cannot be erased. CD is an example of a *write once, read many times* (WORM) medium.

27.6 Other Optical Storage Systems

Since the advent of CD several other optical storage systems have been developed which combine the advantages of CD with the ability to erase, record and re-use the medium. Some combine optical technology with magnetic technology and are known as *magneto-optical*. In these a laser beam locally heats the medium during recording and then a magnetic field changes the crystalline structure of its surface which affects its reflective properties. Others use only a laser beam to record on the medium. In both types playback is essentially the same process as in CD; indeed systems which use 5-inch media can usually also read standard CDs. The capacities of such media vary widely, but 5-inch media hold typically 600 Mbytes to 1.3 Gbytes. These systems are used almost exclusively with computers for the distribution and storage of large collections of software and for back-up purposes. One system, however, is marketed primarily for use in automobile and portable audio systems, as a recordable alternative to CD. The 2½-inch discs have a capacity of 125 Mbytes and use data compression techniques to store up to 75 minutes of stereo audio.

27.7 Solid-state Systems

All of the storage systems described so far require the storage medium to move relative to the reading/writing equipment; the medium is either a spinning disc or a winding tape. And wherever parts are required to move, there is wear and the risk of mechanical failure. The ideal storage system therefore would have no moving parts at all; it would be completely electronic.

In fact the quest for a completely electronic storage system is not so farfetched as might be supposed. It already exists and has done for some years, but for reasons of expense it is usually limited to small quantities of data (a few kbytes). To take a pair of everyday examples, on the back of some commercial vehicles an audible warning system is fitted which issues a spoken message such as, 'Caution! Vehicle reversing' from a loudspeaker when reverse gear is selected. On electronic telephone exchanges a misdialled number no longer results in an impersonal 'unobtainable tone', but in a friendly voice saying, 'The number you have dialled is unobtainable; please check the number and try again.'

These spoken messages are not being reproduced from tape or phonograph records; they are digitised audio stored in an integrated circuit, a so-called *read-only memory* (ROM). To create the recording analogue speech is converted to digital using one of the techniques described in Section 26.12 and the resulting numerical data are stored in successive memory locations in the ROM. Reproduction of such memory-stored sound is a purely electronic process, an oscillator in the circuitry determining the rate at which the data is read from the memory and presented to a digital-to-analogue converter.

For messages lasting a few seconds and which do not need to be changed this medium is inexpensive and, because there are no moving parts, it is very reliable. To put 80 minutes of high-quality stereophonic sound into ROM, however, would be prohibitively expensive at present, although technically it is perfectly feasible. To manufacture an all-electronic hi-fi system in which ROM cartridges can be interchanged as easily as CDs or tape cassettes would introduce a few further complications, although these are by no means insurmountable. Interchangeable ROM cartridges having smaller capacities have been used with handheld computer-based games consoles for some years. Such a medium is viewed by many as the logical next step in hi-fi.

Let us begin by looking at the various types of semiconductor memory; *memory* is, after all, only another word for storage medium.

Most of the personal computers on sale at the time of writing are equipped with 4 to 32 Mbytes of *random access memory** (RAM). This is the memory used to store current programs and also the data that the machine is processing. The valuable feature of RAM is that its contents, every single bit in them, can be changed as often as required. The contents of ROM, in contrast, are usually fixed in production and are not so easy to change. ROM contents can of course be read as often as required, but attempts to change them in the same way as changing RAM have no effect. Most computers contain a small amount of ROM which carries the instructions which the central processing unit executes at start-up (at which time the RAM is of course empty). In some computers frequently used routines and applications are stored in ROM so that they are always instantly accessible.

Most RAM is said to be *dynamic*, that is, each bit is stored as a charge (or as the absence of a charge) in a capacitor. The charge (if any) tends to leak away and so dynamic RAM must be regularly 'refreshed'; it is examined at intervals and each bit that carries a charge is topped up. For this reason, dynamic RAM holds its data only while the host device is switched on. It is said to be *volatile*, meaning that its contents disappear if the host device is switched off

---

* The random access capability distinguishes this form of memory from archive storage systems such as hard discs, in which access is, strictly speaking, serial rather than random. Although an individual byte on a hard disc can, theoretically, be addressed, there is inevitably a (brief) delay in getting the read/write heads to the required track and waiting for the destination sector to reach the heads. In contrast, every byte of RAM is accessible instantly.

or loses its power supply for some other reason. It is therefore only suitable for temporary storage; any valuable data held in RAM must be copied to a non-volatile medium such as hard or floppy disc before the computer is switched off. Dynamic RAM is therefore not a suitable candidate for our electronic storage medium.

But there are other kinds of memory that offer greater promise as a purely electronic storage medium. *Static* RAM consists of bistables (described in Section 25.6) composed of insulated-gate f.e.t.s (see Section 10.11). These need no refreshing, but are not strictly non-volatile since they do need a power supply. They are, however, so miserly in their current consumption (except when actually changing state) that a small battery can maintain the contents of a bank of static RAM for many hours. Credit-card-sized packs of battery-backed static RAM offering storage capacities of up to 80 Mbytes are indeed used as a storage and distribution medium. The format is convenient, but at present is too expensive for all but a few specialised purposes. It is used, for instance, to transfer programs and data to industrial computers operating in very harsh environments, such as adjacent to industrial motors or arc welding equipment, where the electromagnetic interference would render conventional magnetic storage systems unusable.

ROM does not need a battery. Moreover, it is not quite as read-only as the designation Read-Only Memory suggests. Most ROM i.c.s are of a type known as *erasable programmable read-only memory* (EPROM). Using a special kind of igfet, EPROM is both *erasable* and *programmable* meaning that its present contents can be cleared and new contents inserted. Because these processes require special equipment, however, EPROM is usually confined to read-only applications where no change in its contents is needed and, indeed, any change in its contents might be catastrophic! It is sometimes used as the basis of a storage system for portable data processing equipment. The host system contains the special circuitry needed to allow material to be stored in empty memory locations. In this role it resembles recordable CD (Section 27.5) in that it is a WORM system—items can be stored at various times, but once stored they cannot be changed or deleted; consequently after a while the device becomes full and nothing more can be stored in it. The whole EPROM i.c. can, however, be erased by exposing its circuitry (via a window in the package) to ultra-violet light for a few minutes. The resulting recombination of electrons and holes resets the memory contents which can now be used again.

Electrically erasable programmable read-only memory (EEPROM) is an even stronger contender for our wholly electronic storage system. This is similar to EPROM but stored data can be erased and the space it occupied re-used, as in RAM or in a magnetic medium. Its only disadvantage, apart from expense, is that present EEPROM becomes unreliable after several erasures and re-recordings. And an unreliable memory may be worse than no memory at all.

Technology is advancing steadily and nowhere more rapidly than in semiconductor electronics. Developments in EPROM and EEPROM could eventually replace CD-ROM, floppy discs and audio cassette tapes as a reliable wholly electronic demountable storage medium. No moving parts would be needed to retrieve the stored material. The only wear would be in the miniature plugs and sockets used to connect the medium with its host device.

27.8 The Future

Some experts have suggested that storage systems will become less important in the future, at least for domestic audio/video purposes. Broadcasting stations and programme producers will always need professional-quality recording equipment. The need to record material at home, however, could diminish as cable-based on-demand systems come into more widespread use.

Cable television systems which provide 40 or more channels of television and/or radio using fibre-optic cables are spreading rapidly in many countries. Some cable systems now provide other facilities such as telephone services. Feasibility studies have been undertaken into further extensions of the system to allow interactive services, such as so-called video-on-demand systems. A subscriber could, for instance, call up the service provider and request a particular film or musical performance; the service provider's equipment would locate the appropriate recording in a central store and replay it, exclusively to that subscriber, over the cable, immediately. Clearly such a system could eliminate the need, or some of the need, for the subscriber to purchase his own recording or even the equipment to replay it. Much would depend on the comparative costs of subscribing to the cable service and of purchasing or hiring recordings and playback equipment.

For computers, however, local storage will always remain critical. Most modern personal computers will not operate unless a hard disc having a certain critical amount of free space is present. On-demand networks could supply computer software and provide high-speed connections with other computers. In fact on-demand services using low-speed telecommunications lines have been popular with computer users for many years. Many organisations supply software on demand over telephone lines; data transfer is currently limited to approximately 3 kbytes/s and can be considerably slower if the lines are noisy or if the server (the source computer) is providing other services at the same time. Nevertheless, remote back-up services exist which back up their clients' machines automatically overnight for storage on the service provider's equipment. And the growth in popularity of wide-area networks (computers linked over distances using telecommunications lines) has been one of the most widely reported phenomena of the last decade of the twentieth century. The Internet, a global network connecting (at the time of writing—June 1996) an estimated 40 million computers around the world, has dramatic implications for communications, including the facilitating of low-cost international telephony and, to a limited extent, cable broadcasting.

As electronics develops, the media may—and certainly will—change, but data storage will always remain a fundamental function of electronics, an integral part of its role in radio transmission and reception.

Chapter 28 Power Supplies

28.1 The Power Required

In the preceding chapters the power supplies needed to make things work have been assumed. Let us now look at them in detail. Power for valves and cathode-ray tubes can be divided into three classes: cathode heating, anode supplies, and grid bias.

So far as cathode heating is concerned, it is purely a matter of convenience that it is done electrically. In principle there is no reason why it could not be done by a Bunsen burner or by focusing the sun on it with a lens. But there are overwhelming practical advantages in electric heating. Nearly always it is done, for convenience, by a.c. Cathode heating power has to be supplied all the time the valves are required for work, whether or not they are actually doing any.

Power supplied to the anode is more directly useful, in that a part of it is converted into the output of the valve. This is generally not true of power taken by auxiliary electrodes, so valves are designed to reduce that to a minimum.

Not all valves need any grid bias; and, of those that do, not all have to be provided with it from a special supply. And only in large sending valves is any appreciable power consumed, the reason being that with this exception care is taken to keep grid current as nearly as possible down to nil (Section 10.6). This restriction, desirable for most purposes, would unduly limit output from high-power valves.

Transistors have the great advantages of needing no heater power at all and in general needing much lower voltage supplies than valves, e.g., 9 V. Often the sole source of power is a small dry battery, or (where available) a car battery. Larger, permanently installed transistor equipment can more conveniently and economically be supplied (like most valve equipment) from the mains.

28.2 Batteries

A battery consists of a number of cells, and although a single cell is commonly called a battery it is no more correct to do so than to call a single gun a battery.

A cell consists of two different sorts of conducting plates or electrodes separated by an 'exciting' fluid or electrolyte. The e.m.f. depends solely on the materials used, not on their size, and cannot greatly exceed 2 V. The only way of obtaining higher voltage is to connect a number of cells in series to form a battery.

Other things being equal, a small cell has a higher internal resistance than a large one, so the amount of current that can be drawn from it without reducing its terminal voltage seriously is small. To obtain a large current a number of cells of equal voltage must be connected in parallel, or (preferably) larger cells used.

Many types of cell have been devised, but only two are commonly used. They represent the two main classes of batteries—primary and secondary.

Primary cells cease to provide current when the chemical constituents are exhausted; whereas secondary cells, generally called *accumulators*, can be 'recharged' by passing a current back through them by means of a source of greater e.m.f.

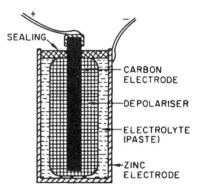

Figure 28.1 Cross-section of a Leclanché dry cell, showing the essential parts

The most commonly used primary batteries are the so-called dry batteries—a badly chosen term, because a really dry battery would not work. The name refers to cells in which the electrolyte is in the form of a paste instead of a free liquid. The essential ingredient is ammonium chloride or 'sal-ammoniac' and the plates are zinc (−) and carbon (+). The e.m.f. is about 1.4 V per cell. When current is drawn, each coulomb causes a certain quantity of the zinc and electrolyte to be chemically changed. Towards the end of its life, the e.m.f. falls and internal resistance rises. Ideally, the cell would last indefinitely when not in use, but in practice a certain amount of 'local action' goes on inside, causing it to deteriorate; in time, therefore, a cell becomes exhausted even if never used. That time is called 'shelf life'.

If only the ingredients named above were included in a dry cell, its voltage would drop rapidly when supplying current, owing to the formation of a layer of hydrogen bubbles on the surface of the carbon. To reduce this layer a *depolariser* is used, consisting largely of manganese dioxide around the carbon electrode. Figure 28.1 shows a section of the cylindrical form of dry cell. Dry batteries of a rectangular shape are built up of interleaved flat plates.

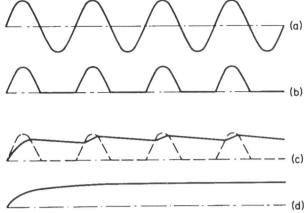

Figure 28.2 (a) Original a.c.; (b) after rectification; (c) the effect of a reservoir capacitor; (d) after passing through a smoothing filter

Where the need for compactness and long life (as in hearing aids and watches) justifies the greater cost, the mercury type of cell is preferred. The positive electrode is steel, the negative zinc, the electrolyte potassium hydroxide and the depolariser mercuric oxide. The voltage remains remarkably constant at about 1.2. These tend to be displaced by lithium cells (up to 3 V) and zinc-air cells (1.65 V).

The most commonly used secondary cell is the lead-acid type, in which the plates consist of lead frames or grids filled with different compounds of lead, and the electrolyte is dilute sulphuric acid. Accumulators have an e.m.f. of 2 V per cell and possess the great advantages of very low internal resistance and of being rechargeable.

Against this there are certain drawbacks. Lead is heavy. The acid is very corrosive, and liable to cause much damage if it leaks out or creeps up to the terminals. If the cell is allowed to stand for many weeks, even if unused, it becomes discharged, and if left in that condition it 'sulphates', that is to say, the plates become coated with lead sulphate which cannot readily be removed and which permanently reduced the number of ampere-hours the cell can yield on one charge. If the terminals are short-circuited, the resistance of the accumulator is so low that a very heavy current—possibly hundreds of amps—flows and is liable to cause permanent damage.

The size, weight and the presence of free acid makes the lead-acid battery unsuitable for some applications where a rechargeable battery would be useful, e.g., in portable television receivers. A better contender is an alternative type of secondary cell, available in sealed construction, which has electrodes of nickel and cadmium, and an electrolyte of potassium hydroxide. The e.m.f. is 1.2 V, less than that of a lead-acid or Leclanché cell, so that more cells are needed to give a desired voltage, but the cells are small and light enough for use in portable equipment.

28.3 D.C. from A.C.

Power from the mains costs not more than about one-fortieth that of power from dry batteries, so there is a strong incentive to use it except where very little power is needed or portability is essential. Almost everywhere mains power is in the form of a.c. Except for cathode heating, a.c. cannot be used as it is, because the requirement is for constant current or voltage in one direction. So we have to convert from a.c. to d.c. and therefore an essential part of a power unit is a *rectifier*.

If the a.c. is sinusoidal (as it usually nearly is) it can be represented as in Figure 28.2a. The result of connecting a rectifier in series with it is to abolish half of every cycle, as shown at b. The average of this is only about one-third of the peak value (Section 18.3) and in any case it would be useless in this form because it is not continuous. By using a *reservoir capacitor* (Section 18.6) this defect is removed and at the same time the average is brought nearly up to the peak voltage.

The conditions are not quite the same as in a detector, because the current drawn is usually much larger and the frequency much lower. Unless the reservoir capacitance is enormous, therefore, it has time to discharge appreciably between one cycle and the next, as in Figure 28.2c. This is described as d.c. with a ripple, or unsmoothed d.c., which would cause a hum if used to feed a receiver. The next requirement, then, is a *smoothing circuit*, which is a low pass filter (Section 18.10) designed to impede the a.c. ripple while allowing the d.c. to pass freely. If the filter is effective the result is as in Figure 28.2d, practically indistinguishable from current supplied by a battery.

In some applications, especially those in which varying amounts of current are drawn and passing through the resistance of the rectifier and filter cause undesirable variations in voltage, there is a need for *stabilisation* (Section 28.9) to keep the voltage automatically constant.

Although in many domestic receivers, particularly television sets, the rectifier is connected straight to the mains, this is not suitable if the voltage required is much different from the mains voltage, and even when it is not the use of an intervening transformer is better practice because it improves safety by isolating the circuits from the mains, enabling them to be earthed.

28.4 Types of Rectifier

At one time vacuum diodes (Section 9.4) were used not only as signal rectifiers (detectors) but also as power rectifiers. To handle the relatively large currents in power units, correspondingly large cathodes (consuming much power to heat) and anodes were needed; and to withstand high voltages the spacing between cathode and anode had to be relatively large, resulting in a high internal resistance and hence loss of voltage and power. So they have given place to semiconductor diodes, which have no heaters to waste power or delay starting while heating up. The voltage drop across them seldom exceeds 1.5 V, so the loss of power here is far less, and because there is much less heat to dissipate they can be much smaller. And they are reasonably cheap and practically unbreakable.

Copper oxide, selenium and germanium rectifiers have been used, in that historical order, and to some extent still are, but the tendency is for silicon to prevail. Although its voltage drop is about three times that of germanium it is still quite small, being of the order of 1 V, and the reverse voltage it can withstand is higher: up to about 1.5 kV, according to type. For higher voltages, diodes are connected in series, usually integrally. Another advantage of silicon over germanium is the smaller reverse current, or alternatively the higher allowable temperature and therefore output. The rectifiers are made not only in very small sizes for domestic equipment but also to provide many kilowatts for senders and industrial uses, and even these rectifiers are surprisingly small.

Thyristors are silicon rectifiers provided with a third electrode, the gate, which is capable of exercising a limited amount of control. Its symbol is shown in Figure 28.3. If the gate is kept at or near the same potential as the cathode,

Figure 28.3 General symbol for a thyristor

the thyristor behaves like a reversed rectifier in both directions; whatever the polarity of the voltage applied between anode and cathode, only leakage current will flow. But if the gate is made a few volts positive so that a small gate current flows, the thyristor behaves as an ordinary silicon rectifier. Although the gate can be used to turn the rectifier on, it is powerless to turn it off; the original state is restored only when forward anode voltage is taken off. This normally happens to a rectifier during half of every cycle. The usefulness of the gate is that it enables forward current to be prevented during any desired proportion of the other half-cycle and so enables the rectified output to be controlled, a facility used, for example, to control the brightness of mains-driven lamps.

28.5 Rectifier Circuits

There are various ways in which rectifiers can be connected in power unit circuits, and one of the simplest is indicated in Figure 28.4a. A transformer T is shown for stepping the mains voltage up or down as required, and a single rectifier—the symbol denotes any type—is connected in series with it and a reservoir C_1. A simple filter circuit consisting of a choke L and a capacitor C_2 completes the apparatus.

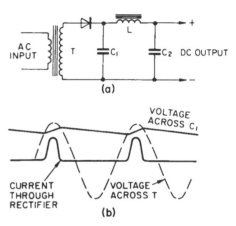

Figure 28.4 The voltages and rectifier current in a simple half-wave rectifier power-unit circuit (a) are shown in diagram (b)

This is called a *half-wave* rectifier circuit, because only one-half of each a.c. cycle is utilised, the other being suppressed (Figure 28.2b). If, as is desirable, the capacitance of C_1 is large enough for the voltage not to drop much between half-cycles, the proportion of each cycle during which the alternating voltage exceeds it is very small; and, as the recharging of C_1 has to take place during these brief moments, the rectifier is obliged to pass a current many times greater than that drawn off steadily at the output. If a large output current is needed, then the reservoir capacitance must be very large, the rectifier must be able to pass very heavy peak currents, and there are difficulties in designing the transformer to work under these conditions. So this circuit is used mainly for low-current high-voltage applications. Note that at the moment when the input voltage is at its negative peak it is in series with the rectified voltage across C_1, which is nearly as great and in the same direction. So the *peak inverse voltage* (p.i.v.) across the diode is nearly twice the peak input voltage, or getting on for three times the input r.m.s. voltage. This fact has to be remembered when choosing rectifiers.

In Figure 28.5 the half-wave circuit (a) is shown along with more elaborate arrangements; and for comparison the

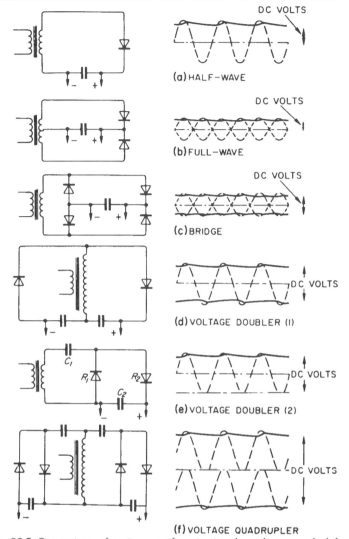

Figure 28.5 Comparison of various rectifier circuits, the voltage supplied from the transformer being the same in all. Note the comparative output voltages

transformer is assumed to supply the same total peak voltage in all of them.

(*a*) Half-wave. The no-load output voltage (i.e., the voltage when no current is being drawn) is very nearly equal to the peak input voltage; but, for the reasons given above, the voltage tends to drop considerably on load.

(*b*) Full-wave. By centre-tapping the transformer secondary, two type *a* circuits can be arranged, in series with the a.c. source, to feed the load in parallel. For the same voltage from the whole secondary, the output voltage is halved; but for a given rectifier rating the current is doubled, and at the same time is steadier, because each rectifier takes it in turn to replenish the reservoir. The resulting ripple, being at twice the a.c. frequency, is easier for the filter to smooth out; but it is more audible.

This is sometimes described as a two-phase circuit (the transformer gives two phases 180° apart); and though less suitable than *a* for high voltage is better for large current.

(*c*) Bridge. Circuit *b* yields a terminal which is either positive or negative with respect to the centre tap by a voltage slightly less than half the total transformer peak voltage. By connecting a second pair of rectifiers in parallel with the same transformer winding, but in the opposite polarity, a terminal of opposite polarity is obtained. The total voltage between these two terminals is therefore twice that between one of them and the centre tap; i.e., it is nearly equal to the peak voltage across the whole transformer. The outputs of the two pairs of rectifiers are effectively in series. Unless a half-voltage point is wanted, the centre-tap can be omitted.

This arrangement gives approximately the same no-load output voltage as *a*, but with the advantages of full-wave rectification.

(*d*) Voltage Doubler: first method. In contrast to *b*, the rectifiers are fed in parallel off the source to give outputs in series. As compared with *a*, the voltage is doubled and the ripple is twice the frequency. The voltage drop on load tends to be large, because each reservoir is replenished only once per cycle.

Like *c* (with centre tap), this circuit gives two supplies, one positive and one negative, from one transformer winding.

(*e*) Voltage Doubler: second method. R_1 and C_1 act, as explained in connection with Figure 18.6, to bring the negative peaks (say) to earth potential, so that the positive peaks are twice as great a voltage with respect to earth. R_2 and C_2 employ this as the input of a type *a* system, giving an output which, on no-load at least, rises to almost twice the transformer secondary peak voltage. Compared with *d*, one output terminal is common to the source, which may be convenient for some purposes. As $R_2 C_2$ is a half-wave rectifier this system suffers from the disadvantages of that type, as well as the losses in R_1, so is confined to high-voltage low-current power units.

(*f*) Voltage Quadrupler. By connecting a second type *e* circuit in parallel with the transformer, to give an equal voltage output of opposite polarity, the total output voltage tends towards four times the peak of the supply. The conversion of *e* to *f* is analogous to the conversion of *a* to *d*.

28.6 Filters

We have already used a simple filter circuit (R_F and C_F in Figure 18.10) to smooth out the unwanted r.f. left over from the rectification process in a detector. The same circuit is often used in power units, but the values of R_F and C_F are greatly different. The frequency of the ripple to be smoothed following a half-wave rectifier is usually 50 Hz, and 100 Hz with full-wave. For effective smoothing, the reactance of C_F at this frequency must be low compared with the resistance R_F. And because of the relatively heavy current in a power unit, R_F must itself be quite low if there is not to be an excessive wastage of volts.

For this reason an iron-cored choke coil is often substituted. It has to be carefully designed if the core is not to be saturated by the d.c. (Figure 4.8) which greatly reduces the inductance and therefore the effectiveness of the coil for smoothing. A typical filter, LC_2, is seen in the half-wave rectifier circuit in Figure 28.4.

Although a reservoir capacitor brings the rectified voltage almost up to the peak input voltage when no current is being taken, the voltage falls off fairly steeply with increase of load current. In technical language, the regulation is not very good. For purposes where it is important that the voltage should remain steady in spite of a fluctuating load current—for example, high-power Class B amplifiers—what is called a choke-input filter is used. In the circuit diagram this looks the same as, say, Figure 28.4*a*, except that the reservoir capacitor is missing. This is not the only difference, however, for the choke has to be of a special kind, known as a swinging choke. Instead of being designed so that its inductance varies as little as possible, its inductance increases substantially as current falls, with the aim of maintaining a nearly constant voltage across it. The output voltage falls very rapidly up to, say, one-tenth full load current (below which the current is not allowed to fall) and therefore is relatively steady. An advantage is that the current through the rectifier is relatively constant, instead of having peak values much greater than the mean as in Figure 28.4*b*.

A choke having high reactance at low frequencies such as 100 Hz is large, heavy and costly. One way of avoiding the need for such components is to use a high ripple frequency. If, for example, the frequency can be raised to 25 kHz, the inductance can be reduced by a factor of 1/250 whilst maintaining the same reactance. Similarly the size of the smoothing capacitors can be reduced in the same ratio. A method of obtaining a high ripple frequency is described in Section 28.11. An alternative method of avoiding the use of a bulky choke at low ripple frequencies is to use a resistor for smoothing whenever possible, and for this reason electrolytic capacitors of very high capacitance—1 mF or more—are now customary. Even so, their impedance to the signal-frequency currents flowing through the power source may not be negligible, and a potential difference set up across this impedance by (say) the signal current in the output stage is passed on to all the other stages fed from the same supply and may seriously upset the working, possibly even causing continuous oscillation. We have, in fact, a form of feedback.

28.7 Decoupling

To obviate such undesirable conditions, *decoupling* is used. It is simply an individual filter for those stages liable to be seriously affected. As those stages are generally the preliminary ones, taking only a small proportion of the total current, a resistor having sufficient impedance at the frequencies concerned can be used. The loss in voltage may be tolerable, as these stages are often required to run at lower voltages than the power output stage; if so, a single cheap component is made to serve the double purpose of a voltage-dropping resistance and (in conjunction with a capacitor) a decoupler. This is the essence of good commercial design.

In Figure 28.6, which is an outline circuit diagram of part of an a.f. amplifier, *R* and *C* are decoupling components. They serve as a filter, preventing signal voltages set up at point X by signal currents flowing through the power source from reaching the previous stage and then being amplified. In this example the output stage is fed direct from the reservoir capacitor, so there will be significant hum voltage at X. This may not matter to the output stage because it is push-pull, which tends to balance out hum, but it could be serious if it came in via the previous stage, so the extra smoothing given by *R* and *C* is needed for this reason too. Because of the comparatively small current flowing through *R* its resistance can be made very much larger and therefore more effective than could

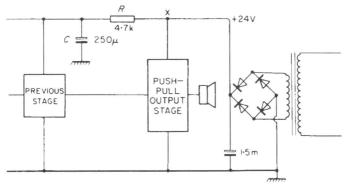

Figure 28.6 Example of the use of decoupling

possibly be used in the main supply line. Signal currents in the stage fed by it find an easy path via *C* and so are kept out of other parts of the amplifier.

It is the extensive use of decoupling and filtering that makes multi-stage circuit diagrams look so complicated. But once their purpose has been grasped it is easy to sort them out from the main signal circuits. Note the use of a detached 'chassis' symbol at the foot of *C* to avoid drawing a confusing crossing wire to the base line. Considerable clarification of complicated circuits can be obtained in this way.

28.8 Bias Supplies

From time to time we have considered various methods of providing bias, and we now review them.

The most obvious way is by means of a battery or other source in series with the circuit connecting the two electrodes concerned, as in Figs. 10.6 and 11.2. A slight modification is the parallel-feed system (Figure 13.4 and many others) where R_B is a resistance so high that it causes negligible loss of the signal applied via *C*. A choke coil is sometimes used instead of high resistance if the bias voltage must not be subject to drop due to current through it.

Obviously it would be possible to replace a bias battery by a mains power unit. Except for high-power equipment this is not done, because the power required is so small and there are more economical alternatives.

With valves and f.e.ts, which normally are biased opposite in polarity to the anode or drain, one of those alternatives is to insert a resistor as explained in Sections 13.5 and 13.6. The required amount of bias is obtained by choice of resistance. To some extent this source of bias is self-adjusting; if an excessive current flows the bias voltage is correspondingly large and the rise in current is restrained. Also it is convenient for providing each stage with its own most suitable bias. Strictly it is cathode or source bias that is supplied, at the expense of anode or drain voltage, and this loss may sometimes need to be taken into account.

If only a simple resistor is used, it carries not only the steady (or z.f.) component of current but also the signal current, which produces a voltage in opposition to the input. We have, in fact, negative feedback (Section 19.15). If that is wanted, well and good; but if not, then something has to be done to provide a path of relatively negligible impedance for the signal currents, without disturbing the z.f. resistance. A by-pass capacitor, C_s in Figure 13.8, serves the purpose; for r.f. a value of 0.1 µF or even less is enough, but for the lowest audio frequencies 50 µF or 100 µF may be needed.

Sometimes it is convenient to have a bias voltage that is equal to the peak value of the incoming signal. This is obtained by using the clamping principle (Section 25.8). With a.f. signals it is sometimes used for the stage immediately following a detector. Although a separate diode detector is commonly used, the gate-source or grid-cathode diode can be made to serve this purpose.

A very common application of this type of bias is for oscillators (Section 14.10); the fact that the bias voltage is nearly equal to the peak oscillation voltage means that it increases and so acts as a governor to prevent the amplitude of oscillation from increasing excessively. If the positive feedback in the oscillator is very large and the time constant equal to the duration of many cycles, the first burst of oscillation may generate such a large bias that the device so biased is cut off, so quenching the oscillation, which cannot resume until some of the charge has had time to leak off the grid. Oscillation then restarts, is again quenched, and so on. This phenomenon of intermittent oscillation is known as *squegging*, and is sometimes usefully employed. Elsewhere, it is a nuisance, which can be cured by reducing the positive feedback or the time constant of the biasing circuit or both. The blocking oscillator (Section 25.4) is sometimes considered as an extreme form of squegging.

28.9 Stabilisation

Often there is a need to maintain the terminal voltage of a power unit constant despite changes in the amount of current drawn. Sometimes the dual problem arises: to keep the output current constant regardless of the voltage. In the first case the desired result would be obtained if the internal resistance of the power unit was reduced to zero; in the second, if it was increased to infinity. We have seen (Section 19.17) that the tendency of negative feedback is to do one or other of these things, according to how it is

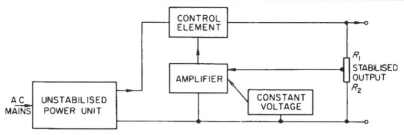

Figure 28.7 Showing the principles of most voltage-stabilised power units

applied, and in fact many stabilised power units can be regarded as negative-feedback systems. But it is perhaps rather clearer to consider them from the following angle.

The general principle of a voltage stabiliser, illustrated in Figure 28.7, is to compare its output voltage (or a certain fraction of it) with a known constant voltage to which it should be equal. If it is not, the difference voltage is amplified and used to vary the output voltage in the direction that brings it toward the desired level. The purpose of the amplification is to ensure that an extremely small 'error' in the actual output voltage is sufficient to exercise effective control. Besides rendering the output voltage almost unaffected by variations (within limits) in the impedance of the load, it does the same with regard to fluctuations in mains voltage.

An essential requirement of such a stabiliser is a source of constant voltage. A much-used device for obtaining it is a special kind of diode, known as a Zener diode, which is distinguished by a very constant (and usually low) breakdown voltage (Section 9.11). One takes great care, with other types of diodes and transistors, not to apply a reverse voltage sufficiently large to risk breakdown, for that would suddenly destroy nearly all resistance and the device would probably be ruined instantly by the uncontrolled current flow. Zener diodes, however, are used in a perpetual state of breakdown. They are no more immune from sudden death than any other kind; survival depends on their being used with sufficient resistance in series to limit the current to a safe amount. This resistance serves also to absorb changes in the voltage of the supply. Figure 28.8 shows a typical Zener characteristic curve, together with a load line, R. With 12 V applied to the diode (note the modified symbol to indicate that the diode is a Zener type) and R in series, R being 1 kΩ, the voltage across R is shown to be 6.3 V and the current through it therefore 6.3 mA.

The amount of impurity ('doping') in this particular silicon Zener diode is such that its voltage at 6.3 mA is 5.7 V, which happens to be about the least affected by temperature so is often chosen although Zeners are made for lower and much higher voltages. If the supply voltage fell even as much as 2 V, so that the load line took up the dashed position, the Zener voltage would fall by only 0.02 V. It therefore gives a 100-fold stabilisation. Fed from a higher voltage through a higher resistance, the ratio would be even greater. And any hum ripple in the supply is greatly reduced. So the value of Zener diodes as a source of known constant voltage is obvious.

The simple arrangement shown in Figure 28.8 can be used by itself as a stabilised power unit. Suppose 2 mA is drawn from across the diode; the current from the source remains practically at 6.3 mA because there is negligible change in voltage across R. So the diode current drops to 4.3, but this has hardly any effect on the voltage across it. Obviously this particular design is very limited in the power it can provide, but Zener diodes are available that can pass 100 mA at 75 V and supply many watts. One disadvantage is that the stabilised voltage is fixed by the particular Zener voltage. Another is that the supply voltage

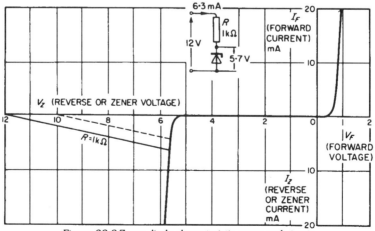

Figure 28.8 Zener diode characteristic curve and circuit

must be much greater and a lot of power is wasted in the series resistor as well as in the diode. But in the role of constant-voltage device in Figure 28.7 it is not required to supply more than a few milliwatts. And because its voltage is compared with a fraction $R_2(R_1 + R_2)$ of the output voltage, any convenient Zener voltage can be used to stabilise almost any output voltage. Further, by making the R_1R_2 potential-divider tapping variable the output voltage can be varied as required.

The design of voltage and current stabilisers has attracted an extraordinary amount of attention, and thousands of books and articles have been written on the subject, in which details of the contents of the boxes in Figure 28.7 can be found.

This section will now be rounded off with a simple current-stabilising device, Figure 28.9. The Zener diode Z, in conjunction with R_B which is chosen to pass a reasonable current through it, corresponds to the 'constant voltage' box in Figure 28.7. The current to be kept constant, passing through R_E, provides the voltage V_E to be compared with

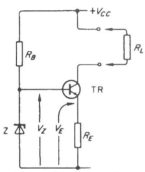

Figure 28.9 Simple type of current stabiliser circuit

V_Z. The difference between them is applied to TR. Because of the high mutual conductance of the transistor chosen for this role, a very small change in V_{BE}, caused by any change in I_E, brings into play a very steep correcting action. Changes in load resistance, R_L, cause corresponding changes in V_{EC}, but we know (from Figure 11.7, for instance) that this has very little effect on I_C (and hence I_E) and such effect as it has is corrected as just explained. But of course there is a limit to the load resistance through which any chosen current can be passed. It occurs when the voltage across R_L nearly equals V_{CC} and the transistor is bottomed. So a fairly high value of V_{CC} (but not exceeding the rated maximum for the transistor) is desirable. The higher V_Z, the stronger the correcting action. The amount of current to be stabilised can be controlled by varying R_E.

28.10 E.H.T.

Cathode-ray tubes in television and radar equipment need a voltage which is measured in kilovolts; 25 kV is a usual figure for colour television receivers. Fortunately the amount of current needed is quite small, of the order of 1 mA. A step-up mains transformer to develop such a voltage in the conventional way by a circuit such as Figure 28.4 would be impracticably large, heavy and expensive, and also lethal. These objections are only partly removed by using the voltage quadrupler circuit (Figure 28.5f). So, as mentioned in Section 23.10, use is made of the very high rate of change of current in the line-scanning transformer, seen in Figure 23.20. The voltage developed across an inductor being proportional to the rate of current change, far fewer turns of wire are needed than with the very slow rate obtained with 50 Hz sinusoidal current for equal peak values. And the 15 625 Hz frequency, combined with the smallness of the current, enables all smoothing components to be dispensed with, apart from the reservoir capacitor formed by the conducting coatings on the c.r. tube.

It is of course not necessary to have line-scanning equipment in order to make use of the same convenient principle for generating e.h.t. for other purposes. All that is needed is a blocking oscillator (Section 25.4) of suitable frequency, with an extra winding on its transformer to step up the voltage, and a rectifier and reservoir capacitor.

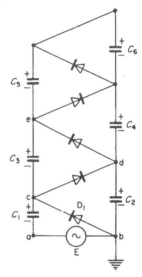

Figure 28.10 Cockcroft-Walton voltage multiplier circuit

It can sometimes work out more economically to use a lower step-up ratio (fewer turns, of thicker wire) and to multiply the voltage by what is known as the Cockcroft-Walton circuit. This has the additional advantage of providing lower (but still high) voltages for focusing, etc. It is simply an extension of the principle of the voltage-doubler circuit, Figure 28.5e. A voltage sextupler of this type is shown in Figure 28.10. The first part of it, *abc*, is a straightforward type of rectifier in which a steady voltage very nearly (when the current is small, as it is in this type of equipment) equal to the input voltage E is maintained across C_1. The potential of c (relative to earth, b) therefore alternates between 0 and $2E$. This is the input to the second rectifier circuit, *bcd*, but, the peak voltage being twice that in the first, C_2 is charged to $2E$. The third rectifier circuit, *cde*, similarly charges C_3 to $2E$. And so on, finally giving an output voltage of nearly $6E$. Note that the potentials across the capacitors in both left- and right-hand columns remain practically constant, but relative to the right-hand (output) column the left-hand one as a whole rises and falls in potential between $-E$ and $+E$ owing to the generator.

28.11 Switch-mode Power Suppliers

A useful development in power suppliers is the 'switch-mode' circuit in which the output from a mains rectifier feeds an astable pulse generator such as a multivibrator. This regularly chops the rectifier output to form an alternating voltage which is transformed, rectified and smoothed to give the required z.f. output voltage. This may seem an unnecessarily complicated method of obtaining a z.f. supply, but it has important advantages. As already pointed out (Section 28.6), if the multivibrator operates at a high frequency (e.g., 25 kHz) effective smoothing can be achieved with quite small inductors and capacitors. Moreover the output voltage of the supplier can be controlled by adjusting the mark/space ratio of the multivibrator (which determines the fraction of each multivibrator cycle during which the high-frequency rectifier conducts) and this can be used in an elegant method of stabilising the output voltage. A sample of the output voltage is compared with a standard constant voltage (as in Section 28.9) and the difference, after amplification, is used to control the mark/space ratio. Power suppliers of this type are very efficient and compact, and are often used in television receivers.

In fact the design of power supply circuits for television receivers poses a number of interesting problems. For example, all the circuits belonging to the receiver proper must be capable of being earthed so that connections to the aerial, headphones and external equipment such as video recorders and hi-fi amplifiers can be made without fear of electric shock.

In the past this has been possible by connecting the receiver to the mains via a transformer which provided the necessary isolation. But mains transformers are bulky, heavy and expensive so that it is not surprising that manufacturers like to eliminate them. One way of doing this is to connect the mains supply to a rectifier and to use the smoothed output to supply a switch-mode power supplier. The chopped d.c. so generated feeds the primary of a chopper transformer with a number of secondary windings proportioned to provide the various z.f. voltages required by the receiver. Thus one transformer is disposed of, but another is required! But the new one can operate at, say, 45 kHz so needs only 1/900th of the inductance and the rectifiers on the secondary side need only 1/900th of the smoothing capacitance. Moreover the chopper transformer can be designed to have very good insulation between primary and secondary windings so that the secondaries can be earthed. The components at mains voltage can be contained within a protected enclosure.

28.12 Cathode Heating

As stated in Section 9.3, the cathode of a valve or c.r. tube may be either directly or indirectly heated. The power for either type can be obtained from a battery, but in practice directly heated valves are then used because they require only about one-tenth of the power needed by the indirectly heated types, and therefore a given battery will run them for much longer, or alternatively a much smaller battery can be used. These points are important, because battery power is so much more expensive than that drawn from the mains.

The vast majority of public electricity supplies are a.c., which can be stepped down to any convenient voltage for valve heating by means of a transformer. If an attempt were made to run directly heated valves in this way it would be unsuccessful, because the slight variation in temperature between peak and zero of the a.c. cycle, and also the variations of potential between grid and filament from the same cause, would produce variations in anode current, which would be amplified by the following stages and cause a loud low-pitched hum. A possible exception is a carefully arranged output stage, which of course is not followed by any amplification.

The temperature of the relatively bulky indirectly heated cathode changes too slowly to be affected by 50 Hz alterations, and as it carries none of the a.c. its potential is the same all over. Incidentally this improves the valve characteristics, and so does its larger area compared with a filament. Further, its greater rigidity allows the grid to be mounted closer to it, resulting in still greater mutual conductance.

A usual rating for a.c. valve heaters is 6.3 V. This odd voltage is to enable these valves to be run alternatively from 6 V or 12 V car batteries, which average 6.3 V or 12.6 V under working conditions. They are usually connected in parallel like battery valves, but if they are all designed to take the same current, so that they can be run in series, the rated voltage is anything from 6.3 to 40 according to type. A common practice is to connect all the valve heaters and c.r.t. heater of a set in series, together with sufficient resistance to enable them to be run direct from 200–250 V mains, either a.c. or d.c.

Appendix A Algebraic Symbols

A.1 Letter Symbols

If a car had travelled 150 kilometres in 3 hours, we would know that its average speed was 50 kilometres per hour. How? The mental arithmetic could be written down like this:

$$150 \div 3 = 50$$

That is all right if we are concerned only with that one particular journey. If the same car, or another one, did 12 kilometres in a quarter of an hour, the arithmetic would have to be:

$$12 \div \tfrac{1}{4} = 12 \times 4 = 48$$

To let anyone know the speed of any car on any journey, it would be more than tedious to have to write out the figures for every possible case. All we need say is, 'To find the average speed in kilometres per hour, divide the number of kilometres travelled by the number of hours taken'.

What we actually would say would probably be briefer still: 'To get the average speed, divide the distance by the time'. Literally that is nonsense, because the only things that can be arithmetically divided or multiplied are numbers. But of course the words 'the number of' are (or ought to be!) understood. The other words—'kilometres per hour', 'kilometres', 'hours'—the units of measurement, as they are called—may also perhaps be taken for granted in such an easy case. In others, missing them out might lead one badly astray. Suppose the second journey had been specified in the alternative form of 12 kilometres in 15 minutes. Dividing one by the other would not give the right answer in kilometres per hours. It would, however, give the right number of kilometres per minute.

In electronics, as in other branches of physics and engineering, this matter of units is often less obvious, and always has to be kept in mind.

Our instruction, even in its shorter form, could be abbreviated by using mathematical symbols as we did with the numbers:

Average speed = Distance travelled ÷ Time taken

Here we have a concise statement of general usefulness; note that it applies not only to cars but to aircraft, snails, bullets, space capsules, and everything else that moves.

Yet even this form of expression becomes tedious when many and complicated statements have to be presented. So for convenience we might write:

$$S = D \div T$$

or alternatively

$$S = \frac{D}{T}$$

or, to suit the printer, $S = D/T$

A.2 What Letter Symbols Really Mean

This is the stage at which some people take fright, or become impatient. They say, 'How *can* you divide D by T? Dividing one letter by another doesn't *mean* anything. You have just said yourself that the only things that can be divided are numbers!' Quite so. It would be absurd to try to divide D by T. Those letters are there just to show what to do with the numbers when you know them. D, for example, has been used to stand for the number of kilometres travelled.

The only reason why the S, D and T were picked for this duty is that they help to remind one of the things they stand for. Except for that there is no reason why the same information should not have been written as:

$$x = \frac{y}{z}$$

or even

$$\alpha = \frac{\beta}{\gamma}$$

so long as we know what these symbols were intended to mean. As there are only 26 letters in our alphabet, and fewer still in the Greek, we cannot allocate any one of them permanently to mean 'average speed in kilometres per hour'. x, in particular, is notoriously capable of meaning absolutely anything. Yet to write out the exact meaning every time would defeat the whole purpose of using the letters. How, then, does one know the meaning?

Well, some meanings have been fixed by international agreement. There is one symbol that means the same thing every time, not only in electronics but in all the sciences— the Greek letter π (read as 'pie'). It is a particularly good example of abbreviation because it stands for a number that would take eternity to write out in full—the ratio of the circumference of a circle to its diameter, which begins: 3.141 592 653 5 . . . (The first three or four decimal places give enough accuracy for most purposes.)

Then there is a much larger group of symbols which have been given meanings that hold good throughout a limited field such as electrical engineering, or one of its subdivisions such as electronics, but are liable to mean something different in, say, astronomy or hydraulics. These have to be learnt by anyone going in for the subject seriously.

Lastly, there are symbols that one uses for all the things that are not in a standard list. Here we are free to choose our own; but there are some rules it is wise to observe. It is common sense to steer clear as far as possible from symbols that already have established meanings. The important thing is to state the meaning when first using the symbol. It can be assumed to bear this meaning to the end of the particular occasion for which it was attached; after that, the label is taken off and the symbol thrown back into the common stock, ready for use on another occasion, perhaps with a different label—provided it is not likely to be confused with the first.

Sometimes a single symbol might have any of several different meanings, and one has to decide which it bears in that particular context. The Greek letter µ (pronounced 'mew') is an example that occurs in this book. When the subject is a radio valve, it can be assumed to mean 'amplification factor'. But if iron cores for transformers are being discussed, µ should be read as 'permeability'. And if it comes before an upright letter it is an abbreviation for 'micro-', meaning 'one millionth of'.

A.3 Some Other Uses of Symbols

The last of these three meanings of µ is a different kind of meaning altogether from those we have been discussing. Until then we had been considering symbols as abbreviations for quantities of certain specified things, such as speed in kilometres per hour. But there are several other uses to which they are put.

Instead of looking on '$S = D/T$' as an instruction for calculating the speed, we can regard it as a statement showing the relationship to one another of the three quantities, speed, time and distance. Such a statement, always employing the 'equals' sign, is called by mathematicians an *equation*. From this point of view S is no more important than T or D, and it is merely incidental that the equation was written in such a form as to give instructions for finding S rather than for finding either of the other two. We are entitled to apply the usual rules of arithmetic in order to put the statement into whatever form may be most convenient when we come to substitute the numbers for which the letters stand.

For example, we might want to be able to calculate the time taken on a journey, knowing the distance and average speed. We can divide or multiply both 'sides' of an equation by any number (known, or temporarily represented by a letter) without upsetting their equality. If we multiply both sides of $S = D/T$ by T we get $ST = D$ (ST being the recognised abbreviation for $S \times T$). Dividing both sides of this new form of the equation by S we get $T = D/S$. Our equation is now in the form of an instruction to divide the number of kilometres by the speed in kilometres per hour (e.g., 150 kilometres at 50 km/h takes 3 hours).

When you see books on electronics (or any other technical subject) with pages covered almost entirely with mathematical symbols, you can take it that instead of explaining in words how their conclusions are reached, the authors are doing it more compactly in symbols. It is because such pages are concentrated essence, rather than that the meanings of the symbols themselves are hard to learn, that makes them difficult. The procedure is to express the known or assumed facts in the form of equations, and then to combine or manipulate these equations according to the established rules in order to draw some useful or interesting conclusions from those facts.

This book, being an elementary one, explains things in words, and only uses symbols for expressing the important facts or conclusions in concise form, or for rearranging them.

A.4 Abbreviations

Another use for symbols is for abbreviation pure and simple. We have already used one without explanation—km/h—because it is known that this means 'kilometres per hour'. The kilometre-per-hour is a unit of speed. '£' is a familiar abbreviation for the pound sterling. The unit of electrical pressure is the volt, denoted by the abbreviation V. Sometimes it is necessary to specify very small voltages, such as 5 millionths of a volt. That could be written 0.000 005 V. But a more convenient abbreviation is $5\,\mu V$, read as '5 microvolts'. The list in Appendix D gives the abbreviations commonly used in electronics.

Still another use for letters is to point out details on a diagram. R stands for electrical resistance, but one has to guess whether it is intended to mean resistance in general, as a property of conductors, or the numerical value of resistance in an equation, or the particular resistance marked R in a diagram. Often it may combine these meanings, being understood to mean 'the numerical value of the resistance marked R in Figure So-and-so'.

Attaching the right meanings to symbols probably sounds dreadfully difficult and confusing. So do the rules of a new game. The only way to defeat the difficulties is to start playing the game.

Here are a few more hints about symbols.

A.5 How Numbers are Used

One way of making the limited stock of letters go further is to use different kinds of type. It has become standard practice to distinguish symbols for physical quantities by italic (sloping) letters, leaving the Roman (upright) ones for abbreviations; for example, 'V' denotes an unspecified amount of electrical potential difference, reckoned in volts, for which the standard abbreviation is 'V'. 'R' is the resistance of a resistor R. And so on.

Another way of making letters go further is to number them. To prevent the numbers from being treated as separate things they are written small near the foot of the letter ('subscript'). For example, if we want to refer to several different resistances we can mark them R_1, R_2, R_3, etc. Sometimes a modification of a thing denoted by one symbol is distinguished by a tick or dash; A might stand for the amplification of a receiver when used normally, and A' when modified in some way.

But on no account must *numbers* be used 'superscript' for this purpose, because that already has a standard meaning. 5^2 (read as '5 squared') means 5×5; 5^3 ('5 cubed') means $5 \times 5 \times 5$; 5^4 ('5 to the 4th power') means 5

× 5 × 5 × 5; x^2 means the number represented by x multiplied by the same number.

x^{-2} seems nonsensical according to the rule just illustrated. It has been agreed, however, to make it mean $1/x^2$ (called the *reciprocal* of x squared). The point of this appears most clearly when these superscripts, called *indices* (singular: *index*), are applied to the number 10. $10^4 = 10\,000$, $10^3 = 1000$, $10^2 = 100$, $10^1 = 10$, $10^0 = 1$, $10^{-1} = 0.1$, $10^{-2} = 0.01$, and so on. The rule is that the power of 10 indicates the number of places the decimal point has to be away from 1; with a positive index it is on the right; negative, on the left. Advantage is taken of this to abbreviate very large or very small numbers. It is easier to see that 0.000 000 002 6 is 2.6 thousand-millionths if it is written 2.6×10^{-9}. Likewise 2.6×10^{12} is briefer and clearer than 2 600 000 000 000.

$x^{1/2}$ also needs explanation. It is read as 'the square root of x', and is often denoted by $\sqrt{x}$. It signifies the number which, when multiplied by the same number, gives x. In symbols, $(\sqrt{x})^2 = x$.

Note that $10^6 \times 10^6 = 10^{12}$, and that $10^{12} \times 10^{-3} = 10^{12}/10^3 = 10^9$. In short, multiply by adding indices, and divide by subtracting them. This idea is very important in connection with decibels (Section 19.2).

Appendix B   Graphs

B.1   What is a Graph?

Most of us when we were in the growing stage used to stand bolt upright against the edge of a door to have our height marked up. The succession of marks did not convey very much when reviewed afterwards unless the dates were marked too. Even then one had to look closely to read the dates, and the progress of growth was difficult to visualise. Nor would it have been a great help to have presented the information in the form of a table with two columns—Height and Date.

But, disregarding certain technical difficulties, imagine that the growing boy had been attached to a conveyor belt which moved him horizontally along a wall at a steady rate of, say, one foot per year, and that a pencil fixed to the top of his head had been tracing a line on the wall (Figure B.1). If he had not been growing at all, this line would, of course, be straight and horizontal. If he had been growing at a uniform rate, it would be straight but sloping upwards. A variable rate of growth would be shown by a line of varying slope.

In this way the progress over a period of years could be visualised by a glance at the wall.

B.2   Scales

To make the information more definite, a mark could have been made along a horizontal line each New Year's Day and the number of the year written against it. In more technical terms this would be a time scale. The advantage of making the belt move at uniform speed is that times intermediate between those actually marked can be identified by measuring off a proportionate *distance*. If one foot represented one year, the height at, say, the end of May in any year could be found by noting the height of the pencil line 5 inches beyond the mark indicating the start of that year.

Similarly a scale of height could be marked anywhere in a vertical direction. It happens in this case that height would be represented by an equal height. But if the graph were reproduced in a book, as here, although height would still represent height it would have to do so on a reduced scale, say half an inch to a foot.

To guide the eye from any point on the information line to the two scales, it is usual to plot graphs on paper printed with horizontal and vertical lines so close together that a pair of them is sure to come near enough to the selected point for any little less or more to be judged. The position of the scales is not a vital matter, but unless there is a reason for doing otherwise they are marked along the lines where the other quantity is zero. For instance, the time scale would be placed where height is nil; i.e., along the foot of the wall. If, however, the height scale were erected at the start of the year 1 AD there would be over three-quarters of a kilometre of blank wall between it and the pencil line. To avoid such inconvenience we can use a 'false zero', making the scale start at or slightly below the first figure in which we are interested. A sensible way of doing it in this example would be to reckon from the time the boy was born, as in Figure B.1. The point that is zero on both scales is called the *origin*.

False zeros are sometimes used in a slightly shady manner to give a wrong impression. Figure B.2a shows the sort of graph that might appear in a company-promoting prospectus. The word 'Profit' would, of course, be in big type, and the figures indistinct, so that the curve would seem to indicate a sensational growth in profits. Plotted

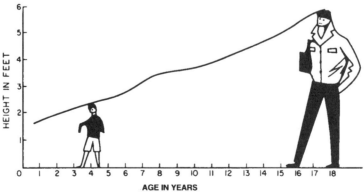

Figure B.1 Simple (but inconvenient) system for automatic graph plotting of human growth. The equipment consists of a conveyor belt moving at the rate of one foot per year, a pencil, and a white wall

without a false zero, as in Figure B.2b, it looks much less impressive. Provided that it is clearly admitted, however, a false zero is useful for enabling the significant part of the scale to be expanded and so read more precisely.

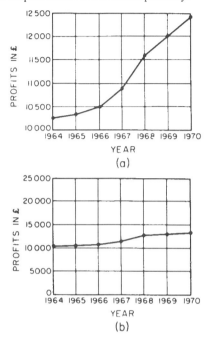

Figure B.2 (a) Typical financial graph arranged to make the maximum impression consistent with strict truthfulness. (b) The same information presented in a slightly different manner

Figure B.2, by the way, is not a true graph of the sort a mathematician would have anything to do with. Profits are declared annually in lumps, and the lines joining the dots that mark each year's result have no meaning whatever but are there merely to guide the eye from one to another. In a true graph a continuous variation is shown, and the dots are so close together as to form a line. This line is technically a curve, whether it is what is commonly understood as a curve or is as straight as the proverbial bee-line.

B.3   What a 'Curve' Signifies

Every point on the curve in Figure B.1 represents the height of the boy at a certain definite time (or, put in another way, the time at which the boy reached a certain height). Since it is possible to distinguish a great number of points along even a small graph, and each point is equivalent to saying 'When the time was T years, the height was H feet', a graph is not only a very clear way of presenting information, but a very economical one.

Here we have our old friend T again, meaning time, but now in years instead of hours. It is being used to stand for the number of years, as measured along the time scale, represented by any point on the curve. Since no particular point is specified, we cannot tell how many the corresponding height may be, so have to denote it by H. But directly T is specified by a definite number, the number or value of H can be found from the curve. And vice versa.

In its earlier role (Appendix A, Section A1) T took part in an equation with two other quantities, denoted by S and D. The equation expressed the relationship between these three quantities. We now see that a graph is a method of showing the relationship between two quantities. It is particularly useful for quantities whose relationship is too complicated or irregular to express as an equation—the height and age of a boy, for instance.

B.4   Three-dimensional Graphs

Can a graph deal with three quantities? There are three ways in which it can, none of them entirely satisfactory.

One method is to make a three-dimensional graph by drawing a third scale at right angles to the other two. Looking at the corner of a room, one can imagine one scale along the foot of one wall, another at the foot of the other wall, and the third upwards along the intersection between the two walls. The 'curve' would take the form of a surface in the space inside—and generally also outside—the room. There are obvious difficulties in this.

Another method is to draw a number of cross-sections of the three-dimensional graph on a two-dimensional graph. In other words, the three variable quantities are reduced to two by assuming some numerical value for the third. We can then draw a curve showing the relationship between the two. A different numerical value is then assumed for the third, giving (in general) a different graph. And so on.

Let us take again as an example $D = ST$. If any numerical value is assigned to the speed, S, we can easily plot a graph connecting D and T, as in Figure B.3a. When the speed is zero, the distance remains zero for all time, so the '$S = 0$' curve is a straight line coinciding with the time scale. When S is fixed at 1 kilometre per hour, D and T are always numerically equal, giving a line ('$S = 1$') sloping up with a 1-in-1 gradient (measured by the T and D scales). The '$S = 2$' line slopes twice as steeply; the '$S = 3$' line three times as steeply, etc.

If a fair number of S curves are available, it is possible to replot the graph to show, say, S against T for fixed values of D. A horizontal line drawn from any selected value on the D scale in Figure B.3a (such as $D = 10$) cuts each S curve at one point, from which the T corresponding to that S can be read off and plotted, as in Figure B.3b.

Each point plotted in a corresponds to an S curve in b, and when joined up they give a new curve (really curved this time!). This process has been demonstrated for only one value of D, but of course it could be repeated to show the time/speed relationship for other distances. Figure B.3b gives numerical expression to the well-known fact that the greater the speed the shorter is the time to travel a specified distance.

Graphs of this kind are important in many branches of electronics; for example, with transistors, whose relationships cannot be accurately expressed as equations.

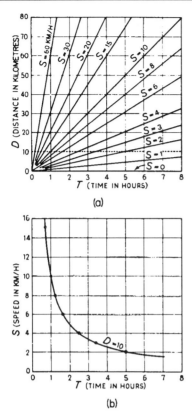

(a)

(b)

Fig B.3 (a) One way of graphing the relationship between speed, time and distance of a journey. Graphs of speed against time for any given distance can be derived from it (b)

shows the distance travelled in 2 hours at 30 kilometres per hour, 3 hours at 20 kilometres per hour, etc. No one would bother to draw a graph for such a simple equation, but this is quite a useful technique for more complicated ones. Figure 20.3 is an example.

Lastly, two (or more) variables can be grouped so as to reduce the total to two. In our $D = ST$, D can be plotted against ST, which in mathematical language is called the product of S and T. The resulting very simple graph is shown as Figure B.4. A point on the ST scale, say 60,

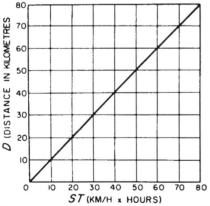

Figure B.4 A third method of graphing three variables is to combine two of them; in this example the product of S and T is used as one of the two variables in the graph

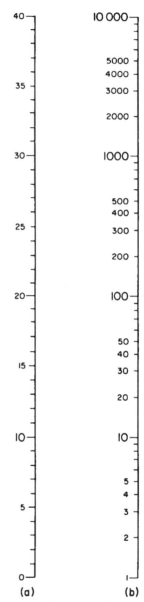

(a)   (b)

Figure B.5 Comparison between uniform or 'linear' scale (a), and logarithmic scale (b)

B.5 Significance of Slope

Even from simple two-dimensional graphs like Figure B.1 it may be possible to extract information that is worth replotting separately. It is clear that a third quantity is involved—*rate of growth*. This is not an independent variable; it is determined by the other two. For it is nothing else than rate-of-change-of-height. As we saw at the start, when the curve slopes upward steeply, growth is rapid; when the curve flattens out, it means that growth has ceased; and if the curve started to slope downwards—a negative gradient—it would indicate negative growth. So it would be possible to plot growth in, say, inches per year, against age in years. This idea of rate-of-change, indicated on a graph by a slope, is the essence of no less a subject than the differential calculus, which is much easier than it is often made out to be. If you are anxious to get a clearer view of any electrical subject it would be well worth while reading at least the first few chapters of an elementary book on the differential calculus, such as S. P. Thompson's classic *Calculus Made Easy* (Macmillan).

B.6 Non-Uniform Scales

Finally, although uniform scales for graphs are much the easiest to read, there may quite often be reasons which justify some other sort of scale. The most important, and the only one we need consider here, is the *logarithmic* scale. In this, the numbers are so spaced out that a given distance measured anywhere along the scale represents not a certain *addition*, but a certain *ratio* or *multiplication*. Musicians use such a scale (whether they know it or not); their intervals correspond to ratios. For example, raising a musical note by an octave means doubling the frequency of vibration, no matter whereabouts on the scale it occurs. Users of slide-rules were familiar with the same sort of scale. Readers who are neither musicians nor slide-rule pushers can see the difference by looking at Figure B.5, where a is an ordinary uniform (or *linear*) scale and b is logarithmic. If we start with a and note any two scale readings 1½ inches apart we find that they always *differ* by 10. Applying the same test to b we find that the difference may be large or small, depending on whether the inch and a half is near the top or bottom. But the larger reading is always 10 times the smaller.

One advantage of this is that it enables very large and very small readings to be shown clearly on the same graph. Another is that with some quantities (such as musical pitch) the ratio is more significant than the numerical interval. Figure 20.7 is an example of logarithmic scales.

Appendix C   Alternative Technical Terms

The reader of books and articles on radio is liable to be confused by the use of different terms to mean the same thing. In the following list the first alternatives to be mentioned are those preferred in this book. The associated terms are not necessarily exact equivalents. Terms distinctively American are printed in italics. The section numbers refer to where the terms are defined or explained.

Term	Section
Accumulator—Secondary battery	28.2
Amplification—Gain	11.5
Anode—*Plate*	9.4
Anode a.c. resistance—Anode incremental resistance—Valve impedance—*Plate impedance*	10.4
Anode battery—H.T.—*'B' battery*.	9.4
Antenna—Aerial.	17.7
Astable—Multivibrator	25.5
Atmospherics—Strays—Xs—*Static*.	17.13
Audio frequency (a.f.)—Low frequency (l.f.)—Speech frequency—*Voice frequency*	1.4
Automatic gain control (a.g.c.)—Automatic volume control (a.v.c)	22.11
Bistable—Eccles-Jordan relay	25.6
Blumlein integrator—Miller integrator	25.2
Bottom bend—Cut-off	11.9
Bottoming—Saturation	11.9
Capacitance—Capacity	3.2
Capacitor—Condenser	3.3
Characteristic resistance—Characteristic impedance—Surge impedance	16.3
Coaxial—Concentric	16.1
Common-anode valve—Cathode follower	12.9
Common cathode (or grid, anode, emitter, base, collector, source, gate, drain)—Earthed cathode (etc.)—*Grounded cathode (etc.)*	12.9
Common-collector transistor—Emitter follower	12.9
Compound pair (of transistors)—Darlington pair—Super-alpha pair—Double emitter follower	19.14
Detection—Rectification—Demodulation (British usage originally reserved this term for a different phenomenon; Sections 18.1, 22.10)	18.1
Dielectric—Insulating material	3.3
Dynamic resistance—*Antiresonant impedance*	8.10
Earth—*Ground*	2.14
Filament battery—L.T.—*'A' battery*	9.4
Frame antenna—*Loop antenna*	17.19
Frequency—Periodicity	1.4
Frequency-changer—Mixer—First detector	21.2
Grid battery—Bias battery—G.B.—*'C' battery*	10.6
Harmonic—Overtone—Partial	15.9
Hum—Ripple	28.3
Igfet—Insulated-gate field-effect transistor—M.O.S.T.—MOSFET	10.11
Image interference—Second-channel interference	21.7
Inductor—Coil	4.6
Integrated circuits—Solid state circuits—S.I.C.s—silicon chips—microchips	24.7
Interference—Jamming	20.5
Jugfet—Junction-gate field-effect transistor	10.11
Long-tailed pair—Emitter-coupled amplifier—Cathode-coupled amplifier	24.26
Loss—Attenuation	19.2
Monostable—One-shot multivibrator	24.9
Moving coil (of loudspeaker)—Speech coil—*Voice coil*	19.9
Moving-coil loudspeaker—*Dynamic loudspeaker*	19.9
Mutual conductance—Slope—*Transconductance*	10.3
Negative feedback—*Degeneration*—*Inverse feedback*	19.15
Noise—Machine interference—*Man-made static*	19.20
Parallel—Shunt	2.8
Peak value—Crest value—Maximum value	5.4
Permittivity—Dielectric constant—Specific inductive capacity (s.i.c.)	3.3
Phasor—Vector—Complexor	5.7
Picofarad—Micromicrofarad	3.3
Q—Q factor—Magnification—Storage factor	8.4
Quality (of reproduced sound)—Fidelity	19.4
Radar—Radiolocation—RDF	23.13
Radio—Wireless	Preface
Radio frequency (r.f.)—High frequency (h.f.)	1.12
Reaction—Retroaction—Positive feedback—*Regeneration*	14.5
Reaction coil—*Tickler*	14.5
Reflection coefficient—Return loss	16.5
Screen—*Shield*	15.7
See-saw circuit—Floating paraphase	24.6
Sender—Transmitter	1.6
Siemens (unit of conductance)—Mho	2.12
Telephones—Phones—Earphones—Headphones—Headset	1.8
Thyristor—Controlled silicon rectifier (c.s.r.)	28.4
Time base—Sawtooth generator—Scanning generator—Sweep	23.4
Tuned circuit—*LC circuit*—Resonant circuit—Tank circuit	8.3
Tunnel diode—Esaki diode	14.8
Turning frequency—Break frequency—Roll-off frequency	19.16
Valve—*Vacuum tube*—*Tube*	9.1

Appendix D — Symbols and Abbreviations

D.1 General Abbreviations

a.c.	alternating current
a.f.	audio frequency
a.g.c.	automatic gain control
a.m.	amplitude modulation
a.v.	alternating voltage
c.r.t.	cathode-ray tube
c.s.r.	controlled silicon rectifier
d.c.	direct current
d.l.	diode logic
d.s.b.	double sideband
d.v.	direct voltage
e.h.t.	extra-high tension
e.m.f.	electromotive force
EPROM	erasable, programmable, read-only memory
f.e.t.	field-effect transistor
f.m.	frequency modulation
h.f.	high frequency
h.t.	high tension
i.c.	integrated circuit
i.f.	intermediate frequency
igfet	insulated-gate field-effect transistor
i.r.	infra-red
i.s.b.	independent sideband
jugfet	junction-gate field-effect transistor
l.f.	low frequency
l.i.c.	linear integrated circuit
l.s.	loudspeaker
p.c.b.	printed-circuit board
p.d.	potential difference
p.d.a.	post-deflection acceleration
p.i.l.	precision in line
p.i.v.	peak inverse voltage
PROM	programmable read-only memory
RAM	random access memory
r.f.	radio frequency
r.g.b.	red, green, blue
r.m.s.	root-mean-square
ROM	read-only memory
s.a.w.	surface acoustic wave
s.c.r.	silicon controlled rectifier
s.s.b.	single sideband
s.w.r.	standing wave ratio
t.l.	transistor logic
t.w.t.	travelling-wave tube
u.h.f.	ultra-high frequency
u.v.	ultra violet
v.c.r.	video cassette recorder
v.d.u.	visual display unit
v.f.	video frequency
v.h.f.	very high frequency
v.s.b.	vestigial sideband
z.f.	zero frequency

D.2 Greek Letters

Letter	Name	Usual meaning
α	alpha	current amplification factor (of common-base transistor)
β	beta	current amplification factor (of common-emitter transistor)
ε	epsilon	permittivity
θ	theta	an angle
λ	lambda	wavelength
μ	mu	(1) permeability (2) voltage amplification factor (of valve) (3) one-millionth (as prefix to a unit symbol)
π	pi	$\frac{\text{circumference}}{\text{diameter}}$ of circle ($= 3.14159...$)
ρ	rho	resistivity
σ	sigma	conductivity
φ	phi (small)	angle of phase difference
Φ	phi (capital)	magnetic flux
ω	omega (small)	$2\pi f$
Ω	omega (capital)	ohm

D.3 Quantities and Units (S.I.)

Quantity	Symbol	Unit	Abbreviation for unit
Time	t	second	
Time period for one cycle	T	second	
Frequency	f	hertz cycles/second	Hz
Wavelength	λ	metre	m
Electromotive force	E	volt	V
Potential difference	V	volt	V
Current	I	ampere	A
Quantity of electricity, or charge	Q	coulomb	C
Power	P	watt	W
Capacitance	C	farad	F
Self inductance	L	henry	H
Mutual inductance	M	henry	H
Resistance	R	ohm	Ω

continued

284 Scroggie's Foundations of Wireless and Electronics

continued

Reactance {Capacitive, Inductive}	$X \begin{cases} X_C \\ X_L \end{cases}$		ohm	Ω
Impedance	Z		ohm	Ω
Conductance	G		siemens	S
Susceptance {Capacitive, Inductive}	$B \begin{cases} B_C \\ B_L \end{cases}$		siemens	S
Admittance	Y		siemens	S
Magnetic field strength	H		ampere/metre	
Magnetic flux	Φ		weber	Wb
Magnetic flux density	B		tesla	T
Q factor, X/R	Q			
Signal gain or loss	A		decibel	dB

Note: Small letters (e, v, i, etc.) are used to indicate instantaneous values.

D.4 Unit Multiple and Submultiple Prefixes

Symbol	Read as	Means
G	giga-	one thousand million ($\times 10^9$)
M	mega-	one million ($\times 10^6$)
k	kilo-	one thousand ($\times 10^3$)
m	milli-	one thousandth ($\times 10^{-3}$)
μ	micro-	one millionth ($\times 10^{-6}$)
n	nano-	one thousand-millionth ($\times 10^{-9}$)
p	pico-	one billionth ($\times 10^{-12}$)

Examples:

$$1 \text{ MHz} = 1\,000\,000 \text{ Hz}$$
$$1 \text{ k}\Omega = 1000 \text{ }\Omega$$
$$1 \text{ mA} = 0.001 \text{ A}$$
$$1 \text{ }\mu\text{H} = 0.000\,001 \text{ H}$$
$$1 \text{ pF} = 0.000\,000\,000\,001 \text{ F}$$

D.5 Transistor Abbreviations and Symbols

Bipolar		Field effect	
e emitter	C-E common-emitter	s source	
b base	C-C common-collector	g gate	
c collector	C-B common-base	d drain	

C-E	C-C	C-B	Unspecified	
h_{fe}	h_{fc}	h_{fb}	h_f	forward current transfer ratio with input short-circuited
h_{ie}	h_{ic}	h_{ib}	h_i	input impedance with output short-circuited
h_{oe}	h_{oc}	h_{ob}	h_o	output admittance with input open-circuited
h_{re}	h_{rc}	h_{rb}	h_r	reverse voltage transfer ratio with input open-circuited

The above are small-signal values. D.C. values are distinguished by capital letter subscripts: e.g., h_{FE}. This also applies to such symbols as I_C (collector d.c.), I_C (collector r.m.s. signal current), i_C (collector instantaneous signal current), V_C (static voltage at collector), etc. To define the voltage more precisely, V_{CE} means collector voltage with respect to emitter. A doubled subscript (V_{CC}, for example) means supply voltage (Figure 13.1). For I_{CBO} and I_{CEO}, see Figure 13.3. For elements of hybrid-π circuit, see Figure 22.4.

f_{hfe}	frequency at which h_{fe} is 3 dB less than at low frequencies
f_1	frequency at which $h_{fe} = 1$
f_T	$f_{hfe} \times$ low-frequency $h_{fe} \approx f_1$
g_m	mutual conductance

D.6 Valve Abbreviations and Symbols

k	cathode
g	grid
g_1	first grid (nearest cathode)
g_2	second grid; and so on
a	anode
μ	voltage amplification factor
r_a	anode a.c. resistance
g_m	mutual conductance
g_c	conversion conductance (of frequency-changer)

Note—Symbols are frequently combined, thus:
I_a = anode current
V_{g2} = voltage at second grid
R_a = resistance connected externally to anode
c_{gk} = internal capacitance from grid to cathode

Capital letters are used for associated items outside the valve; small letters for items inside the valve itself.

D.7 Special Abbreviations used in this Book

f_r	frequency of resonance
f_o	frequency of oscillation
f'	frequency off-tune
f_t	turning frequency
A	voltage amplification
B	feedback factor
R_L	load resistance
R_G	generator resistance

Appendix E

Decibel Table

The decibel figures are in the centre column: figures to the left represent decibel loss, and those to the right decibel gain. The voltage and current figures are given on the assumption that there is no change in impedance.

Voltage or current ratio	Power ratio	− dB +	Voltage or current ratio	Power ratio
1.00	1.000	0	1.000	1.000
0.989	0.977	0.1	1.012	1.022
0.977	0.955	0.2	1.023	1.047
0.966	0.933	0.3	1.035	1.072
0.955	0.912	0.4	1.047	1.096
0.944	0.891	0.5	1.059	1.122
0.933	0.871	0.6	1.072	1.148
0.912	0.832	0.8	1.096	1.202
0.891	0.794	1.0	1.122	1.259
0.841	0.708	1.5	1.189	1.413
0.794	0.631	2.0	1.259	1.585
0.750	0.562	2.5	1.334	1.778
0.708	0.501	3.0	1.413	1.995
0.668	0.447	3.5	1.496	2.239
0.631	0.398	4.0	1.585	2.512
0.596	0.355	4.5	1.679	2.818
0.562	0.316	5.0	1.778	3.162
0.501	0.251	6.0	1.995	3.981
0.447	0.200	7.0	2.239	5.012
0.398	0.159	8.0	2.512	6.310
0.355	0.126	9.0	2.818	7.943
0.316	0.1000	10	3.162	10.00
0.282	0.0794	11	3.55	12.6
0.251	0.0631	12	3.98	15.9
0.224	0.0501	13	4.47	20.0
0.200	0.0398	14	5.01	25.1
0.178	0.0316	15	5.62	31.6
0.159	0.0251	16	6.31	39.8
0.126	0.0159	18	7.94	63.1
0.100	0.0100	20	10.00	100.0
3.16×10^{-2}	10^{-3}	30	3.16×10	10^3
10^{-2}	10^{-4}	40	10^2	10^4
3.16×10^{-3}	10^{-5}	50	3.16×10^2	10^5
10^{-3}	10^{-6}	60	10^3	10^6
3.16×10^{-4}	10^{-7}	70	3.16×10^3	10^7
10^{-4}	10^{-8}	80	10^4	10^8
3.16×10^{-5}	10^{-9}	90	3.16×10^4	10^9
10^{-5}	10^{-10}	100	10^5	10^{10}
3.16×10^{-6}	10^{-11}	110	3.16×10^5	10^{11}
10^{-6}	10^{-12}	120	10^6	10^{12}

Index

Abbreviations, 276, 283, 284
A.C. power supplies, 267–271
Acceptor circuit, 62
Acceptor impurity, 70
Accumulator, 26, 266, 267
Active device, 84
Adder, 250–251
Admittance, 48
Aerial. *See* Antenna
After-glow, 210, 226
Air traffic control, 226
Alternating current, 35–41
   addition and subtraction, 39
   frequencies of, 35
   meters, 37
Alternating voltage, 35
   addition and subtraction, 39
Ammeter, 12, 20
Ampère, 12, 13, 21
Amplification, 85, 86
   a.f., 162–180
   calculation, 87
   r.f., 196–209
   v.f., 229–236
   voltage, 77, 86, 88
Amplification factor, 75, 88, 276
   current, 81
   voltage, 75
Amplifier, 8
   a.f., 168–173
   aperiodic, 196
   balanced, 235
   cascode, 202
   Class A, 89, 168
   Class B, 89, 124, 169–173, 178
   Class C, 89, 124
   common-base, 100, 106
   common-base preselector stage, 202
   common-collector, 98, 99, 106
   common-emitter, 97, 98, 106
   d.c., 231, 235, 243, 246
   differential, 235
   direct-coupled, 231, 235, 243, 246
   i.f., 200, 201
   inverter, 235, 249, 250
   long-tailed pair, 235
   microwave, 121, 208–209
   non-inverting, 235
   operational, 234–236, 238, 240, 256
   pulse, 229–236
   r.f., 162, 201–203, 208, 209
   u.h.f., 203
   v.f., 229–236
   voltage, 77, 86
   wide-band, 229–236
Amplitude, 2
Amplitude modulation, 123, 124, 126, 127, 189, 191

Analogue computer, 236, 247
Analogue-to-digital conversion, 255–257
AND gate, 249
Anode, 66
Anode a.c. resistance, 76
Anode conductance, 76
Anode power supply, 266
Antenna, 142–148
   coupling, 206–208
   dipole, 142–148
   directional characteristics, 143
   director, 144
   effective height, 148
   folded dipole, 147
   frame, 149
   gain, 145
   horn, 149
   inductor, 149
   input circuit, 137, 138
   loop, 149
   Marconi or quarter-wave, 146
   microwave, 149
   parasitic, 144
   reflector, 144
   slot, 144, 145
   tuning, 148
   Yagi, 144, 148
Antinode, 137
Aperiodic coupling, 196
Application development, 254
Arithmetic and logic unit, 253
Aspect ratio, 214
Astable system, 240
Atomic number, 67
Atoms, 10
Attenuation distortion, 162
Audio frequencies, 3, 35, 54, 85, 162–180
Automatic frequency control, 180
Automatic gain control, 205–206
Auto-transformer, 55, 115, 116
Avalanche effect, 73
Average value of a sine wave, 36, 37
Azimuth, 225, 226

Backing up, 261, 262
Baffle (loudspeaker), 167
Balun, 147
Bandpass filter, 187, 201
Bandwidth, 128, 129, 182–184, 201, 222
Base (transistor), 78
Baseband, 129
Base bias, 108, 111, 173, 242
Base characteristic, 80
Battery, 11, 13, 19, 42, 266
   charging, 19, 26
Beat frequency, 189
Beating, 189, 190, 194

Bel, 162, 163
Bell, A. 162
Bias automatic, 120
   base, 108, 111, 173, 242
   gate, 110–111
   grid, 77, 111
Bias signal, tape recording, 260
Bias supplies, 111, 271
Binary
   addition, 248–250
   scale, 130, 247–248
Bipolar transistor, 77–81. *See also* Transistor, bipolar
Bistables, 240, 243, 244, 251–253
Bit, 253
Bitmap, 254
Blocking capacitor, 115, 155, 156, 172
Blocking oscillator, 241, 242, 271, 273
Blumlein integrator, 240
Boltzmann's constant, 179
Bootstrapping, 178
Bottom bend, 89
Bottoming, 89
Bridge circuit, 269
Brightness control, 213
Broadcasting channels, 128, 129
Buffer memory, 262
Byte, 253

Cable broadcasting, 265
Camera tube, 224
Capacitance, 22–27
   active-device internal, 196
   analysis, 23
   definition, 22
   in a.c. circuits, 42–48
   in parallel and series, 44
   in parallel with inductance, 59
   in parallel with inductance and resistance, 63, 64
   in parallel with resistance, 47
   in series with inductance, 55
   in series with inductance and resistance, 56, 57
   in series with resistance, 46
   stray, 57, 63, 116, 157, 175, 203, 233, 241, 242
   unit of, 22
Capacitor, 24
   blocking, 115, 156, 172
   bypass, 271
   charge and discharge of, 25
   electrolytic, 24, 270
   ganged tuning, 193, 194, 202, 203, 208
   microphone, 126
   neutralising, 124, 201
   voltage-controlled, 74, 124, 125, 137, 161, 194

Index

Carrier wave, 123, 124, 125–129, 141, 189
  amplitude-modulated, 123, 124, 125–128, 189–191
  frequency-modulated, 124–128, 158–161
Cascode, 202
Cassette tape, 260–261
Cathode, 65, 67
  follower, 100, 104, 177
  heating, 266, 274
Cathode-ray oscilloscope, 213–214
Cathode-ray tube, 5, 126, 210–217
  alternatives to, 227–228
  description of, 210, 214–216
  electric focusing and deflection, 211, 212
  in computers, 227
  in radar, 225–227
  in television, 214–216, 222–224
  magnetic deflection and focusing, 212, 213
  power requirement, 266
  power supplies, 273, 274
Cavity resonator, 64, 121, 138
CD. *See* Compact disc
CD-ROM. *See* Compact disc
Cell, 266, 267
  primary, 266
  secondary, 267
Centre frequency, 126, 160
Ceramic filter, 201, 236
Channel separation, 128, 129
Characteristic curves
  bipolar transistor, 79, 80
  germanium diode, 156, 157
  silicon diode, 73
  thermionic diode, 66
  transistor, 79, 80
  triode, 75–77
  Zener diode, 73
Characteristic resistance, 132, 136
Charge carriers, 68, 70–72
Charge-coupled device, 252
Chemical effects of electricity, 19
Choke, 203, 204, 270
  control, 124
  input filter, 270
Chopper, 246
Chrominance, 222
Cinematography, 5, 258
Circuit
  closed, 11, 17
  differentiating, 237–239
  integrating, 237–239
  integrated, 160, 236, 242, 246, 248, 249, 255, 256, 263–265
  short, 18
Circuit diagrams, 13–15
  crossing wires, 13
  layout, 14
  warning, 14
Circular polarisation, 142
Clamping, 244, 245

Class A amplification 89, 168
Class B amplification 89, 124, 169–173, 178
Class C amplification, 89, 124
Clipping circuit, 244, 245
Closed circuit, 11, 17
Cockcroft-Walton voltage multiplier, 273
Collector, 78
Collector characteristic, 79, 80
Colour burst, 223
Colour-difference signals, 222
Colour television, 6, 129, 222–224
  camera, 225
Colpitts oscillator, 115, 116, 121, 161
Common-mode rejection, 235
Common-mode signals, 235
Communication, natural, 1
Compact cassette, 260
Compact disc, 262–263
Comparator, 256
Compatibility, 222
Compensation, impurity, 70
Compiled language, 254
Complementary pair, 172, 178
Compound pair, 173
Computer, 227, 247, 248, 253–255, 261–265
  analogue, 236, 247
  digital, 247–255
  languages, 254
  network, 255
  program, 253
Computer-aided design, 255
Computer-aided manufacture, 255
Computer memories or stores, 253, 254, 261–265
Concertina circuit, 172, 177
Condenser, 24
Conductance, 17, 23
  conversion, 192
  mutual, 76, 81, 82, 104, 110, 192, 233
Conduction, intrinsic, 69, 71, 74
Conductor, 11
  linear, 13
  non-linear, 66, 76, 156
Constant current generator, 92, 108
Constant-voltage generator, 92, 109
Control unit, computer, 253
Conversion conductance, 192
Conversion gain, 193
Corkscrew rule, 30, 142
Cosine, 237
Coulomb, 22
Counter discriminator, 161
Counter circuit, 252–253
Counterpoise, 147
Coupling antenna, 206–208
  aperiodic, 196
  critical, 186
  reactance, 185
  tuned-circuit, 185
Covalent bonds, 67
Critical coupling, 186

Critical damping, 113
Cross-modulation, 204, 207
Crossing wires in circuit diagrams, 13
Crossover distortion, 171, 173, 178, 179
Crossover filter, 167
Crystal, 67–69, 71
Crystal microphone, 126
Crystal oscillator, 121
Current amplification factor, 81, 91
Current carrier, 68
Current dumping, 178
Current feedback, 177
Current gain, 98, 99, 101, 103
Current generator, 91, 93
Current leakage, 72, 79, 107
Current meter, 19, 20
Current stabiliser, 271–273
Curve, 279
Cut-off frequency, transistor, 200
Cycle, 3

Damping, 113
Dark current, 83, 252
Darlington pair, 173
Data storage, 258–265
D.C. amplifier, 231, 235, 243, 246
D.C. power supplies, 267
D.C. restorer, 218–220, 245
Decibel table, 285
Decibels, 162–163
Decoupling, 178, 270
De Forest, 76
Delayed a.g.c., 206
Demodulator, 150
Depletion layer, 71, 73, 74, 81, 194,
Depletion mode, 78, 82
Depolariser, 266
Detection, 65, 150–161
Detector
  crystal, 151
  distortion in diode, 156, 157
  f.m., 158–161
  loading effect of diode, 154, 155
  quadrature, 160
  ratio, 159, 160, 206
  series diode, 154, 155
  shunt diode, 157, 158
  television diode, 158
Dielectric constant, 24
Dielectric losses, 63
Dielectrics, 24, 73
Differential amplifier, 235
Differentiating circuit, 237–239
Differentiation, 236
Diffusion, 71
Digital audio tape, 261
Digital Compact Cassette, 261
Digital data storage, 261–265
Digital techniques, 247–257
Digital-to-analogue conversion, 255–257
Digital Video Disc, 263
Diode, 65–74, 268
  Esaki, 118
  mixer, 190, 227

Diode (*continued*)
  germanium, characteristic curve, 156, 157
  light-emitting, 83
  photovoltaic, 83
  reverse-biased, 71
  silicon, 73
  tunnel, 118
  use of, 65
  varactor, 74, 125, 137, 161, 194
  Zener, 73, 271–273
Diode characteristics, 73, 74
Diode detector, 150–160, 245
Diode rectifier, 151, 268–270
Dipole antenna, 142–148
Direct-coupled amplifier, 231, 235, 243, 246
Director, 144
Discriminator
  counter, 161
  Foster-Seeley, 158–159
  phase, 150–160
  phase-locked-loop, 160, 161
  quadrature, 160
  ratio, 159–160, 206
Disc storage, 261–265
Dish, 149
Diskette, 261
Displacement current, 27, 69
Distortion, 124
  amplitude/frequency, 162
  attenuation, 162
  crossover, 171, 173, 178, 179
  diode detector in, 156, 157
  gain/frequency, 162, 164, 166, 173, 174, 182
  harmonic, 164–166, 169
  intermodulation, 165
  loudspeaker, 167
  non-linearity, 120, 162, 164, 166, 182, 260, 261
  oscillator, 120
  phase, 166, 174
  transient, 177
  waveform. *See* Distortion, non-linearity
Domain, magnetic media, 259
Donor, 69, 70
Doping, 71, 78
Doppler effect, 226
Doppler radar, 226
Drain, 81
Drain current, 110, 112–114
Drain current/drain voltage curves, 82
Driver stage, 170, 172
Dry battery, 266
Dual phasor diagram, 93, 94, 175
Duality, 92, 95, 97, 117, 144
Duty cycle, 227
Dynamic limiter, 159
Dynamic random access memory, 264
Dynamic resistance, 60, 62, 137, 181, 188
Dynamo, 11

Ear, 4, 8
Earth connection, 14, 18–19
Eccles-Jordan relay, 243–244
Edison, T. A., 258
Eddy currents, 62
Effective height of antenna, 148
Effective radiated power, 145, 149
Efficiency,
  Class A amplifier, 169
  Class B amplifier, 170
E.H.T. power supplies, 273–274
Electrical power, 20, 26
Electric charge, 10, 22, 26
  unit of, 22
Electric current, 10, 12, 13
  direction of, 11, 23
  effects of, 19
  in capacitive circuits, 26, 42
  in inductive circuits, 32, 49
  measurement of, 19–20
Electric field, 10, 18, 23, 24, 26, 66–68, 71, 74, 113, 140–142, 145, 149, 204, 211
Electric lines of force, 139
Electric space constant, 24
Electric waves, 6, 7
Electricity, 10
  chemical effects, 19
  heating effect, 19, 36
  magnetising effect, 19
  nature of, 10
  negative, 10
  positive, 10
Electrode, 65
Electrolyte, 266
Electromagnet, 28, 259–260
Electromagnetic waves, 141
Electromotive force, 11, 18–19,
  induced, 29
Electron gun, 210
Electronic devices, 65
Electronics, 65
Electrons, 10–11, 17, 19, 65–72
  thermionic emission of, 65, 66
  valence, 67–70, 74
Electrostatic loudspeaker, 168
Elements, 67
Emitter, 78
  follower, 100, 109, 177
Energy, 21, 26
Enhancement mode, 78, 82, 83
Envelope, 3
  of modulated r.f. wave, 123
ε, 23, 24
EPROM, *See* Erasable programmable read-only memory
Equalisation, 163, 182
Equation, 275–277
Equipotential lines, 211
Equivalent circuit, 15, 53, 54, 86, 91–105
Equivalent generator, 86, 87, 91, 93, 118, 196
Erasable programmable read-only memory, 264–265

Esaki diode, 118
Ether, 6
Exponential curve, 25, 33
Extremely-high frequencies, 35

Fall time, 231
Farad, 22
Faraday, Michael, 22, 30
Feed, 84, 106
Feedback, 95, 270
  coil, 114
  current, 177
  effect on gain/frequency distortion, 173–176
  effect on input impedance, 176, 177
  effect on output impedance, 176, 177
  effect on phase shift, 174–176
  negative, 173–180, 240, 271
  positive, 114, 117, 181, 184, 241, 242, 271
  shunt-injected, 176
  voltage, 99, 173–177
Feeder, 131
  coaxial, 131, 147
  parallel-wire, 131, 147
Feed-forward error compensation, 179
Ferrite, 62, 149
Ferromagnetic materials, 28, 30
Field (television), 215
Field-effect transistor, 81–83. *See also* Transistor, field-effect
Filament, 65
Filter, 117, 155, 156, 160, 166, 206, 238, 270
  bandpass, 187, 201
  ceramic, 201, 236
  choke input, 270
  crossover, 167
  lowpass, 160, 161, 267
  optical, 126
  surface-acoustic-wave, 201
Flicker, 215
Flicker noise, 179
Floppy disc, 261
Fluorescence, 210
Flux, 28, 33, 52, 53
Flyback, 213, 215, 219, 221
Flywheel synchronisation, 220
Forward a.g.c., 206
Forward bias, 72, 79, 106, 152, 246
Forward resistance, 152, 154, 159
Foster-Seeley discriminator, 158
Fourier principle, 164, 230
Frequency, 2, 3, 6, 7, 8
  alternating current, 35
  audio, 3, 35
  bands, 35
  beat, 189
  carrier, 129, 145
  centre, 126
  cut-off, 200
  main divisions of, 35
  microwave, 35
  resonance, 57, 58, 61, 62

Frequency (continued)
  radio, 6, 35
  tuning, 175, 183, 200
  video, 126, 129, 158, 215, 216
Frequency-changer, 189, 191, 192, 217
  as modulator, 191
  types of, 191, 192
Frequency characteristics, 163, 164
Frequency divider, 241
Frequency-division multiplex, 129
Frequency modulation, 124, 125, 128, 158–161
Frequency multiplier, 165, 242
Frequency off-tune, 182–184
Frequency-shift keying, 125
Frequency spacing, 128–129
Frequency-time relationship, 230–231
Fundamental, 164
Fuse, 19

Gain
  antenna, 145
  current, 98, 99, 101, 103
  voltage, 86, 97–99, 101, 103–105
Gain/frequency distortion, 162–164, 166, 173, 174, 182
Ganged tuning, 193, 194, 202, 203, 208
Gate
  f.e.t., 81, 82
  linear, 246
  logic, 248–251
Gate bias, 110
Gate insulation, 82
Gated a.g.c., 206
Generator, 11, 21, 36
  equivalent, 86–87, 91, 93, 118, 196
  pulse, 240–244
  sine-wave, 113–118
Germanium, 67, 70, 73
Germanium diode, 156, 157
Germanium transistor, 80, 107, 108
Graphical user interface, 254
Graphic equaliser, 164
Graphics, 255
Graphs, 278–281
  three-dimensional, 279–280
Greek letters, 283
Grid, 75
Grid bias, 77, 266
Ground wave, 146
Gun, electron, 210

h parameters, 96, 98, 198
Half-adder, 250–251
Hard disc, 253, 261
Harmonic distortion, 164–166, 169
Harmonics, 164, 165, 195, 229, 230
Hartley oscillator, 115, 116, 120, 161
Hartley's law, 222, 230
Heater, 65
Heating effect of electricity, 19
Heising modulation, 124
Henry, 30
Hertz, H., 6, 7, 64

Heterodyning, 189
High tension, 66
Holes, 67–69
Horn antenna, 149
Hue, 222
Hum, 267
Hybrid-π equivalent circuit, 198
Hysteresis, 62

Iconoscope, 224
Igfet, 82, 252
Image interference, 194
Image orthicon, 224–225
Impedance, 46, 47, 48
  matching, 135, 136, 208
  transformation, 54
Impurity conduction, 69–71
Indices, 277
Inductance, 28–34, 112, 232–234
  analysis, 30
  in a.c. circuits, 49–55
  in parallel with capacitance, 59
  in parallel with capacitance and resistance, 63, 64
  in parallel with resistance, 51, 53, 54
  in series and in parallel, 50
  in series with capacitance, 56
  in series with capacitance and resistance, 56, 57
  in series with resistance, 51
  leakage, 53
  mutual, 33, 50, 185, 186
  self-, 30, 50, 139
Induction, 29
Induction field, 141
Inductive circuit, electric current in, 32
  power in, 33
Inductor, 31, 194
Inductor antenna, 149
Input characteristic, 80
Input circuit, 84
Input device, 254–255
Input resistance, 91, 94, 95–97, 99, 101–105, 176, 177
Input signals, 85
Input unit, 253
Instruments for measuring current, 19, 20
Insulators, 11
Integrated circuits, 160, 236, 242, 246, 248, 249, 255, 256, 264
Integrating circuit, 237–239
Integration, 236
Intercarrier, 218
Interference, 187, 194, 195
  second-channel or image, 194
Interlacing, 215
Intermediate frequency, 189, 193
Intermodulation, 165
Internal resistance, 19
Internet, 255, 265
Interpreted language, 254
Intrinsic conduction, 69, 71, 74
Inversion of signal, 86
Ionisation, 67

Ions, 11
Iron core, 62
  in transformer, 33, 52

Johnson noise, 179
Joule, 26
Jugfet, 81, 82

Keying, frequency-shift, 125
Kinetic energy, 21, 33
Kirchhoff's laws, 17, 18, 23, 26, 27, 41, 47, 80, 105
Klystron, 64, 122, 125, 208, 209

ladder network, 256
λ, 4
Laser, 262
Laser disc, 263
Lead, 13
Leakage current, 72, 79, 107
Leakage flux, 30
Leakage inductance, 53
Lenz's law, 30
Light-sensitive devices, 69, 83
Light-emitting diode, 83, 227–228
  matrix, 227–228
  seven-segment, 227
Limiter, 158–160, 206
Lines of force, 23, 140
Liquid crystal display, 228
Load, 21, 36
  conductance, 97
  current in transformer, 52
  line, 85, 88, 89, 119, 168, 169
  resistance, 85, 87, 96
  use of term, 84
  Local area network, 255
Logarithmic scale, 280
Logic circuits, 248–251
Logic convention, 249
Logic operations, 248–251
Long-tailed pair, 235
Loop antenna, 149
Loop coupling, 138
Loudspeaker, 8, 167–169
  baffle, 167
  distortion, 168
  electrostatic, 168
  horn, 167
  line source, 167
  moving-coil, 167
  vented-enclosure, 167
Low-level modulation, 124
Low-tension, 66
Lumière brothers, 258
Luminance, 222

Magnet, 28
  permanent, 29
Magnetic coupling, 50
Magnetic deflection, 212
Magnetic field, 28, 29, 33, 37, 50, 112, 139–142, 144, 149, 204, 212
Magnetic flux, 28, 29, 30, 31, 52, 53

Magnetic focusing, 212
Magnetic saturation, 31
Magnetic storage, 259–262
Magnetising current, 52–54
Magnetising effect of electricity, 19
Magneto-optical disc, 263
Magnification of circuit. *See Q*
Mainframe computer, 255
Majority carriers, 70, 74
Marconi antenna, 146
Mark/space ratio, 239, 240, 243, 274
Matching, 88, 135, 136, 147, 208
Maximum-power law, 88, 89
Maximum ratings, 85, 169
Maxwell, Clerk, 6, 27
Measuring instruments, symbol, 14
  moving-coil, 20, 29, 37
Memory, 253–254, 263
  computer 253–254
  video, 227
Meter, 37
  current, 19, 20
  moving-iron, 19
Mho, 17
Microcomputer, 255
Microfarad, 24
Microphone, 4, 8, 9, 125, 125
  capacitor, 126
  crystal, 126
  electromagnetic, 126
  piezo-electric, 126
Microprocessor, 255
Microwave antenna, 149
Microwave frequencies, 35
Microwave oscillator, 121
Miller effect, 197, 201, 203
Miller integrator, 240
Milliammeter, 14, 19, 20
Milliampere, 13
Minority carriers, 70, 74
Mixer, 192
  self-oscillating, 192
Modem, 255
Modulation, 123–130
  amplitude, 123, 124, 126, 128, 189, 190, 191
  choke control, 124
  cross-, 204, 205, 207
  depth of, 123
  factor, 123, 182
  frequency, 124, 125, 128, 158–161, 182, 184
  Heising, 124
  methods of amplitude, 124
  methods of frequency, 125
  negative, 216
  pulse systems, 130
  quadrature, 223
  sinusoidal, 123, 124
  velocity, 121
Modulator, 8, 191
  frequency-changer as, 191
  reactance, 125
'Monkey chatter', 129

Monophonic reproduction, 130
Monostable systems, 240
Monostables, 244
Mosfet, 82
Moving-coil loudspeaker, 157
Moving-coil meter, 19, 29, 37
Moving-coil microphone, 126
Moving-iron meter, 19, 37
μ 28, 52, 276
Multimeter, 37
Multiplex, 129
Multiplier, 20
  frequency, 165, 242
Multi-sessional, 263
Multivibrator, 130, 242–244, 274
Mutual conductance, 76, 81, 82, 104, 110, 192, 233
Mutual inductance, 33, 34, 50, 185, 186
Mutual reactance, 185

NAND gate, 249
Negative electricity, 10
Negative feedback, 161, 173–180, 232–236, 271
  current, 177
  shunt-injected, 176
  voltage, 177
Negative modulation, 216
Negative resistance, 117, 118
Network, computer, 255
Neutralisation, 201, 202
Node, 137
Noise, 125, 179, 193, 202
Noise factor, 179
Noise figure, 179
Noise reduction, tape recording, 260
Non-linearity, 66, 76, 151, 156, 162, 191, 205, 213
Non-linearity distortion, 120, 162, 164–166, 182
NOR gate, 249
NOT gate, 250
Nucleus, 10

Offset voltage, 80, 152, 178
Ohm, 12
Ohmmeter, 20
Ohm's law, 12, 15–17, 25, 32, 36, 38, 41, 43, 44, 46, 66, 76, 86, 87, 92, 134, 151, 230
ω, 44
On-demand service, 265
Operational amplifier, 240, 256
Optical character recognition, 254
Optical fibre, 138
Optical filter, 126
Optical storage, 262–263
Opto-isolator, 83
OR gate, 248
Orthicon, 224
Oscillation, 112–122
  amplitude, 118–120
  distortion, 120

Oscillation (*continued*)
  frequency, 113
  self-, 114
Oscillator, 181, 189
  automatic biasing, 120
  bias, 271
  blocking, 241, 242, 271, 273
  Colpitts, 115, 116, 121, 161
  crystal-controlled, 121
  frequency, constancy of, 120, 121
  frequency-changer, in, 191
  Hartley, 115, 116, 120, 161
  microwave, 121
  relaxation, 240–244
  resistance-capacitance, 117
  variable frequency, 121
  voltage-controlled, 160, 161
Oscillator harmonics, 195
Oscillatory circuit, 112, 113
Oscilloscope, 213
Output characteristic, 79, 80
Output circuit, 84
Output conductance, 97, 99, 100, 102
Output resistance, 176–177
Output stage, 168–173
Output unit, computer, 253
Over-coupling, 187, 208
Overshoot, 230

Padder, 194
Parallel connection, 15–17, 44, 47, 50, 51, 54, 59, 63
Parallel feed, 116
Parallel resonance, 61–62
Parallel signal, 252
Parameter, 76, 80, 91
  transistor, 80, 81
  valve, 75, 76
Passive device, 84
Peak inverse voltage, 268
Peak value of a sine wave, 36, 37, 39
Pentode, 80, 89, 104
Period, 3
Permanent magnet, 29
Permeability, 28, 31, 32
Permeability tuning, 201
Permittivity, 24, 132, 134
  absolute, 24
  relative, 24
Phase, 38
Phase angle, 46, 53, 117
Phase delay, 231
Phase discriminator, 158, 159, 161
Phase distortion, 166, 174, 231, 232
Phase lag, 135
Phase-locked-loop detector, 160
Phase shift, 166
  with negative feedback, 174
Phase splitter, 172, 177
Phasor, 38
Phasor diagram, 38–41
  amplitude modulation, 127
  capacitance and resistance in parallel, 47

Phasor diagram (*continued*)
  capacitance and resistance in series, 46
  capacitive circuit, 45
  direction signs, 39
  dual, 93, 94, 175
  inductance and capacitance in parallel, 59, 60
  inductance and capacitance in series, 55
  inductance and resistance in parallel, 51
  inductance and resistance in series, 51
  inductance, capacitance and resistance all in parallel, 63
  inductance, capacitance and resistance all in series, 56
  inductive circuit, 51
  subscript notation, 40
  transformer, 52, 53
  transistor, 94, 97
  transistor amplifier, 100, 101, 198, 199
  transmission line, 133
  tuned circuit, 54, 56, 59, 63, 182
  valve, 87
  voltage and current, 39–41
Φ, 28, 52
φ, 40, 47, 51–53
Phonograph, 258
Phosphor, 210
Phosphorescence, 210
Photo-CD, 263
Photoconductive devices, 83
Photodiodes, 83
Photoelectric devices, 83
Photoemissive devices, 83
Phototransistors, 83
Photovoltaic diode, 83
π, 39, 59
Picofarads, 24
Picture elements, 5, 214
Picture signal, 5
Picture tube, 5, 8, 210, 212, 222, 224, 232
Piezo-electric microphone, 126
Piezo-electric effect, 121
Pinch-off voltage, 82
Pixel, 227
Plan position indicator, 225, 226
p-n junction, 71, 72, 74, 77–79, 81, 94
Polar diagram, 143
Polarisation, 142
Poles, magnetic, 28, 29
Positive electricity, 10
Positive modulation, 216
Post-deflection acceleration, 213
Positive feedback, 114, 117, 181, 184, 241, 242, 251
Potential barrier, 77
Potential difference, 11, 15, 16–18, 19, 26
Potential divider, 17, 100, 115, 119, 157, 173, 174, 236
Potential effect, 19

Potential energy, 21
Potentiometer, 17
Power, electrical, 20, 26, 36
  in capacitive circuit, 26, 27, 45
  in inductive circuit, 33, 50, 52
  in mixed circuit, 47
Power amplifier, 77, 168–173
Power curve, 36, 45
Power factor, 47
Power gain, antenna, 145
Power ratio, 162, 163
Power supplies, 266–274
  bias, 271
  cathode heating, 274
  d.c., 267–270
  e.h.t., 273–274
  stabilised, 271–273
  switch-mode, 274
  picture tube, 224
Prefixes, multiple and sub-multiple, 284
Preselector, 193–195, 202, 203, 205, 217
Primary cell, 267
Printed circuit, 236
Probe coupling, 138
Process control, 255
Product detector, 160
Program, computer, 253
Programmable memory, 263–265
Programming language, 254
Proximity effect, 62
Pulse code modulation, 130
Pulse generator, 240–244
Pulse modulation systems, 130
Pulse recurrence frequency, 226
Pulse shaping, 239, 240
Pulse waveforms, 229, 237
Push-pull circuit, 89, 169, 170, 178
Pythagoras's theorem, 46

Q, 57–58, 62, 113, 120, 121, 154, 181, 182, 200, 202, 208
  and selectivity, 181–185
  of tuned circuit, 64
Quadrature, 38
Quadrature detector, 160
Quadrature modulation, 223
Quantising, 256
Quantities and units, 284
Quantity of electricity, 22, 42
Quarter-wave resonator, 136, 137
Quarter-wave transformer, 136
Quartz crystal, 121
Quiescent point, 169

Radar, 7, 138, 149, 192, 225–227
  Doppler, 226
  power supplies, 273, 274
  receiver, 227
Radians, 39
Radiation, 141
Radiation field, 141
Radiation resistance, 143
Radio, nature of, 1
Radio frequencies, 6, 35

Radio sender. *See* Transmitter
Radio telegraphy, 7, 125
Radio telephony, 7
Radio waves, 6, 8
Ramp generator, 256
Ramp waveform, 213, 229, 237, 240
Random Access Memory, 254, 264–265
Raster, 215
Rated frequency deviation, 125, 128
Ratio detector, 159, 160, 206
Reactance, 44, 46, 50, 87, 88, 94, 136, 174
Reactance coupling, 186
Reactance modulator, 125, 180
Reaction coil, 114, 181
Read-only memory, 264–265
Real time, 254
Receiver, 4
  superheterodyne, 189–195
  television, 217–224, 242, 245
Reciprocal, 277
Recombination, 69–72, 74
Rectification, 65, 67, 151–153
Rectifier, 37, 72, 151, 192, 267–270
  bridge, 267, 269
  circuits, 268, 269
  diode, 151, 268, 269
  full-wave, 269
  half-wave, 268, 269
  resistance, 152
  semiconductor diode, 72–74
  types, 268, 269
  vibrator, 151
  voltage-doubler, 270
Reflection coefficient, 135
Reflector, 144
Reflex klystron, 121, 122, 125
Regeneration, 114
Rejector circuit, 62, 195
Relative permeability, 28
Relaxation oscillators, 240–244
Repeller, 122, 125
Reluctance, 204
Reservoir capacitor, 152, 153, 267, 270
Resistance, 12, 21, 23
  analysis of, 17
  characteristic, 132, 133, 135, 136
  dynamic, 60, 62, 137, 181, 188
  electrical power expended in, 21
  h.f., 63
  in parallel, 15, 16
  in parallel with capacitance and inductance, 63, 64
  in parallel with inductance, 51, 54
  in series, 15
  in series with capacitance, 47
  in series with capacitance and inductance, 56, 57
  in series with inductance, 51
  negative, 117, 118
  of coil, 62
  of parallel *LC* circuit, 59, 60
  radiation, 143

Resistance (*continued*)
　rectifier, 152
　temperature dependence, 17
Resistance-capacitance oscillator, 117
Resistivity, 17
Resistor, 14
Resonance, 57, 58, 61, 62, 181–184
　frequency of, 58, 61
　parallel, 61, 62
　series, 57, 61
Resonance curve, 57, 58, 61, 182–188
Resultant, 39
Retroaction coil, 114
Return loss, 135
Reverse a.g.c., 206
Reverse bias, 77, 78, 83, 106
Reverse voltage feedback ratio, 96
ρ, (resistivity), 17
　(return loss), 135
Rheostat, 65
Rhumbatron, 64
Ribbon microphone, 126
Ringing, 229
Rise time, 229, 231, 234
Rochelle salt, 126
Root-mean-square (r.m.s.) value, 37–39

S numbers, 182–184
Sampling, 256, 262
Satellite, 145
Saturation, colour, 222
　current, 66, 89
Sawtooth waveform, 213
Scales, 278, 279, 281
Scanning, 5, 126, 215, 260
Scanning circuits; 220–222
Schmitt trigger, 244
Schottky diode, 249–250
Screening, principle of electric, 201–203
　principle of magnetic, 203, 204
Secondary cell, 266
Secondary emission, 210
Second-channel interference, 194
See-saw circuit, 235
Selectivity, 58, 129, 181–188, 202, 208
　and Q, 181–185
Selectivity curve, 182–188
Self-inductance, 30, 50, 139
Self-oscillating mixer, 192
Semiconductor, 11, 65, 67
　i-type, 71
　metal-oxide-semiconductor, 82
　metal-oxide-silicon, 82
　n-type, 69, 74
　p-type, 69, 74
Semiconductor devices, 65
Semiconductor diode, 72, 118, 268
Sender. *See* Transmitter
Sensitivity, 181
Serial/parallel conversion, 252
Serial signal, 252
Series connection, 15–17, 19, 22, 44, 46, 50, 51, 56, 57, 62
Series-parallel combinations, 16

Series resistance of a tuned circuit, 181
Seven-segment display, 227–228
Shadow-mask picture tube, 224
Shift register, 251–252
Shock excitation, 113
Shelf life, 266
Shift register, 251, 252, 256, 257
Short circuit, 18
Short-circuited-output current gain, 199
Shot effect, 179
Shunt, 20
Shunt-injected negative feedback, 176
Shunting, 15
SI units, 17, 22, 24, 30, 31
Sidebands, 126–129, 182–188, 190, 217
　definition, 128
Siemens, 17
σ, 17
Sign convention, transistor equivalent generator, 93, 94
Signal/noise ratio, 179, 193, 202
Signal, 84, 85
　input, 84, 85
　inversion, 86
Silicon, 67, 69, 70, 72
Silicon diode, 73
Silicon transistor, 79–81
Sine wave, 35, 38, 164
Single-sideband system, 129
Sinusoidal modulation, 123, 124, 126–129
Skin-effect, 62
Skip distance, 146
Slope, 76, 280, 281
Slot antenna, 144, 145
Smoothing, 267, 270
Solid-state. *See* semiconductor
Solid-state storage, 263–264
Sound generator, 3
Sound waves, 1, 2, 6, 7, 35, 125, 167
Source, of an f.e.t., 81
　of a signal, 84, 85
Source follower, 104
Space charge, 66, 71, 72, 75, 80
Space-charge-limited current, 66
Speed-up capacitors, 244
Spider, 167
Square root, 277
Square wave, 36, 37, 42, 49, 229, 237–239
Squegging, 271
Stabilisation, 267, 271–274
Stagger tuning, 187, 203
Standing wave, 135, 136
Standing-wave ratio, 136
Static random access memory, 264
Statistical process control, 255
Store, computer, 253–254, 258–265
Stray capacitance, 57, 63, 116, 157, 175, 203, 233, 241, 242
Stub, 137
Subcarrier, 129, 222
Subscripts, 276
Substrate, 82, 236

Superheterodyne principle, 189–195
Superposition theorem, 230
Superscripts, 276
Surface-acoustic-wave filter, 201, 218
Surge impedance, 132
Susceptance, 48, 60
Sweep unit, 213
Swinging choke, 270
Switch, 14
　electronic, 246
Switch-mode power supplies, 274
Symbols, algebraic, 275–277
　binary logic components, 250
　circuit diagram, 13, 15
　letter, 275, 276
　mathematical, 276
Sync. separator, 218, 219
Synchronising circuits, 214, 218–220
Synchronising signals, 5, 160, 215, 216, 241
Synchronous detector, 223

Tape
　recorder, 259–261
　speed, 260
　storage, 259–262
Target, 126, 224
Technical terms, equivalents, 282
Telecine, 126
Telegraph, 4
Telegraphy, 125
　radio, 7
Telemetry, 126
Telephone, 4, 247
Telephony radio, 7
Television, 5, 126, 129, 145, 206, 214–224, 231, 233, 234
　aspect ratio, 214
　camera, 5, 8, 126, 212, 224, 252
　camera tube, 224–225
　colour, 6, 222–224
　diode detector, 158
　frequency allocations, 129
　405-line system, 216
　625-line system, 215–217
　819-line system, 216
　picture tubes, 5, 8, 210, 212, 222–224, 232
　power supplies, 273–274
　receivers, 187, 203, 217–224
　scanning circuits, 220–222
　signal characteristics, 216, 217
　sound, 217
　synchronising circuits, 218–220
Temperature-limited current, 66
Tesla, 31
Tetrode, 80
Thermal-agitation noise, 179
Thermionic emission of electrons, 65, 66
Thermocouple, 19
Thermocouple meter, 19, 37
Thermojunction meter, 37
Thyristor, 268
Time base, 213, 214

Index 293

Time constant, 26, 33, 57, 153, 154, 158, 159, 206, 230, 234, 238, 240, 241, 243–245
Tonal value, 5
Tone control, 164
Touch sensor, 194
Tracking in superhet. receivers, 193, 194
Transconductance, 76
Transducer, 201, 236
Transfer parameters, 91
Transformer, 30, 33, 34, 52–55, 92, 267, 268–270
  a.f., 54, 171, 172, 174
  balance-to-unbalance, 187
  dot convention, 52
  efficiency, 54
  iron core, 33, 34, 52
  line-scanning, 273
  load current, 52, 53
  losses, 53, 54
  quarter-wave, 136
  step-up, 240, 273
Transient, 26, 177
Transient distortion, 177
Transistor, bipolar, 77–81
  a.f. amplifiers, 168–179
  base biasing, 108, 109, 111
  bistables, 243–244, 252
  blocking oscillator, 241
  characteristics, 79, 80
  common-base configuration, 100–103
  common-collector configuration, 98–100, 101–103
  common-emitter configuration, 95–98, 101–103, 197–199
  current stabiliser, 273
  equivalent output circuit, 96–97
  frequency-changer, 192, 203
  gating circuits, 246
  graphical symbols, 79
  internal capacitances, 196
  i.f. amplifiers, 200, 201, 205, 206
  leakage current, 107, 108
  limiting frequency of operation, 199–200
  line output stage, 220–222
  load lines, 89, 107
  operational amplifiers, 234–236
  oscillators, 115–117
  parameters, 80–81
  phasor diagrams, 94, 97, 100, 101, 104, 198–200
  pulse shaping circuits, 239, 240
  pulse and video amplifiers, 232–234
  quadrature detector, 160
  ramp generator, 240
  reactance modulator, 180
  r.f. amplifiers, 201–203
  sync separator, 218, 219
Transistor, field-effect, 81–83
  biasing, 110, 120
  characteristics, 82–83
  equivalent circuit, 104–105
Transistor, field-effect (*continued*)
  frequency-changer, 192
  oscillator circuit, 112, 114, 115, 120
  r.f. amplifier, 202
Transistor/Transistor Logic, 249
Transmitter, 4, 7, 8, 19, 121, 123, 125–130, 131, 137, 145, 147, 149, 180, 181, 190, 195, 217, 222, 225, 227, 268
Triode valve, 75–77, 85–87, 124
  characteristics, 75, 76
  load line, 85
Truth table, 248
Tubes, 65
Tuned circuit, 56–64, 174, 181–188
  coupled, 185–188
  parallel, 59–63
  Q, 181–185
  series, 56–59, 62
Tuning, 7, 58, 181–188
  antenna, 148
  ganged, 193, 194, 202, 203, 208
  permeability, 201
  stagger, 187, 203
Tunnel diode, 118
Turning frequency, 175, 183, 200

Ultra-violet waves, 6
Under-coupling, 208
Unilateralisation, 201, 203
Unipolar transistor. *See* Transistor, field-effect
Unit multiple and submultiple prefixes, 284

Valves, 65–67, 75–77, 85–87
  abbreviations and symbols, 284
  biasing, 77, 110, 271
  diode, 66, 268
  parameters, 76
  pentode, 80, 89, 104
  power requirements, 266
  triode, 75–77, 85–88, 124
Varactor, 74, 125, 137, 161, 194
Variable-frequency oscillator, 121
Vectors, 39
Velocity modulation, 121
Vented enclosure, 167
Vertical helical scan, 260
Vestigial sideband, 129, 217
Vibration, 3
Video cassette recorder, 260–261
Video frequencies, 126, 129, 158, 215, 216
Video memory, 227
Video signals, 5, 8, 215, 260
Video still camera, 252
Video tape recording, 260–261
Virtual earth, 235, 236, 239
Visual display unit, 227, 255
Volatile store, 264
Volt, 12
Volt coil, 37
Voltage, alternating, 35
  addition and subtraction, 39

Voltage amplification, 86–88
Voltage amplification factor, 75
Voltage amplifier, 77, 86
Voltage-controlled device, 77
Voltage doubler, 270
Voltage drop, 18
Voltage feedback, 100
Voltage gain, 86, 97, 98, 99, 101, 103, 105
Voltage generator, 91, 93
Voltage inverter, 235
Voltage multiplier, 273
Voltage-operated device, 77
Voltage quadrupler, 270, 273
Voltage sextupler, 273
Voltage stabiliser, 271–273
Voltmeter, 14, 213, 247
  electrostatic, 20, 37

Watt, 20, 21, 27
Wattmeter, 37
Wave reflection, 134
Wave transmission along a line, 133, 134, 139
Wavebands, 35
Waveform, 2, 229–231
  analysis, 328
  non-sinusoidal, 229–236
  saw-tooth, 213, 229, 237, 240
  shaping, 239, 240
  sinusoidal, 35, 36, 38, 164
  synthesis, 229–230
Waveform distortion. *See* distortion, non-linearity
Waveform generator, 240–244
Waveguides, 137, 138
Wavelength, 3, 35
Waves, electric, 6, 7
  electromagnetic, 35, 141
  radio, 6, 7, 35
  sine, 35, 38, 164
  sound, 1, 2, 6, 7, 35, 125, 167
  square, 36, 37, 42, 49, 229, 237–239
  standing, 135, 136
  ultra-violet, 6
Wear in storage systems, 259, 262
Weber, 30
Whiskerless diode, 74
Whistle, 194, 195
White noise, 179
Wireless, nature of, 1
Word, 248, 251–253
Work, 20, 26
Working point, 85, 106–111
Write once, read many times, 263, 264

X-rays, 6
X-shift, 213

Yagi antenna, 144, 148
Y-shift, 213

Zener diode, 73, 272, 273
  characteristic curve, 272